INTRODUCTION TO GEOGRAPHICAL INFORMATION SYSTEM

About the Authors

Prithvish Nag, (*b.* 1951), M.Sc. (Gold Medalist), Ph.D., is the Director of the National Atlas and Thematic Mapping Organisation (NATMO) of the Ministry of Science and Technology, Government of India. He has been the President of the Indian National Cartographic Association, Indian Association of Special Libraries and Information Centre, Institute of Indian Geographers, Indian Institute of Geomorphologists, Professor S.P. Chatterjee Memorial Foundation and Indian Geographical Foundation. At the international level, he was the Chairman of the International Cartographic Association (ICA), Commission on Population Cartography, UN/PCGIAP Committee on Institutional Strengthening for Asia and the Pacific, International Steering Committee on Global Mapping (ISCGM), Working Group on Data Standardization; and Full Member of the International Geographical Union (IGU), Commission on Population Geography, and Member, Editorial Board, International Journal on Geographical Information System.

Dr. Nag was the Visiting Member of the Staff of the School of Oriental and African Studies, London. He is an author and editor of over 35 books and 110 research papers. His well known books are *Geography of India* and *Digital Remote Sensing*.

Dr. Nag was UN Consultant in Cartography in the Sultanate of Oman and the Surveyor General of India.

Smita Sengupta, (*b.* 1956), M.A., Ph.D., had a varied experience of working with universities, institute, government and the GIS industry. At present she is working as GIS Consultant. Dr. Sengupta is a recipient of the UGC and CSIR fellowships. She has brought out books on *Residential Patterns of Suburbs, Geography of India* and *Environment, Population and Development*, apart from over 35 research papers in well known journals.

INTRODUCTION TO GEOGRAPHICAL INFORMATION SYSTEM

Prithvish Nag

and

Smita Sengupta

Concept Publishing Company, New Delhi-110059

Cataloging in Publication Data—DK
Courtesy: D.K. Agencies (P) Ltd. <docinfo@dkagencies.com>

Nag, P. (Prithvish), 1951-
Introduction to geographical information system /Prithvish Nag and Smita Sengupta.
p. cm.
In Indian context.
Includes bibliographical references.
Includes index.
ISBN 8180694399

1. Geographic information systems—India. 2. Geography—India—Data processing. I. Sengupta, Smita, 1956-, joint author. II. Title.

DDC 910.285 22

ISBN-13: 978-81-8069-440-0 (PB)
ISBN-10: 81-8069-440-2 (PB)

First Published 2008

Published and Printed by
Ashok Kumar Mittal
Concept Publishing Company
A/15-16, Commercial Block, Mohan Garden
New Delhi-110059 (INDIA)

Phones : 25351460, 25351794
Fax : 091-11-25357103
Email : publishing@conceptpub.com

Dedicated to
Shri Naurang Rai
Who selflessly promoted
Indian Geography

Preface

After publishing the book on *Digital Remote Sensing*, my publishers were keen on bringing out another book, and this time it was on the Geographical Information System. They were pursing me to do so since 1998. But due to my hectic activities and involvement in several scientific and professional activities, I was not able to concentrate on this volume. Nevertheless, I managed to plan the whole scheme and was successful in convincing my co-author to work with me again. Earlier, we brought out books on *the Geography of India* and *Environment, Population and Development*. But the problems were of different type. Even with fast means of communications available now, it was not always possible to interact easily. I was based in Kolkata and my co-author in Mumbai. Later, my movement to Dehra Dun did not make our life easy. I returned to Kolkata with a determination to complete the pending work. Hence this book.

In the beginning of eighties I came to know about the power of computers and start thinking of its application, particularly in thematic mapping where information from a variety of sources has to be processed for a meaningful topic. Initially my exposure to the digital technologies in the Department of Geography, University of Durham, United Kingdom in 1982 gave a big boost. Professor John I. Clarke and Professor Michael Blakemore were extremely helpful in this regard. Following this, I wrote a paper for the Auto-Carto London 1984 which was accepted but I could not go there for presentation. Nevertheless, it attracted the attention of Professor J.T. Coppock of the University of Edinburgh. He not only approached me to get this paper published in the then newly brought out *International Journal on Geographical Information System* but also to accept the Membership of its Editorial Board. My paper entitled *A proposal for the development of a geographical information system for India* was published in the first volume of this journal. A few years later I joined the Indian Space Research Organisation as Scientist/Engineer. There I was involved in several committees to assess the GIS requirements and to analyze GIS software for various activities of ISRO. I soon returned to the National Atlas and Thematic Mapping Organisation or NATMO and was entrusted with similar activities.

Meanwhile, my co-author, Dr Smita Sengupta, after completing her Ph.D., was on a look out of a thrust area which is exciting. I suggested her to explore GIS related topic for post-doctoral work. Her proposal was accepted by the Council for Scientific and Industrial Research, Government of India. She continued her interest in this subject even after shifting to Indian Institute of Technology, Mumbai and other GIS related industries in the same city. Due to her exposure of the potentials of GIS in industry, we were able to contribute on business opportunities and applications in the realm of GIS.

I was greatly benefited by the experience I gained in carrying out a project of the Technology, Information, Forecasting and Assessment Council (TIFAC) on Government-Industry Co-operation in Cartography in 1997. With this survey I could get an insight of the status of the GIS related activities in the government institutions, and the industry and how the combination works in the country. The support of Shri Y.S. Rajan, then Executive Director of TIFAC was extremely beneficial. Around the same time, I had close connection with several professional bodies related to geography and cartography. These associations gave me an opportunity to know the status of GIS related activities in the universities, institutions and collages. Recently my association with Survey of India and the National Spatial Data Infrastructure allowed me to look into the issue entirely from a different angle, i.e. availability of basic data for the development of GIS. I am sure that the new National Map Policy and acceptance of the NSDI by the government will give a boost to GIS activity in the country.

This book is our combined effort to promote GIS technology in India. The language has been made simple as possible in order to make the subject easily understandable. Some may find it useful as a text book, however, the intention was never to do so. We did our best to get Indian examples of GIS applications in business. We went through the company profiles, their presentations in different conferences and web sites. But we were only partially successful. Firstly, the Indian companies hardly take up local projects; and if they do so, it is difficult to get the details. Hence, a lot of examples have been taken from abroad. Perhaps this will help in exploring the potential of GIS in different businesses. For this purpose, a world wide search was launched through the web, which was time consuming and cumbersome as well. By the time we could finalize the manuscript, many web sites were withdrawn. However, we have included the case studies here. Nevertheless, after going through the book, one could gather that the scope of GIS is mind boggling. Another objective of this book is also to promote entrepreneurship in the field of GIS, not only of the projects in the country but also of abroad. With our success in Business Process Out-sourcing (BPO), which basically deals with non-spatial data, there is no reason why India can not become hub for GIS BPO as well. The man-power, technology and infrastructure are getting ready to take up such a challenge as well. We recently brought out a book focusing on this subject. This book is a modified version prepared to meet the requirements of the students' community of the country.

This book could not have been brought out without the support of our friends and colleagues in the government, industries and institutes. My association with the Department of Science & Technology, National Atlas & Thematic Mapping Organisation, Survey of India, different institutions under the Department of Space and ISRO including NNRMS, Indian National Cartographic Association; and my coauthor's association with IIT and industry is worth mentioning. We are extremely grateful to Shri A.K. Malik, Dr A.K. Dasgupta, Dr G.C. Debnath, Shri A.J. Kurrian, Dr R. Siva Kumar, Dr P.S. Acharya, Dr P. Venketachalam, Brig Girish Kumar, Dr B. Nagarajan, Shri H.B.Madhwal, Dr. Devi Dutt Chuniyal, Dr M.Kudrat, Dr N.C. Gautham, Dr K.V. Sundarum, Shri Sanjay Kumar, Shri Ayon Tarafdar and Dr A. Perumal.

Our special thanks are to Dr T. Ramasami, Secretary, Professor V.S. Ramamurthy, former Secretary of the Department of Science & Technology for their constant encouragement. Dr K. Kasturirangan, Hon'ble Member of Parliament and former Chairman, ISRO has been constantly in touch with our scientific and academic pursuits. We are grateful to him as well.

Shri Ashok Kumar Mittal of the Concept Publishing Company, New Delhi has always supported us in our endeavors and this is no exception. Unfortunately we could not bring out this book during the life time of his father, Shri Naurang Rai. However, we would like to dedicate this book to him.

Prithvish Nag

Kolkata
19 October 2006

Content

List of Figures

List of Tables

1

Introduction

Tradition of map-making is as old as human civilization. Geographical knowledge traces its historical background from very beginning of civilization. Knowledge of maps as a science grew with time and religious thinkers also started delving into its mysteries. In the ancient literature, there are descriptions of the territories or *Bharatavarsha* or *Jambu-Dvipa* or different regions of India, named as *janapadas, khanda*, and *dvipas*. The descriptions are fairly correct and have relevance even today, particularly to understand India as a concept. There is an ancient Indian concept of the spherical world with the *Jambu-dvipa* (*i.e.* India) as the central part and Mount Meru (Kailas) as the centre of the world. In *Vedas, Puranas,* Buddhist and Jain literature, there are references about time, longitudes, latitudes, rhumb lines and the like. In *Puranas* (500 BC to 700 AD), particularly *Brahman Purana* the concept of map scale, and generalization of geographical features were realized. Preservation of collection of maps in the form of atlases has been mentioned in *Padma Purana.*

Man's first conception of the earth's figure was a plane. Greek philosopher thought it a sphere. Later, Sir Issac Newton showed it must be spheroid. However, development of Science of Surveying and Mapping of earth's surface in an orderly manner is ascribed to Eratosthenes (276-198 BC) who based the compilation of his comprehensive map of the world upon his famous determination of the circumference of the earth. Claudius Ptolemy, a mathematician, astronomer and geographer produced a map of India in the 2nd century AD which was subsequently made use of and commended in 18th century. The golden age of Indian renaissance started when in 5th century, genius Aryabhat who calculated the earth circumference to be 25,080 miles. The next contributors to Indian geography were Chinese and Arab travellers who left the interesting detailed accounts of their journey.

In Kautilya's *Arthasastra* (4th century BC), the significance of village boundaries and methods for land measurements have been described. The accounts given by him about the *Siva Purana* and Panini's *sutras* indicate the importance given to geodesy, cadastral surveys, plans and sketch maps for army movements.

References to regions and places with the help of astronomical calculations were included in traditional Indian system. Ujjain was considered the oldest prime meridian much before Ptolemy. The main source of traditional Indian cartography is *Sata Patha Brahmana* which gives account of the concept of earth in the 5th and 6th centuries. Aryabhat's astronomical contributions are well known, particularly in calculating the diameter of the earth, its axial rotation, position of planets in space, and planetary orbits in 5th century AD. The land surveying manuals were part of *Sulba Sutras* (science of mensuration). In Vedic texts, apart from the description of physical features like mountains, rivers and plants, there are *sutras* for trigonometry, co-ordinate and solid geometry, analytical conics, calculus (differential and integral), equation of straight lines and even Pythagoras theorem. Some of the above *sutras* of Vedic mathematics have relevance in modern surveying and mapping as well. It is said that *trikonamiti* is what we understand in the form of 'triangulation' today. This forms the basis of modern map-making, including geodesy.

The rich Arab traditions and their knowledge of stars and related information, the skills of Portuguese cartographers and experiences from Egypt, China and Japan (particularly calligraphy and artistic designing), Mesopotamia (Iraq), and other colonial powers contributed to the Indian cartographic traditions as well. Navigational charts, route charts for land and sea were commonly used in this part of the world. Compass lines and related navigational details were included in such charts. Nevertheless, the Indian knowledge of sea, oceans, other neighbouring countries and coasts, navigation skills etc., was excellent. *Markandeya* and *Bhagavata Puranas* have references to sea-borne trade and commerce. In the *Rigveda* and *Satapatha Brahmana,* there are references to navigation in *purva* and *paschim samundra* (Bay of Bengal and Arabian Sea). Raghuvamsa records the presence of a strong naval force. The Cholas, Oriyas, Marathas and Gujaratis used their own navigational charts. Indians had direct port to port established routes in the Indian Ocean Realm. Trade and commercial links with Oman including Muscat, the East African coast including Zanzibar, Sofala coast and different islands and territories in south and Far East Asia are well known. When the Portuguese appeared in the Indian Ocean, they were surprised to see that the Indian pilots were using compass and charts. These charts were good enough to navigate through the ocean. Such charts or maps included relevant information, such as islands, reefs, shoals, ports, cities, pole star, latitude co-ordinates, depth of water, lighthouse, and distances expressed in *zams*. But, with the emergence of colonial powers in the Indian Ocean, this tradition got affected.

The age of great exploration started with the quest for uncorrupted truth. The major exploration got underway at the end of 15th century with Columbus crossing the Atlantic in 1492, Vasco da Gama circumnavigated Africa in 1497 and the Magellan's expeditions circumnavigated the world between 1519-1522. The Dutchman Snell (1591-1626) carried out the first accurate triangulation and made the rigorous study of refraction. The French Clergyman Picard, in 1670 made the measurement of size of earth. His result of 6,275 km for the radius of earth was the first improvement on Eratosthenes determination of the size of the earth. Newton's theory, that the earth should be oblate because of the centrifugal force, caused by the spin was validated by the measurements two meridian arcs, one at the equator and the other closer to pole measured by two survey expeditions organized by French Academy of Sciences.

The dissemination of geographical knowledge was limited till 15th century before the invention of printing technique due to the labour and skill involved in duplicating graphic data. Once, however, lithographic techniques were discovered map making got a stimulus, particularly in Europe where adventurous men were out to conquer new worlds for commerce. Father Monserrate, a Jesuit Missionary brought out a map after a visit to Akbar's court. This was a first of those maps which are based on measured routes and astronomical observations.

During the early Pathan period, Sher Shah Suri was instrumental in preparing revenue maps in northern India. Later, Raja Todar Mal, a Minister in Emperor Akbar's court, made valuable contributions in preparing revenue maps for a larger part of the country. Sher Shah Suri and Todar Mal's revenue map, based on regular land survey systems were well known in the medieval period and continued to be in practice during the mid-eighteenth century. These are available information of the surveys instituted of Akbar during 16th century; measurements being made by a hempen rope which was replaced by a "Jarib" of bamboos joined by iron rings. The other noted cartographers of the time also published their vision of maps of India, but in 1752 the French geographer, Jean-Baptiste Bougigon d'Anville published map of India and put Indian geographical knowledge on a scientific footing. In the south, Rajaraja, the Chola king (985-1001 AD), carried out revenue assessment survey of his territory. Further, plans, sketches, religious maps, concept maps, route maps, charts, *parikrama* maps and the like were very popular. The layouts of palaces, forts, cities, tombs and gardens were based on topography, directions and astronomical calculations. Sun light, position of moon, wind directions, seasonal variations, calculations of geographical co-ordinates (as mentioned in *Ain-i-Akbari*), etc., were included in the plans which cannot be done without proper knowledge of the geographical space. Attempts were made from very early times to establish the geographical locations of important places. Raja Jai Singh, Sawai of Jaipur of pre-British period established astronomical observatories at Jaipur, Delhi, Mathura, Ujjain and Varanasi and determined geographical positions in India. Even today we can find these traditional materials in old cities like Allahabad, Mathura and Varanasi. A lot of materials have made their way to the libraries and collections in the West.

The pioneering work done by French expeditions showed that terrestrial measurements (angles and distances) are the main tools for the task of relative positioning. The technique of triangulation, astronomical determination of positions and azimuths as well as levelling were started in mid-seventeenth century. In the early days of the influx of foreigners in India, the geographical knowledge of southern parts of the country was greatly increased during the period of wars of political supremacy but northern parts of the country (Bengal, Uttar Pradesh etc.) was sketchy.

During the middle of 18th century, latitude could be easily determined by observing the meridian altitude of the sun or star. For longitude determination one had to wait for favourable phenomenon such as an eclipse of sun, moon or satellites in clear sky with similar observations to be made at some known places. In 1787, astronomical observations for latitude and longitude at various places were made covering India. The proper astronomical centres were established in due course of time using sextants, chronometer and telescopes. The results of these instruments were imperfect because of their accuracy level and faulty mathematical tables. It is interesting to note that because of difficulty in determining longitude early maps of India do not indicate longitude with reference to Greenwich meridian but are referred to arbitrary meridian passing through Madras or Calcutta.

Most of the early maps were local surveys carried out by cursory method. In 1776, comprehensive instructions for preparing the maps on the scale two mile to an inch formulated: in which distances were measured by parambulator rather with chains and bearing to the conspicuous hills. Short base lines laid out and measured and distant points to be fixed by triangulation.

Development of the Indian Cartographic Empire

The history of Survey of India is a part of the saga of nation building. It was perhaps never realized that the foundation of modern India is synonymous with the activities of this department. But its contribution has never been emphasized—not even by the *Survey* itself. How scientific foundation or development initiatives could have been possible without the anticipatory actions taken by this department. The process of understanding of the priorities in development and defence has been the forerunner in all respect.

All such developments were of course not without price and efforts. The scientific measurement of the country had several ramifications. Surveyors had to traverse from regions to regions, waiting for an opportune time free from natural, man-made and logistic problems in order to continue their task. Hazards, diseases, snakebites, dacoity, hostile local people, battles and the like came in the way of their mission. Surveyors penetrated the jungles, climbed mountains, crossed rivers and fixed poles, stations and control points through out the length and width of the country. There was no respite whether it may be on the slopes of Western Ghats, or swampy areas of Sundarbans, or ponds and tanks, ox-bow lakes and meandering rivers of Bengal, Madurai or Ganga basin. The deserts, mighty Himalayas, *Rann*, river Chambal or Gandak, *terai* or *dooars* were also not spared. The survey has been associated with the waves of Arabian Sea or Bay of Bengal, dust storms of Rajasthan, cyclones of Eastern Coast, cold waves of the north and the widespread monsoons and heat all over the country.

It was against this price, determination and missionary zeal that the foundation of the country was laid. The knowledge assets built over years and generated with whatever technology then available proved to be valuable. The best part of the whole process is that a new information set which may be based on latest technology like aerial photography or global positioning system take benefits of the earlier generated information. No piece of information is wasted and has relevance even after decades. How such an empire of knowledge has been built up? It is not only a technological feat that is most important but it is an emblem of pursuance of one of the longest scientific experiment carried out in the world, i.e. mapping the nation against all odds. The legacy and traditions continue. It is a matter of great interest to know how such a superstructure of information has been built ? How this great scientific experiment is carried out for so long. What were the compulsions and apparent benefits ? Why the colonial rulers and later the independent government with limited resources continued their interest in this expensive exercise. These are some of the questions that will keep on bothering the people interested in the history of science of this country.

The Survey of India traces its birth starting from Major James Rennell appointed as Surveyor General by the East India Company on the 1st January 1767. In those days, the urgent need was for a picture of the country showing general course of main rivers and the situation of principle towns. This task was taken up with speed and produced serviceable maps of the area of Bengal and Bihar in less than twelve years, an area of over 1,50,000 sq. miles, producing a continuous and uniform set of maps. These maps, however, could lay no claim to accuracy of details but were sufficient to meet the needs of the time. Rennell also produced a map "Map of Hindoostan" in 1783 after relinquishing the post of Surveyor General (Fig. 1.1).

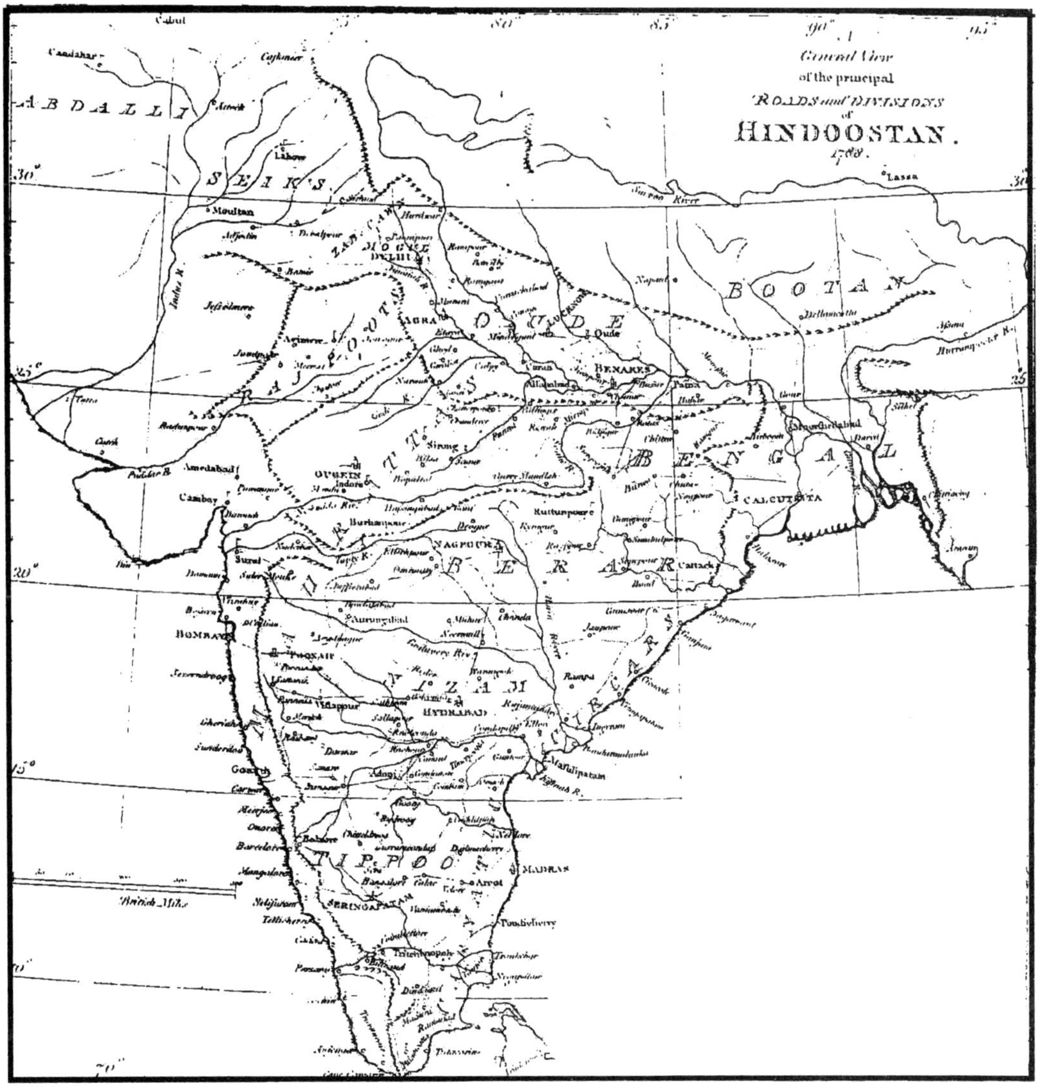

Fig. 1.1: Map of Hindoostan by Rennell

The progress of topographical surveys in Madras and Bombay presidencies was more or less independent and uncoordinated with the work of Bengal presidency till the beginning of 19th century. This was not a satisfactory state of affairs and retarded the progress considerably. It was only in 1787 that an accurate survey of the sea coast from Madras to southernmost extremity of the peninsula was taken up by running a 300 mile line of triangles along the coast with a view not only to ascertain the actual line of the sea coast but a complete survey of the peninsula of India. This survey was the first Indian survey to have base on triangulation.

Towards the close of 18th century, primitive theodolites had been brought in use. The angles and bearing were measured with theodolite and pocket compass for direction of the road etc. The technique of plane tabling was first used in 1792. Plane tabling survey was subsequently developed into an art and has been extensively used down to present time for topographical surveys in all types of terrain. Even now, though the modern techniques of surveying are coming up, yet this simple technique is widely used in various surveys such as large scale mapping, engineering survey, cadastral survey etc.

Beginning of Scientific Surveys

The period of piecemeal surveys was over by end of 18th century. A new era began with the William Lambton and George Everest which signalled the consolidation of coordinated efforts. The foundation of a truly scientific Survey of India was laid which was the beginning of a stupendous work and occupied the lifetime score of noble and devoted surveyors. A network of primary triangles was established by the trigonometrical surveys. It was a magnificent scheme, timely conceived and brilliantly executed. Although, technique of triangulation, astronomical determination of positioning and azimuths as well as levelling were started in mid-seventeenth century but the scientific procedures started only by the end of 18th century when a project for the measurements of an arc of the meridian through a network of trigonometrical surveys covering the Indian peninsula was formulated.

The actual work of the Great Trigonometrical Survey was commenced on 10th April 1802 by the measurement of a base line near Madras, (now Chennai). Madras base line was established using steel chain which consists of 40 links of 2½' each, measuring in the whole 100 feet. The baseline was measured with the aid of coffers (long boxes) as it was required for the triangulation of the "Great Arc" where utmost possible accuracy was aimed at. From this base line a series of triangles was carried up to the Mysore and the second base was measured near Bangalore in 1804. To start with primary reference station of origin was the Astronomical Observatory at Madras. Having connected the two sides of the Peninsula, Lambton devoted much of his labours to the measurement of an arc of meridian. The series measured for the purpose is known as the *"Great Arc Series"*. In addition to the measurements of Great Arc Series, webs of triangle extended for establishing the positioning of main cities. This idea of web of triangles was replaced due to cost effectiveness by an All India Grid composed of cris-crossing 'chains' or 'bars' of triangle centred on the Great Arc. The holes on the grid could be filled later by cheaper and less rigorous topographical surveys. This idea gave birth the name "grid-iron" (Fig. 1.2).

The grid-iron layout consisted generally of an outer frame of two extreme meridional and two longitudinal series closing at each junction on a measured base line. To bring the Great Arc across the plains, masonry tower stations were built which were about 50 feet high. The first essentials of every observation station, whether on hill top or tower, or otherwise was the stability of the instrument and immovability of the mark over which instrument and signals were centred. Since these marks were established after putting hard work, therefore marks were handed over to the civil authorities when all connections had been completed. In 1866, it was ordered that all stations of Great Trigonometrical Survey should be placed under the official protection of district magistrates and visited periodically. This practice is still in vogue till the present time for all primary Great Trigonometric (GT) stations and benchmarks.

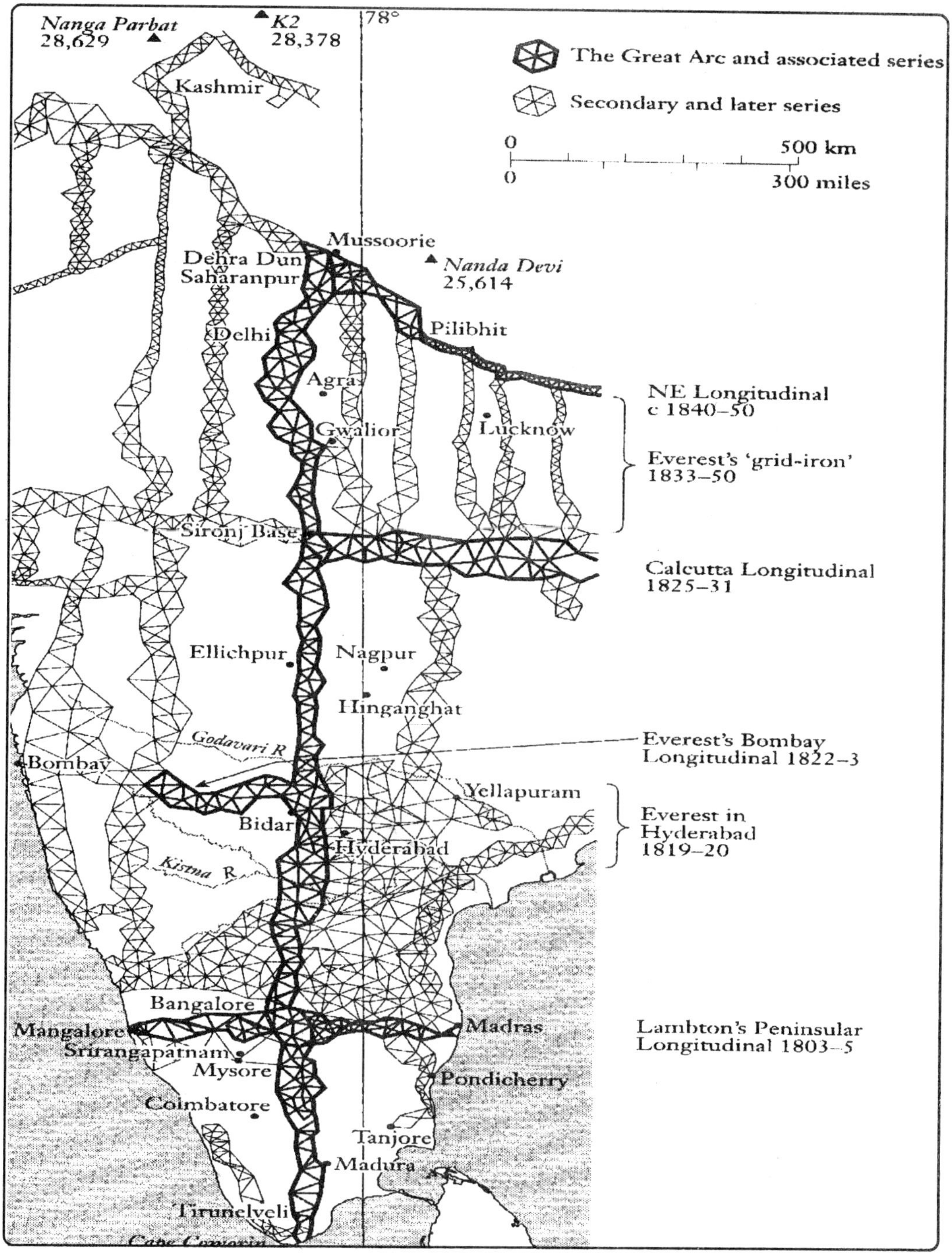

Fig. 1.2: The Great Arc

Lambton's main instrument referred as the Great Theodolite which was a marvel workmanship of those days (Fig. 1.3). The horizontal circle was 36 inches in diameter and the vertical circle 18 inches; each was read by two microscopes. This theodolite was used by Lambton and his assistant and then by Everest and others till 1866. Various other theodolites were used for observations for meridional series, viz. 36 inch theodolite, which was built up from Lambton great theodolite, 34 inch theodolite, 24 inch, 18 inch and 15 inch theodolites. For laying out the series and running secondary and minor triangulation, small theodolites were used i.e. 14", 12" and 7". For baseline measurements, compensation bars and other baseline apparatus were used. The compensation bars remained, however, the only means available for measuring the base line of the main triangulation framework—Vizagapatnam and Cape Comorin base were measured during 1862 to 1869. In 1856 'standard yard' arrived from England and the following year a special room at Dehra Dun was set aside where subsidiary standard could be laid off or compared by microscope as and when marking of staves for levelling operations required. Standard spirit levels were used during those days.

Fig. 1.3: The great theodolite

Triangulation and levelling computation were made to a regular routine, adapting rules and formulae to the requirement of the department. Computation forms were lithographed at Calcutta under the direction of the Chief Computer. One of the greatest contributions made to the Great Trigonometrical Survey was the preparation and publication of a set of tables to be used with departmental formulae and the computation forms by Babu Radhanath Sickdhar, Chief Computer. The first edition entitled *Tables to facilitate the computation of a Trigonometrical Survey and the Projection of Maps* was published in 1851. Auxiliary tables to facilitate the calculations of the survey department were published in 1868. The first official lists of geographical coordinates were published in 1842.

For the dispersal of triangular error, method followed by Everest was tested against new method devised by Gauss. Radhanath first tried two simple figures and obtained results closely agreeing with old method. He tried this on complicated figures that occur in the trigonometrical survey. The results were highly satisfactory, showing that the greatest discrepancy between Gauss's and Everest's methods would not exceed 0.14". The final distribution of errors and reduction of results of the triangulation of the Great Trigonometrical Survey was carried out under the direction of General James Walker whilst he was Superintendent of the Trigonometrical Survey. For this purpose the whole triangulation was divided into five figures—North-West, North-East, South-West, South-East quadrilaterals and the southern trigon.

Lambton and Everest did not go deeply into the subject of heights above sea level. Lambton first connected to the sea at Madras in 1802, but for his great Central Arc he preferred the connection made at Cape Comorin in 1809. From this he brought up his height by vertical angles from station to station. Surveyors in India had no professional interest in the measurement of the vertical rise and fall of tides along the coast except to find the level of the sea from which to calculate their land heights. Lambton followed the practice of his time in calculating his heights from low water. It was only in 1837, when the mean between high and low tide observations for at least for half a month were proved to agree closely from one place to another. Self-registering tide gauges had been invented during 1830-33. Self-registering tide gauge was established at Colaba observatory, Bombay in 1842. Tide gauges at various other places were also connected through spirit level during 1851 to 1860. Indian levelling was started upon correct and scientific lines in 1858. Again it was General Walker who was the founder of Indian levelling for the care and the thought he bestowed upon the subject before he initiated the fieldwork.

Waugh took over in 1843 from Everest and during seventeen years of his administration, triangulation series was advanced eastward to over Ganga valley to Calcutta and extendedup to Assam. Regular observations to the Himalayan peaks were taken. Waugh was responsible for the discovery of the highest mountain in the world, 29002 feet above the sea level, which he recommended should be named after George Everest who had built up the triangulation system by which the discovery was alone made possible (Fig. 1.4).

Regular astronomical observations for azimuth and meridian were continued along both meridional and longitudinal chains of triangle as check against accumulation of errors in direction. Both Lambton and Everest had been well aware that their observations were influenced by visible mountain masses and variation of density. Various other mathematicians and geodesists worked on this subject and attracted wide attention suggesting the value of pendulum observations for the determination of variations of gravity and introduced the theory of compensation or isostacy.

Fig. 1.4: Picture of Mount Everest (Photographed by P. Nag).

Development of Great Arc

The measurement of Great Arc from Cape Comorin to Himalayas was completed by 1843. The 'grid iron' system consists of meridional chains of triangle tied together at upper and lower end by longitudinal chains. Ambitious scheme of triangulation commenced with the Great Arc Series, having Dehra Dun base at the north and Sironj base in central India as the southern end. From Dehra Dun base NW Himalayan series was extended while from Sironj Base the Calcutta longitudinal series was extended up to Karachi base. In 1887 observation of a series from NW Himalayan known as Kashmir series was initiated. The height of stations averaged 17,000 ft. in this series. The second highest peak, next to Mount Everest was found during this triangulation which was 28,290 ft. high and was named K2.

Indian triangulation was first adjusted in 1880. In 1937-38, Bomford attempted to readjust the triangulation after incorporating the new triangulation series observed, Laplace stations to arrest azimuths errors and new baseline measured between 1930 and 1934 using invar wires. He carried out preliminary figural adjustments for the distribution of discrepancies in scale and azimuth in a rational way. The triangulation adjustment carried out in 1937-38 suffered weakness due to ill fitting nature of the ellipsoid, measurement of base lengths by 10 feet bar suspected to have undergone changes in length by 3 ppm, lack of Laplace observations, non-application of various geodetic corrections like deflection, skew normal, geodesics, polar motion etc.

The adjustment of 1937-38 was not considered comprehensive and explicit. Instead of resolving simultaneous normal equations, formed after incorporation of new data, graphical technique of adjustment was employed. Adjustment of 1880 has remained the basis of all India triangulation and mapping except for a constant change of—(2′27″.18) in longitude. This change was essential as the longitude of Madras Observatory was revised in 1905.

The mapping activities was scientifically executed and based on triangulation series adjusted in 1880. Most of the maps were on the scale half inch or larger. Mapping activities opened up new possibilities with the introduction of aerial photographs early years of the war. The science

opened up new possibilities with the introduction of aerial photographs early years of the war. The science of photogrammetry had not developed but the immense value of aerial photos impressed the surveyors. The air survey techniques were, however, not much used in the department up to 1939 except for Training for Military Survey purpose and for survey of inaccessible areas. At the time of independence (1947) about 60 per cent of the country was covered by primary scale of mapping on 1 inch to 1 mile. This primary map scale 1 inch to a mile was adopted in Survey of India in 1905. It was also decided that these maps may be produced in colours and show the relief by rigorous contourings. The requirement of printed maps increased during 1939. The main printing organization at Calcutta could not meet the mapping requirement. Therefore, the present site at Dehra Dun was selected and the printing machines were installed and by middle of 1943, the printing group was in operation with three large high speed machines.

In post-independence era, there was immense increase in the developmental activities which is continued till today. The Survey of India was the only premier organization in surveying and mapping which could take up the survey work for developmental scheme, due to which normal topographic survey was kept to a secondary place.

During the First and Second Five Year Plans, the department faced with demands for survey which required for more survey potential than it had. During the First Five Year Plan (1951-56), 70 per cent of the department's potential employed on developmental surveys and in Second Five Year Plan about 60 per cent potential could be directed for this purpose. The department had diverted work load with the switching over to metric system in 1956, the basic map scale having been changed to 1:50,000 from 1 inch to 1 mile. It was during second five year plan, when advancement in various techniques of mapping were available and there was pressing demands for maps for the development of the country. Various techniques in the field of photogrammetry, cartography, printing and training were adopted in Survey of India.

Geodetic and Geophysical Surveys

The Great Trignometrical Survey triangulation network in India and adjacent country was started in 1802 and by about 1880 sufficient triangulation series had been observed. The responsibility of establishing geodetic and geophysical control in India lies with Geodetic and Research Branch of Survey of India. With the introduction of modern instruments in improving the precision and accuracy of all geodetic and geophysical survey, the activities of this branch have been diversified and today it engaged in number of research and developmental programmes. Theodolites used in the past were replaced by glass arc theodolites in the late forties of twentieth century which read directly to an accuracy of 0.2 second of arc. Geodetic bases of length 10-12 km were established during the period 1831-82 have been measured with India 10 foot bars, while the modern base lines were measured with invar wires. Astronomical azimuths have also been measured at a few stations to keep check on the orientation of the triangulation series. The triangulation network of India for providing first order control by chain of primary and secondary series every 4° (about 400 km) apart was completed by the year 1956. The work for densifications to bring control at 2° (about 200 km) apart and upgradation of the existing series classified as secondary has been taken up by the Survey of India. There has been tremendous growth in the field of space geodesy in the last couple of decades. TRANSIT system adopted in Survey of India in seventies to establish geodetic control. A more rigorous system called NAVSTAR GPS (Navigation Satellite with Time and Ranging Global Positioning System) or simply GPS, a satellite based radio navigation system, providing 3-dimensional position, navigation and time information to suitably equipped users all over the globe at any instant of time, was introduced in the year 1980. GPS technique was adopted in Survey of India around the year 1990 and enormous work has been since been done in production mode. Electronic distance measuring device have been introduced since 1962 and number of bases have been added to the principal Indian triangulation. After incorporation measured base lines, lap lace observations etc. in the Indian Triangulation net, simultaneous adjustment of the primary and secondary series has been taken up to redefine the Indian Geodetic Datum.

Various geodetic and geophysical modern instrumentations are being put in use for geodetic, gravity, geomagnetic observations. Glass Arc theodolites, astrolab, chronograph, crystal clocks for geodetic work, Frost, Worden, Lacoste Romberg and Automated (CG-3M) gravimeters for gravity measurements and QHM (Quartz Horizontal Magnetometer), ZBM (Zero Balance Magnetometer), Proton Magnetometer for geomagnetic observations are in use for geodetic and geophysical surveys. The appearance of the computers is extensively used for better and more homogenous adjustments and data processing.

Topographical Mapping

After establishing the ground control points either by conventional ground method or by space based techniques, the subsequent steps are: survey ground details and verify, annotate and introduce names of rivers, roads, villages, towns etc. and depict topographic features by contour and other suitable symbols. These steps, in the past, were carried out through the ground techniques. With the emergence of aerial photography, Survey of India introduced the technique of surveying aerial photographs. The development of photogrammetry in India has been rapid only since 1954 when the first stereo-plotting machine a Wild Autograph A7, was processed and installed, though multiplexes were purchased as early as 1950s. The photogrammetric Techniques available became the standard procedures in quite a few units of the department. The introduction of photogrammetric techniques has expedited many developments in the country.

In Survey of India, the process of map making with aerial photograph starts with planning of aerial photograph. The Ground Contour Points (GCPs) at selected locations are connected with existing geodetic control by field survey and their positions are captured. With the photographs and minimal ground contour and their position identified, the stage is set for undertaking Independent Model Photogrammetric Triangulation and subsequent computation provides requisite control points in each model. The model properly oriented is then used for generating cartographic data which are verified on ground before preparing the final map.

By early fifties of twentieth century precision analogue photogrammetric instruments were used for topographical mapping. Analytical plotters captured the data in digital form, which had added advantage for acquiring directly digital map. The SPOT, LANDSAT and IRS imagery for topographical mapping on small scale are being used in the department. Digitally generated data are being used for creating cartographic data base in various scale, viz. 1:250,000 and 1:25,000. In addition to this data base for district planning map series on scale 1:250,000 is being carried out. Digital Terrain Model (DTM) being generated for various developmental activities in the country.

Survey of India as a part of its activities to keep abreast with latest technology in the science of map making, made digital map making programme in early eighties with a view to adopt digital technology in the country. Survey of India adopted Computer Asserted Cartography (CAC) system, the "AUTOMAP" in 1981. In this system the geographical and other data are stored in digital form and processed on demand to draw the required map output by using computer operated drafting table. Survey of India has taken a giant leap forward in 1986 when integrated digital map production systems were installed at 3 locations, i.e. Modern Cartographic Centre (MCC), Dehra Dun, Digital Mapping Centre (DMC), Dehra Dun and Digital Mapping Centre (DMC), Hyderabad. Having overcome the initial difficulties of new technology induction in early nineties, the department aimed to achieve flexibility in satisfying users demands by establishing structured Nations Cartographic Data Base (NCDB) and Geographic Information Systems (GIS).

Recently, a major initiative has been taken in Survey of India. Development and business potentials have been included in the activities of the organisation. As a result, reorganization has taken place in 2003. Directorates for each state, known as Geospatial Data Centres (GDCs), have been created which are equipped with digital and GPS technologies. DMCs and MCC have been converted to new directorates, namely National Geospatial Data Centre and GIS and Remote Sensing. A new Business and Publicity Directorate has been opened.

With the completion of 1:50,000 Topographical Survey of whole country by 1982 and mapping by 1985, India has joined a select group of countries which have complcted map cover on the national scale. The department now has an ambitions programme of covering the whole of India by maps on 1:25,000 scale. The country covers 394 maps on scale 1:250,000 and 5,106 sheets on scale 1:50,000. In 2003 all the 1:50,000 scale maps were digitized. With the new National Map Policy, these maps are now being converted to WGS84 datum and UTM map projection. In the process, a seamless map data is being generated (Appendix I).

Responsibilities and Challenges

Survey of India is the premier national survey and mapping organization in the country. In its assigned role as a national principal mapping agency, SOI bears a special responsibility to ensure that the country's domain is explored and mapped suitably to provide base maps for expeditious and integrated development of the nation. In the process, SOI has been producing all purpose topographical and various other series maps required by defence, general administration, internal security, developmental needs, irrigation, watershed management, resource management and various types of engineering projects. It is also responsible for establishing precise planimetric control, heights above MSL, gravity, geomagnetic and tidal prediction as prerequisite for mapping activities and other scientific applications. The department is committed to provide technical expertise to other countries in the field of geodesy, surveying, cartography and survey education.

With the introduction of digital technology in the department, three digital centres are generating Digital Topographical Database for entire country for use in various planning processes and creating GIS. Its specialized Directorates such as Geodetic and Research Branch, Research and Development Directorate and Survey Training Institute have been further strengthened to meet the growing requirement of user community. The department has also been immensely contributing to a number of multi-institutional scientific programmes related to the field of geophysics, remote sensing, glaciology, study of seismicity and seismotectonics, scientific expedition to Antarctica and digital data transfer.

India is fast moving into information and knowledge society. Emphasis is increasingly placed on IT driven transparent e-governance. The nation has been generating voluminous field/digital data, i.e. information through systematic topographical surveys, geological surveys, soil surveys, cadastral survey etc. Access and availability of such information to the citizen, society, private enterprise and government are of immense importance. As a part of this vision NSDI is being involved through partnership approach among various data generating agencies to facilitate integration, easy access and networking of databases. NSDI has been conceived as a national system that synergistically combines the resources and infrastructure of various players with the power of IT and enabling information support for decision-making in government, industry, academia and other organisation besides serving the public needs. NSDI strategy and action plans have been formulated through multi-institutional approach involving Department of Science and Technology (DST), Survey of India (SoI), Indian Space Research Organisation (ISRO), Geological Survey of India (GSI) and National Atlas and Thematic Mapping Organisation (NATMO) etc.

Atlas and Thematic Mapping

At the time of independence it was felt that different types of maps are required for development, planning and reconstruction of the newly formed country. These maps should focus on development issues in accordance with the priorities of the country as envisaged by the Five Years Plans. The progress of cartography in India for its development and reconstruction process was then initiated. The earlier approach for capture, control and colonization was given reorientation. From Table 1.1, the planning priorities in different Five Year Plans may be gathered. Initially, attention was more on the development of agriculture and related activities that were necessary following the Second World War and the partition of the country. In fact, a better understanding of the newly born country was required with all its facets. Hence, the preparation of the national atlas was felt necessary. Apart from maps on the physical aspects, the Hindi version or the first *National Atlas of India* included plates on soils, forests, land use, irrigation, crops, livestock, power and industries. There were on social aspects as well, such as on population and health (Massey, 2002). The Second Five Year Plan was meant to give a shape to the dreams of the first Prime Minister of the country, Pandit Jawaharlal Nehru. Iron and steel, heavy machinery and industry was given more priority. This required power and appraisal of the available resources in different parts of the country. Considering the size of the country and the varieties, it became imperative to prepare the enlarged version of the *National Atlas of India* (Fig. 1.5).

During the Third Plan, planning and development at regional and state level got impetus. This lead to the preparation of atlases at sub-national level. One of the most important one was *The Planning Atlas of the Damodar Valley Region*. Following this, several regional atlases came into existence. These atlases followed the state framework however, the focus was not always the same. As a result, we have statistical, administrative, resources, planning or simply

Fig. 1.5: National Atlas of India: Physiography

Note: Prepared for National Commission for Integrated Water Resources Development

state atlases. If we compare the contents, we will find there is a great degree of commonality though they have come out under different arrangements and funding. This process is not yet complete. Earlier, some agricultural atlases at state levels were prepared. In this plan period, the preparation of the *National Atlas of India* consisting of 300 plates was in full swing. It consists of various aspects of the country, which were required for any planning strategy at national and regional levels. This period also witnessed the diversification of the thematic mapping activities and applied geography.

Table 1.1: Planning Priorities and Cartographic Initiatives

Plan	*Priority sectors*	*Objectives*	*Cartographic initiatives*
First Plan (1951-56)	Agriculture, irrigation and power projects	To correct disequilibrium in economy caused by Second World War and partition of the country	Hindi edition of the *National Atlas of India* and state atlases.
Second Plan (1956-61)	Iron and steel industry, heavy machinery and industry	To achieve a socialistic pattern of society	*National Atlas of India;* State and regional resources and planning atlases.
Third Plan (1961-66)	Agriculture and industry	To achieve self-sustaining growth	DVC Atlas, Census atlases, DCHs.
Fourth Plan (1969-74)	Employment and education, especially for the poor	To combine equity with growth	Forest, Agricultural, Irrigation and Tourist Atlases.
Fifth Plan (1974-79)	Infrastructure relating to irrigation, power and transport	To go by self-reliance and control inflation	Land use, Geomorphological, mapping, urban studies.
Sixth Plan (1980-85)	Technological upgradation of agriculture, industry and energy	To work towards satisfaction of basic needs and removal of poverty	Land use and landform maps of Development blocks and districts. Modernization in cartography. Water Resource Development Atlas.
Seventh Plan (1985-90)	Productive employment	To work towards satisfaction of basic needs and removal of poverty	*Indian Ocean Atlas, Students' Reference Atlas and Socio-Economic Atlas.* Digital mapping system, wasteland maps. RS based mapping.
Eighth Plan (1992-97)	Human development-employment, fertility control, education and health	To upgrade efficiency through restructured economic management system and redefined role of government	District planning maps. Digital topographical maps. Forest maps by FSI. Block maps. Environmental sensitive maps.
Ninth Plan (1999-04)	Growth with social justice and equity	Priority to agriculture and rural development for productive employment and eradication of poverty; providing minimum basic services; environmental sustainability; promotion and developing people's participatory institutions like Panchyati Raj, co-operatives and self-help groups; and building self reliance.	Panchyat maps. Block and taluk profiles. Large scale environmental maps for formulation of polices. Digital maps for urban management. Cartographic base for decision support system.

Source: Based on Nag (1997).

The focus of the Fourth Plan (1969-74) was on employment and education. Though several new mapping activities were initiated but they did not have any direct bearing on this focal issue. But a close scrutiny will indicate that our contribution was to prepare a larger base that would ultimately help in generating employment and education. NATMO took up the preparation of the *Tourist Atlas of India* and several resource atlases on water, forest and agriculture. On behalf of the Irrigation Commission of India, Irrigation Atlas was also take up during this period.

The Next Plan, i.e. Fifth Plan (1974-79) will be considered to be a watershed in the history of thematic cartography in the country. The present name of NATMO came into existence in order to include thematic mapping in addition to atlas cartography. Apart from the preparation of the National Atlas, following two objectives were included in the manifesto:

(a) Research studies on environmental aspects and their impact on social and economic development; and
(b) Preparation and compilation of land use and land capability maps on 1: 1,000,000 scale from data collected by the relevant agencies as a pilot scheme to begin with.

The above modifications paved the way for taking up research work and large scale mapping, such as landuse mapping, geomorphological survey and mapping and urban studies. Obviously training was required to meet the new challenges. Hence training programmes were strengthened. At the instances of different bodies, project work on Dandakaranya, Sundarban, maps for land use planning of CMDA area were taken up.

The Sixth Plan (1980-85), the focus was on technological upgradation of agriculture, industry and energy. This objective could only be met by providing a sound spatial base in the form of maps. To meet the requirements at larger scales, landuse and landform maps were prepared for selected districts and development blocks. This was also the time when we start getting the satellite imagery, particularly from Landsat. Hence, satellite imagery and air photographs were widely used for such large-scale mapping. On the other hand, water was considered important for agricultural development. Due to this reason, initiative was taken to prepare the Water Resource Development Atlas as well as to revise the Irrigation Atlas of India.

The focus of the Seventh Plan (1985-90) was on more socio-economic aspect, i.e. on productive employment. Several proposals were suggested to the Department of Science and Technology and through then to the Planning Commission. Finally, three projects were accepted which were basically to strengthen the organisation infrastructure. These are:

Establishment of Thematic Atlases Division with the preparation of the following atlases:

(i) Establishment of Research, Development and Training Division
(ii) Establishment of Digital mapping and Geographical Information System

The Eighth Plan was having a focus on human development-employment, fertility control, education and health. Obviously the attention was shifting to micro level. Government took a lot of initiatives for planning and development at the local level by providing employment, authority of decision-making and infrastructure development. The 73rd and 74th Amendment of the Constitution is a marked development in this regard. Though, no plan was sanctioned *per se*, but the following activities will indicate the priorities.

District Planning Map Series (DPMS)

The Department of Science and Technology (DST) of the Government of India mooted the idea of having a Mega project on District Planning Map Series for the whole country as developmental tools for resource mobilization at grassroots levels. Both the organisations, National Atlas and Thematic Mapping Organisation (NATMO) and the Survey of India are to participate in this National Project by sharing the districts of the country in equal numbers and in phases during the Eighth Plan period. This project provides comprehensive parameters for districts and its region as scientific and technical inputs for development, thus producing for the first time in the country a Mega Project map series for spatial planning, environmental consciousness and education. These maps also give stress on ecological balance by depicting forest cover characteristics and land utilization. In addition, the maps highlight themes that aid the planners, development authorities or agencies, NGOs and decision-makers in planning. The format of the map has been designed in such a manner that variety of natural resources data like relief and slope, climatic conditions, soil, general landuse and cropping pattern, irrigation and hydrogeology. Resources like population, rocks and minerals and associated industries etc., that are related with the district are also shown in most convenient way either in main maps or in inset maps to cater the needs of users from various disciplines and levels (Fig. 1.6).

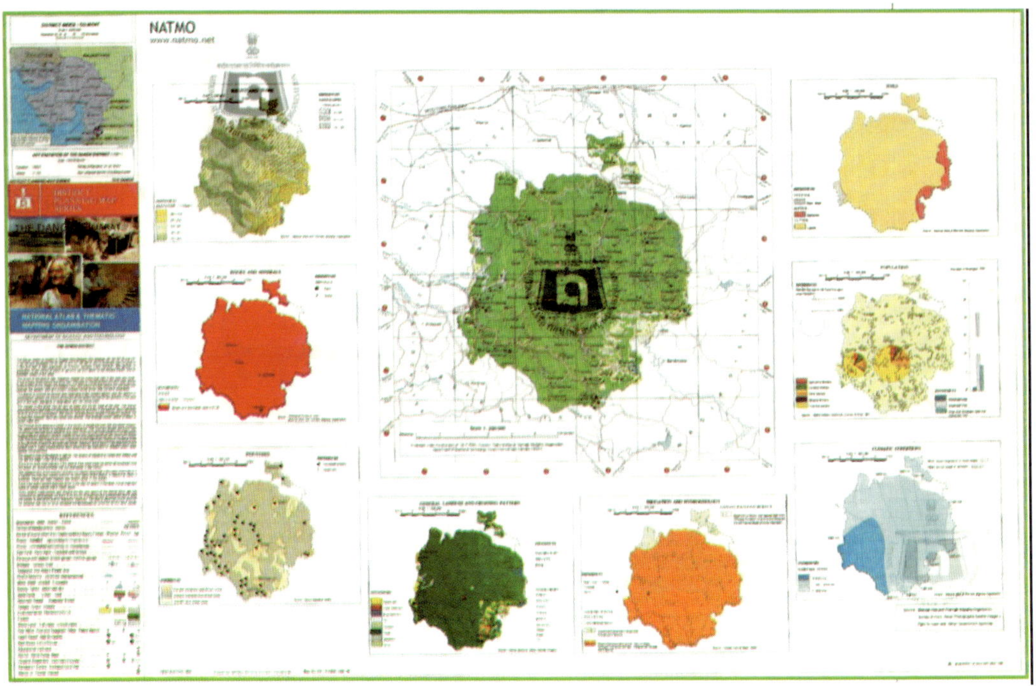

Fig. 1.6: District Planning Map Series of Dehra Dun (*Source:* NATMO)

In the above, if necessary, specific themes pertaining to the district can also be represented which may have direct bearing on the socio-economic inputs for development of special districts. These maps may further give a base for systematic generation of GIS.

Challenges in Environmental Mapping

Challenges are more prominent in the field of thematic cartography. Theme for a map has got a specific endurance. If a map is not produced within a reasonable time, the theme becomes inconsequential or of historical significance only. NATMO is bound to take up steps in order to circumvent the delays.

Thematic Cartography is facing new challenges of different kinds as well. Due to the demand of users, this field of specialisation is undergoing a serious methodological challenge. Users would like to have thematic maps for precise applications, like maps on industrial siting or on mapping of forest species. Here separate maps on different themes like slope, forest coverage, rainfall etc. are no longer acceptable. The requirements are for mapping of different components from these thematic maps that are relevant for specific applications. For example, a map on tea industry should show a certain range of slope, a required rainfall zone, persons employed in manufacturing in tea industry, and infrastructure for processing and marketing of tea and the like. Earlier thematic maps used to have such features in several maps on the concerned themes, and the users were left to interpret on their own; and obviously to have their own opinion. Another example of Zoning Atlas for Industrial Siting is given in Table 2.2.

Table 2.2: Details of the Maps for Environmentally Sensitive Zones

—Forests : Reserved, protected, unclassified
—National parks/Wildlife sanctuaries,
—Wetlands
—Flood prone areas : Protected
—Monuments : National, state, local
—Places of tourist interest/Health resorts
—Agricultural research stations
—Areas above 300m
—Coal fields
—Horizontal seismic coefficient
—Offices of state pollution control boards
—Industrial areas, Industrial clusters, Industrial estates,
—Growth centres
—Boundaries : International, state, district
—Names : State, district headquarters, important place, other sites
—Railways with stations, bridges and zonal names
—National highways with numbers, other roads,
—Rivers, canal, lakes.

Against this backdrop, the new approach in thematic cartography is challenging. We not only have to understand the spatial interrelationship between two or more themes, but also the relationship between a sub-set or sub-order with other sub-set or sub-order of another theme. Since, several such themes are to be shown on one map, thematic mapping has virtually become a multi-dimensional approach. Obviously, we have to wake up to meet this challenge in thematic cartography. A lot of research and development work has to be put in to make thematic maps more meaningful (Fig. 1.7).

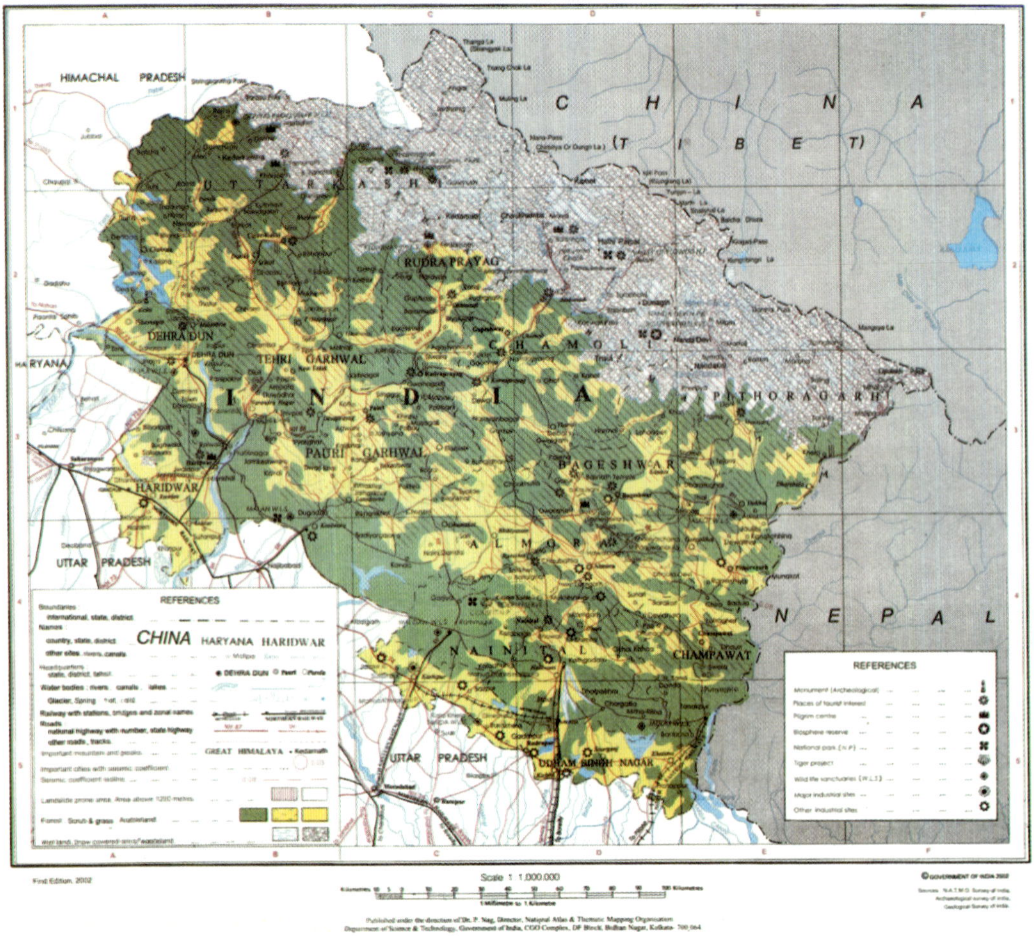

Fig. 1.7: Environmentally sensitive Zone of Uttaranchal (*Source*: NATMO)

In collaboration with the Central Pollution Control Board (CPCB), Mapping of Environmentally Sensitive Zones and Industrial Sites of several states have been carried out. NATMO will soon bring out a publication entitled *India: Environmental Status* which will contain these maps. The digital data of these are also very much in demand. Again, in collaboration with the Central Pollution Control Board, NATMO has brought out an atlas on environment. It consists of maps on physical environment, sensitive areas, pollution sources, environmental quality and pollution control. This atlas is at the national level.

Ninth Plan Priorities

The Ninth Plan is yet to get a shape. But from the documents available, some ideas about its objectives can be gathered. The focus is on *Growth with Social Justice and Equity*. The specific objectives of this plan arising from the greater reliance on market forces and the imperative policy and action are as follows:

1. Priority to agriculture and rural development with a view to generating adequate productive employment and eradication of poverty.
2. Accelerating the growth rate of the economy with stable prices.
3. Ensuring for all, particularly, the vulnerable sections of society.
4. Providing the basic minimum services of safe drinking water, primary health care facilities, universal primary education, shelter, and connectivity to all in a time bound manner.
5. Containing the growth rate of population.
6. Ensuring environmental sustainability of the development process through social mobilization and participation of people at all levels.
7. Empowerment of women and socially disadvantaged groups such as scheduled castes and scheduled tribes and other backward classes and minorities as agents of socio-economic change and development.
8. Promoting and developing people's participatory institutions like Panchyati Raj institutions, co-operatives and self-help groups.
9. Strengthening efforts to build self-reliance.

At least some of the above objectives are meant for (a) infrastructure development at the grassroots level with the help of Panchyati Raj and other institutions, (b) identification of priority areas for socio-economic change and development, and (c) mapping of environment-related issues. Regarding the former issue, the block development maps would be useful (Fig 1.8).

Further development of cartography or cartography for development in the country will depend on providing detail maps for *Panchayats*. This requires maps at 1:10,000 scale. Fortunately, IRS panchromatic data at 1:12,500 scale is available which can be used for this purpose. With the availability of Cartosat-1 2.5m data, mapping at this scale will be strengthened. Cartosat-2 is also due to be launched soon having better than 1m data. This will be beneficial for the cartographic activities and GIS-related activities in the country.

The development of thematic cartography is a feature of post-independence period. Several government institutions and departments have made valuable contributions in this regard. The office of the Registrar General of India produced a series of publications containing maps, exclusively or partially. The Census atlases of the country and states, administrative atlases and maps contained in the district census handbooks proved to be very handy for all sorts of users though they were not very high quality cartography work. The other survey institutions were also not lacking behind. Geological Survey of India, All India Soil and Land Use Survey, National Bureau of Soil Science and Land Use Planning, Forest Survey of India did produce good theme specific maps.

Fig. 1.8: Darjiling District Block Development Block Map

Department of Space

The Department of Space and the Indian Space Research Organisation (ISRO) were established with several objectives basically focusing on ways and means to utilize space for development. One of the major activities is on remote sensing. Some of the major institutions came over the national scene much before the establishment of the National Natural Resource Management System (NNRMS) in 1984. These institutions are the Indian Institute of Remote Sensing (earlier Indian Photo-interpretation Institute), National Remote Sensing Agency and Space Applications Centre—all under the umbrella of the Department of Space. These institutions are engaged in space borne data based cartography. They have best equipped technologically as well. Under the aegis of NNRMS, five Regional Remote Sensing Service Centres were established and several state and university centres were supported. But the real boost was following the successful of the IRS series of satellites, starting from IRS-1A in 1988. There are now five orbiting Indian satellites providing remote sensing data, the best in the world among the civilian satellites. As mentioned earlier, another IRS series, known as Cartosat, has been in action.

Other Government Initiatives

The other departments under the Union Government also have different activities related to cartography. The Department of Science and Technology (DST), apart from having two major mapping institutions (Survey of India and NATMO), has a major cartographic activity under its programme on Natural Resources Data Management System (NRDMS). Apart from developing a GIS software known as Gram++, it sponsors projects related to cartography on the management of natural resources. Similarly, the National Informatic Centre under the Planning Commission has countrywide set up for data collection and transmission. They are also involved in the GIS activities.

The state governments are now having state councils for science and technology. Through these councils' projects are granted which are exclusively or partly related to cartography. In some states, there are institutions funded by the state governments which are engaged in mapping activities. Further, the Directorates of Land Records in each state are also involved in map preparations. Some state governments are trying to modernize their set up. Under the state governments, there are planning authorities, municipalities and *panchayats*. Some of them do have very rich collections of maps meant for infrastructural development. State Ministries of Urban Development and Ministries of Rural Development and *panchayats* are supporting cartographic activities. With the 73rd and 74th Amendments of the Constitution, infastructural development at the grassroots level has been given high priority where maps are unavoidable.

The universities do have departments or centres on geography, geology, remote sensing, civil engineering or forestry which are involved in the preparation of maps. But they are basically meant for training and development of expertise. However, some centres under the university set up do take up major cartographic jobs, like the Institute of Remote Sensing in Anna University, Chennai or Visservaria Technological University of Karnataka or Jawaharlal Nehru Technological University, Hyderabad. Nevertheless, there are no university level courses specifically for the discipline of cartography. Though there are special papers on this subject in geography, but not a single university awards a degree in cartography. However, some universities are awarding M.Tech. degrees in GIS and Remote Sensing which do contain several aspects of cartography as well.

GIS Activities in Industries

The scenario on the other side, i.e. industry, is entirely different. Before 70s the cartographic activities were limited to some publishing houses engaged in producing school atlases and some wall maps. Only a very few publishing houses were serious in this matter. On the other hand, there were land surveyors who carried out specific surveys for layout plans of housing colonies, buildings, plantations such as tea estates, area development and the like. The maps produced by them very close to cadastral maps which not necessarily be linked to the geographical co-ordinate system. The surveyors were either passed from Industrial Training Institutes spread all over the country or were ex-Survey of India personnel.

The real boost in the cartographic activities in the Industry in India came with the introduction of the computers. The systems capable for digital mapping, image processing and GIS were having multi-national backing. The principals were located in the US or in Europe or even in Australia and Japan. The selling of the hardware and software lead them to explore in the service sectors. This sector also got benefit of the cheap-trained manpower in India. As a result, bulk jobs from abroad started coming to India, however, these very basically conversion of hard copy maps to vector data. In the major cities in the southern India like Hyderabad and Bangalore, a good number of industries are engaged in this activity. They do employ a good number of skilled persons. With the advent of internet and web technologies, the transfer of digital data has become easy. The time difference between India and the Western World is an additional advantage. However, these digital conversion centres are looking towards the west for their survival and steady flow of projects. The Indian jobs or projects are optional in their business strategy, which is very unfortunate. This is due to bureaucracy, delay in accepting tenders and payments, complicated procedures and non-availability of existing information. Their interaction with the government is negligible but they look forward to the government for jobs. The industry has clearly indicated that they are well equipped to take up projects of national importance, but the occasions are rare. Due to the delays in granting of jobs, the infrastructure cost becomes exorbitant. The cost of space, equipment and trained manpower is very high and cannot be sustained by the industry for a long time if they do not diversify their activities. This has resulted in to two major trends: (a) the small scale industry may not survive in long run, and (b) bigger players are taking up GIS projects as their side business along with supplying hardware, software or exporting services through wire.

The healthy sign on the industrial front is the Industry-Industry co-operation. With the liberalization of the economy, the multi-nationals and NRI investments have multiplied. They require maps or information in GIS format for the Indian operations, which includes retail-selling strategies. In such situations, the accuracy and reliability is not high on agenda. Further, in case of telecommunication and infrastructural development the government is heavily dependent on the industry. They require maps, preferably in digital form, or GIS data as well. Hence there is growing co-operation within the industry.

The third major force, which is likely to emerge, is the Non-Governmental Organisations (NGOs). Almost in all major and medium size cities, several NGOs have emerged who have some activities in the field of GIS. In fact their outfits can orient them selves to take up medium size GIS projects. But, as for the industry, they suffer from non-availability of existing data. Perhaps, the formation of the National Spatial Data Infrastructure or NSDI will be useful to them. But for them it is slightly easier to obtain government jobs. Due to this reason, some of the industries have a NGO outfit as well. Retired government officials, university teachers and unemployed youths are actively engaged in GIS related activities under the banner of NGOs. Further, there is a series of professional bodies who have been registered as NGOs but engaged in the promotion of such activities, such as the Indian National Cartographic Association (INCA), Institute of Surveyors, Indian Society of Geomatics, and the Indian Society of Remote Sensing. They hold conferences, workshops and training courses every year.

The GIS scenario in the country regarding the government and the industry is quite confusing. Several surveys reveal that there is definitely a scope for co-operation provided certain barriers are removed. There is strength on both sides, which has to be exploited for taking up jobs of national and international importance. The growth of the country can no longer be staggered for desirable co-operation in this specialized field. The weaknesses have to be identified and measures are to be taken for removing the bottlenecks.

Conclusion

If we glance through the planning, environmental or business priorities and the cartographic or GIS initiatives, we will find both the activities are not always compatible. Sometimes cartographic or GIS activities are late in translating their action in tune with the government or industry requirements. This is not only due to the lack of visualization, but the procedure for preparing the maps is lengthy and it takes time to bring out the product. Fortunately, digital mode, in the form of GIS, is changing this scenario. In lot of cases, the cartographic initiatives have got a boost. Nevertheless, such a healthy situation has not always been beneficial because by the time the digital data could actually be used, the information contained is already old.

Another trend, which is very apparent, is that the concept of single theme maps is becoming outdated considering the application of maps in planning and decision-making. The example is the series of maps being prepared on Environmental Sensitive Zones and Industrial Sites. These maps contain selective information from a variety of thematic maps. Hence, the situation is ripe to collect and collate spatial information where queries can be made for decision-making. In other words, GIS has to become an essential part of the development or business initiatives.

Since the planning and retail business priorities are becoming more area specific, the spatial unit of planning and development or even operational unit is becoming smaller. Hence, from national and state atlases, and later district maps, the priorities are now at block and panchyat level mapping. Similarly, in urban areas, digital maps for planning, development, management and taxation are being preferred. This makes a case for a variety of GIS applications in the day to day life. In a country like India, which is so diversified, the scope of GIS applications is still under explored.

Lastly, maps are not being always being prepared as a solution or a blue print for planning. The digital technologies have provided other options as well. It can provide digital base and software options to play around with the spatial information and work out initiatives required for given set of priorities. Decision support system with the help of GIS can look into the planning options specific to the area concerned. Obviously, such demands and trends make a case for the promotion of GIS-related activities in the country, not only for government-sponsored activities but also for business promotion. India is poised for a developed country from fifteen years from now. Both, government and business, have to be better organized for the quantum jump based on the better organized and managed information.

REFERENCES

Audet, Richard and Gail Ludwig (2000), *GIS in Schools,* ESRI Press, Redlands.

Annonymous (1999), Highlights of the Ninth Five Year Plan, *Yojana,* May, pp. 29-49.

Arunachalam, B. (1997), Cartography of the Chola administrative hierarchy, *Indian Cartogarpher*, pp. 3-8.

Chandrasekhara, C.S. and S.P. Chatterjee (1972), Planning Regions of India, in S.P. Chatterjee and S.P. Das Gupta, *op. cit.*

Dastidar, Sachi G. and Shefali S. Dastidar (1991), *Regional Disparities and Regional Development—Planning of West Bengal,* Firma KLM Pvt. Ltd., Calcutta.

Dutt, A. K. (1984), The National Atlas of India, *Geographical Review*, Vol. 74, pp 94-100.

Dutt, G. K. (1979), Thematic mapping in India—contribution of the National Atlas Organisation, *Landscape System*, Vol. II, pp 6-12.

Dutt, G. K. and P. Nag (1985), *Life and Work of Professor S. P. Chatterjee*, NATMO, Occasional Monograph-1, Calcutta.

Dutt, G. K. and P. Nag (1987), *NATMO Scenario by 2001 AD*, NATMO Occasional Monograph-4, Calcutta.

Gogate, Prasad P. and B. Arunachalam (1997), Two eighteenth century Maratha maps of Administration, *Indian Cartographer*, pp. 30-34.

Gupta (eds), *Regional Geography and Cartography*, Selected Papers Vol. 4, 21st International Geographical Congress Publication, Calcutta, pp. 49-67.

Krishan, Gopal (1997), From centralised planning to liberal economy: The development experience of India since independence, in Abha Lakshimi Singh (ed), *Land Resource Management*, B.R. Publishing Corporation, Delhi, pp. 93-105.

MacEachren, Alan (1994), *Some Truth with Maps: A Primer on Symbolization and Design*, Association of American Geographers.

Massey, Jack (2002), A tribute to social GIS in India, *Geospatial Today*, Vol. 1, issue 4, pp. 75-77.

Monmonier, Mark (1993), *Mapping It Out*, University of Chicago Press.

Monmonier, Mark (1996), *How to Lie with Maps*, University of Chicago Press.

Muehreke, Philip and Julina Muehke (1992), *Map Use—Analysis—Interpretation*, J.P. Publication.

Nag, P. (1997), Cartography and administration, *Indian Cartographer*, pp. 299-305.

Nag, P. and M. Kudrat (1998), *Digital Remote Sensing*, Concept Publishing Company, New Delhi.

Nag, P. (1992), Role of NATMO in promoting Indian Geography, in R.L. Singh and Rana P. B. Singh (eds). *The Root of Indian Geography*, NGSI Publications, Varanasi, pp. 301-10.

Nag, P. (1992), Professor S. P. Chatterjee (1903-1989): A Life Sketch, in Singh and Singh (eds), *op. cit.*, pp. 5-8.

Nag, P., A. K. Kundu and D. Saha (1992), Changing concept of national atlas in P. Nag (ed), *Thematic Cartography and Remote Sensing*, Concept, New Delhi, pp 39-49.

Nag, P. (1997), NATMO: His Dream and our action, *Manchitran*, No. 2, pp. 1-5.

Nag, P. (2003), *The Great Arc,* Department of Science and Technology, New Delhi.

Nath, V. (1972), Geography and development planning in India, in S.P. Chatterjee and S.P. Das Gupta (eds), *Regional Geography and Cartography*, Selected Papers Vol. 4, 21st International Geographical Congress Publication, Calcutta, pp. 148-61.

Rao, K. Narshima, V. Bhanuja Kumari, N.S. Sekhar, V.S.S.S.R. Murthy and S.V.L. Bhavani (1997), Relevance of cartography in planning process, *Indian Cartographer*, pp. 18-26.

TIFAC (1999), *Report on the Government-Industry Co-operation in Cartography*, Technology Information, Forecasting and Assessment Council, Department of Science and Technology, New Delhi.

2

Basic Concepts

Geographic Information System or GIS is so versatile that it is difficult either to define or to imagine or even to visualize it. By the end of the 1970s the term GIS had emerged in recognition both of common technological requirements and of the opportunity to build the system that could potentially satisfy all of various applications. Nevertheless, GIS is information about geography, i.e. information tied to some specific set of locations on the earth surface, including the zones immediately adjacent to the surface and subsurface, ocean and atmosphere. 'Spatial' is often used synonymously with or even in preference to 'geographical' in the context, although in principle it might be taken to include information that is tied to frames other than the earth surface, such as the human body (as in medical imaging) or a building as in architectural drawings (Faust *et al.,* 1991). Goodchild (1992) identifies two distinct primitive types of geographical information: (a) field information and (b) discrete information.

Field Information

Field information in which geography is conceived as a set of spatially continuous functions, each having a unique value everywhere, and information about discrete entities, where the world is conceived as populated by geometric objects that either and otherwise empty space and are characterize by attributes, such that any point in space may lie in any number of discrete entities. The field or object dichotomy underlies many areas of GIS including its data models, data qualities, and analysis and modelling.

Discrete Information

Modern GIS has enhanced the utility of the map by replacing it with a large number of coverage that can be automatically analyzed and their themes combined to give meaningful answer for decision-makers (Demers, 1999). Today there is an ever increasing recognition of the need to perform large scale mapping and map analysis operation for wide variety of traditionally manual task. Some of the examples are:

- Foresters are interested to keep an updated inventory of their timber resources. GIS is an efficient management tool for their day-tc-day operations.
- Fire departments have a need for GIS to enhance their routing capabilities to ensure rapid response in emergencies.

- Military or defence forces could use GIS to determine appropriate battle places and to organize troops movements.
- Local governments may employ GIS to develop growth and development plans and to modify zoning regulation to account for increasing population pressures.
- Businessmen are in use of GIS to market products and even to develop mailing list based on selected spatial criteria.
- Real estate companies are beginning to use GIS to isolate available housing on the basis of customer criteria such as proximity to school, type of neighbourhood or access to highway.
- Police departments are currently using GIS to compile information to characterize the movements and operational settings of suspected serial killers
- Academic disciplines such as geography, biology, geology, landscape architecture and wildlife management now have the capability to employ the technology to develop and test hypothesis concerning pattern of natural phenomenon on the earth.

The potential users of GIS are limitless and the types and number of users are growing at a logarithmic pace. More examples have been given in the following chapters.

The modern GIS has the utility of the maps by replacing it with a large number of mapped 'coverages', each with an interrelated theme. These coverages can be automatically analyzed and the themes are combined to give meaningful answers for decision-makers. The GIS is gradually changing the way we do think with maps, the way we think about geographic Information, even the way in which the geographic data is collected and compiled. The task, which was impossible with the traditional maps, is now common place. To answer such question requires access to the geographical information, which is characterized by its multidimensional nature (x, y, z coordinates and time), its large volume and high processing cost. Even to answer apparently simple questions needs that data from several sources must be integrated into consistent form and be analyzed.

In the late twentieth century, the awareness of the importance for data on topography and specific themes of the earth surface, such as natural resources, has increased greatly. Aerial photography and satellite based remote sensing data ease the means of mapping large areas with a greater degree of accuracy. The same technology also gave the resource scientist enormous advantage for reconnaissance and semi detailed mapping. The resulting thematic maps were the sources of useful information for resource exploitation and management.

GIS has been defined by its objective, as in Cowen's definition of GIS as spatial decision support system (1988). In the context of remote sensing, GIS tends to be reserved for system that integrate remotely sensed data with other types or processed data that have already been cleaned and transformed. Similarly, an atmospheric scientist or oceanographer will tend to associate GIS with system used more for multidisciplinary work and policy study, and will use other software environments for modelling and analyses within the confines of his or her own discipline. In short, because GIS implies generalized software environment which is exclusive to geographical information implying that there is a tendency for it to be most strongly associated with multidisciplinary, integrative work and applications in more narrowly defined environments.

As mentioned earlier, the term GIS is difficult to define and represents the integration of many subject areas. It is becoming hybridized and modified in order to conform to intellectual, cultural, economic and political objectives. Geographic Information Systems have three major components. These are as follows:

- Computer hardware
- Sets of application software
- Proper organizational context including skilled persons

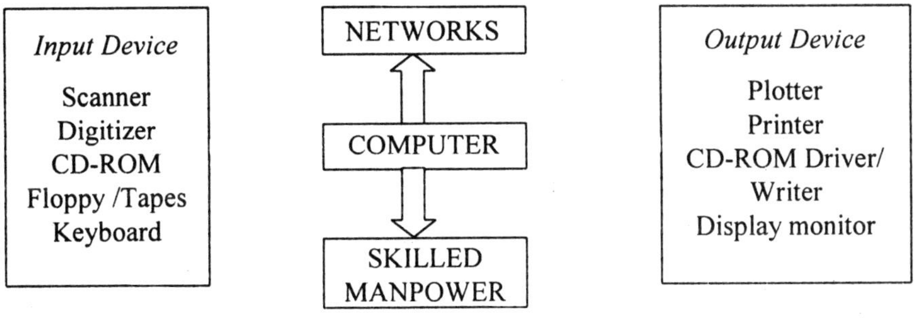

Fig. 2.1: Components of GIS

The general hardware components of GIS are presented in the above Fig. 2.1. The computer has a hard disk for storing data and operational software package (programmes), but extra storage can be provided via network or by digital tape, optical CD-ROM and other devices. A digitizer and or scanner is linked with a personal or desktop computer or work station in order to convert maps into digital form so that they can be used by the specific software. A plotter or printer or any other kind of display device or monitor is used to present the results of the data processing. Inter—local and global electronic networks using special data lines with optical fibers or over ordinary telephone lines by using a device is known as modem which provide computer communication. The user controls the computer and the peripherals (a general term for plotter, printer, digitizer and other apparatus linked to computer), via the computer screen and keyboard, added by a mouse.

Software for GIS

The software for a GIS may be categorized into five functional groups:

- Data inputting and verification or checking
- Data storing and data base management
- Data transformation and projection
- Data processing, modelling and analysis
- Interactions with users

Data Input and Verification

Data input covers all the aspects of capturing spatial data from existing maps, field observations, and sensors (aerial photography, satellite, lazar mapping and recording instruments) and covering them to a standard digital forms. Many tools are available including the interactive computer screen, mouse, and digitizer or scanner, word processors and spread sheet programmes, and devices necessary for reading data already written on a media (CD-ROM, magnetic tape etc.). Once the data have entered, there is a need to check them for errors—possible inaccuracies, omission, and other problems. This is to be done prior to the linkage of the spatial and attribute data.

Data Storage and Data Management

This means the retrieval of data from the data base and display on the computer screen. This also includes how the results of analysis are reported to the users. Data may be presented in the form of maps, tables and figures (graphs and charts) in a variety of ways ranging from the ephemeral image on the computer display screen, through hard copy output drawn on the printer, plotter or information recorded on an electrical media (CD-ROM, data etc.) in digital form.

Data Transformation

Data transformation embraces two types of operations, namely: (i) transformation is needed to remove errors from the data or to bring them up-to-date or to match them to other data sets; and (ii) the large array of analytical methods that may be applied to the data in order to solve the problems in question. Transformation can operate on spatial, topological and non-spatial aspects of the data, either separate or in combination. A few examples are given below:

- Those associated with changing of scale
- Those fitting data to new projection systems
- Logical retrieval of data
- Computation of area, perimeters, length and so on.

Many of these transformations are of general nature that one should expect to find them in every kind of GIS softwares in one form or another. Other kinds of manipulation may be extremely application specific and their incorporation into a particular GIS may be only to satisfy the particular users of that system. The kind of transformations methods available, their optimum use and misuse, the way in which sets of simple transformations may be combined in order to achieve certain types of geographical or spatial modeling are the major themes of GIS. Certain spatial transformations are necessary to ensure the integrity of a database.

Data Processing, Modelling and Analysis

It is long felt that the requirements of users to retrieve and transform data are unlimited. Therefore, most systems provide a range of interfaces by which the user can interact with the system. The simplest one is menu-driven commands that can be selected by simply pointing, checking and clicking with the mouse and this is an efficient way of performing complex function for ordinary users. An alternative one is for typing the simple commands with the help of command language. This is very much similar to word processors. Many commands in GIS are written in straightforward, modular language so that the users can understand exactly what is to be done. There are options for the users to create reproducible sets of commands that carry out well-defined operations linking several modular commands together. For this there is no need that the users must be very skilful and expert in programming.

Of course, menu driven GIS may not be ultimate solution for all the GIS problems. In many situations users may need to write some programme to meet their specific needs. Now a days, some GIS are equipped with so called macro languages—simplified and formal programming languages that may be used to link many basic applications together. The latest development, particularly for simulation modelling with GIS is to provide the user with a high level compliable programming language in which the models can efficiently be written. In other situations, the GIS is used to assemble the data for a complex model that is programmed outside the system using standard computer language (Burrough and McDonnell, 1998). After the modelling, the results of the models can be transferred back to GIS for display and query.

Interactions with Users

Geographic information comprises data about the surface, subsurface, and atmosphere of the earth, interpretations and explanations applied to those data, and an organizational framework for understanding the information. Commonly, geographic information is considered that which comes from maps, but the concept embraces any type of information, which may be positioned, on or relative to the earth. It may be acquired through direct measurement, remote sensors, *in situ* observations; defined through survey or iegal description; interpreted from data analysis or simulation. Geographic information obtains meaning from both spatial and non-spatial aspects of the data, i.e. "where" and "what".

The availability of computerized geographic data has led to profound changes in how we use and understand information about the earth (Fig. 2.2). Map-making in its traditional form has evolved into systems for enabling user-directed views of selected information. Cartographic principles are essential to successful automation of geographic information—the concepts of abstraction, classification, delineation, and interpretation is equally applicable to a computerized system. Other map-making concepts such as projection and symbology have taken new forms as coordinate algorithms and database attributes, respectively. Although earlier systems for computerizing geographic data focused on faithful automation of the components of map sheets, such as a geometric shape with a label, newer approaches consider the underlying problem of effectively modelling real-world conditions, such as a bounded area associated with its description.

Other realms of geographic information have also been transformed by the increased prevalence of automation. Remotely sensed data and direct measurements are frequently acquired as digital data straightaway, and traditional aerial photography is being exploited using new scanning technologies. Even information not traditionally considered as purely 'geographic', such as reports about a location or videotape of an event, are often incorporated into databases of geographic information.

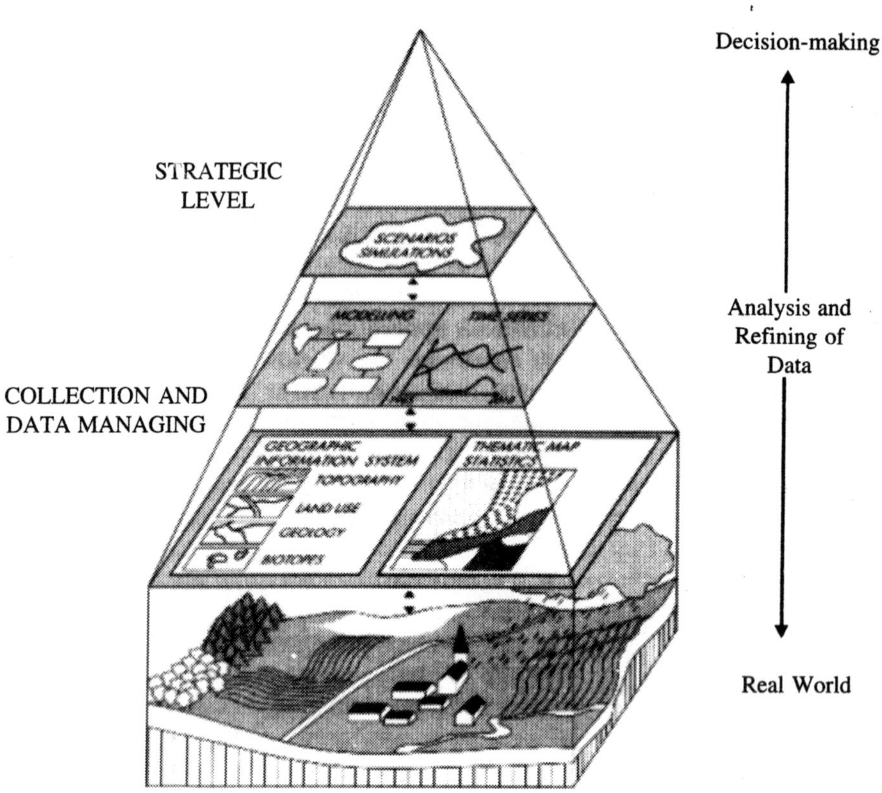

Fig. 2.2: By using GIS, a simplified world can be brought into the computer. GISs are in the process of filling the upper half of the pyramid.

The primary value of computerized geographic information is integration of disparate data types and sources. In fact, this is really the objective of GIS, i.e. to provide a systematic structure for managing complex, diverse collections of geographic information, as well as the tools and functions for display, query, map algebra, simulation, and so forth. What GIS has wrought is a new way of *thinking spatially*. Spatial analysis goes beyond simple retrieval, and allows us to explore spatial relationships and process (Fig. 2.3).

System Components

GIS have five primary components, namely, (a) *People,* computer staff, experts from various science disciplines, GIS operators, GIS experts, applications developers; (b) *Data,* which may be of type *spatial*, *temporal*, or *attribute*; (c) *Engines* that perform various data storage, retrieval, analysis, reporting, and communication functions; (d) *Interfaces* such as GUIs having widgets based on toolbcxes such as X-Windows or MOTIF, Microsoft Windows; and (e) *Hardware,* including workstations and networks, disk and tape storage, digitizers, plotters, and communications devices (Fig. 2.4).

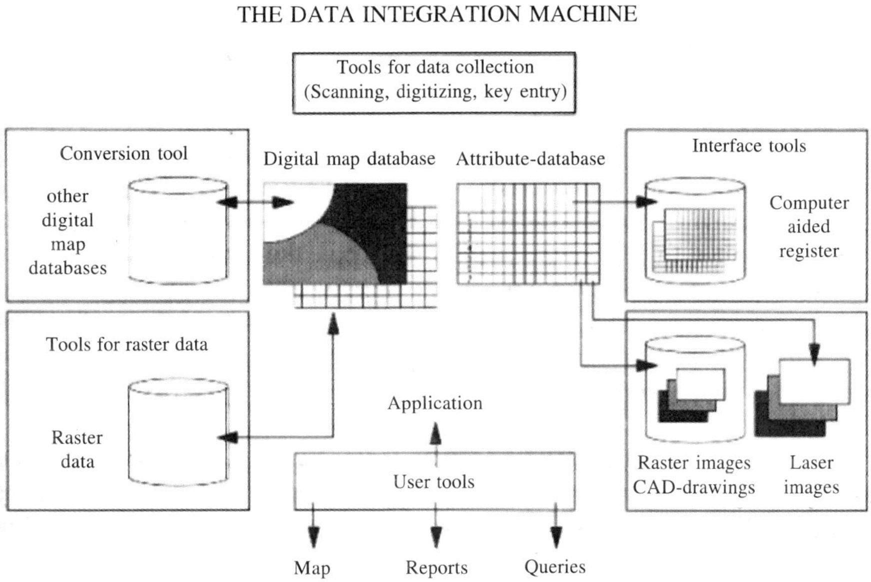

Fig. 2.3: GIS data integration machine

Fig. 2.4: Elements of GIS (*Source*: ISRI)

People

Whether a GIS is operating in a university environment or in a business or government organization, it is not in a vacuum. Inside the system there are interactions with the people responsible for day-to-day operations of the GIS, with those who operate the system, and with those responsible for project management and oversight (Fig. 2.5).

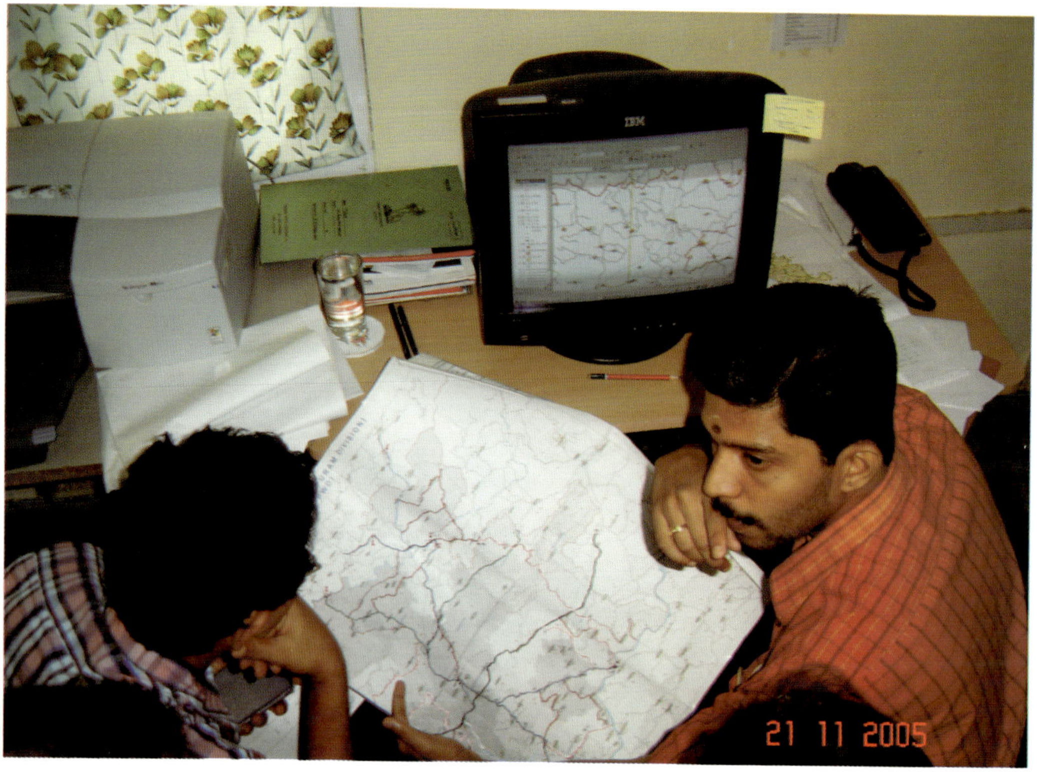

Fig. 2.5: Qualified and trained people are success key of every GIS (*Courtesy*: RSI, Hyderabad).

The system users—those who will use the GIS to solve spatial problems—are most often people who are well trained in GIS, perhaps in a specific GIS. Digitizing, error checking, editing, analysis of the raw data, and output of the final solutions to queries are the primary tasks of the system users. Whether trained in university courses or on the job, the system users will require additional training because of the constant changes both in software and in the demands of the system for new analytical techniques. In many GIS operations, this group is likely to be the most transitory, with new employees replacing those who have moved on to different positions.

System operators are responsible for the day-to-day operations of the system, more often performing tasks that allow the system users to function efficiently. Their work involves troubleshooting when programmes hang up or the complexity of the analysis requires additional insight. They are also responsible, in most cases, for training the users. In many cases system operators have experience as system users. They also know enough about hardware and software configurations to be able to make adjustments for upgrades. In addition to acting as system administrators, they frequently act as database administrators, keeping track of the security and integrity of the database to prevent possible loss or corruption of data.

GIS suppliers are responsible for providing software support and updates of the software as new and improved methods are put into the system. They may work as concert with hardware development companies to provide bundled packages of software and hardware. In many cases,

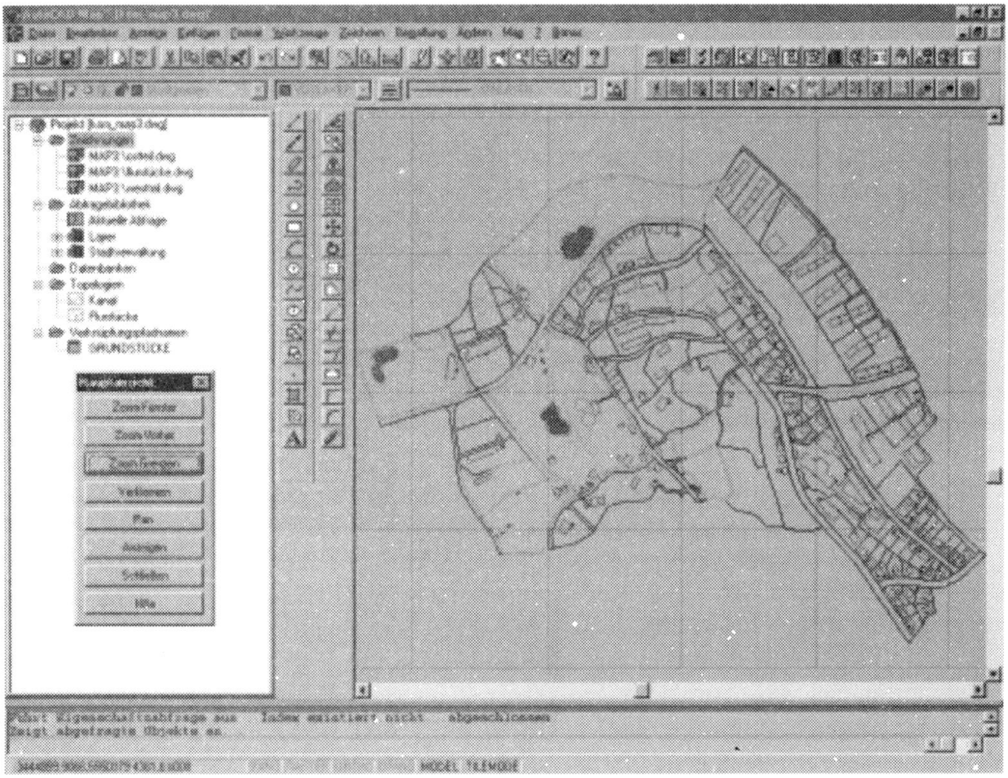

Fig. 2.6: Map production in AutoCAD Map R3

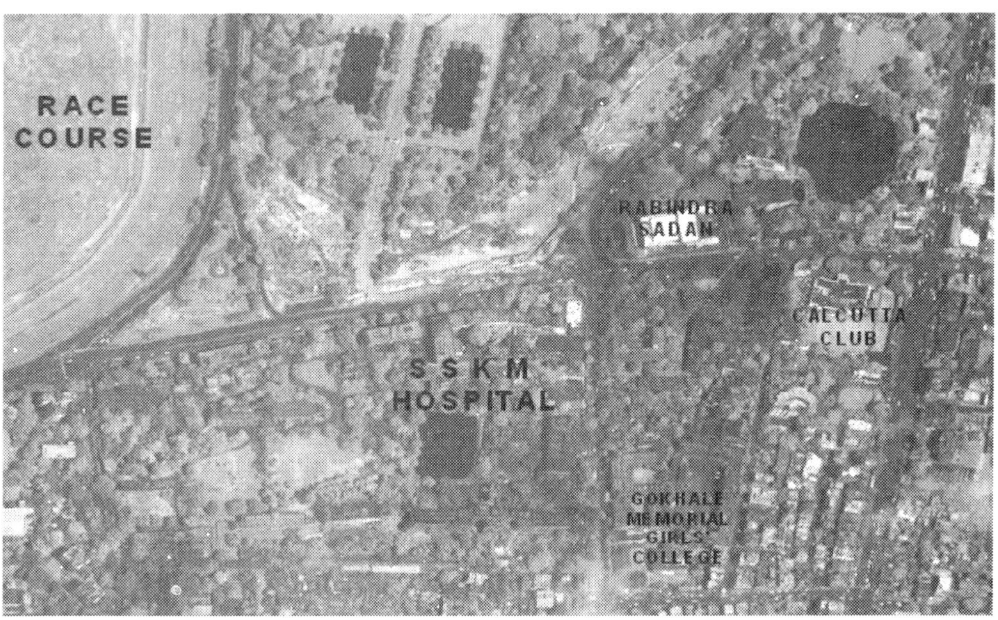

Fig. 2.7: Satellite image of the Part of Kolkata

training is provided for GIS users, either on site or at the supplier's facility, through contacts with the system administrator.

The data supplier could be either private or industry or public. The private company or industry may provide internally generated data or data obtained from public agencies modified to better fit needs expressed by the user community. Public agencies, primarily federal government agencies, provide data for large portions of the country. In many cases these data are designed for use by the agencies themselves, but the data are available to external users as well. Some public agencies offer data at cost, or in some cases at no cost at all, provided the use is for non-profit GIS operations.

Application developers are generally trained programmers who will provide user interface to reduce the reliance on specialized GIS professionals to perform common tasks. In many cases the programming is done in macro languages provided by the GIS supplier to support applications development without the need for interfacing with traditional computer languages. While in the past, most application developers were internal personnel who had years of experience with the analytical capabilities of the GIS software, today there are many small consulting firms specializing in the development of applications for third-party clients.

GIS systems analysts are group of people specialise in the study of systems design. Most often system analysts are part of a team of professionals responsible for determining the goals and objectives of the GIS system within an organisation, fine tuning the system so that it provides the right analytical techniques, and assuring the successful integration of the system in to organizational framework. Commonly the systems analysts are part of a larger consulting firm specializing in the implementation of GIS for third-party users, but in some cases they are employees of the GIS suppliers.

Data

The every GIS system should be able to absorb data in a variety of formats, not just in the native format of the particular GIS. For example, an outline map of India may be available as an AutoCAD DXF format file. The GIS should at a minimum be capable of absorbing the DXF file without further modification (Fig. 2.6). Similarly, attributes may already be stored in DBF format, and should be absordable either directly or through the generic ASCII format. Support for data formats is important to a GIS when data are to be brought in from outside (e.g. public-domain data from the internet). Ideally, the GIS software should be able to read common data formats for both raster (DEN, GIFF, TIFF, JPEG, EPS) and vector (TIGER, HPGL, DXF, Postscript, DLG). Some GIS packages have import functions only into a single data structure, usually either an entity-by-entity structure or a topological structure (for details *vide* Chapter 3: Data Structure).

For three-dimensional data, these systems usually support only the triangular irregular network. Others support only raster structures based on the grid, including the quad tree, and either convert all data into this structure or just ignore it. Some GIS packages continue to support only data in a proprietary format, available only at cost from the software vendor. A rather critical GIS function is the ability to convert between raster and vector data, an absolutely essential feature for the integration of multiple data sources such as global positioning system or GPS data and satellite images (Fig. 2.7).

Of increasing interest in recent years has been the development of GIS functions that support data in standard exchange formats. At the national and international levels, several data transfer standards have now been developed, such as the Spatial Data Transfer Standard (SDTS) and DIGEST (for details see Chapter 3). Much of the power of GIS software comes from the ability to manage not just map data but also attribute data. Every GIS is built around the software capabilities of a Database Management System (DBMS), a suite of software capable of storing, retrieving selectively, and reorganizing attribute information. The database manager allows us to think that all the data are available, that the data are structured in a simple flat-file format, and that they constitute a single entity. In fact, the database manager may have partitioned the data between files and memory locations and may have structured it in anyone of several formats and physical data models.

GIS Functions and Capabilities

A GIS is often defined not for what it is but for what it can do. This functional definition of GIS is very revealing about its use, because it shows us the set of capabilities that a GIS is expected to have. The functional capabilities can be grouped in the six categories. These are capabilities for data capture, data storage, data management, data retrieval, data analysis and data display. These six functions must always be present for the software to qualify as a GIS. For detail analysis of the GIS software features offered by different vendors, Appendix VI may be seen.

Data Capture

Getting the map into the computer is a critical first step in GIS and a main task of data capture functions. Before a map can be digitized it needs to be prepared. If the package supports scanning, the map needs to be clean, fold-free, free of handwritten annotation and marks, and on a stable base. GIS software could have support for digitizing and sophisticated editing systems for detecting and eliminating digitizing errors. Level of this support depends on software vendor. One of the essential data capturing functions is editing of the maps after they have been captured. For a vector data set, at a minimum we should be able to delete and reenter a point or a line. For raster data set, we should be able to modify the grid by selecting subsets, changing the grid spacing, or changing a specific erroneous grid value. Other functions typical of an editor are node snapping, in which points which are close to each other that should indeed be the same point, such as the endpoints of a line segment, are automatically placed into the graphic database with the identical coordinates; dissolve, when duplicate boundaries or unnecessary lines are eliminated automatically or manually; and mosaicing or zipping, in which adjacent map sheets scanned or digitized separately are merged into a seamless database without the unnecessary discontinuities caused by the lack of edge matching of the paper maps.

Another important editing ability is the ability to deal with map generalization (Fig. 2.8). Many digitizing modules of GIS systems, and certainly scanning, generate far more points than are necessary for the use of the GIS. This extra detail can complicate data reformatting and display, slow the analysis process, and lead to memory problems on the computer. Many

GIS packages allow the user to select how much detail to retain in a feature. Most will retain points that have a minimum separation, and snap together all points within a fuzzy tolerance. For point data sets, most GISs will eliminate or average duplicate points with the same coordinates. Some will allow line generalization, using any one of many algorithms that reduce the number of points in a line (Joao, 1998). Area features can be eliminated if they become too small, or can be grouped together, a process many GISs call clumping.

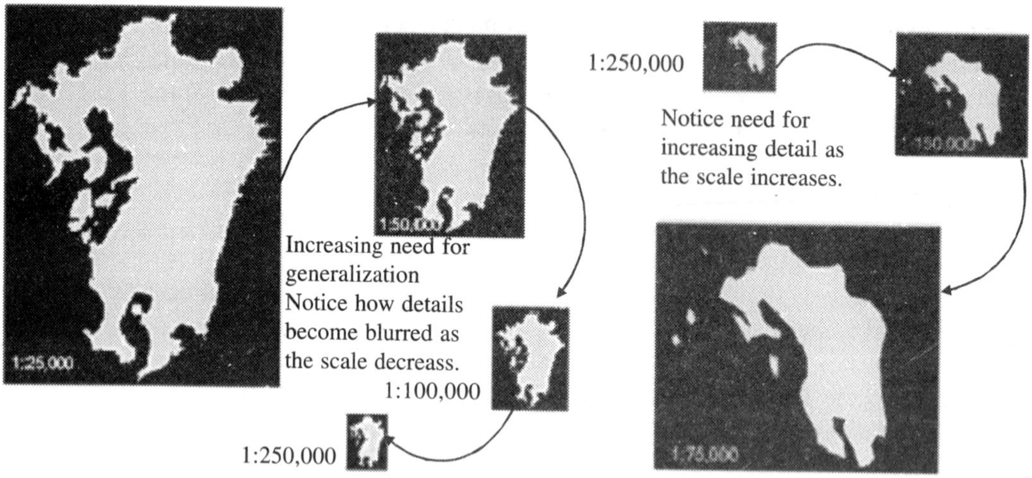

Fig. 2.8: Effect of generalization

Topological data based GIS can check automatically to make sure that the lines meet at nodes and that the entire map area is covered by polygons without gaps or overlaps. Beyond simply checking, they allow automatic cleaning of topology, snapping nodes, eliminating duplicate lines, closing polygons and eliminating slivers. The GIS user should be careful when using automatic cleaning, since the tolerances may eliminate important small features or move the features around in geographic space without accountability. Some GISs have ability for direct GPS data conversion, with survey-type data from COGO (coordinate geometry) systems, or with remotely sensed imagery.

Essential to *geocoding* capabilities, since GIS allows maps from many sources to be brought into a common reference frame and overlaid, is the *geocoding* software's ability to move between coordinate systems and map projections. Most GIS packages accomplish these operations using affine transformations. Affine operations are related to plane geometry; they manipulate the coordinates themselves by scaling the axes, rotating the map, and moving the coordinate system's origin. In some cases, when no good control is available, maps must be statistically registered together, especially when one layer is a map and one an image or photograph. The statistical method known as *rubber sheeting* or *warping* is used for this and is a function inside many GIS packages.

In order to be useful, a GIS must provide tools above and beyond the editor to check the characteristics of the database. Checking the attributes is the responsibility of the database manager. The database system should enforce the restrictions on the GIS that are specified during the data definition phase of database construction and stored in the data dictionary. Most of this checking is done at data-entry time. It checks to determine that values fall within the correct type and range.

Data Storage

Data storage within a GIS has historically been an issue of space: (a) usually how much disk space the system requires and access, or (b) how flexible a GIS is in terms of making the data available for use. The massive reductions in the cost of disk storage, new high-density storage media, and the integration of compression methods into common operating systems have made the former less critical. Although disk storage is less critical than in the past, it can still be a constraint. GIS software on a micro-computer can occupy tens of megabytes even without data, and on a workstation perhaps hundreds of megabytes. As data become higher resolution, as more raster layers are used, and as finer and finer detail becomes available, many GIS data sets can easily move into the gigabyte range size.

Database Management System

Database management system (DBMS) allows data entry, data editing and supports tabular or other list types of output, sometimes independent of the GIS. Many operations on data are very important from a mapping perspective. For example, very often maps captured from different sheets must be merged together, or sometimes a mask must be placed over the data to exclude features entirely from the GIS. Even more complex, sometimes line features such as latitude/ longitude grid, a river, or a political boundary must be sectioned up or have points added as new features or layers are introduced. This feature, called dynamic segmentation, can be done automatically by the GIS.

Data Retrieval, Analysis and Display

Another major area of GIS functionality is that of data retrieval. GIS supports the retrieval of features by both their attributes and their spatial characteristics. Geographic analysis requires a close association between the spatial elements and their attribute data. Previously these issues have been held and managed separately within automated cartography packages and databases respectively. GIS provides the technology to perform more sophisticated analysis which makes use of the links between the two. Queries to a GIS can thus be graphics driven or data driven. Graphics driven queries involve spatial based searches for objects and retrieval of the associated attribute data, or point and query functions to select displayed features and retrieve the associated information. Data driven queries involve the use of data values selectively to display the matching spatial features or the use of attribute values to determine shading pattern or colour coding of the relevant spatial elements. Operations to retrieve, summarize, selectively display and analyze both the alphanumeric and graphical data are usually embedded within a GIS. Examples of the types of spatial analysis usually available within a GIS include (Fig. 2.9).

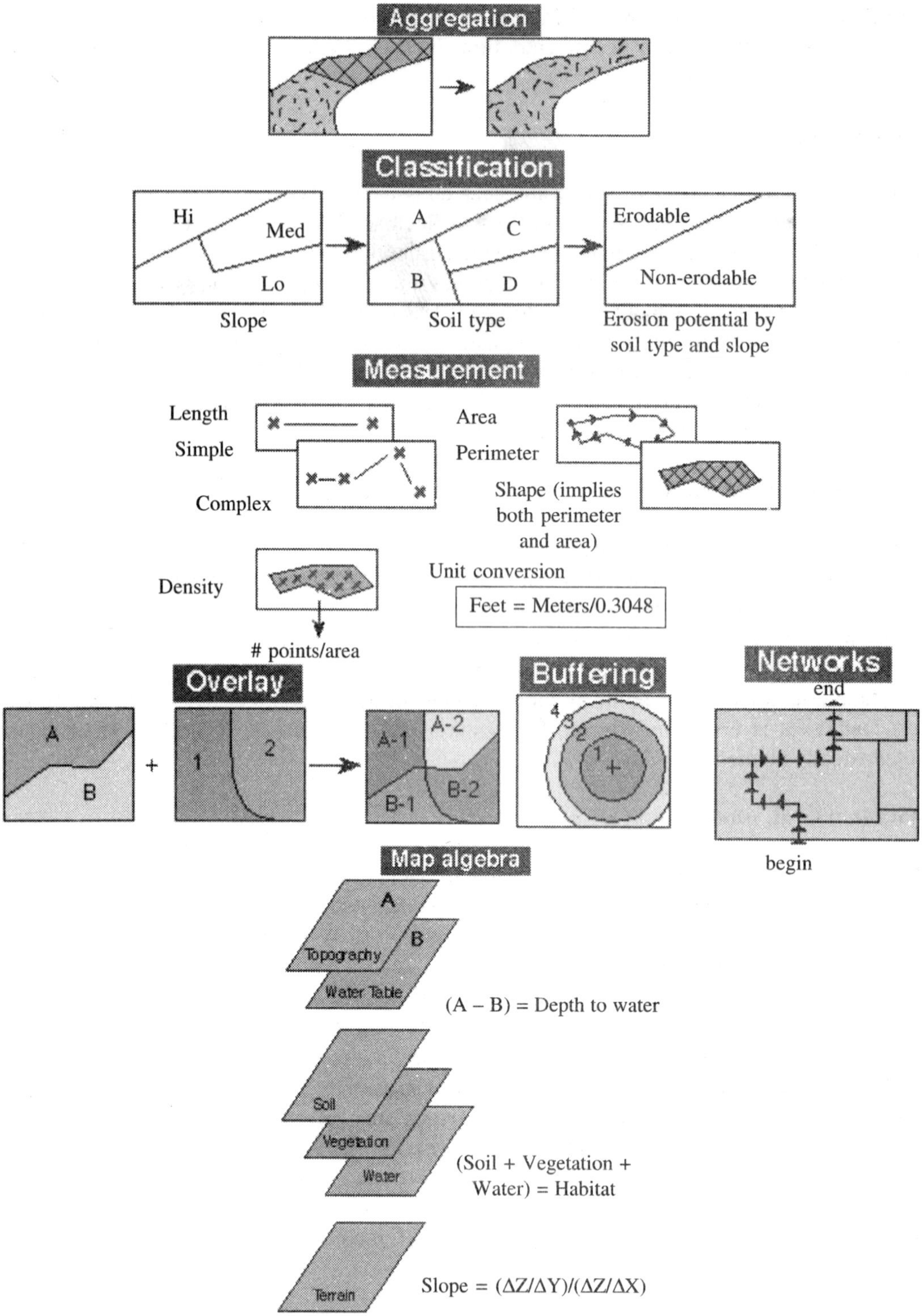

Fig. 2.9: GIS functions

Choropleth Mapping

Area features are classified according to their attributes. A legend acts as a look-up table with each range of attribute values being associated with a particular colour or shading pattern, e.g. shading enumeration districts according to population density.

Buffer Generation

Boundaries are created around points, lines or areas at an equal distance in all directions. The result being circular, corridor or polygon shaped features, respectively, which represent areas at set distances from the original object, e.g. creation of buffers around an airport to calculate zones of different noise intensity.

Polygon Overlay

Area features on one data layer are overlaid onto those of other data layers in order to calculate areas which have a certain combination of attributes or lack certain values, e.g. overlaying soil, drainage and slope data layers will provide information on the land most suitable for crop production.

Contouring

The interpolation of lines of the same elevation from spot height data, e.g. the calculation of height contours from which to generate slope and elevation data for land surfaces or any other statistical surfaces (Fig. 2.10).

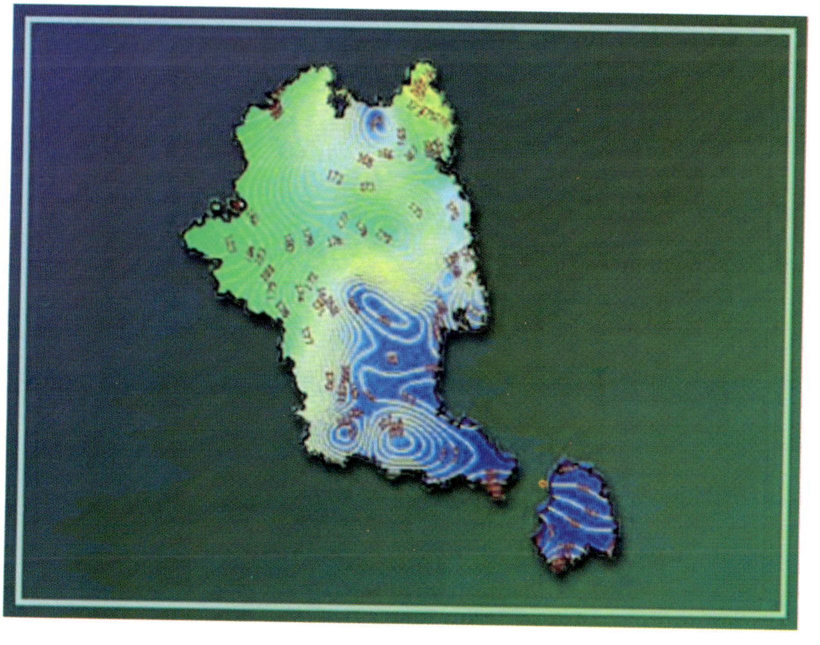

Fig. 2.10: Contours (*Source*: NRDMS)

Terrain Analysis

The creation of three dimensional views or Digital Terrain Models (DTM) from height data, e.g. a three-dimensional view of a landscape can be used to assess the visual impact of an afforestation programme or for watershed planning (Fig. 2.11).

Fig. 2.11: Digital terrain Model Created from Contours (*Source*: Survey of India)

Network Analysis

Tracing through a network of connected line features in order to simulate flows of traffic or water. Often used to calculate shortest or quickest paths between two points in a network, e.g. use of a network to calculate optimum routes for emergency vehicles (Fig. 2.12).

Area and Length Calculations

The calculation of the area of polygon features or the length of linear features. Often these simple calculations can be constrained to include only those features with attributes satisfying certain selection criteria, e.g. calculation of the area of parkland within a city which needs to be maintained or calculation of the length of pipeline within a sewage network which needs replacing. Examples of how the attribute information may be manipulated within a GIS include the retrieval of data values for selected features, the computation of attribute means and the calculation of statistical summaries and samples.

Output from a GIS can be in the form of an interactive computer display of the graphic and/or attribute data or as traditional paper maps, tables, graphs and reports. Automated cartography techniques are employed to produce the graphical output from GIS. Commonly used output devices include pen plotters, inkjet printers, electrostatic plotters, laser printers and screen copiers. Export to other systems such as statistical packages, word processors and multi-media will usually require the GIS to be able to convert internal data into other standard formats.

Interfaces

GIS was heavily influenced by the types of operating systems in use. Early operating systems were quite unsophisticated but were used with GIS nevertheless. Among these were IBM mainframe operating systems, MSDOS by Microsoft and DEC VMS.

In the microcomputer environment, the GUI-based operating systems include Windows and Windows NT, now updated as Windows 2000XP, Apple Macintosh GUI and IBM OS/2. These operating systems added two critical elements to the microcomputer's capabilities, multitasking and device independence, meaning that plotters and printers could be taken out and assigned to the operating system instead of the GIS package, in somewhat the way that printing and screen fonts are handled centrally, rather than duplicated in every Windows package.

One system that had encompassed these capabilities since its inception, and that swept the workstation environment, was UNIX. It has been the dominant workstation environment for two reasons: first, because it has complete integrated network support, and second, because several full GUIs exist for UNIX in the public domain, the most important being the X-Windows system.

X-Windows implementations of most leading GUIs exist, including OpenLook and the Open Software Foundation's MOTIF interface. In many UNIX systems, the user can switch the GUI to suit particular needs or applications. A full GUI programming toolkit, including such tools as Xt, Xview, Tcl/Tk and the X-Windows libraries Xlib are part of the X-Windows release.

As a final benefit, several versions of UNIX and all of the GUI systems run extremely efficiently on microcomputers, including shareware UNIX releases such as Linux, not only outperforming the Windows type GUIs, but being available free or as shareware on the Internet or CD-ROM. A key element here has been the Free Software Foundation's releases, including GNU (GNU is NOT UNIX) versions of virtually every key element of UNIX.

Thus two main avenues for GISs have evolved as far as operating systems are concerned. On the microcomputer platform a lingering set of DOS applications is rapidly being rewritten for the updated versions of Microsoft's Windows. In this GIS environment, the number of systems installed, the mobility of laptop and subnotebook computing, and low cost of software have been major strengths. On the workstation platform, UNIX and X-Windows, often with MOTIF as the GUI, reign supreme. This work environment has led to high end applications, large data sets, networking, depth software, and high quality graphics. Both are healthy and prospering workplaces for GIS.

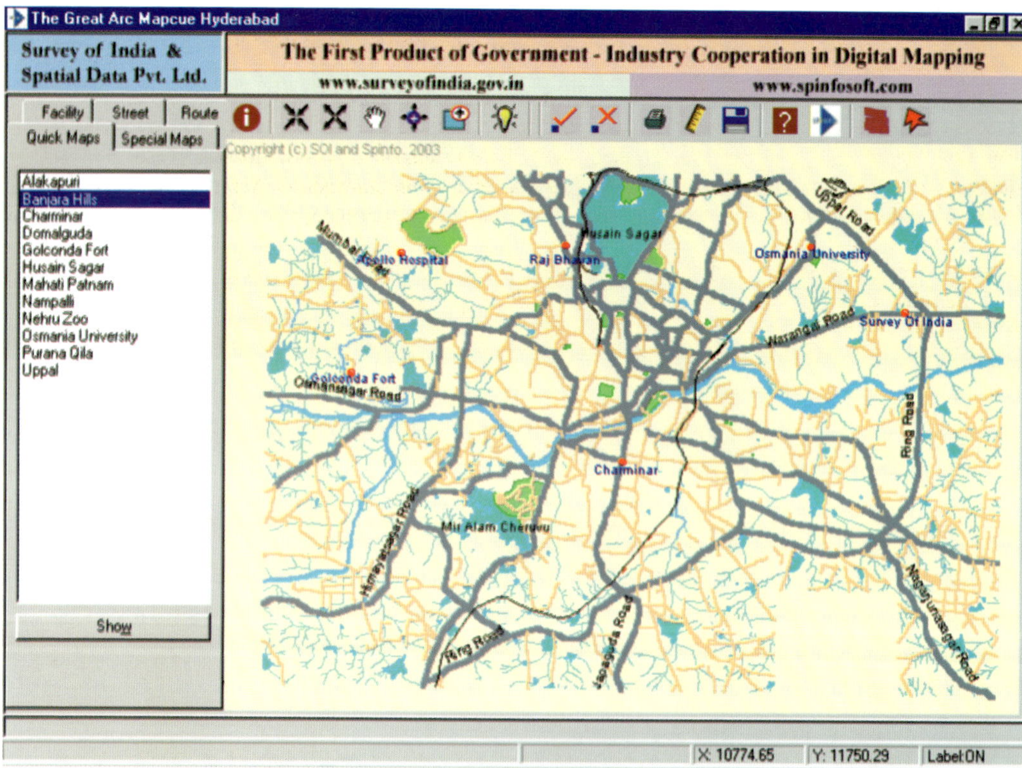

Fig. 2.12: Network of roads in Hyderabad (*Source*: Survey of India and Spinfo, Bangalore)

Summary

In any set up, big or small, the geographical information system or GIS will be having different combinations of the above hardware. As mentioned earlier, the choice of hardware is software dependent as well. The prices of hardware and even software are falling. Further, in case of the latter, the licencing policy also determines the cost. In addition, newer items are coming to the market everyday. Hence, the decision to have hardware, software and networking is a complex, technically and financially. In the following chapters we will find several examples of such combinations procured by different agencies to bring out GIS based business solutions.

REFERENCES

Burrough, P.A. (1992), Development of intelligent geographical information system, *International Journal of Geographical Information System*, Vol. 6, No. 1, 1-12.

Burrough, P.A. and McDonnell, R.A. (1998), *Principles of Geographical Information Systems*. Oxford University Press Inc. New York.

Cowen, D.J. (1988), GIS *versus* CAD *versus* DBMS: What are the differences ?, *Photogrammetric Engineering and Remote Sensing*, Vol. 54, pp. 1551-5.

Demers, M.N. (1999), *Fundamentals of Geographic Information Systems*, 2nd Ed., John Wiley and Sons. Inc.

ESRI (u.d.), *ESRI Map Book*, ESRI Press, Vol. 19, Redlands.

Faust, N.L., W.H. Anderson and J.L. Star (1991), Geographic information systems and remote sensing future computing environment, *Photogrammetric Engineering and Remote Sensing*, Vol. 57, Issue 6, pp. 655-68.

Goodchild, M.F. (1992), Geographical information science, *International Journal of Geographical Information System*, Vol. 6, No.1, pp. 31-45.

Joao, Elsa Maria (1998), *Causes and Consequences of Map Generalisation*, Taylor and Francis.

Maguire, D.J.,Goodchild, M.F. and Rhind, D., (1991), *Geographic Information System*, Longman, London.

Reddy, M. Anji (2001), *Remote Sensing and Geographical Information System*, 2nd. Ed. BS Publication, Hyderabad.

Sengupta, Smita (1992), Geographical information system, in P. Nag (ed), *Thematic Mapping and Remote Sensing*, Concept Publishing Company, New Delhi, pp. 69-78.

3

Data Structure

Geographic database management systems are more complex than database management systems used for banking, library searches, airline bookings and medical records. Three general features of the data within a GIS must be maintained are: (a) information on the position of the feature being stored, (b) topological information on the spatial relationships of the features (topology is the way in which geographic features are connected and provides a mechanism to identify the positional relationships among features), and (3) attributes of the feature (Burrough, 1986). The spatial component of geographic data describes the location of a feature and the possible topological relationships among features. The attribute component describes various attributes of the feature. There are five kinds of data to be represented in a GIS (Figs. 3.1 to 3.5).

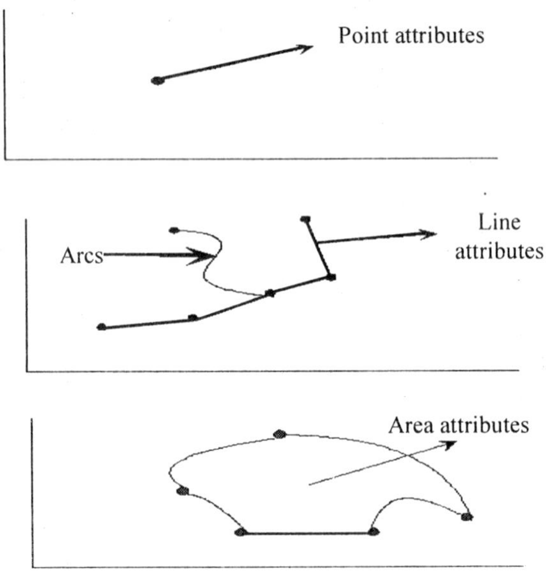

Fig. 3.1: Simple GIS display

1. ***Point features,*** e.g. location of soil samples, boreholes, manholes, rain gauges, burst water mains, pumping stations, trees, buildings. The points consist of a number of nodes with no thickness and are often referred to as zero dimensional. One method to store a point feature in a GIS is as a table in the data base management system.

Point ID	*X Coordinate*	*Y Coordinate*	*Pointer to attribute data*
1	30123.6	19782.4	Point 1
2	30167.3	19745.7	Point 2
3	34952.2	19648.1	Point 3

Where the pointer to attribute data is a link into another database table when other data about that point is kept, for example that it represents an access chamber to a sewer system and that it has properties such as date of construction, condition, size, material etc. This is a link into a full database management system so further relations are permitted from this point.

2. Linear features, e.g. roads (on small scale maps), rivers, pipelines, power lines, elevation contours. The nodes are linked with arcs, each with a number of vertices (the simple arc is a straight line). Between vertices the arc is usually considered a straight line but curved links are possible. Line data can either be non-branching lines, or tree or network structures. In a network there is more than one route between two nodes. This data has one dimension, i.e. it does not have thickness and care must be taken in the definition of the system that a loop is not confused with a polygon. A simple structure for a line feature or network is:

Line reference
Attribute Pointer
arc 1
arc 2
...
etc.

Arc reference	X coordinate	Y coordinate
Node 1	30123.6	19782.4
Node 2	30167.3	19745.7
Vertex 3 etc.	30952.2	19648.1

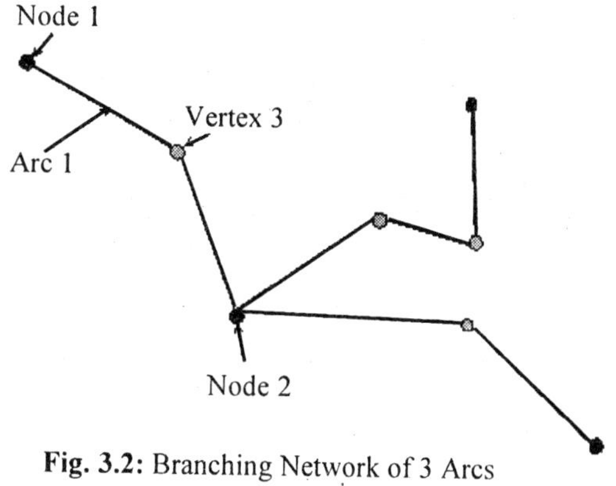

Fig. 3.2: Branching Network of 3 Arcs

3. Areas or polygons with common properties, e.g. pressure zones, catchments, contributing areas, soil association mapping units, climate zones, administrative district areas, buildings and other land cover. The polygon consists of a number of arcs or linear features that form a closed loop without crossing over one another. The arcs are usually straight between vertices but may be curved.

Polygon Reference	*Attribute Pointer*
3	Point 1
X Coordinate	*Y Coordinate*
30123.6	19782.4
30167.3	19745.7
30952.2	19648.1
30123.6	19782.4

Simple Polygon Structure

The simple polygon representation shown above where a quadrangle is represented, as used in CAD (or DXF format), is of little use in GIS. The three major problems with simple polygons are:

(a) The boundary between two polygons needs to be stored twice. There is always a possibility that the nodes for each boundary polygon are in slightly different positions resulting in artificial gaps between polygons, or slivers where an area is assigned to two or more polygons (Fig. 3.3). These problems are particularly acute with manually digitizing.

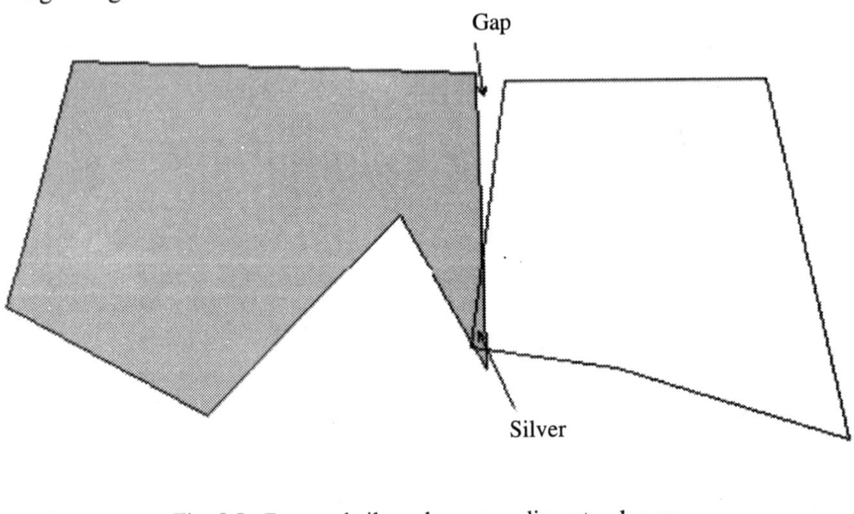

Fig. 3.3: Gaps and silvers between adjacent polygons

(b) When manually digitizing it is possible to accidentally cross over from one polygon to another creating a totally false polygon or to pass from one node to another in incorrect order giving rise to a weird polygon (Fig. 3.4).

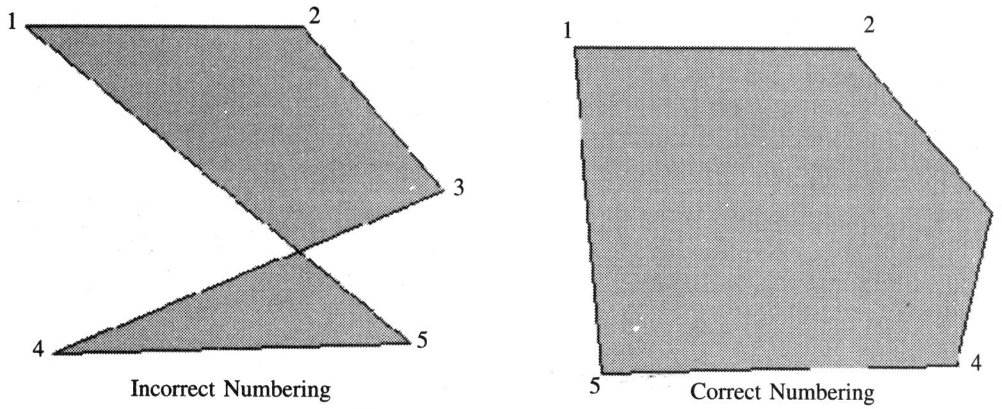

Fig. 3.4: Weird polygons

(c) Complex geographical objects are difficult to represent, for example islands or disjointed polygons (Fig. 3.5). If we consider an example from urban drainage where a garden area is completely surrounded by car park, for example, at a prestige office complex, then if we calculate the area of simple polygons the area of the car park will include, erroneously, the area of the gardens and grossly overestimate the impermeable area.

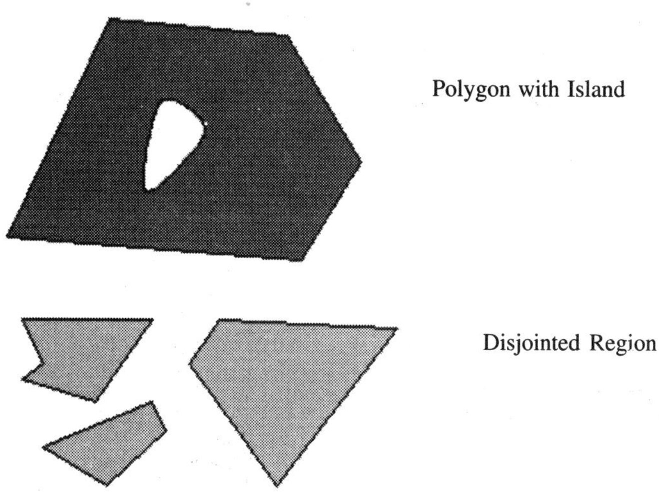

Fig. 3.5: Polygons with Islands and disjointed regions

In GIS, therefore, area data is represented as topological structure in one of a number of ways. The ArcInfo method of storing this information is shown in Fig. 3.6. A separate list is used to hold information about islands and disjointed structures. Different themes can be represented on the same coverage and there is no requirement that polygons do not overlap. For example, a single coverage may contain polygons representing land cover, whereas another

other polygons may contain the contributing areas to inlet nodes of a storm water drainage system. The polygons naturally overlap and the intersections of these polygons provides one of the main uses of GIS and is known as overlay to reflect the graphical process of overlaying one theme upon another (Fig. 3.6).

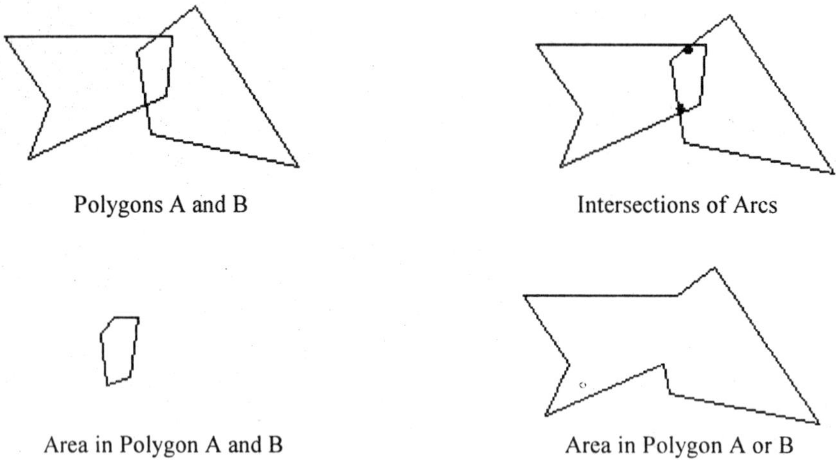

Fig. 3.6: Intersections of polygons

4. Actual or potential surfaces, e.g. ground elevation, variation of mean annual temperature, spatial distributions of rainfall, population densities.

5. Temporal elements, e.g. changes in land use over time, changes to a pipe network, rainfall records or stream flow records. These are not well represented in current GIS technology, but newer object oriented GIS should make this more readily available

Representation of Spatial Data

Spatial data are represented in GISs in two very different ways. Fig. 3.7 shows the two different ways that a stream could be represented in a GIS. Spatial data can be represented as either rasters or vectors. In raster format, a grid is used to represent the study area. The location of features in the study area is depicted by the values in the cells overlaying the feature. Vector data represent geographic features by coordinates of points, lines, and polygons. Points represent small features such as wells, towers, or nest locations. Linear features such as roads and streams are represented by lines. Areas such as cities, forests, wetlands, and soil units are represented by polygons. Polygons are bounded on all sides by a series of straight line segments.

Raster Data

Raster data are stored in the computer as a matrix. The cells are referenced by lines and elements (Fig. 3.8). In the simplest form, each line is a computer record. Each record will contain the values for all elements in the line. Any cell not containing a feature would have the value of "0". In the simplest raster system, the value stored for each cell is the attribute component of the geographic data. In Fig. 3.8(a) cells with value of "1" are forests, cells with value of "2" are croplands, and cells with value of "3" are rangelands. In more sophisticated raster systems, the cell value is a label that will link to records as an attribute file. In the above example, cells labeled as "1" could have many attributes, such as species composition, age of forest stand, and estimated volume of marketable timber.

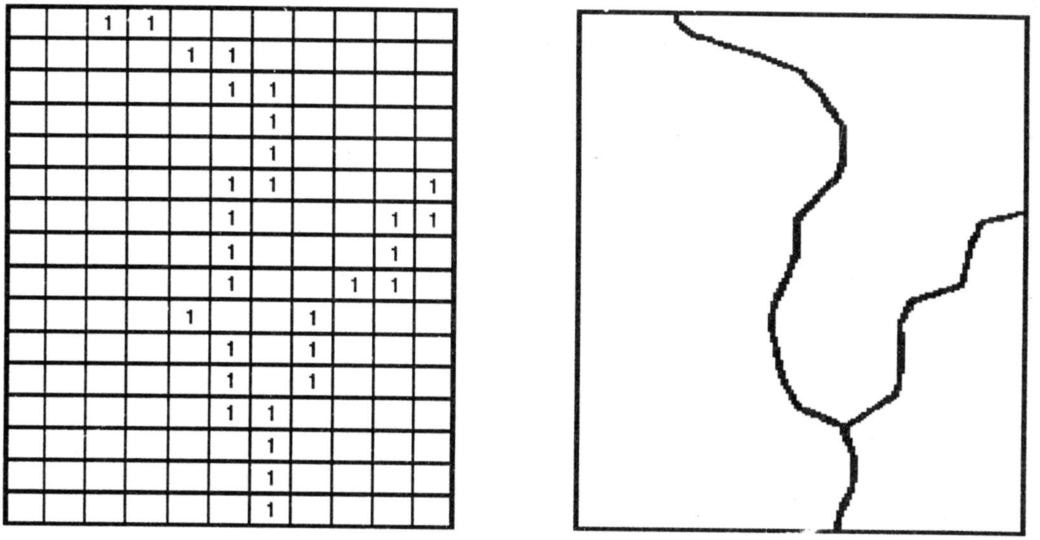

a b

Fig. 3.7: Raster and vector data format

Note: A stream can be represented in a GIS either by a raster (a) or vector (b) format.

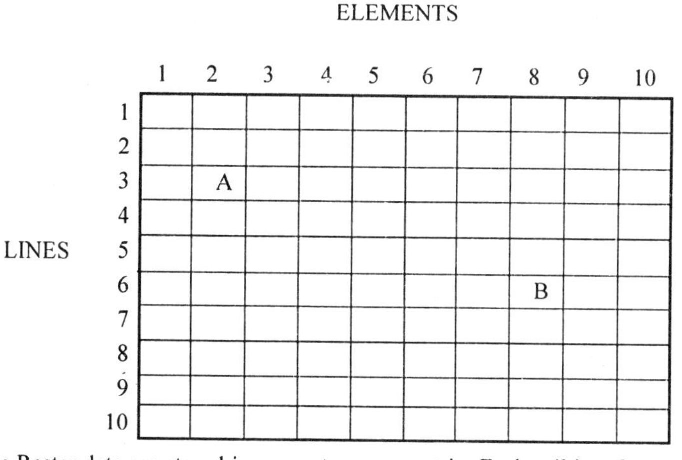

Fig. 3.8 (a): Raster data are stored in computers as a matrix. Each cell is referenced by its line and element number. The example shown is for a small file with 10 lines and 10 elements. Cell A is located at line 3, elements 2. Cell B is located at line 6, element 8.

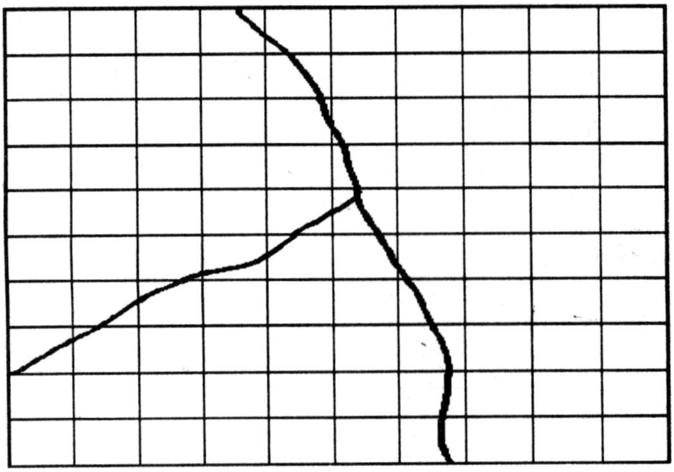

Fig. 3.8 (b): Raster and vector data storage

Because the raster system is strictly a two-dimensional matrix, various types of geographical data are stored as different layers or overlays in the GIS (Fig. 3.9). One layer may contain land use/land cover, another layer may contain wetland data, and another layer may contain information on the transportation system. The raster data formats are as follows:

Table 3. 1: Raster data formats

- Raster ADRG—ARC digitized raster graphic
- ADRI—ARC interchange raster image file
- BIL—Band interleaved by line
- BIP—Band interleaved by pixel
- BMP—Bitmap file
- BSQ—Band sequential
- CADRG—Compressed ADRG
- CEOS—Radarsat image file
- CIB—Controlled image base file
- CGM—Computer graphics metafile
- DEM—Digital elevation model
- DIB—Device independent bitmap
- DOQ—Digital orthophoto quad
- DTED—Digital terrain elevation data
- ECW—ERMapper compressed image file
- EPS—Encapsulated postscript
- GeoTIFF—TIFF image with georeferencing
- GIF—Graphic interchange format
- GIS—ERDAS Imagine file
- IMG—ERDAS Imagine native format
- JFIF—JPEG file interchange format
- JPEG—Joint photographic experts group format
- LAN—ERDAS Imagine 7.x file
- MrSID—MrSID compressed file
- NITF—National imagery transmission format
- PCX—Bitmap graphics file
- PICT—Apple's graphics file
- PNG—Portable network graphics
- PSD—Adobe Photoshop image file
- RLC—Run length encoding compressed format
- RPF—Raster product format
- SDTS—Spatial data transfer standard
- SUN/RAS—Sun raster file
- TGA—Targa graphics file
- TIFF—Tagged image file format

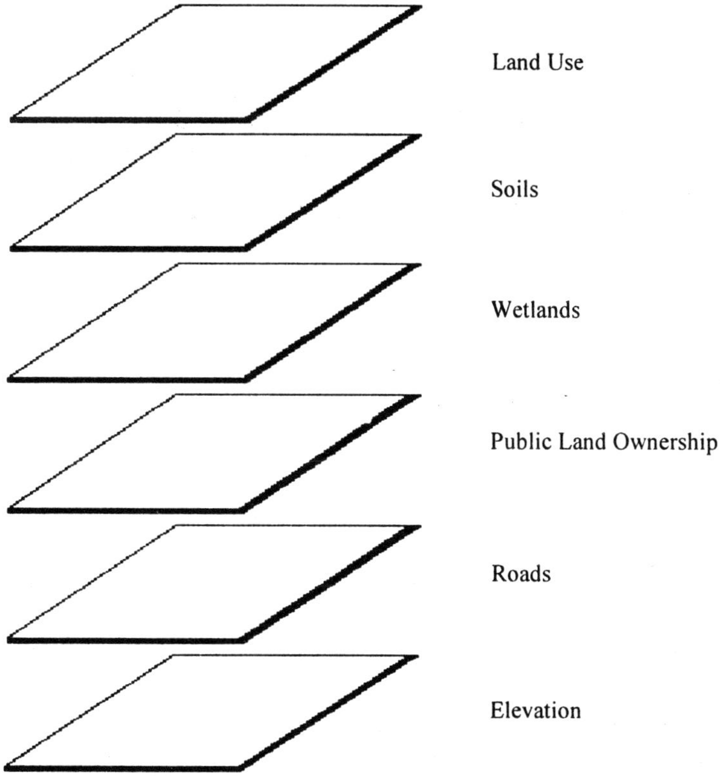

Fig. 3.9: GIS data layers

The user of a raster system must determine the size of the cells to be used. This size is referred to as spatial resolution. The cell size can vary tremendously depending upon the size of the study area and the objectives for the GIS. Cell size as large as 20 ha for state or regional planning may be adequate. For a wildlife management area, a cell size of 0.05 ha or smaller might be required, depending upon the application of the GIS and the size of the wildlife management area. Storage requirements increase drastically as the cell size is reduced. Reducing the cell size by one-half will increase the data storage requirements by a factor of four. Conversely, as cell size increases, the precision of the representation of the land feature is reduced. Choosing the appropriate cell size for a particular GIS application is a compromise between cost of data storage and computer time and reliability of the representation of the land feature.

Vector Data

Vector data provide for high precision in representing the location of features. Aronoff (1989) described how vector data can be used to define the location of a point, a line, and an area. A point is represented by a simple pair of coordinates. The line is represented by an ordered list of pairs of coordinates. The area is represented as a polygon with ordered pairs of coordinates that close the polygon (the first and last pair being the same).

The coordinates can be any arbitrary units but usually are stored as UTM, state plane, or latitude and longitude coordinates. The first vector system used simple techniques to store the X and Y coordinates for polygons. In this simple system the coordinates for the common boundary between two areas were stored twice, once for the first area and again for the adjacent area. These duplicate storage techniques, simplified computations and plotting but wasted storage space and, more importantly, provided no information as to adjacency or connectivity of geographic features (topology). Most vector systems now use topological models for representing the location of areas (Aronoff, 1989).

Topology deals with the spatial relationships between connecting or adjacent features (e.g. arcs, nodes, polygons and points). For example, the topology of an arc includes its from— and to-nodes, and its left and right polygons. Topological relationships are built from simple elements into complex elements: points (simplest element), arc (sets of connected points), areas (set of connected arcs) and routes (sets of sections, which are arcs or portions of arcs). Redundant data (coordinates) are eliminated because an arc may represent a line feature, part of the boundary of an area feature, or both. Topology is useful in GIS because many spatial modelling operations do not require coordinates, only topological information. For example, to find an optimal path between two points require a list of the arcs that connect to each other and the cost to traverse each arc in each direction. Coordinates are only needed for drawing the path after it is calculated.

In topological models (Fig. 3.10), a polygon is defined by a series of arcs. Arcs begin and end at nodes, which occur wherever two or more arcs meet. Each arc is defined by a series of coordinates, starting with the coordinates for the beginning node and ending with the coordinates for the ending node. Topological relationships are stored in three tables. The polygon topology table describes the arcs that bound each polygon, the node topology table describes the arcs that end at each of the nodes, and the arc topology table describes which end points (nodes) occur on each arc and which polygons are to the left and right of each arc. These three topology tables provide the tools required to efficiently determine the positional relationships of one feature to other features. A coordinate table defining the coordinates for each arc also is used in topological models. In addition to these topological databases, the attributes for the features are stored in an attribute database.

The vector data formats are as follows:

Table 3.2: Vector data formats

- AOI—ERDAS Imagine area of interest
- ARC—ARC export file
- ARCE—ArcInfo export format
- ARCG—ArcInfo generate format
- BNA—Atlas GIS file
- CDF—Cartographic data format
- CGM—Computer graphics metafile
- DAT—MapInfo native data format
- DFAD—Digitized feature analysis data
- DGN—Microstation design file
- DLG—Digital line graph file
- DWG—AutoCAD drawing file
- DXF—Drawing exchange format
- E00—ArcInfo export format
- EPS—Encapsulated postscript file
- GDF—Geographic data file
- GRID—ArcInfo GRID files
- IGDS—Interactive graphic design software export format
- IGES—Initial graphics exchange standard format
- MGE—Intergraph file format
- MIF/MID—MapInfo interchange format
- OVR—ERDAS Imagine annotation layer
- PICT—Apple's graphics file
- SHP—ESRI shapefile
- SIF—Standard interchange format
- SDTS—Spatial data transfer standard
- TAB—MapInfo format
- TIGER—Topologically integrated geographic encoding and referencing/line census file
- VPF—Vector product format
- VFR—Vector relational format
- WMF—Windows metafile format

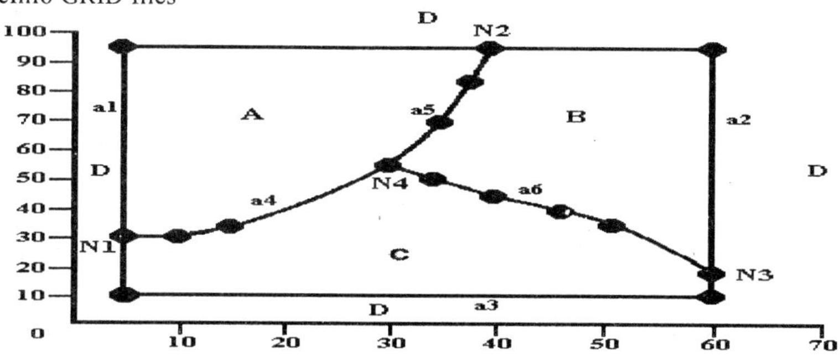

Polygon Topology	
Polygon	**Arcs**
A	a1, a4, a5
B	a2, a5, a6
C	a3, a4, a6
D	outside study area

Node Topology	
Node	**Arcs**
N1	a1, a3, a4
N2	a1, a2, a5
N3	a2, a3, a6

Arc Topology				
Arc	**Start Node**	**End Node**	**Left Polygon**	**Right Polygon**
a1	N1	N2	D	A
a2	N2	N3	D	B
a3	N3	N1	D	C
a4	N1	N4	A	C
a5	N4	N2	A	B
a6	N3	N4	C	B

Fig. 3.10: Typological model

Raster *versus* Vector Systems

Early GISs were either raster or vector systems. Various advantages and disadvantages of vector and raster data systems are mentioned below (Burrough, 1986; Aronoff, 1989). Both approaches are equally valid ways of representing spatial data. The advantages and disadvantages of raster and vector systems have been heavily debated. Most modern GISs handle both raster and vector data but usually are designed primarily for one data type. The complete integration of raster and vector data capabilities will be common in GISs of the future (Faust *et ai.*, 1991). These new GISs will quickly and efficiently convert among rasters, vectors, and other data structures most appropriate for the application being performed. GISs must function equally well with both raster and vector data. Comparison advantages and disadvantages of vector and raster methods as revised from Nag and Kudrat, Burrough (1986) and Aronoff (1989).

Raster Method

Advantages

- Data structure is simple.
- The method is compatible with remotely sensed or scanned data.
- Procedures for spatial analysis are simple.

Disadvantages

- Greater disk storage is often required.
- Topological relationships are difficult to represent.
- Unless extremely small cell sizes are used, the graphic output is often aesthetically less pleasing.
- Projection transformations are more difficult.

Vector Method

Advantages

- Compact data structure requires less disk storage.
- Topological relationships are readily maintained.
- Graphic output is aesthetically more pleasing and more closely approximates hand-drawn maps.

Disadvantages

- Data structures are complex.
- Overlaying multiple vector maps is often time consuming.
- Output graphics may take hours to draw on plotters.
- Some spatial analysis procedures are difficult.
- Software and hardware for vector systems are often more expensive.
- The method is not as compatible with remote sensing data.

Database Planning and Design

The primary purpose of this phase of the GIS development process is to specify "how" the GIS will perform the required applications. Database planning and design involves defining how graphics will be symbolized (i.e., colour, weight, size, symbols, etc.), how graphics files will be structured, how non-graphic attribute files will be structured, how file directories will be organized, how files will be named, how the project area will be subdivided geographically, how GIS products will be presented (e.g., map sheet layouts, report formats, etc.), and what management and security restrictions will be imposed on file access. This is done by completing the following activities:

- Select a source (document, map, digital file, etc.) for each entity and attribute
- Set up the actual database design (logical/physical design)
- Define the procedures for converting data from source media to the database
- Define procedures for managing and maintaining the database

The database planning and design activity is conducted concurrently with the pilot study and/or benchmark activities. Clearly, actual procedures and the physical database design cannot be completed before specific GIS hardware and software has been selected while at the same time GIS hardware and software selection cannot be finalized until the selected GIS can be shown to adequately perform the required functions on the data. Thus, these two activities (design and testing) need to be conducted concurrently and iteratively.

In many cases, neither database design matters nor hardware and software selection are unconstrained activities. First, the overall environment within which the GIS will exist must be evaluated. If there exist "legacy" systems (either data, hardware or software) with which the new GIS must be compatible, then design choices may be limited. Both GIS hardware and software configurations and database organizations that are not compatible with the existing conditions should be eliminated from further consideration. Secondly, other constraints from an organizational perspective must be evaluated. It may, for example, be preferable to select a specific GIS or database structure because other agencies with whom data will be shared have adopted a particular system. Finally, assuming that the intended GIS (whether it will be large or small) will be part of a corporate or shared database, the respective roles of each participant need to be evaluated. Clearly, greater flexibility of choice will exist for major players in a shared database (e.g., county, city, or regional unit of government) than for smaller players (town, village, or special purpose GIS applications). This does not mean that the latter must always go with the majority, but simply that the shared GIS environment must be realistically evaluated. In fact, one way for the smaller participants in a shared GIS to ensure their needs are considered, is to fully document their needs and resources using procedures recommended in these guidelines.

Finally, with the completion of both the database planning and design and the pilot study and benchmark activities, sufficient detailed data volume estimates and GIS performance information will be known to calculate reliable cost estimates and prepare production schedules. This becomes the final feasibility check before major resources are committed to data conversion and GIS acquisition.

What is Already Known about the GIS Requirement

Prior phases of the GIS development process should have produced the following information which is needed at this time:

- A complete list of data, properly defined and checked for validity and consistency (from the master data list, data model and metadata entries).
- A list of potential data sources (maps, aerial photos, tabular files, digital files, etc.) catalogued and evaluated for accuracy and completeness (from the available data survey). This inventory would also include all legacy data files, either within the agency or elsewhere, which must be maintained as part of the overall shared database.
- The list of functional capabilities required of the GIS (from needs assessment).

Selecting Sources for the GIS Database

This activity involves matching each entity and its attributes to a source (map, document, photo, digital file). The information available for this task is as follows:

- List of *entities and attributes* from the conceptual design phase

Table 3.3: Master data list

Master Data List Entity	*Attributes*	*Spatial Object*
Street_segment	name, address_range	Line
Street_intersection	street_names	Line
Parcel	section_block_lot#, owner_name, owner_address, sites_address, area, depth, front_footage, assessed_value, last_sale_date, last_sale_price, size (owner_name, owner_address, assessed_value (as of previous January 1st))	Polygon
Building	building_ID, date_built, building_material, building_assessed_value	Footprint
Occupancy	occupant_name, occupant_address,occupancy_type_code	None
Street_segment	name, type, width, length, pavement_type	Polygon
Street_intersection	length, width, traffic_flow_conditions, intersecting_streets	Polygon
Water_main	type, size, material, installation_date	Line
Valve	type, installation_date	Node
Hydrant	type, installation_date, pressure, last_pressure_test_date	Node
Service	name, address, type, invalid_indicator	None
Soil	soil_code, area	Polygon
Wetland	wetland_code, area	Polygon
Floodplain	flood_code, area	Polygon
Traffic_zone	zone_ID#, area	Polygon
Census_tract	tract#, population	Polygon
Water_District	name, ID_number	Polygon
Zoning	zoning_code, area	Polygon

- The list of *surveyed data sources from the Available Data Survey* and their recorded characteristics in the metadata tables *Source Documents, Entities Contained in Source, and Attributes by Entity.*

Table 3.4: Source documents

Source Document Name:	Smarts Map
Source ID #:	1
Source Organization:	Kolkata Municipal Corporation (KMC)
Type of Document:	Map
Number of Sheets (map, photo, etc.):	100
Source Material:	Paper maps
Projection Name:	UTM
Coordinate System:	State Plane
Date Created:	5-Oct-1910
Last Updated:	8-Nov-95
Control Accuracy Map:	National Map Accuracy Standard
Scale:	Variable; 1:600 or 1" = 50 feet
Availability:	Current
Reviewed By:	NATMO
Review Date:	18 August 1997
Spatial Extent:	Original Calcutta Municipal Corporation area
File Format:	N/A
Comments:	Covers about 60% of today KMC area

Table 3.5: Entities contained in source

Source ID #:	1
Entity Name:	Parcel
Spatial Entity:	Polygon
Estimate Volume Spatial Entity:	100 per map sheet
Symbol:	None
Accuracy Description Spatial Entity:	Earlier Map Accuracy Standard
Reviewed By:	NATMO
Review Date:	02-Jan-94
Scrub Needed:	Yes
Comments:	

Table 3.6: Attributes by entity

Source ID #:	1
Entity Name:	Parcel
Attribute Name:	Property Number
Attribute Description:	Borough, Ward, Street and plot Number
Code Set Name:	N/A
Accuracy Description Attribute:	N/A
Reviewed By:	NATMO
Review Date:	08-August-97
Comments:	

If there is a choice between sources, that is, two or more sources are available for a particular entity attribute, then criteria for deciding between them will be needed. In general, these criteria will be:

- Accuracy of resulting data
- Cost of conversion from source to database
- Availability of the source for conversion
- Availability of a continuing flow of data for database maintenance.

Occasionally, alternative sources may result in different representations in the database, such as a vector representation *versus* a scanned image. In this situation, the ability of each representation to satisfy the requirements of the GIS applications will need to be evaluated. Once a source has been selected, the metadata tables that record source data information need to be completed as appropriate. These are:

- Data object information
- Attribute information
- Spatial object information
- Source document information

To complete the accuracy information, the accuracy expected from the conversion process will need to be determined. This accuracy target will also be used later in the database construction phase by the quality control procedures. The metadata tables that need to be completed at this time are shown below:

Table 3.7: Data Object Information

Data Object Name	Parcel
Type:	Simple
Data Object Description:	Land ownership parcel
Spatial Object Type:	Polygon
Comments:	

Table 3.8: Attribute Information

Data Object Name:	Parcel
Data Attribute Name:	Property Number
Attribute Description:	Borough, Ward, Street and Plot Number
Attribute File Name:	Parcel.PAT
Codeset Name/Description:	N/A
Measurement Units:	N/A
Accuracy Description:	N/A
Comments:	

Table 3.9: Spatial Object Information

Data Object Name:	Parcel
Spatial Object Type:	Polygon
Place Name:	Amherst
Projection Name/Description:	UTM
HCS Name:	State Plane Coordinate System
HCS Datum:	NAD83
HCS X-offset:	1000000
HCS Y-offset:	800000
HCS Xmin:	25
HCS Xmax:	83
HCS Ymin:	42
HCS Ymax:	98
HCS Units:	Feet
HCS Accuracy Description:	National Map Accuracy Standard
VCS Name:	
VCS Datum:	
VCS Zmin:	0
VCS Zmax:	0
VCS Units:	
VCS Accuracy Description:	
Comments:	

Table 3.10: Source Document Information

Data Object Name:	Parcel
Spatial Object Type:	Polygon
Source Document Name:	Parcel Map
Type:	Map
Scale:	Variable: 1" = 50 feet To 1" = 200 feet
Date Document Created:	17-Nov-89
Date Last Updated:	05-Oct-94
Date Digitized/Scanned:	24-Apr-95
Digitizing/Scanning Method Description:	Manual digitized with Wild B8
Accuracy Description:	90% of all tested points within 2 feet
Comments:	

For some of the above tables, information will be available for only some of the entries. The remaining entries will be completed later as the database is implemented. The examples shown are from the metadata portion of the GIS design software package that accompanies these guidelines. This package is a Microsoft Access programme that runs "stand-alone" (user does not need a copy of Microsoft Access) on a regular PC. Where the same information is needed for multiple tables, this information is only entered once. The information is then automatically transferred to the other tables where it is needed.

Logical/Physical Design of the GIS Database

This activity involves converting the conceptual design to the logical and physical design of the GIS database (hereafter referred to as the physical design). The GIS software to be used dictates most of the physical database design. The structure and format of the data in a GIS, like ARC/INFO, Intergraph, MapInfo, System 9 etc. have already been determined by each vendor respectively. If one separates the conceptual entity and its attributes from the corresponding spatial entity and its geometric representation, it can be seen that the physical database design for the spatial entity has been completely defined by the vendor and the GIS designer does not need to do anything more for this part of the data. The attributes of the entities may, however, be held in a relational database management system linked to the GIS. If this is the case, the GIS analyst needs to design the relational tables for the attribute information. It will not always be the case where one entity from the E-R diagram translates into a single layer. More complex representations will be needed. Generally this will involve two or more entities forming a single layer with, possibly, several relational database tables.

The water main segments, the valves and the fire hydrants have been placed together in one layer as line segments, and two sets of nodes. However, each entity has its own relational table to record its respective attributes. The relationship is maintained by unique keys for each instance of each entity. Every entity shown on the E-R diagram must be translated to either a GIS layer, a relational table(s), or both, as indicated by the information to be included. In addition, every relationship of the type "relationship represented in database" (single line hexagon on the E-R diagram) must be implemented through the primary and secondary keys in the tables for the entities represented.

The entity "parcel" may "contain" the entity "building." The table for each entity would have its own primary key (ID#), however, the table for building must also have a secondary key (parcel ID#) to maintain the relationship in the database. The completed physical database design must account for all entities and their attributes, the spatial object with topology and coordinates as needed, and all relationships to be contained in the database. The remaining items on the E-R diagram, the two types of spatial relationships, must be accounted for in the list of functional capabilities, that is, the implied spatial operations must be possible in the chosen GIS software.

Procedures for Building the GIS Database

Developing a GIS database is frequently thought of as simply replicating a map in a computer. As can be inferred by the nature and detail of the activities recommended upto this point in these guidelines, building a GIS database involves much more than replicating a map. While substantial portions of the GIS database will come from map source documents, many other sources may also be used, such as aerial photos, tabular files, other digital data, etc. Also, the map representation is only part of the GIS database. In addition to the map representation and relational tables, a GIS can hold scanned images (drawings, plans, photos), references to other objects, names and places, and derived views from the data. The collection of data from diverse sources and its organization into a useful database requires development of procedures to cover the following major activities:

- ***Getting the Data*** which may include acquiring existing data from both internal and external sources, evaluating and checking the source materials for completeness and quality, and/or creating new data by planning and conducting aerial or field surveys. Contemporary GIS projects attempt to rely on existing, rather than new, data due to the high cost of original data collection. However, existing data (maps and other forms) were usually created for some other purpose and thus have constraints for use in a GIS. This places much greater importance on evaluating and checking the suitability of source data for use in a GIS.

- ***Fixing any problems in the data source***, often focused only on map source documents, this activity has been called *map scrubbing*. Depending on the technology to be used to convert the map graphic image into its digital form, the source documents will have to meet certain standards. Some conversion processes require the map to be almost perfect which other processes attempt to automate all needed *fixes* to the map. What is required here is for the GIS analyst to specify, in detail, a procedure capable of converting the map documents into an acceptable digital file while accounting for all known problems in the map documents. This procedure should be tested in the pilot project and modified as needed.
- ***Converting to digital data***, the physical process of digitizing or scanning to produce digital files in the required format. The major decision here is whether or not to use an outside data conversion contractor or to do the conversion within the organization. In either case, specifications describing the nature of the digital files should be prepared. In addition to including the physical database design, specifications should describe the following:

  —Accuracy requirements (completeness required, positional accuracy for spatial objects, allowable classification error rates for attributes).
  —Quality control procedures that will be conducted to measure accuracy.
  —Partitioning of the area covered by the GIS into working units (map sheets) and how these will be organized in the resulting database (including edge matching requirements).
  —Document and digital file flow control, including logging procedures, naming conventions, and version control.

- ***Change control***, most map series are not static but are updated on a periodic basis. Once a portion of the map has been sent to digitizing (or whatever process is used), a procedure must be in place to capture any updates to the map and enter these into the digital files.
- ***Building the GIS Database,*** once digitizing has been completed, the sponsoring organization has a set of digital files, not an organized database. The system integration process (a subsequent guideline document) must take all the digital files and set-up the ultimate GIS database in a form that will be efficient for the users.

Procedures for Managing and Maintaining the Database

Because the physical world is constantly changing, the GIS database must be updated to reflect these changes. Once again, the credibility of the GIS database is at stake if the data is not current. Usually, the effort required to maintain the database is as much as, or more than that required creating it. This ongoing maintenance work is usually assigned to in-house personnel as opposed to a contractor. The entire process should be planned well in advance. Once again, the equipment and personnel must be ready to take over the maintenance of the database when the data conversion effort and database building processes are complete.

Database maintenance requires two supporting efforts: (a) ongoing user training and (b) user support. Ongoing user training is needed to replace departing users with newly trained personnel. This will enable the data maintenance to be carried out on a continuous and timely basis. It is also important to offer advanced training to existing users to provide them with the opportunity to improve their skills and to make better use of the system. GIS is a complicated technology, making operating problems inevitable. User support will help users solve these problems quickly. It will also customize the GIS software to enable them to execute processing tasks more quickly and more efficiently. User support is usually provided by in-house or contract programmers. It requires knowledge of the operating system and macro programming language as well as troubleshooting common command and file problems.

GIS Data Sharing Cooperatives

The establishment of data sharing cooperatives within the public sector is a cost-effective means of database development and maintenance which is encouraged. Cooperative multi-participant database projects allow for data exchange, and the opportunity to create new means for developing, maintaining, and accessing information. The sharing of data in the public sector, especially between government agencies and offices which are funded by the same financial resources, should be expected. It does not make fiscal sense for public funds to be utilized in the development of two GIS databases of the same geographic area for two different agencies. Benefits of data sharing thus would include: the development of a much larger database, for far less cost; the development of more efficient interaction between public agencies; and through the utilization of a single, seamless database the availability of more accurate information, since all agencies would share the same, up-to-date information (see Chapter 8: Spatial Data Infrastructure).

The goal of a data sharing strategy is to maximize the utility of data while minimizing the cost to the organization. It is important that all parties involved have clear and realistic expectations as well as common objectives to make the data sharing work. Under any circumstance, however, database management and maintenance will require us to redefine our relationships with those we routinely exchange data with, whether they are within an organization or part of a multi-participant effort including outside agencies. Work flow and information flow must be reviewed and changed, if necessary. Procedures and practices for the timely exchange and updating of data must be put in place and data quality standards adhered to, whether it is hard copy data which must be converted for inclusion or digital files which might be available for importing to our system. Systematic collection and integration of new and/or updated data must be employed in order to safeguard the initial investment, maintain the integrity of the database and assure system reliability to meet function needs.

Database Structure and Database Management

Basics of Computer File/Database Structures

The essential features of any data storage system are that they should allow data to be accesses and cross referenced quickly. There are several ways of achieving this, some of which are more efficient than others. Unfortunately, there seems to be no single 'best' method that can be used for all situations. A goal of GIS is to represent and store the graphic entities of mapped information along with relevant attributes in such a way to make all the data easily retrievable and scope for manipulation. This is done by taking advantages of the ways computers handle data in logical fashion through file and database structures. A brief overview of the ways computers can handle data is offered here.

Simple list—The most basic file structure is a *simple list*. In this file structure, there really is no absolute ordering of the data. The data occur in the file is essentially the same way in which they were entered. Simple list may start out in logical fashion but whenever modifications are made, they rapidly get out of order because new data are appended to the end of the list. It is very easy to add data to this system but retrieving is data is inefficient.

Ordered sequential files can be thought of as a rolodex in which we keep everything in alpha-numeric order. As new data are added, the file is restructured (sorted) to maintain that order. Very often ordered sequential files are accessed by binary search procedure.

Indexed file structures provide pointers to more efficiently search data. The most efficient system is to develop an index that is based on a commonly searched attribute in the database. In many applications, particularly in geographical information, the individual items (pixels, points, lines or areas) will not have any key attribute such as an identification number or a name, but will also carry information about the associated attributes. Very often it is the information of the associated attributes that is required. With *indexed files* original data file can be speeded up in two ways. If the data items in the files themselves provide the main order of the file, then the filers are known as *direct files*. The location of items in the main file may also be specified according to topic, which is given in second file, known as *inverted file*.

Types of Database Systems

A database is a collection of files that contain interrelated information, and a Database Management System (DBMS) is a program that lets the user add, delete and modify records in the database. All GIS databases include links between graphic (map) features and tabular data that provides descriptive information for the graphic objects. If it were not for this link, a GIS would not exist but would be similar to computer aided design or computer mapping systems. The tabular data provides the *information* in GIS. These days most GIS's provide links to a variety of databases allowing users a choice between a hosts of commonly used systems. Therefore, the GIS can be a combination of two distinct systems generated individually, but used for a common purpose. Database systems come in three main varieties:

- Hierarchical database systems,
- Network database systems, and
- Relational database systems.

Of these three types of database systems, the hierarchical is the fastest but the most restrictive for the user. This system is appropriate if you can naturally organize the information in the database into the same structure as a hierarchy chart. The network system is more flexible than the hierarchical system but more difficult for a user than the relational database system. The relational system is rapidly becoming the most popular of the three. It is the most flexible and easiest to use. But this flexibility comes at the cost of low speed compared to other database systems.

Hierarchical Database Systems

A hierarchical database system is one in which the DBMS supports a hierarchical structure of records organized in files at various logical levels with connections between the levels. A record at a particular level contains data common to a set of records at the next lower level. There are no connections between records of the same level. Each record contains a field defined as the key field, which organize the hierarchy.

Records of the same type are collected in files known as elements. Several types of elements may reside at the same logical level. Starting from the highest-order element, the hierarchy permits a set of elements to be accessed at the next lower level. Any one element may retain connections to only one set of lower level elements and itself may be a member of only one such set. Construction of a hierarchical structure begins with a main object at its top. The main object has a range of characteristics which can be collected at the various levels of hierarchy (Fig. 3.11).

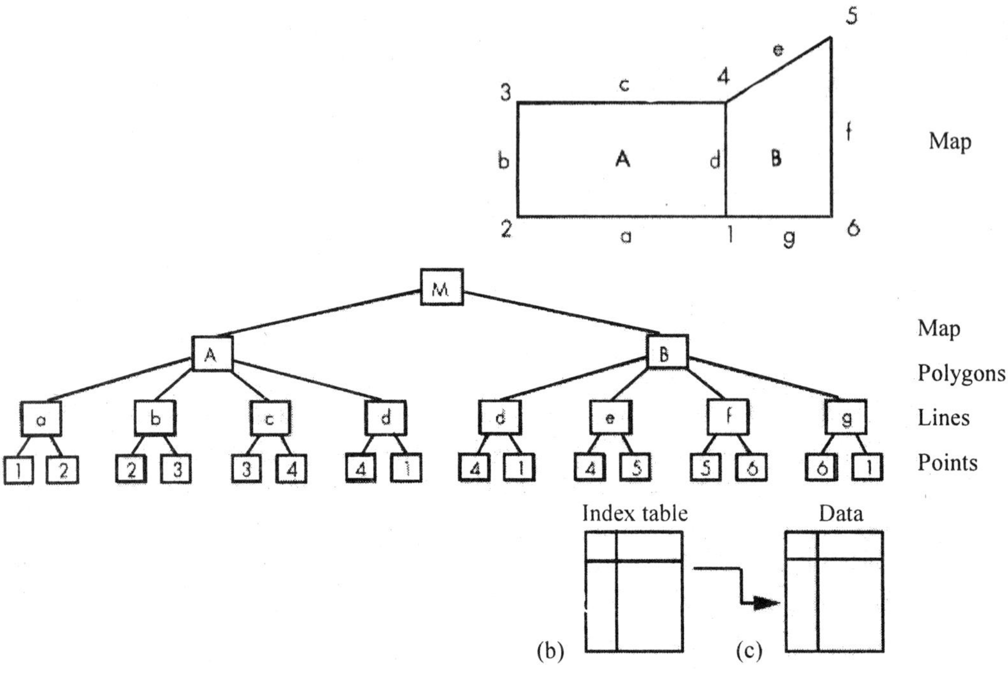

Fig. 3.11: Map data stored in a hierarchical database system.
Note: The double storage of lines and points (Bernhardsen, 1992)

Geographic data which is organized hierarchically fit into this model. Geographical data may be stored in a hierarchical structure in a manner that reflects the real world. An example is when the levels of a hierarchical model correspond to the levels of real-world administration, for example in a country, county, township or town/city. A good example of spatial data that can fit into this model is the data of the Census of India in which information for one enumeration area has no lateral relation to another enumeration area, but is related to the higher levels in the hierarchy to city, *tahsil* or *taluk*, district and the state. In the hierarchical model an owner record (file) can have many members, but the member records can only be owned by one owner record.

Hierarchical database systems are easily expanded and updated. However, they require large index files, must be frequently maintained and are susceptible to multiple entries. Searches are rapid, but search routines are fixed and constrained by the structures. The elements or the structure are related only through one-to-many connections. This constraint imposes the presupposition that all queries are known in advance and accounted for in structuring and entering data. This constraint is not always natural or suitable in GIS applications. As a result, hierarchical database systems are usually restricted to storing digital map data in GIS.

Network Database Systems

A network database system is one in which the DBMS supports a network type of organisation. Each element, or collection of like records, has connections to several different level elements. Interconnections are made in the hierarchical organisation, and a characteristic may be associated with two main objects. The resultant network structure more closely represents the complex interrelationships which often exist between real-world geographical objects. The elements of the structure may be related through one-many, many-one and many-many connections. The purpose of the network structure is to improve the flexibility and reduce the multiple entries of the hierarchical structure. Points and lines are entered only once. Searches need not pass through all levels, but can take shortcuts. However, the volume of index data is greater than that of the hierarchical structure (Fig. 3.12).

A network structure permits rapid connection between data which physically are stored in different disk sectors. Maintaining data stored in a network structure is complex, so although a network structure is better suited to geographical data than is a hierarchical structure, it is infrequently used in GIS applications.

Comparison between Hierarchical and Networks Data Structure

A record type represents each entity type in the E-R diagram. A 1:N relationship is represented as hierarchy type where the record type corresponding to the one side of the relationship is the parent (a 1:1 relationship is a special case of the 1:N relationship). A week entity can be represented as a separate record type. This record type becomes a dependent in a hierarchy where the record type, corresponding to the strong entity in identifying relationship, is the parent. An M:N relationship is not easy but it requires introducing duplicate record types or using multiple hierarchies and introducing virtual records.

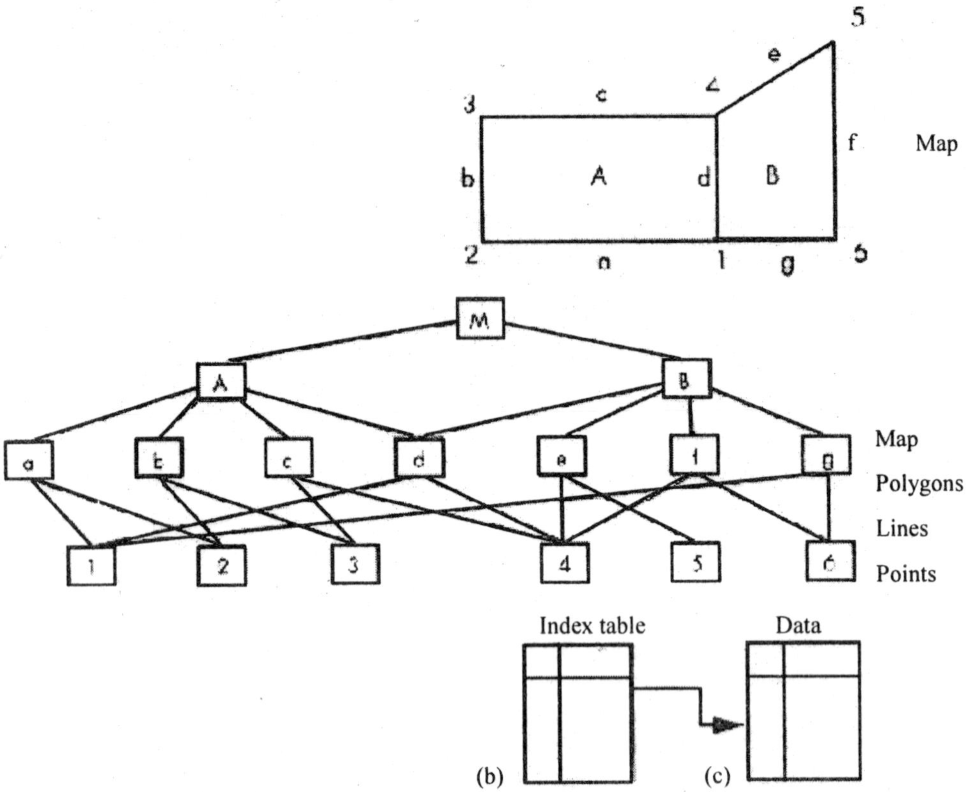

Fig. 3.12: Storage structure in a network-based database system.
(a) shows a map section; (b) shows polygons stored in the network;
(c) is a pointer table which specifies where the map data
(d) are in fact stored (Bernhardsen, 1992)

Each entity type in E-R diagram is represented by record type and each instance of the entity is represented by record occurrence. A week entity can be represented as a separate record type. In the case of networking data model the identifying relationship is represented as a set type wherein the record type corresponding to the week entity type forms the member and the record type corresponding to the strong entity is owner. A 1:1 or 1:N relationship is represented as a set type. An M:N relationship requires introducing an intermediate record type. This record type is a common member in two set type, one of which is owned by each record types involved in M:N relationship.

Merits and Demerits

The major limitation of hierarchical system cannot handle many to many relationship and GIS data bases. have, in addition to one to one and many to one relationship, many to many relationships in which a single entry may have many attributes, and each attribute is linked explicitly to many entities. However, network system is the improvement over hierarchical structure for GIS. Networking system can handle many to many relationships. In network system each piece of data may be associated with an explicit computer structure called a pointer which directs it to all of the other pieces of data to which it relates. Unlike hierarchical data structure it is not restricted to the branching tree structure, each individual piece of data can be linked directly anywhere in the database without any parent-child relationships. The major drawbacks is that in very complex GIS database, the number of pointers can get quite large, coming to comprise a substantial portion of storage space. In addition, although linkages between data elements are more flexible, they must still be explicitly defined with the use of pointers. The numerous possible linkages may become an extremely tangled web, often resulting in confusion and missed and incorrect linkages. This limitation may be frustrating for the novice GIS users. The disadvantages of the network data model are: the query language is procedural and requires the user to navigate through the database by specifying sets, owners, and members. This in turn means that the user has to be cognizant of the data structure of the database.

Relational Database Systems

This has become the most popular form of a database model since it has the most flexible approach to linkages between individual databases. There is no hard structure to a relational model and the form of the model is determined by the needs and concepts of the user. In a relational database each record (file) has a set of attributes and a set of rows (entries) that form the table. Separate tables are linked, or related, to one another through a common attribute. Any attribute in the table can be used as the key link between that table and another table. Links can be formed in any direction and between any nodes (table). This database model allows maximum flexibility since ownership is not defined as it is in the hierarchical and network models (Fig. 3.13).

Building I.D.	Property	Owner	Year	Type
589	44 110			
610	44 50			
955	44 99			

Property I.D.	Owner	Area	Address
44 99			
44 50			
44 110	Nils Nilsen	6,51	9999 Toppen

Fig. 3.13: A relations database.
Note: Each field in a table can be the key to locating data in another table (Bernhardsen, 1992)

There are no pointers in the data tables. Internal index tables are used to direct inter-table communications, which means that the system must open at least one index table for each connection between data tables. This can result in relatively large databases and slow access. Moreover, as there are no pointers, searches for records or fields must be sequential through the tables. The result is that relational database systems are inherently slower than hierarchical or network database systems.

A field may be located quickly in a given table, and the data accessed may be manipulated. The simple structures of relational database systems have permitted the development of standard query languages, one of which is Standard Query Language (SQL). Relational algebra may be performed using two classes of storage and retrieval operations. The set of operations include union, intersection, difference and product. The relational operations include selection (accessing rows), projection (accessing columns), joining and dividing. Joining creates a new table from data retrieved from various tables. The new table need not physically be stored in the database.

Relational DBMS are now the type used most frequently in GIS applications, primarily because of their simple, flexible structures. Another reason is that they support the complex relationships common among real-world geographical objects. Multiple entries are less frequent than in hierarchical or network database systems, although search times tend to be longer. This is particularly noticeable in specific operations on digital map data, for instance when forming polygons and with automatic text placement (Fig. 3.14).

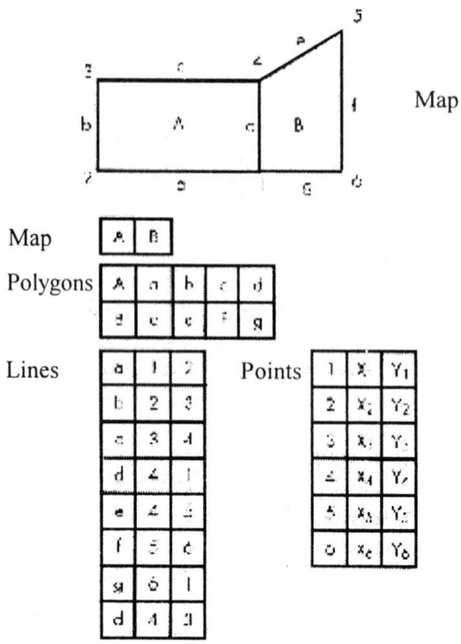

Fig. 3.14: Map data stored in a relations database. (Bernhardsen, 1992)

What is SQL?

SQL is an abbreviation for Structured Query Language. It is a tool for organizing, managing, and retrieving data stored in a database. It works with a specific type of database model, a relational database.

History of SQL

SQL is a language used for accessing relational databases. As with relational databases, the origin of SQL dates back to 1970 when Dr. Codd proposed a relational model for databases. Codd's article triggered research on relational databases. IBM started a research project called System/R in 1974. The goal of the project was to prove the workability of the relational concept and to provide some experience in actually implementing a relational DBMS. In 1978, System/R was distributed to a number of IBM customers for evaluation. With the successful implementation of System/R, IBM proved that relational databases were feasible, and could be the basis for commercial products. SQL was born during the implementation of System/R. In 1979, Relational Software Inc. (now Oracle Inc.) came to market with a relational Database Management System (DBMS) based on SQL. IBM worked hard to produce commercial relational products. In 1982, IBM introduced its first commercial product SQL/DS, and in 1983, DB2, DB2's SQL became the *de facto* standard database language.

In 1986, the American National Standards Institute and the International Standards Organization (ANSI/ISO) published a standard for SQL, and gave SQL official status as a standard. The standard is largely based on DB2's SQL. The standard was revised and expanded in 1989, and is called SQL89 (SQL1) standard. The ANSI committee continued its work on the standardization of SQL, and in 1992 a new more rigorous standard SQL92 (SQL2) was published. After some extensions to SQL2, ANSI published standard SQL99 (SQL3) in 1999.

The ANSI has published three standards over the years: SQL89 (SQL1), SQL92 (SQL2), and SQL99 (SQL3). Today most relational database products are designed to meet the SQL92 standard. However, none of them implement all features of the SQL2 standard and many of the advanced features of SQL92 have yet to be implemented. SQL3 addressed more advanced issues in SQL, but no vendor has implemented the standard at the present time. The majority of the language has not changed through these updates. All of the core functions of SQL, such as retrieving, adding, and modifying data, are the same. The SQL statements given in this tutorial are the core part of SQL. They are common to most available relational products today.

Object Oriented Database Systems

Hierarchical, network and relational database systems are essentially intended for administrative tasks. Consequently, they are not particularly well suited to representing a conceptual data model of geographical reality. All these systems are record oriented, which is to say that the data they contain are filed record by record. For instance, in a relational database, all the elements which together comprise a particular map object may reside in several records in various tables. Furthermore, geometrical data and attribute data are often physically separated, each residing in a separate database. Topological storage of vector data is object oriented, but the relevant database structure remains a problem. Object-Oriented Database Systems (OODS) attempt to overcome these difficulties by representing more faithfully the real world, which comprises homogeneous, through complex, objects with varying internal and external relations. There are several different object-oriented technologies, all sharing some of the same concepts and philosophical principles, but the direct benefits of OODS come from the following ideas:

(a) feature classes (a *shema*) in which each feature type has several attributes;
(b) active data-features that have their own behavior ("methods") stored in the repository not in application programmes; and
(c) inheritance between the classes of data (attributes) and behaviour (methods).

An active object database allows geographic feature data to be grouped into classes that represent similar entities. This technology approach allows that spatial data could be stored in complex, richly structured *shema*—complex features or complex feature classes. Some other important features of object-oriented approach are:

Integrity control—active data can check their own integrity, the validity of their relationships with other features, topological constraints, etc.;

Version control including virtual check-in and check-out of repository segments—spatial object repositories make it easy to provide novel versioning mechanism so each user can see a virtual copy of all data while editing only a part of them;

Advanced topology support—geometric topology results from the sharing of geometry—poles placed at cable intersections, rivers forming parts of administrative boundaries. Spatial object repositories can identify spatial coincidence and dynamically model them in the database. This ability exploits two important spatial object database features: the ability to hold all types of data and the ability to effectively manage a vast web of connections among objects;

Rich representation—display methods allow objects to display themselves, providing important information about the object (a valve/switch can indicate whether it's open or closed). Display method is centralized as part of the feature, and appearance of objects can be made consistent across a utility—local authority, conforming to standards;

Multiple presentation from a master dataset—spatial objects can be associated with several geometries, and display methods can use any of the geometries. The same objects can appear differently in different contexts.

As the first commercial GISs built upon the object-oriented technology were sold more than ten years ago (in 1991) OO GISs are not widely used (U.S. Geological Survey, U.K. Ordnance Survey, Land Information New Zealand, Czech Land Survey Office, some other national mapping agencies, marine and air chart producers, military spatial data users and automated mapping/facilities management spatial resource planning utilities) but initial success is proof that OO GIS is a future requirement, not an option.

REFERENCES

Aronoff, Stan (1989), *Geographic Information System: A Management Perspective,* WDL Publication, Ottawa.

Bernhardsen, T. (1992), *Geographic Information Systems,* VIAKIT IT, Norway.

Burrough, Pater and R. McDonnell (1998), *Principles of Geographical Information Systems,* Oxford University Press, Oxford.

Burrough, P.A. (1986), *Principles of Geographic Information Systems for Land Resources Assessment,* Oxford University Press, Oxford.

Faust, N.L., W.H. Anderson and J.L. Star (1991), Geographic information systems and remote sensing future computing environment, *Photogrammetric Engineering and Remote Sensing,* Vol. 57, Issue 6, pp. 655-68.

Nag, P. and M. Kudrat (1998), *Digital Remote Sensing,* Concept Publishing Company, New Delhi.

4

Software and Hardware Options

In last few years a huge volume of spatial information has been available, stored and managed in the form of Geographical Information System (GIS) for various applications which has opened a new vista for business opportunities. International Geo-sphere Biosphere research, digital earth, digital regions and several global and national projects related to digital information have given a boost to a new type of applications. These applications or projects are based on geo-spatial data and knowledge about such databases. The management of such data, information or knowledge depends on efficient and user-friendly GIS. The GIS technology now permits us to collect basic common sense knowledge linked with geo-spatial data, and also allows this knowledge to be closely associated with such data. The association of knowledge with spatial data makes the option of having feature-oriented objects in the GIS environment. The new applications look towards the database which can produce intelligent levels. Such a combination of data and knowledge has opened several types of applications not only in the developed world but also in the developing world. Hence, there has been a greater urgency to understand the spatial or geospatial data infrastructures and their various facets.

We live in information age which has become essential to solve problems of national development. Further, spatial information has proved to be an important element underpinning decision-making for many disciplines because it deals with location. In the sophisticated digital environment, it has expanded to include databases, satellite positioning, communication networks including internet, web technology and wireless applications. It draws experiences from the geographic information systems, computer science, land administration, geography, surveying and mapping, legal and public administration disciplines. GIS is so versatile that it is more difficult to define than to imagine or visualize. By the end of the 1970s the term GIS had emerged in recognition both of common technological requirements and of the opportunity to built the system that could potentially satisfy all of the various applications. GIS is information about geography, i.e. information tied to some specific set of location on the earth surface (including the zones immediately adjacent to the surface and subsurface, ocean and atmosphere.

The potential users of GIS are limitless and the types and number of users are growing at a logarithmic pace. Further, the modern GIS has the utility of the maps by replacing it with a large number of mapped coverages, each with an interrelated theme. These coverages can be automatically analysed and these themes combined to give meaningful answers for decision-makers. The GIS is gradually changing the way we do think with maps, the way we think about geographic information (GI), even the way in which geographic data collected and compiled. The task, which was impossible with the traditional maps, is now common place. To answer such question requires access to the geographical information, which is characterized by its multidimensional nature (x, y, z coordinates and time), its large volume and high processing cost. Even to answer apparently simple questions, the data from several sources must be integrated into consistent form and be analyzed.

In the late twentieth century, an awareness of importance for data on topography and specific themes of the earth surface, such as natural resources, have accelerated greatly. Aerial photography and satellite-based remote sensing data ease the means of mapping large area with great accuracy. The same technology also gave the resource scientist enormous advantage for reconnaissance and semi-detailed mapping. The resulting thematic maps were the sources of useful information for resource exploitation and management.

GIS has been defined by its objective, as in Cowen's definition of GIS as spatial decision support system (Cowen, 1988). In the context of remote sensing GIS tends to be reserved for system that integrate remotely sensed data with other types or processed data that have already been cleaned and transformed. Similarly, an atmospheric scientist or oceanographer will tend to associate GIS with system used more for multi-disciplinary work and policy study, and will use other software environments for modelling and analyses within the confines of his or her own discipline. In short, because GIS implies generalized software environment that is exclusively to geographical information, i.e. a tendency for it to be most strongly associated with multi-disciplinary, integrative work and applications, in more narrowly defined environments less general solutions may be adequate.

The term GIS is difficult to define and represents the integration of many subject areas. The term itself is becoming complex and modified to conform to intellectual, cultural, economic and political objectives. The terminology has in fact become extremely elastic resulting in an increasingly confusing jargon due to new definition that constantly creeps into both scientific and popular literature.

A preliminary definition for consideration might be that of David Rhind (1988), who defines GIS as "a computer based system for collecting, checking, integrating and analysing information related to surface of the earth." This definition has some highly worthwhile elements that should be examined. First, it indicates that it deal with the surface of the earth. Although this is not an absolute requirement, the vast majority of GIS applications do deal with the portion of the earth. Moreover, the statement that GIS is used to collect, check, integrate and analyse information enumerates a large number of the necessary groups of operations for any GIS.

Components of Geographical Information Systems

GIS refers to an information technology capable of collecting, storing, analysing and displaying both spatial (geographically-referenced) and non-spatial data. Components of a GIS include both computer hardware and software, spatial data features (raster grid cells, or points, lines, and polygons) and associated tabular attribute data. Utilization of GIS technology occurs in a wide range of applications, from Land Information Systems (LIS), Automated Mapping and Facilities Management (AM/FM), Census and Legislative Redistricting and "transportation" GIS (GIS-T), to natural resource management, environmental modelling and Spatial Decision Support Systems (SDSS). Consequently, each area of specialization has unique needs for data and computing. This standard focuses on general criteria for those aspects of GIS technology specifically associated with computing architecture (e.g., hardware and software). In addition, guidelines are provided for two other GIS standard categories: spatial data transfer format and metadata content. GISs have three major components. These are as follows:

—Soft ware
—Hard ware
—Proper organizational including skilled persons

The computer has a hard disk for storing data and operational software package (programmes), but extra storage can be provided via network or by digital tap, optical CD-ROM and other devices. A digitizer and or scanner are used to convert maps into digital form so that they can be used by the specific software. A plotter or printer or any other kind of display device or monitor is used to present the results of the data processing. Local and global electronic networks using special data lines with optical fibers or over ordinary telephone lines by using a device known as modem provide computer communication. The user controls the computer and the peripherals (a general term for plotter, printer, digitizer and other apparatus linked to computer), via the computer screen and keyboard, added by a mouse. Now a variety of mouse is available.

GIS SOFTWARE

Selection of a GIS software is highly dependent on a number of factors, including the intended use of the chosen application, software and maintenance cost, software "learning curve" and the current technical expertise of agency personnel. The software standards identified are representative of two broad GIS application categories (i.e., professional and desktop). GIS application software should be capable of running under a Microsoft Windows (Windows-95, Windows-98, Windows-NT, Windows-2000XP) or X Windows (e.g., UNIX) environment. While it is recommended that software incorporate both vector and raster data models (Chapter 3), priority should be given to vector data model capabilities. GIS software incorporating a relational database must conform to the Structured Query Language (SQL) standard described in the database management systems (DBMS) for multiuser applications standard. GIS software not using a relational database should be able to export and import SQL data.

Beyond the differing data model specifications described above, GISs may be categorized, compared and contrasted based on their functional capabilities, ranging from basic mapping packages to desktop and professional GIS software applications. Typically, mapping packages are not true GISs, possessing a flat file structure data model with no topology or spatial analysis capabilities; such applications will not be addressed in this standard. Becoming more powerful every year, desktop GIS software applications most often run within a Windows environment, and make use of a relational database management structure. In addition to map creation, desktop GIS functionality typically includes address geocode matching capabilities, proximity analysis (e.g., *what is near?*) and buffering, as well as polygon overlay utilities. Desktop GIS packages are gaining in popularity particularly due to their point-and-click graphical user interfaces and relative ease of use. Professional GIS applications are exemplified by software running under either a UNIX or Windows-NT operating system (OS), incorporating both vector, raster and, potentially, TIN data models, as well as some level of CADD integration capabilities. The best choice for database development and management, professional GISs are capable of performing complex spatial operations from feature overlays to nearest neighbor analysis. Such systems also feature programming capabilities for user-customization of the system.

Two levels of software application standards have been identified based on the *professional* and *desktop* functionality levels described above. For the professional level, the ESRI ARC/INFO suite of products is the standard. At the desktop level, the standard consists of two vendor packages: the MapInfo product suite and ESRI's ArcView GIS and associated extensions. In addition to contrasts in functionality, professional and desktop applications may also be differentiated based on cost and *ease-of-use/learning curve*. Typically, professional level GISs require a higher software investment (up to five times more) than a desktop application. Cost of supporting hardware platforms also tends to be greater when operating on a UNIX-based platform, but is negligible under Windows-NT, etc. Designed as much for the end-user as the analyst or the programmer, desktop packages are generally menu-driven, containing a number of generic query, display and mapping utilities which allow the user to quickly access, manipulate and utilize spatial digital data without extensive specialized training and/or experience. Certified training is available throughout the region for each of the product suites identified below. Nevertheless, the choice of software is generally user oriented. There are standard packages now available off the shelf. The components of a GIS software are more-so-ever standardized. The detail criteria are as follows:

- *Data entry*: digitizing, scanning, automated data capture, interface with existing digital files, key board entry.
- *Analysis*: map overlay, proximity analysis, mathematical analysis, enclosure, buffer, measurements, attribute analysis and interpolation.
- *Surface modelling*: 3-D surfaces, slope analysis, and draping.
- *Network modelling*: route selection, allocate demand to network and facility sitting.
- *Manipulation*: image merging, projection, clip, window, update, generalize and aggregate.
- *Query*: spatial and attribute.
- *Output*: tabular list, map display, digital outputs.

There are a number of key components to data and its management.

- **Database Support Environment**
  - Data extractions
  - Data Integrity
  - Data Version Control
  - Storage Management
  - Schema
  - Data Dictionary
  - Security/authorization
- **Database Management**
  - GIS Data components

Layers
Attributes

—GIS Data modeling

Satisfying queries
Dangers of data modelling
Make data model flexible for future projects

—Develop Data Standards and Procedures

Conventional data standards
Framework Layers
Conventional QA/QC standards
Distribution standards
Backup Standards

—Organization

Hierarchical (tree structure with data dependency defined)
Pros and Cons:
Fast access
Large storage
Not dynamic
Limited sharing
Network (multiple hierarchical, multiple data paths)
Pros and Cons
Increased flexibility
Difficult to update
Limited sharing
Better than hierarchical
Relational (organize data into tables of rows and columns)
Pros and Cons:
Flexible
Standard Query Language (SQL)
Performance is system dependent
Widely used

—Linkages

Combining information contained in existing production systems and files with your GIS application

2 Methods:

Copied files: exporting the data to a file and importing that file into your GIS. This is the most common method and is time consuming when matching data structure of source application to GIS.

Direct link: importing data into the GIS from a live information application. This is least common and very difficult if the data structure of the source application does not match with existing GIS.

—Integration, Ownership, and Sharing

On going data management policy. Should be an organization's IT/IS Policy.

Ownership/stewardship—who's responsible for data maintenance and accuracy?

Sharing—define procedures for distribution and maintenance (updates, corrections, etc).

Cost effective

Geospatial Database Conversion and Creation

—Quality, Quality, Quality

Data is only as good as its source.

Can be costly

Greater the accuracy, higher the cost.

"Measure twice, cut once"—More expensive if data is developed twice.

Costs are coming down (DOQQ base maps)

$$\boldsymbol{cost = area \times density \times features \times process}$$

What level of accuracy to get?

Follow Federal Standards?

Base quality on needs of users?

Base quality on needs of future too!

—Data Conversion and Creation

Define data conversion as taking what exists and making it useful to GIS

Paper maps to digital map

Tabular data joined to geospatial data.

Expense of Conversion/Creation:

Quantity of Conversions

Quality of Conversions

Availability of organizational resources:

Staff, Hardware, Software, Expertise

Conversion/Creation Methods:

Manual Digitizing

Land Surveys

Data Purchases

In order to bring some order to all of this complexity, this chapter summarizes some of the available software products according to their intended application area. Four categories of software are identified: (a) desktop mapping; (b) data viewers; (c) components, and (d) internet mapping. First, the general characteristics and application area of each product is briefly described. Second, a table with a list of vendors and products listed according to each application area is presented. Details related to GIS vendors, applications and capabilities have been included in the book (Appendices III, IV and VI).

Desktop Mapping

Desktop mapping is the traditional software application that is thought of when a product is labeled a GIS. A full featured GIS includes built-in ability to:

- display and query spatial data and non-spatial attribute data,
- perform spatial topological operators such as clip, union, and join,

- alter the coordinate system or projection of spatial data,
- create high-quality cartographic output,
- convert between spatial data formats available from federal and commercial sources,
- framework for implementing customizations either through a proprietary or third generation programming language.

Data Viewer

Data viewers are a more recent development among the increasingly specialized products offered by vendors. Strictly speaking, a data viewer application allows you to browse spatial data without the ability to perform spatial topological operations. Typically, these programs support a limited number of data formats and require that all data to be displayed is in a uniform coordinate system or projection. Functionality relating to map cartography may be present in such applications, but is generally limited in comparison to a GIS.

GIS Software Components

Like data viewers, deployment of GIS component software is also a relatively recent trend in the field. Developments in software engineering during the last 10 years, such as object-oriented programming languages and the availability of dynamically linked libraries, have enabled the increased reuse and improved modularity of code written for commercial software. Customer demand for increasingly specialized applications coupled with the establishment of several software component specifications has enabled vendors to re-sell some of the technology used to rebuild their own software as component technology. Components are particularly well suited for building standalone applications which require GIS functionality, but will not be operated by GIS personnel.

Software Component Overview

The GIS software components shall fully support and exploit the capabilities of the proposed hardware platform and shall provide full functionality for entry, editing, maintenance, analysis, display, and hard copy output of both graphics and tabular data on a continuous and interactive basis. For purposes of this procurement, software component capabilities have been grouped into the functional categories of:

- Database structure
- User interface
- Data entry
- Data editing/maintenance
- Data query and analysis
- Data display/output
- Application development
- Operating system requirements

Data Editing and Maintenance

The vendor shall describe the tools and capabilities of the proposed system to modify and manipulate spatial and attribute data in the GIS for the following categories:

- Interactive Graphic Editing
- Attribute Editing
- File Copying
- Deletion of Features
- Edit Controls
- Rubber Sheeting
- Coordinate Registration and Transformations
- Quality Control/Error Detection
- Merging, Extraction, Edge Matching of Data
- Data Transactions, including the capabilities of the proposed system to translate data into and out of the following formats:

  —GFIS to Proposed System Format (specify how attribute data is addressed)
  —AutoCad DXF (specify how attribute data is addressed)
  —AutoCad CWG (specify how attribute data is addressed)
  —Intergraph IGDS (specify how attribute data is addressed)
  —USGS DLG and DEM
  —TIGER Line Files
  —ArcInfo Export Files
  —Exchange data with KVS Computer Assisted Mass Appraisal (CAMA) System

Data Query and Analysis

The proposed software shall support the following data query and analysis capabilities:

- Graphic Data Query
- Area/Perimeter/Distance Calculation
- Attribute Data Query
- Spatial Aggregation
- Buffer Analysis
- Address Matching
- Polygon Overlay Analysis
- Linear Network Analysis
- Area Districting and Zoning

Data Display and Output

The data display and output tool capabilities that the proposed software shall support including the following:

- Graphic Display

- Tabular Display
- Raster Image Display/Production
- Vector Map Overlay
- Hard Copy Map Production
- Hard Copy Report Production
- Map Plot/Display Relationship with Scale
- Graph/Chart Production
- Interactive Map Composition

Application Development

The vendor shall propose one of more software components that, in a well integrated manner, provide the following capabilities and features:

- Menu Design and Custom Application Development
- Programming Features
- Supporting High-Level (4 GL) Programming
- Subroutine Libraries

Basic Operating System Requirements

The operating system component of the software shall be the primary operating system of the proposed hardware platform and shall provide all of the traditional features of current operating systems as described below:

- Multi-user Support
- Multi-tasking, Multi-threading Support
- Security Management
- File Management
- Memory Management
- Database Backups
- Error Monitoring/Disaster Recover
- System Diagnostics
- Anti-viral Protection
- Electronic Mail (E-Mail)

Network Management Functions

The proposed system shall provide capabilities for monitoring and managing all data and devices on the GIS network as one unified system and support the following capabilities:

- Multi-user Database Access and Maintenance
- Monitoring of Network Activity
- Network Problem Diagnostics
- Print and Plot Management

GIS Database Structure

Database Model

A GIS database model defines the nature and usage of spatial (geographic) data with a database. The proposed software shall support a spatial data model that is capable of creating, managing, and manipulating data sets, defined on the basis of spatial coordinates and associated attribute data sets.

Feature Types: The data model shall support multiple feature types including point, node, line, polygon, and text features.

Data Storage: Features shall be stored as double precision x and y coordinates.

Data Types: The data model shall support multiple graphic and nongraphic data types.

Database Organization: Vendors shall describe strategies for organizing data into logical groups on the basis of data themes, and shall describe the capabilities of the data model for supporting simple and complex feature types.

The geographic data model shall support the creation and maintenance of topological data. Topology shall be created through execution of a software function to structure graphic data sets. Vendors shall describe the ability of the proposed data model to support logical polygons, networks, and user-defined topological structures. Software capabilities that support large-scale engineering and design activities should be outlined as well as specific engineering functions and appropriate modules. It shall support for storage of raster map images (e.g., scanned bluelines, orthophotos) and for raster scanned documents, and the geographic data model shall support the creation and storage of a continuous geographic database. A Relational Database Management System (RDBMS) will be able to maintain a minimum of 30,000 records of parcel ownership information in a single table and shall provide functionality for updating database content, queries, and production of reports. The recommended RDBMS must be either a part of the GIS software or have a direct access capability. Some of the major functions of standard GIS packages are shown in Table 4.1.

Table 4.1: Popular GIS Software and their functions

Vendor Name	*Desktop Mapping*	*Data Viewer*	*Component*	*Internet Mapping*
Autodesk	AutoCAD Map	Map Guide Viewer	N/A	Map Guide
Bentley	MicroStation	N/A	MicroStation/J	Model Server Imager
Blue Marble Geographics	N/A	Geographic Explorer	GeoObjects GeoView	no out-of-box solution see BeyondGeo.com
Environmental Systems Research Institute, Inc	ArcInfo ArcView	ArcExplorer	MapObjects ArcObjects	ArcIMS
Intergraph	MGE (Modular GIS Environment)	GeoMedia Professional	GeoMedia	GeoMedia WebMap
MapInfo, Inc	MapInfo Professional	MapInfo ProViewer	MapX	MapXtreme
Object FX	N/A	N/A	SpatialFX	N/A

N/A: Not Available in Category

The details of the capabilities of industry standard software are mentioned in Appendix VI. The list of free GIS software is enclosed in Appendix V; and the list of GIS vendors and related software developers is resting in Appendix III.

Selection of Software

Initially it is required to evaluate the software independently of hardware. The software will be selected based on the functionality it offers. The hardware selection is based on the selected GIS software and the operating system strategy in the organization uses. It is needed to test the hardware and software together making sure it works uniformly without any hindrances. The nature of hardware and software technology is that it changes. In recent years it has been changing very quickly. Do not let this stop the efforts. It is easy to get intimidated. The important thing to remember is to get a product that has been proven in the market place and continues to have a clear development path. Avoid technology that is outdated or is on the bleeding edge and has not been proven. Software is evaluated on functionality and performance.

Functionality

What is important here is the ability of the software to do the things you need it to do in a straightforward manner. As an example, if the intended users are relatively new to using computers, the software has to have an easy-to-use Graphical User Interface (GUI). If the organisation needs to develop specific applications, the software should have a programming language that allows the software to be modified or customized.

Standards

Standards are a way of making sure that there is a common denominator that all systems can use. This can be in the form of data formats that can import and export data into the system, guidelines used for developing software, supporting industry developed standards that allow different applications to share data. Standards are generally developed by a neutral trade organization or in some cases are defined by the market. There is a group that has formed for the GIS industry called *Open GIS.* This organisation is developing standards for developers to use as they engineer software. *Open GIS* is made up of representatives from the software developer companies.

Performance

The performance of the software is dependent on two factors: (a) how it is engineered and (b) the speed of the hardware it is running on. GIS software is complex and will use a large amount of the system resources (memory, disk, etc.). The more complex the software, the more resources it will need. Performance will be impacted if you have a minimally configured computer. Look for the developer's software specifications to see what configuration is needed to run the software. This will give the minimum requirements. It has to be followed up by getting the recommended specifications from the developer or a user group. These recommendations will give a more accurate idea of the type of configuration one will need.

Expandability

The software needs of today will change over time. More than likely the system will need to expand. Is the software being evaluated able to provide networking capabilities? Will it share data with other applications? Will it grow as the organisation's GIS grows? Evaluate software based on the ability to grow with the organisation. This may mean that there are complimentary products that can be used in conjunction with the package one is evaluating today or the developer has clearly defined plans for added functionality. Discussions with other users to see if the developer has a good track record for providing these enhancements would be beneficial.

Licensing

GIS software is not purchased, it is licensed. There is normally a one-time licence fee with an on-going maintenance fee that provides us with the most current versions of the software as they are released. In large systems this will be spelled out in a licensing agreement with a corresponding maintenance agreement. For desktop software a shrink wrap license is used with subsequent releases being offered to existing users through a discounted upgrade. The maintenance fees and upgrade costs generally run between 15 to 30 per cent of the initial licence fee. The terms in most software packages spell out how the software can and cannot be used. The terms of the licence should be reviewed by an attorney before signing up. This can save hassles later as one is developing and using the system.

HARDWARE

Computer hardware used to support GIS is a highly variable part of the overall system. Users will customize their hardware environment to best meet their own individual needs. In all cases, however, a fully functional GIS must contain hardware to support data input, output, storage, retrieval, display and analysis. During its infancy, GIS processing was supported only on large mainframe computers. Today, a variety of platforms support GIS processing, ranging from large mainframe computers to mini-computers to scientific work stations to personal computers. In many cases, hardware used to support other applications (e.g., payroll or accounting or digital image processing) is also used for GIS. Hardware configurations for GIS span a tremendous range in terms of start-up and acquisition costs. All systems carry associated costs for maintenance and support.

When discussing hardware, there are terms and concepts that need to be understood. However, GIS software selection drives the hardware requirements. Therefore, before launching a full scale evaluation of hardware, the GIS software selection should be made. The GIS hardware standard establishes consistent configurations for GIS-based computing platforms and associated peripherals. PC computing hardware should be Intel-compliant. Intel platforms utilized for GIS should conform to the desktop system with minimum specification of (a) minimum 32MB RAM, (b) minimum 17" monitor size with 1280×1024 by 256 colour with 4MB VRAM, and (c) 101-key Keyboard.

In addition to PC hardware platforms, GIS technology has been widely implemented on workstation computers utilizing a UNIX operating system. Because so many brands (e.g., non-compatible variations) of UNIX exist, an ITCC UNIX standard has not been established. However, in considering purchase of such a workstation, one should consider six major vendors, including (in alphabetical order): DEC, Hewlett-Packard, IBM, Intergraph, Silicon Graphics, and Sun Microsystems. Other hardware specifications to be taken into consideration include RAM requirements (min. 32MB), monitor size and resolution, and hard disk requirements (min. 1 GB in shared (e.g., networked) disk-space environment; 2 GB for stand-alone workstation).

Deciding what hardware to use for the GIS is a difficult yet important task. It will make up the foundation on which one will build his system. There is no clear-cut formula to use to make the selection process easier. When discussing hardware, there are terms and concepts that one need to understand. Hardware has following basic components (Fig. 4.1):

- Operating System
- Processor
- Disk
- Memory
- GIS Hardware Used in Digital Data Conversion
- Communications

Fig. 4.1: Hardware required for GIS

Operating Systems

An operating system is the software that runs the computer hardware. It is this programme that tells the computer what to do and how to do it. One may already be familiar with some of the operating systems that are on the market such as Microsoft's Windows product or various brands of the UNIX operating system.

It is an important to have an Operating System (OS) plan within the organisation. The plan should take into account the departments that will be using the computer system, the type of network being used (or being planned), what operating systems are currently being used, how large the database is and what kind of technical support skills one have access to (in-house or contractor). The GIS will need to fit into the OS plan. This will be important as to add other departments onto the system.

Processor

The processor or CPU (Central Processor Unit) is the part of the computer that actually does the calculations or "processes" the instructions being sent to it. The most common term that describes the processor's capabilities is the clock speed. This is stated in terms of MHz (MegaHertz). The clock speed simply describes how many cycles per second the processor works. The higher the clock speed the faster the processor.

Another description of the processor's capability is how many bits it can access at one time. Many of the new processors are 32-bit processors. This means that the CPU can access or "grab" 32 bits of information during each cycle. Older computers such as a "386" machine where 16-bit machines. There are some machines on the market that manufacture a 64-bit machine (such as Digital Equipment Corporation, Sun Microsystems). These are very fast CPU's but are hampered by the lack of a 64-bit OS that can take advantage of its speed. It is the direction the hardware industry seems to be heading.

Disk

The disk or hard drive is the device used to store the operating and application software. It is also used to store data and images. In working with a GIS one will quickly find out that GIS uses a large amount of disk space. It is not uncommon to have multiple gigabytes of hard drive on a single end-user machine and 10-20 gigabytes on a central data server. Luckily, the prices of hard drives have been coming down and will continue to be affordable.

Memory

Memory or Random Access Memory (RAM) is used as a temporary storage space by the OS and by the application software which is running on the computer. Most applications run better as the amount of memory increases. This is true up to a point. At some point, the performance increases will begin to taper off as additional memory is added. Most software developers can give you configuration data that indicates where this point is.

GIS HARDWARE USED IN DIGITAL DATA CONVERSION

Most contemporary GIS software packages are structured to operate on computer workstations to accomplish digitizing and editing tasks. Four basic types of workstations can be identified:

- A digitizing station, a workstation which is connected to a precision digitizing tablet, which utilizes a high-resolution display terminal, and which also has all of the analysis functions necessary for querying, displaying and editing data.
- An editing workstation, which is used for conducting most of the QA/QC functions of the conversion process, having all the functionality of the digitizing station except for the ability to digitize data via a digitizing tablet.
- Graphic data review/tabular data input workstations are used for displaying and reviewing graphic data, and for the entering of tabular attribute data associated with these features.

- X terminals are the fourth type of workstation and these allow for graphic display and input of data utilizing the X Window system communications protocol.

With the increasing power of today's personal computers, many GIS analysis packages are being designed for PCs. As GIS data files are very large, PC-based GIS packages usually require a PC with minimum requirements including a 486 processor and 16 megabytes of RAM. Hard-drive disk space depends upon how large the datasets are which are being used. A safe bottom-line for hard-drive space with a PC is 500 megabytes. For most data conversion projects, much more hard-drive space will be needed in order to store data as they are converted. Tape storage hardware is also necessary in order to efficiently backup the many megabytes of files created in the conversion process. Just to provide an idea of the storage requirements necessary for basic scanning conversion, the file size of one tax map alone, in (Tagged Image File Format or TIFF) image format, scanned at a 500 dots per inch (dpi) resolution, can range anywhere from 1-3 megabytes alone.

Fig. 4.2: Digitizer

Digitizing hardware requirements vary according to the conversion approach which is applied. For vector conversion, a digitizing tablet will be necessary in usually a manual digitizing process (Fig. 4.2). Another piece of digitizing hardware, a scanner, is used to create raster images (Fig. 4.3). Automatic digitization, through the use of a scanner is a very popular approach for capturing data. Raster data can subsequently be transformed into vector data in most turn-key GIS packages, through the use of raster-to-vector conversion algorithms.

After the conversion of map data into digital form, hardware will be needed for outputting digital data in hardcopy format. When handling a data conversion project, a necessary piece of output hardware is a pen or raster plotter. GIS software allows for the creation of plots at any view scale. The plotter, with its ability to draw on a variety of materials (including paper, mylar and vellum), allows for the creation of quality map plots. Most plotters usually have a minimum width of three feet. Vector and raster plotters are both available on the market. Vector, pen plotters utilize various pens for the drawing of linear features on drawing media.

Pen plotters can handle most plotting jobs, but they do not produce good results in area shading such as in the production of cholorpleth maps. *Raster plotters,* on the other hand, are excellent in producing shading results. *Raster plotters* usually cost more than vector plotters, but are substantially more versatile and have better capabilities (Fig. 4.4).

Fig. 4.3: Scanner

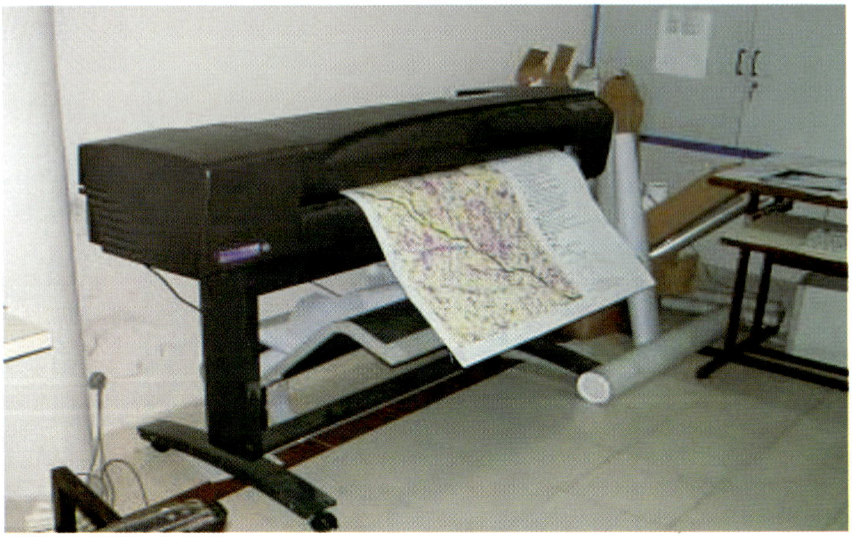

Fig. 4.4: Plotter

Other output devices for the creation of hardcopies of GIS data include: screen copy devices, used for copying screen contents onto paper without having to produce a plot file; computer FAX (facsimilie) transmissions, often used in communications between conversion contractors and clients, produce small letter-size plots, and the fax transmission files (as raster images) can be saved and viewed later; printers are used to output tabular data which is derived from the GIS, and if configured correctly, can produce small letter size plots (Fig. 4.5).

Fig. 4.5: Printer

Communications

The trend in most systems today is to link up users throughout the organization on a network. This is an area in the computer industry that is advancing very rapidly. It is recommended that the organisation retain a competent consultant who works with networks to give the detailed and current information. In simple terms, a network is a connection between computers that allows information to be passed around from computer to computer. In a typical organisation, this is a Local Area Network (LAN). In order to connect a computer to the network it will need a network card for the wiring to plug into and network software to allow the computer to transmit and receive signals over the net (Fig. 4.6). Of course, the physical network (wiring) is also needed.

A small network within a department is inexpensive and can allow the users to share network resources such as printers and database servers. The network can provide services like e-mail, disk sharing and using common hardware facilities. It can also be the entryway into larger networks that go outside the building or campus your organisation is located on.

This is called a Wide Area Network (WAN). A WAN requires a more structured network architecture. It does give users access to more resources (Fig. 4.7).

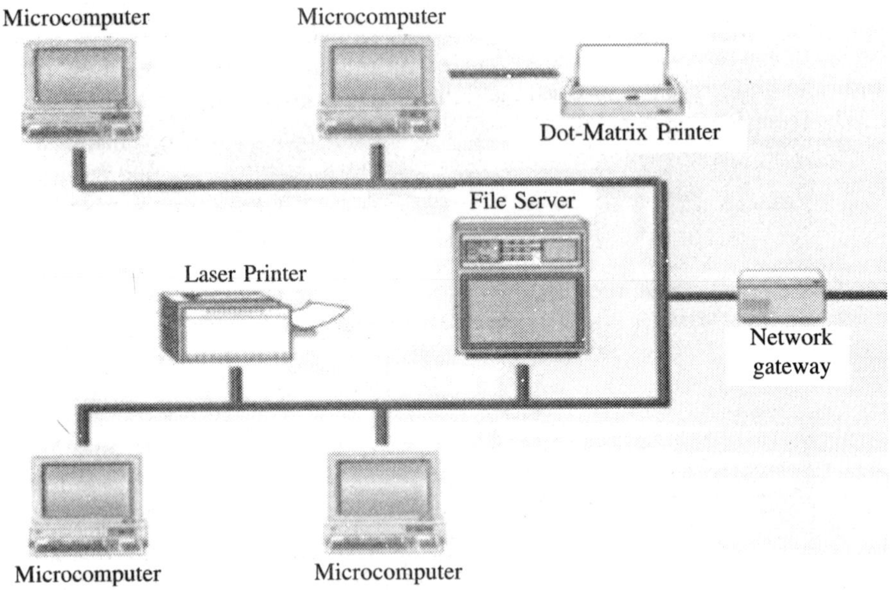

Fig. 4.6: Local Area Network—Intranet

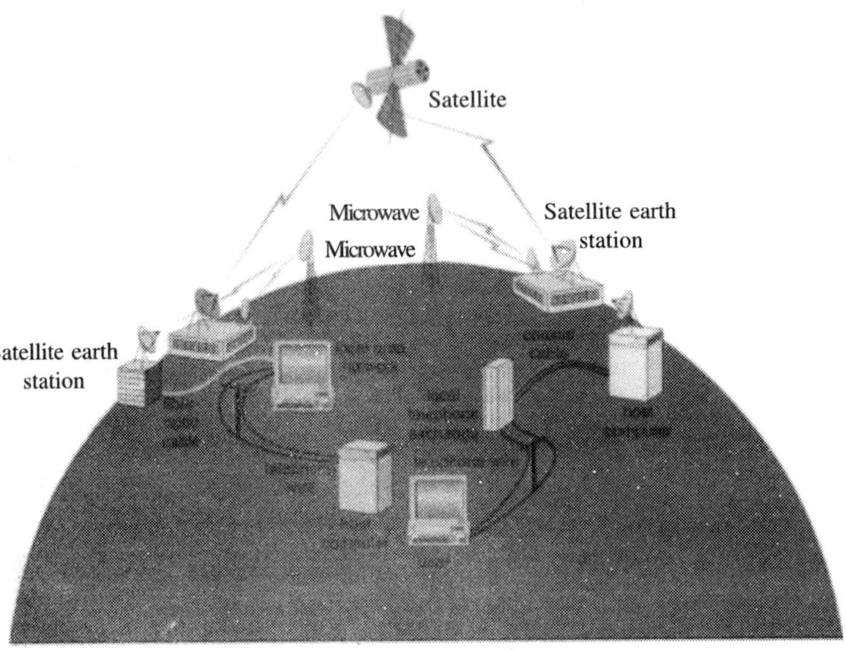

Fig. 4.7: Wide Area Network (WAN) Internet

Another important point to consider is developing access to the internet. This specialized network is growing rapidly and provides an incredible amount of resources for a user. The internet is an area to share ideas in a GIS forum, download data for use in the system, get technical support for a problem, get the latest information on a product from a vendors' home page or develop one of your own. The amount of information is overwhelming and too diverse to list here. The point is that one should seriously be considering getting a connection to the internet. When considering the network requirements, this factor must be considered. Further, internet based GIS has opened new areas for applications. Networks should have features to support storage of data which is accessible by users on the network by specifying particular files, collections of features or attributes, and geographic areas; access security at different levels of the GIS database; ability to support and connect query workstations on the network and to allow database queries directly from workstations on the network; and ability to allow network-wide access to plotters and printers. Further, the network should also be able to perform the following functions:

- Access to data on remote nodes by reference to the node, disk, directory, and file
- Access to programs on remote nodes by similar reference
- Assignment of logical names or aliases for programs or data locations on remote nodes
- Control of peripheral devices from any node on the network
- Passing of mail messages across nodes
- Programme-to-programme communications across nodes
- Monitoring of traffic and errors on the network

The proposal shall include all cabling and devices required to implement all data communication connections, utilizing existing facilities. The proposed system must operate at a minimum raw data speed of 10 megabytes per second.

Benchmarking a System

Benchmarking a GIS can be a very involved process. The level of effort needed for the benchmark should be proportional to the size and complexity of the overall system being developed. A benchmark is the process of testing various combinations of hardware and software and evaluating their functionality and performance. The benchmark is usually part of a tender process and is only done with a limited number of selected vendors (i.e., those that have responded and been short listed). Each combination is tested under similar conditions using a predefined data set that is indicative of the database. This dataset should be used with all of the hardware and software configurations selected for evaluation. When completed, an organisation will have results that can used to objectively evaluate the systems.

Installation

When putting a benchmark together there is strength in numbers. A committee may be appointed for this purpose. The committee will take the burden off of a single person and give the process more objectivity. The members of the committee should be drawn from all the interested departments and agencies within the organisation. A working group of about 8-10 committee members is reasonable. The committee should develop the criteria that will be used to evaluate the systems. These criteria will form the basis of the benchmark. A series of tasks that each vendor will need to complete during the benchmark is to be worked out. The tasks should be measurable (i.e., time, ease of use, can the function be done). Also prepare a form that each of the committee members will use to rate the tasks performed in the benchmark.

In the benchmark one will not only be to rating various aspects of the system, but also going to be rating the vendor. One has to be sure to include some measurement for teamwork, communication, and technical skills of the vendor. It might be useful to work with a consultant that has experience setting up benchmarks or to get advice (and examples of documentation) from another local government who has recently completed a benchmark. Well in advance of the scheduled benchmarks, information that outlines the tasks the vendor will need to perform is to be sent out and any rules they will need to follow (how much time for set up, time given to perform various tasks, how many people can be present for the benchmark, etc.).

Vendor Support

The selected vendor will become an extended team member for the GIS group. The vendor will be a good source of support and information. All vendors provide some type of technical support. Ask current users how it has worked for them. If there have been problems in the past, do existing users see improvement? The GIS industry has been growing very fast over the last few years, there are bound to be some growing pains. What one should be looking for is a vendor who listens to the need and makes improvements based on the user input.

Spatial Data Transfer Formats

The standardization of spatial data transfer formats is intended to provide efficient interchange of data between users. In reality, the methods selected for such interchange is often dependent on limitations imposed by hardware and software resident at both the source and target sites. Currently a number of efforts are ongoing to refine existing standards or develop alternatives, including the Open GIS Consortium or OGC. Spatial Data Transfer Standard (SDTS) components include a conceptual model, specifications for a quality report, transfer module specifications and definitions of spatial features and attributes. Most major GIS software vendors now support conversion utilities to and from the SDTS format. In addition, the U.S. Geological Survey has developed a set of public-domain software tools designed to support, among other tasks, the encoding and decoding of logically compliant SDTS data into and out of the SDTS's ISO 8211-FIPS 123 general information interchange format. Additional information on data transfer standards may be obtained by accessing the following sites on the World Wide Web:

SDTS:ftp://sdts.er.usgs.gov/pub/sdts/standard/latest_draft/pdf
OGIS:http://www.opengis.org/techno/interop/index.htm

Metadata Content

Documentation provides information to users about the development, content and format of digital spatial data. Known as metadata, this information serves to describe the content, quality, condition, and other characteristics of various datasets. The GIS standard for documentation for digital spatial data shall be the Content Standard for Digital Geospatial Metadata, developed by the Federal Geographic Data Committee. Regarding India, the National Spatial Data Infrastructure or NSDI has formulated the metadata standards (Appendix II). The standard specifies the information content of metadata for a set of digital spatial data, including a common set of terminology and definitions for documentation related to these metadata. The standard does not specify how this information is to be organized or transferred within and between computer systems, or how the information is to be transmitted or communicated to users. Additional information on the FGDC metadata content standard may be obtained by accessing the following site: http://www.fgdc.gov/metadata.

Sample Hardware Specifications

Specifications for a system configuration to support GIS development and operational applications follow the system configuration consists of various devices that will be networked together to support data capture, storage, processing and display in both digital and hard copy forms, including:

- mapping/analysis workstations (2)
- colour laser printer (1)
- black and white laser printer (1)
- cartridge tape drive (1)
- colour raster plotter (1)

The proposal shall include technical and functional capabilities of the devices offered to meet these specifications. Provision of the following information should be included for each device:

- manufacturer
- year of production
- model number and version
- capabilities/configuration of each device in comparison to the device specifications
- documentation provided with the device (i.e., manuals: CD or hardcopy)
- features of warranty and guarantee included in the purchase price
- post purchase user support, duration and minimum response time

GIS Workstations

The mapping or analysis workstations will support a widerange of GIS activities, including database development, database quality control, user application development, database maintenance and all GIS applications supported by fully functional GIS software such as cartographic production, geographic database queries, and advanced geographic analysis using both spatial and attribute information. One of the GIS workstations must support high capacity data storage, and multi-user GIS processing, and should perform all GIS operations and applications within acceptable user response times. General specifications for workstations are as follows:

- Mass storage may be configured within the workstations' cabinetry and/or as external drives
- High resolution (1280 × 1024 or greater) color monitor with at least 19" minimum diagonal screen dimension
- Keyboard and a pointing device such as a mouse
- Network-ready and should be capable to connect a local area network (LAN) and supporting a minimum transmission speed of 10 megabytes per second (mbps).
- Multi-user, multi-tasking OS supporting logical security measures such as user name/ password validation, and user access privileges.
- Virtual memory operations, either through a dedicated hardware controllers(s) or through software OS functions.

Specific Details of Workstations

- 32-bit processor supporting both 64-bit address and data buses, minimum of 75 MHz clock speed and may have multiple CPUs on board.
- 128 MB (megabytes) of main memory and shall support 32 MB memory modules and be expandable to 256 MB.
- mass storage disk drive(s) for direct access of data and software functions with a minimum of 3 GB of mass storage.
- CD-ROM drive with one 1.44 MB floppy drive.
- Should support multi-user/multi-tasking operations.
- Options for upgrading the speed and performance.

Small-Format Colour Printer

One colour printer will be used for the production of colour hard copy graphic plots and non-graphic report generation. The colour laser printer should meet the specifications or equivalent described below:

- 300 dots per inch (dpi) resolution
- 100 sheet paper tray
- 4 MB memory onboard with capacity for memory upgrades
- Support for letter, legal size, and 11" × 17" paper sizes
- Built-in postscript compatibility
- Serial and parallel interface

Large-Format Colour Raster Plotter

A colour raster plotter shall be included for the production of high quality, large format cartographic products. This device must provide a high-volume colour plotting capacity. The plotter shall support 36" × 42" plots and produce colour plots at a minimum resolution of 300 dots per inch. The plotter shall be compatible with the proposed LAN hardware and communications protocols and must be accessible by all workstations on the LAN. A sample hard copy output from the proposed plotter shall be included in the proposal. Additionally, the plotter should meet the specifications described below:

- Capable of supporting true colour plotting
- 8 MB memory onboard with capacity for memory upgrades
- Support for all paper sizes, A through E size
- Built-in postscript compatibility
- Serial and parallel interface

Conclusion

A proper combination of software, hardware and communication technology has become a core issue for GIS development in any setup. One has to be very careful in working out the details which is not easy. Fast changing technology, increasing volume of data to be handled, new features, licensing protocols, falling prices and options of integrations with other technologies have made such decision more complex. As mentioned earlier, a consortium approach to decision-making may reduce the pitfalls.

REFERENCES

Burrough, P.A. (1992), Development of intelligent geographical information system, *International Journal of Geographical Information System,* Vol. 6 No. 1, 1-12.

Burrough, P.A. and McDonnell, R.A (1998), *Principles of Geographical Information Systems.* Oxford University Press Inc., New York.

Cadoux-Hudson, J. and Heywood, I. (eds), (1991) *Geographic Information System,* Taylor and Francis, London.

Carstensen, Jr., Laurence, W. and Campbell, James, B. (1991), Desktop scanning for cartographic digitization and spatial analysis, *Photogrammetric Engineering and Remote Sensing,* Vol. 57, No.11, pp. 1437-46.

Cowen, D.J. (1988), GIS versus CAD versus DBMS: What are the differences ? *Photogrammetric Engineering and Remote Sensing,* Vol. 54, pp. 1551-55.

Frank, Andrew U., Egenhofer, Max J. and Kuhn, Werner (1992), A perspective on GIS Technology in the nineties, *Photogrammetric Engineering and Remote Sensing,* Vol. 57, No.11, pp. 1431-36.

Goodchild, M.F. (1991), Geographical information science, *International Journal on Geographical Information System,* Vol. 6, No. 1, pp. 31-45.

Goodchild, B.F. (1992), Geographic data model, *Computer and Geosciences,* Vol. 18, pp. 401-08.

Green, N.P., Finch, S. and Wiggins, J. (1986), The state of art in geographic information system, *Area,* Vol. 17, No. 4, pp. 295-301.

Groff, James R. and Paul N. Weinberg. (1999), *SQL, the Complete Reference.* Berkeley, California: Osborne/McGraw-Hill.

Knapp, W.G.M. van der (1992), The vector to raster conversion: (mis)use in geographical information systems, *International Journal on Geographical Information Systems,* Vol. 6, No. 2, pp.159-70.

Maguire, D.J., Goodchild, M.F. and Rhind D., (1991), *Geographic Information System,* Longman, London.

Mark, David, M. and Gould, Michael, D. (1991), Interacting with Geographic information: A commentary, *Photogrammetric Engineering and Remote Sensing,* Vol. 57, No.11, pp. 1427-30.

Mounsey, H.M. and Tomlinson, R.F., (1988), *Building Data Bases for Global Science,* Taylor and Francis, London.

Nag, P. (1987), A proposed geographical information system for India, *International Journal on Geographical Information System,* Vol. 1, No. 2, pp. 181-87.

Rhind, D.W. (1988), A GIS Research Agenda, *International Journal of Geographical Information System,* Vol. 2, p. 23-28.

Worrall, Les (1990)(ed), *Geographic Information System: Development and Applications*, Taylor and Francis, London.

5

Datum, Projection and Co-ordinate System

Geospatial information, including maps and images, are vital to support decision-making at various level and implementation of action plans. With the availability of space borne imagery, Global Positioning System (GPS) data and the GIS technology, users are now able to process maps–both individually and along with tabular data and crunch them together to provide a new perception—the spatial visualization of information. Space based images has the characteristic to cover vast areas and have larger information quality as well as multi-spectral and repeat observations ability, thus, it is suitable for regional and global environment monitoring. The application of satellite images is not only used for mapping but also to monitor temporal changes in the environment.

Global positioning system is one of the recent and important advances positioning technique. Its effectiveness and high positioning accuracy, the system has been widely applied in many different fields. GIS has now become a tool to integrate different sources to help users to manage earth's resources and environment. It assists decision makers to formulate the best policy for the development and the use of earth's resources. Vast amount of geographical information that was originally scattered in different organizations like maps, photos, statistics, detailed list etc could be integrated in GIS environment.

The science of surveying and mapping encompasses a broad range of disciplines including surveying and mapping, Remote sensing, GIS and GPS. The technology of digital image processing, the GPS and GIS are being used to integrate, processed and analyzed the spatial data for sustainable development of the natural resources. The digital data (e.g. high resolution images, scanned aerial photographs, scanned maps and digital orthophotos) are currently available for this purpose. GIS is powerful system for decision-making tool is now in the hand of cartographers. Beginning with a computerized topographic map as its base, a GIS overlays and integrate graphic and textual information from separate databases. The end result is a customized and reliable tool that can support decision making and problem solving and provides almost instantaneous answers to complex questions.

In view of the present day requirement, Survey of India (SOI) plans to provide digital topographical databases for entire country on 1:50,000 scale initially, in a years time. Topographical data or maps in digital and analogue form on WGS 84 (World Geodetic System-84) is made available for general public without any restriction (existing topographical maps of SOI are on Everest datum). The department uses the state-of-the-art technology in glossarial data acquisition, management and dissemination. High-resolution satellite imagery viz. IRS, SPOT, IKONOS data etc are being used for updating of topographical maps. All these data sets would be useful in developing spatial data infrastructure (Chapter 8) that will better facilitate the availability and access to spatial data for all level including government sectors, commercial sectors, academia and citizens in general.

Reference Ellipsoids

Ellipsoidal earth models are required for accurate range and bearing calculations over long distances. Loran-C, and GPS navigation receivers use ellipsoidal earth models to compute position and waypoint information. Ellipsoidal models define an ellipsoid with an equatorial radius and a polar radius. The best of these models can represent the shape of the earth (Fig 5.1) over the smoothed, averaged sea-surface to within about one-hundred meters. Reference ellipsoids are defined by semi-major (equatorial radius) and semi-minor (polar radius) axes. Other reference ellipsoid parameters such as flattening, and eccentricity are computed from these two terms. Many reference ellipsoids are in use by different nations and agencies.

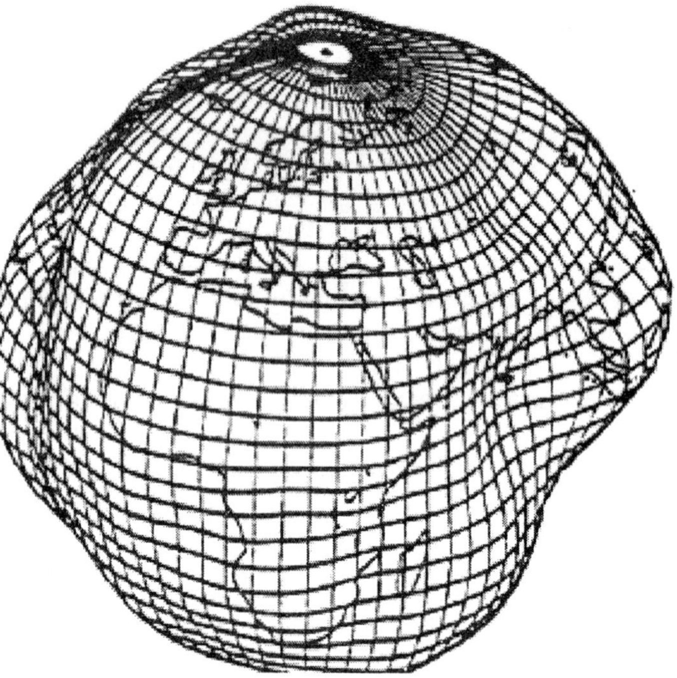

Fig 5.1: Perspective view of the world (*Source*: Rolf A. de By, ITC)

Geodetic Datums

Geodetic datums define the reference systems that describe the size and shape of the earth. Hundreds of different datums have been used to frame position descriptions since the first estimates of the earth's size were made by Aristotle. Datums have evolved from those describing a spherical earth to ellipsoidal models derived from years of satellite measurements.

Modern geodetic datums range from flat-earth models used for plane surveying to complex models used for international applications which completely describe the size, shape, orientation, gravity field, and angular velocity of the earth. While cartography, surveying, navigation, and astronomy all make use of geodetic datums, the science of geodesy is the central discipline for the topic. Referencing geodetic coordinates to the wrong datum can result in position errors of hundreds of meters. Different nations and agencies use different datums as the basis for coordinate systems used to identify positions in geographic information systems, precise positioning systems, and navigation systems (Fig 5.2). The diversity of datums in use today and the technological advancements that have made possible global positioning measurements with sub-meter accuracies requires careful datum selection and careful conversion between coordinates in different datums. Further, two different datums are used for horizontal and vertical coordinates—the horizontal one is an ellipsoid while the vertical one is geoid.

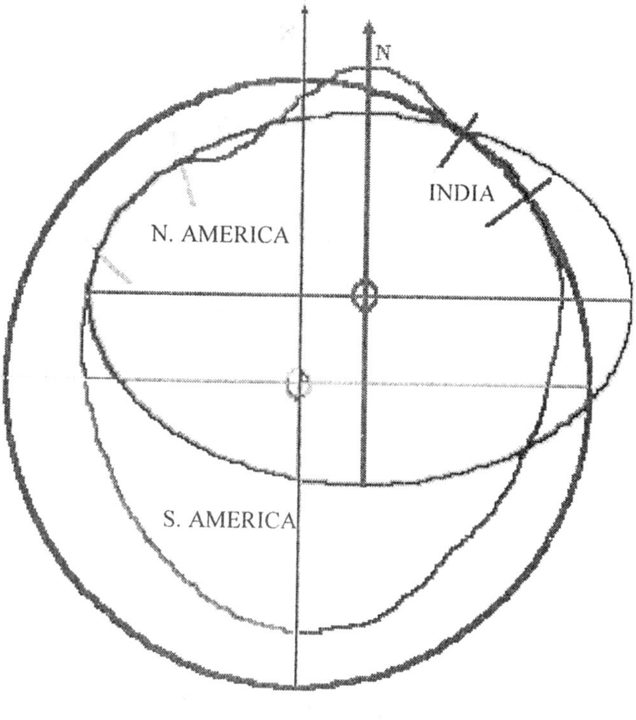

Fig 5.2: Geoid and best fitting ellipsoids

Geodetic datums can be broadly classified into two categories : (a) geocentric datums—World Geodetic System 1984 or (WGS-84), and (b) local geodetic datums—European datums 1987 or ED-87 and Indian geodetic datum 1880 (Fig 5.2). More than hundred local horizontal datums are used in different countries of the world; while GPS provides locations on WGS-84 datum. With the introduction of GPS in the mapping and GIS activities, it has become imperative to establish relationship between the WGS-84 and Everest datum. The New Map Policy of 2005 has further indicated the necessity of such relation to bring out the Open Map Series or OSM (Appendix I). Further, the heights provided by GPS refer to WGS-84 datum. In order to obtain geoidal heights, the ellipsoid-geoid relationship is also to be established with a greater degree of accuracy (Fig 5.3).

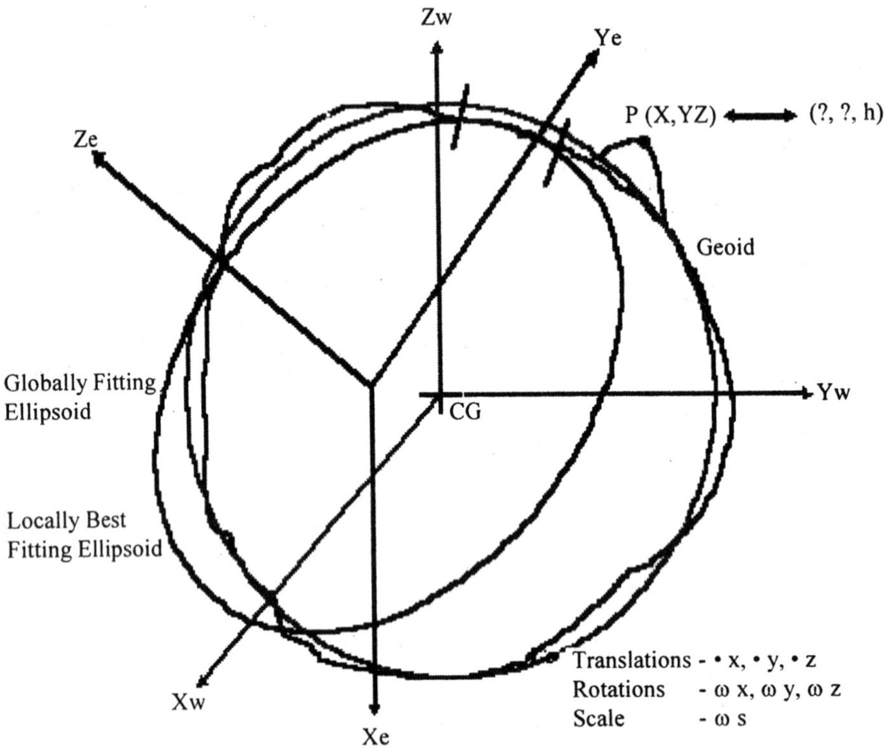

Fig 5.3: Geocentric and locally fitting ellipsoids

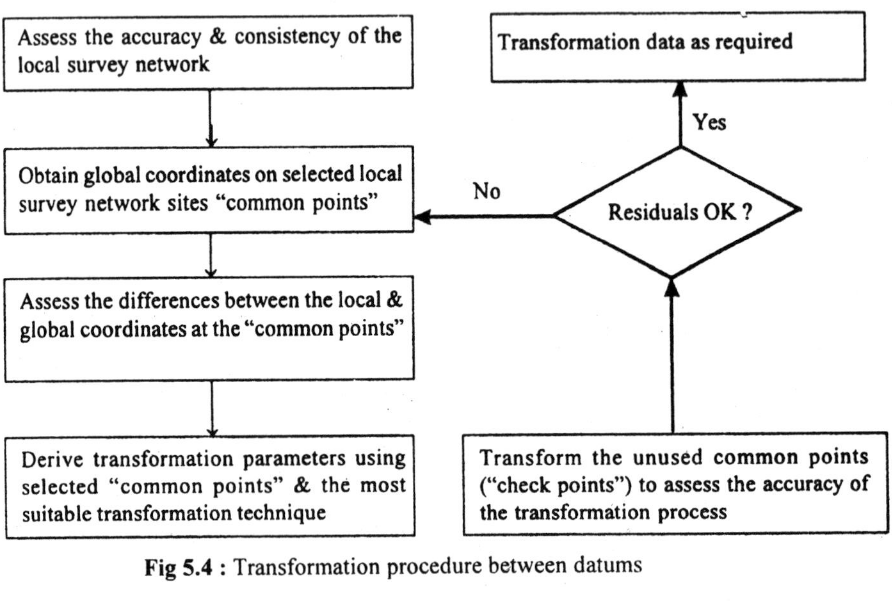

Fig 5.4 : Transformation procedure between datums

The transformation between any two horizontal datums require seven parameters : three due to shift of origin, three for rotation and a scale factor. These datum transformation parameters can be worked out using several points or stations in both the systems. Now software is available which can transform coordinates accurately between two geodetic datums (Fig 5.4 and Fig 5.5).

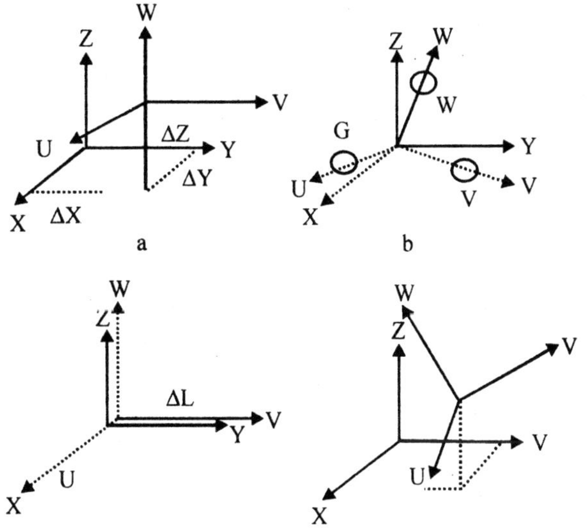

Fig 5.5: Translation and rotations

The Great Trignometrical Survey

The Great Trignometrical Survey or GTS in due course turned out to be a mega scientific activity not only for undivided India, but for the whole world *per se*. It laid the scientific basis for measuring the base line and thereby providing reliable data to find out the shape and size of the earth. The exercise was based on triangulation over the length and breath of the country and hence it is known as GTS. The basic requirement for a map or even GIS is to know the location or geodetic coordinates in terms of longitude, latitude and elevation above the sea level. A framework of control points was established through out the country onto which geographical or topographical features are adjusted. This framework is known as *triangulation network.* The chains of triangles whose lengths and sides are measured for converting the distances along the spherical surfaces of the earth to geodetic coordinates. A series of triangles were measured along the 78°E of Greenwich, i.e. from Kanyakumari to the foothills of Himalayas at Banog. This chain of triangles is now known as *The Great Meridional Arc of India.* With the success of this activity, the surveying technique changed from astronomical to triangulation method which was more accurate and economical as well (see also Chapter 1).

Ultimately, the chains of triangulation covered the whole country and provided a grid iron mesh of points which became the basis for topographical, revenue and cadastral surveys. This work which was initiated by Col. William Lambton and completed by Sir George Everest. This helped in calculating the parameters of the Everest datum or the Indian datum which are as follows :

Semi major axis (a) = 6377301.243m or 20922932 ft
Semi minor axis (b) = 6356100.231m or 20853375 ft
Flattening (f) = 1/300.8017

India has travelled quite far in last two hundred years. i.e. from the days of astronomical methods to GTS and finally to the Global Positioning System or GPS.

The Indian Datum

Since the earth surface is uneven and not exactly suitable for direct mathematical computations, a hypothetical reference surface must be defined which is known as *geodetic reference datum.* The reference datum adopted in India is the Everest Ellipsoid which has been developed after adjustments of the Indian geodetic network existing in 1880. Kalyanpur was taken as the initial point (Latitude : 24°07'11".26; Longitude : 77°39'17".57). Hence there are limitations regarding the availability of data and computational methods. Meanwhile a lot of geodetic data has been collected and advanced computational methods are now available. It is now possible to take up integrated adjustments leading to redefinition of the datum. The Indian geodetic datum or the Everest ellipsoid is based on Great Trignometrical network which was initiated in April 1802. Further, geoid based on the mean sea level is used as datum for finding out height of any point in the country. Survey of India has the primary responsibility in this regard. The first level net and the vertical datum was established in 1909. In order to further refine the datum, a second level net was mooted in 1977.

The South Asian region was earlier divided into five parts : four quadrilaterals (NW, NE,SE and SW) and the southern trigon for the adjustment. Owing to the computational limitations, these quadrilaterals could not be adjusted. In 1937-38, an unsuccessful attempt was made to readjust the network. As a result, Everest Eliipsoid is continued to be used as the geodetic datum for India and neighbouring countries. The Indian topographical maps and all geodetic coordinates are based on this datum. However, serious attempts are being made to overcome this problem and to redefine the Indian geodetic datum. This will certainly improve the accuracy of the datum and thereby higher accuracies in geodetic coordinates, distances, azimuths and gravity anomalies would be possible.

Maps

The procedure from geoid to map projection consists of several steps, including the selection of appropriate reference ellipsoid considering the topography of the country (Fig 5.6). However, a GIS analyst hardly comes across to such intricacies. In most of the cases he is bound by the reference ellipsoid based on which the data is available or chosen by the country or the regional authorities. Nevertheless, in some very specialized application, the choice of ellipsoid becomes significant.

For a user of GIS technology maps are of supreme importance without which it is not possible to progress in this field. The maps are the main source of data for GIS. Hence, the traditions of cartography are fundamentally important to GIS. GIS has roots in the analysis of information on maps, and overcomes many of the limitations of manual analysis. A map is a representation of a small or large area of the earth on a flat piece of paper. It may be regarded as a concept ional picture of the earth's surface as seen from above to which lettering is added for identification. A map always bears some proportions to the actual size of the area on the ground. This proportion is known as scale of the map and is always given to it. From the above definition 4 essentials of perfect map can be deduced:

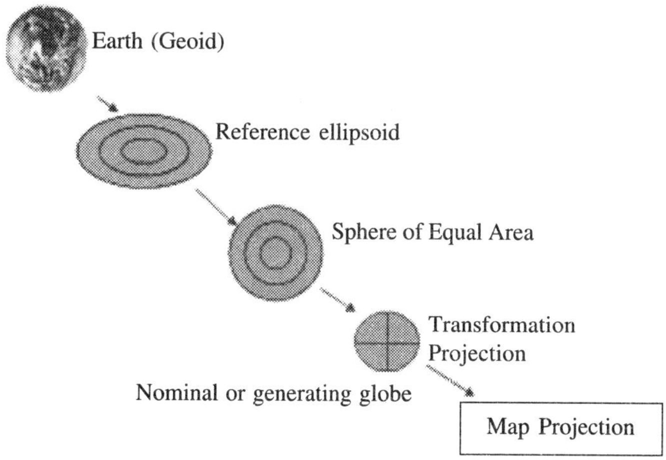

Fig 5.6 : Map Projection Process

Map is conventional representation of earth's surface, which implies that various conventional signs or symbols are used in a map for showing various objects such as roads, railways, canals and trees etc. A key to such conventional signs is always given on the map.

A map is a representation of the ground as seen from above. The objects represented on a map have only two dimensions. For showing three dimensions various methods are to be used like contours etc.

In a map lettering is added for identification. A map must bear title, which state what the map is intended to show. Lettering inside map varies according to the importance of the place.

The map is some fraction of true size of the area on the ground. This fraction or scale forms one of the most important and fundamental essentials of a map. Without scale, any diagram will be only a sketch and not a geographic map.

Types of Maps

Maps are classified either on the basis of their scale or on the basis of the contents and purpose for which they are required. Based on the scale maps have been divided in to following three types:

(1) Cadastral maps
(2) Topographical maps
(3) Atlas maps

Cadastral Maps

Cadastral maps are maps on large scale and are some times known as plans. The very large scale allows full details to be given namely boundaries of properties and individual buildings etc. Therefore, these maps are useful for identification of boundaries of property in legal documents and for detailed affairs of every kind such as local administration, taxation and management estate. In Indian case, the community development block or a police station maps are also included in this category (Fig 5.7).

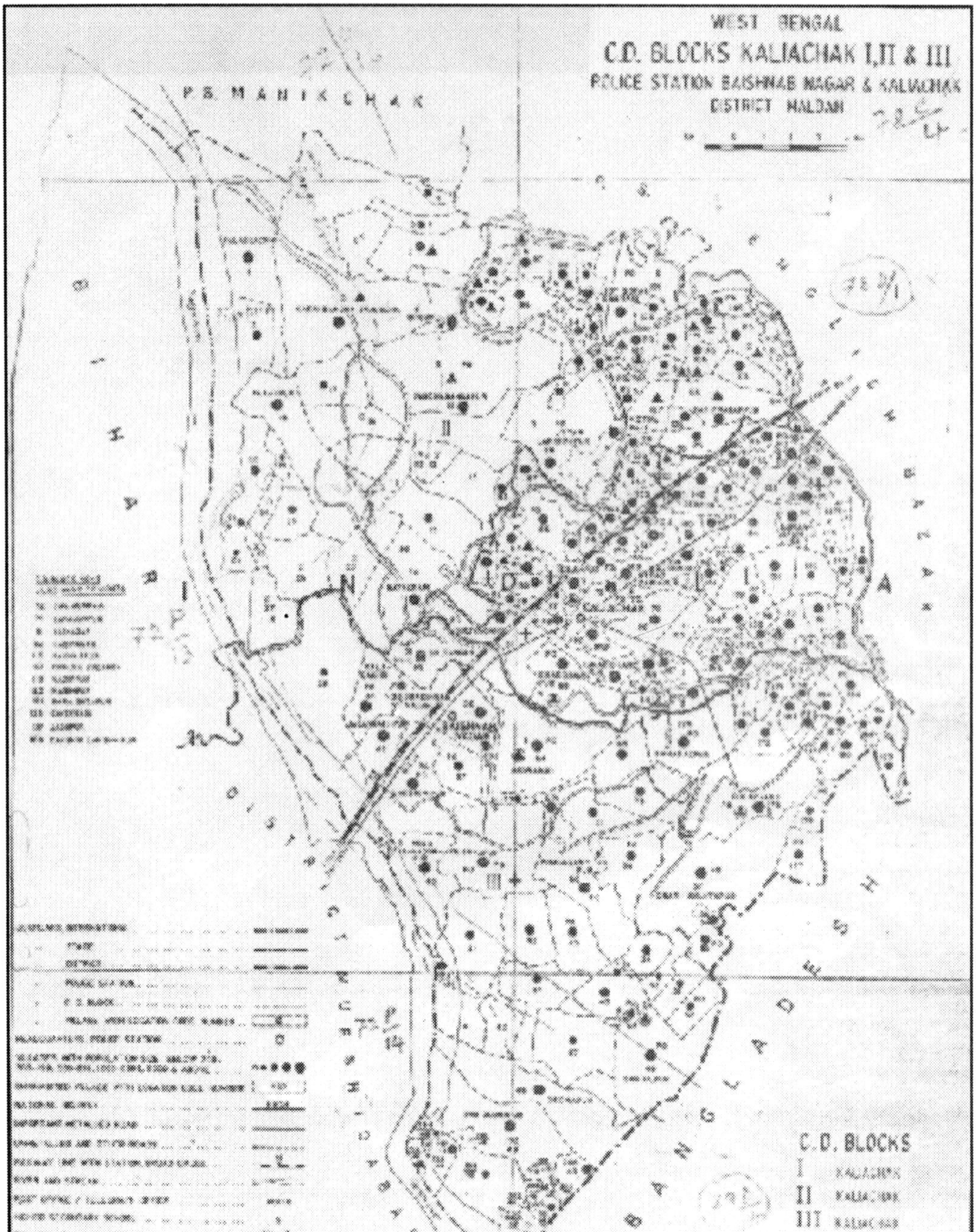

Fig 5.7: Cadastral maps

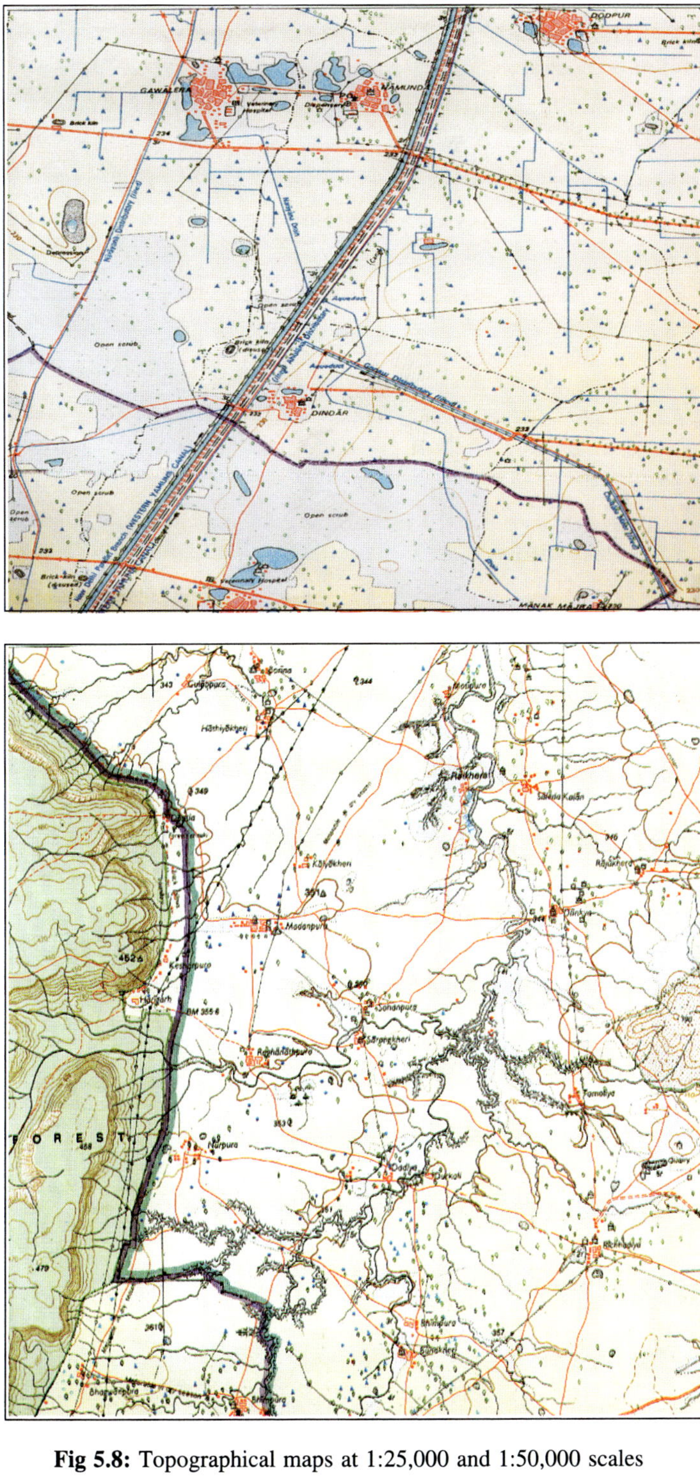

Fig 5.8: Topographical maps at 1:25,000 and 1:50,000 scales

Topographical Maps

Topographical maps of India are prepared by Survey of India. The principal maps are those on a scale of 1: 250,000, 1:50,000 and 1:25,000 respectively (Fig 5.8). Contours are used for showing the altitude of the area. They are large scale maps but their scale is smaller than that of cadastral maps. Since scale is large they show considerable details of natural and man made features such as hills, rivers, forests, towns, roads, building blocks, railways and the like.

Atlas Maps

Atlas maps are small scale maps. Most of the atlas maps are to give general impression of any particular area. Based on the contents and purpose different themes are created to represent data in an Atlas maps. These maps are also known as thematic maps. In India, the National Atlas and Thematic Mapping Organistaion (NATMO), based in Kolkata, is responsible for the preparations of atlases and other thematic maps, including the National Atlas of India. An example of an atlas map is depicted here (Fig 5.9).

Thematic Maps

Several types of thematic map are important in GIS. Standard GIS software packages are very powerful tool to generate thematic maps. Some of the features of such GIS software have been discussed in Chapter 2. Examples of the applications of different techniques for generating thematic maps have been discussed here.

(a) A choropleth map uses reporting zones such as counties or census tracts to show data such as average incomes, percent female, or rates of mortality. An example of the ratio of Child Population to total population in different districts of the country is shown in the figure below (Fig 5.10).
(b) An area class map shows zones of constant attributes, such as vegetation, soil type, or forest species. Example of Soil Map of Banjora development block has been shown in Fig 5.11.
(c) An isopleth map shows an imaginary surface by means of lines joining points of equal value, "isolines" (e.g. contours on a topographic map). Based on topographical data, a colour coded elevation zones in a part of the country has been depicted in Fig 5.12.
(d) *Elevation Models* can now be generated by GIS software. Statistical or appropriate attribute data can be chosen for raising surfaces at different levels. A composite picture can be generated which are generally known as digital elevation model. An example of the distribution of educational institutions in Nadia district has been shown in Fig 5.13.

Basic Map Elements: Information Commonly Needed in a Map

Almost all maps must include certain basic elements that provide the reader with critical information. Among these are the title, scale, legend, body of the map, north arrow, cartographer, neat-line, date of production, projection used, and information about sources. The placement of this information and the style of its depiction will vary greatly from map to map depending on the audience and message to be communicated. Some elements are found on almost all maps no matter what the theme, others are depended heavily on the context in which the map will be read. Let us consider these elements to see how they are depicted and why their importance may vary from map to map.

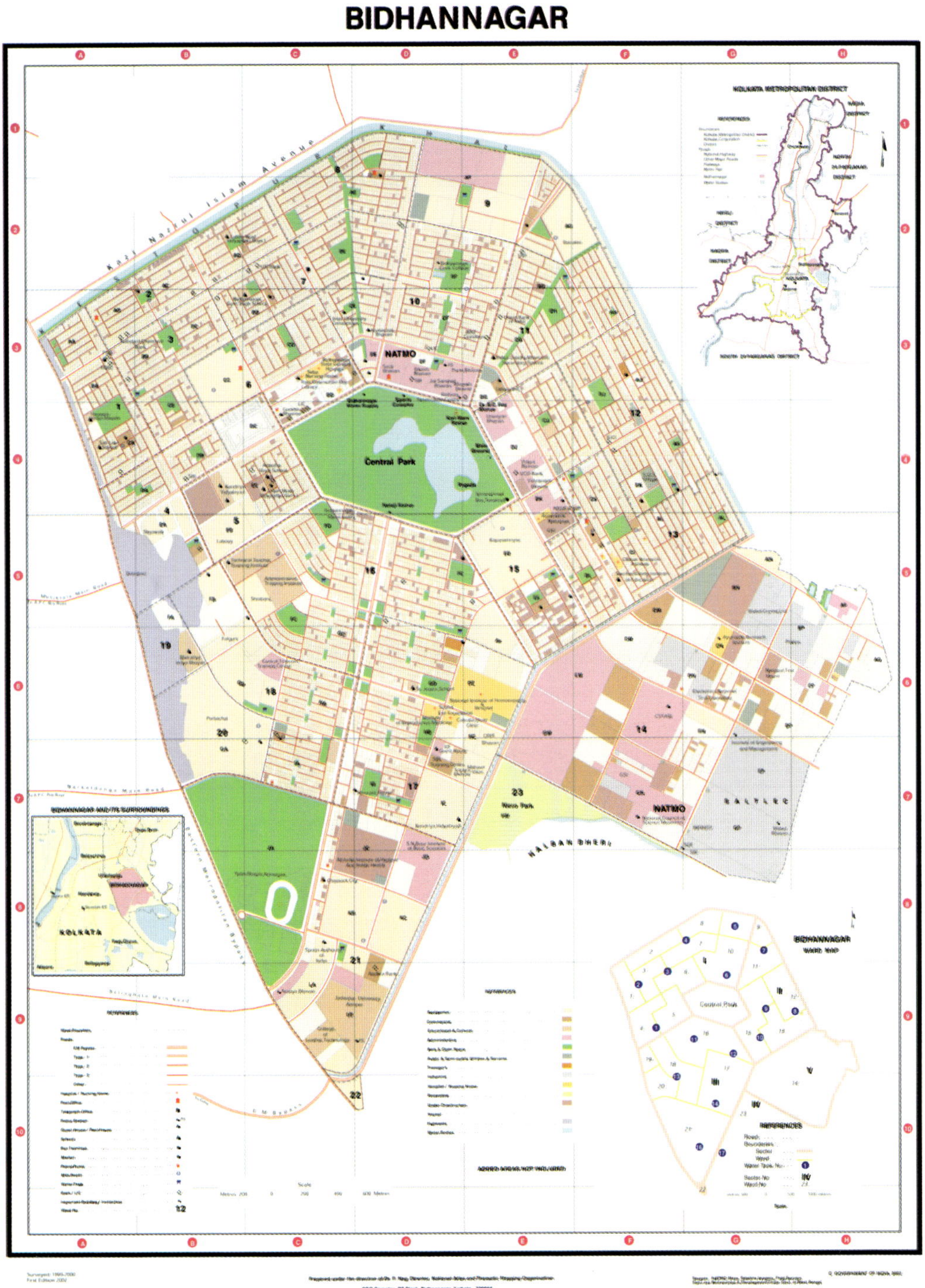

Fig 5.9: Thematic maps

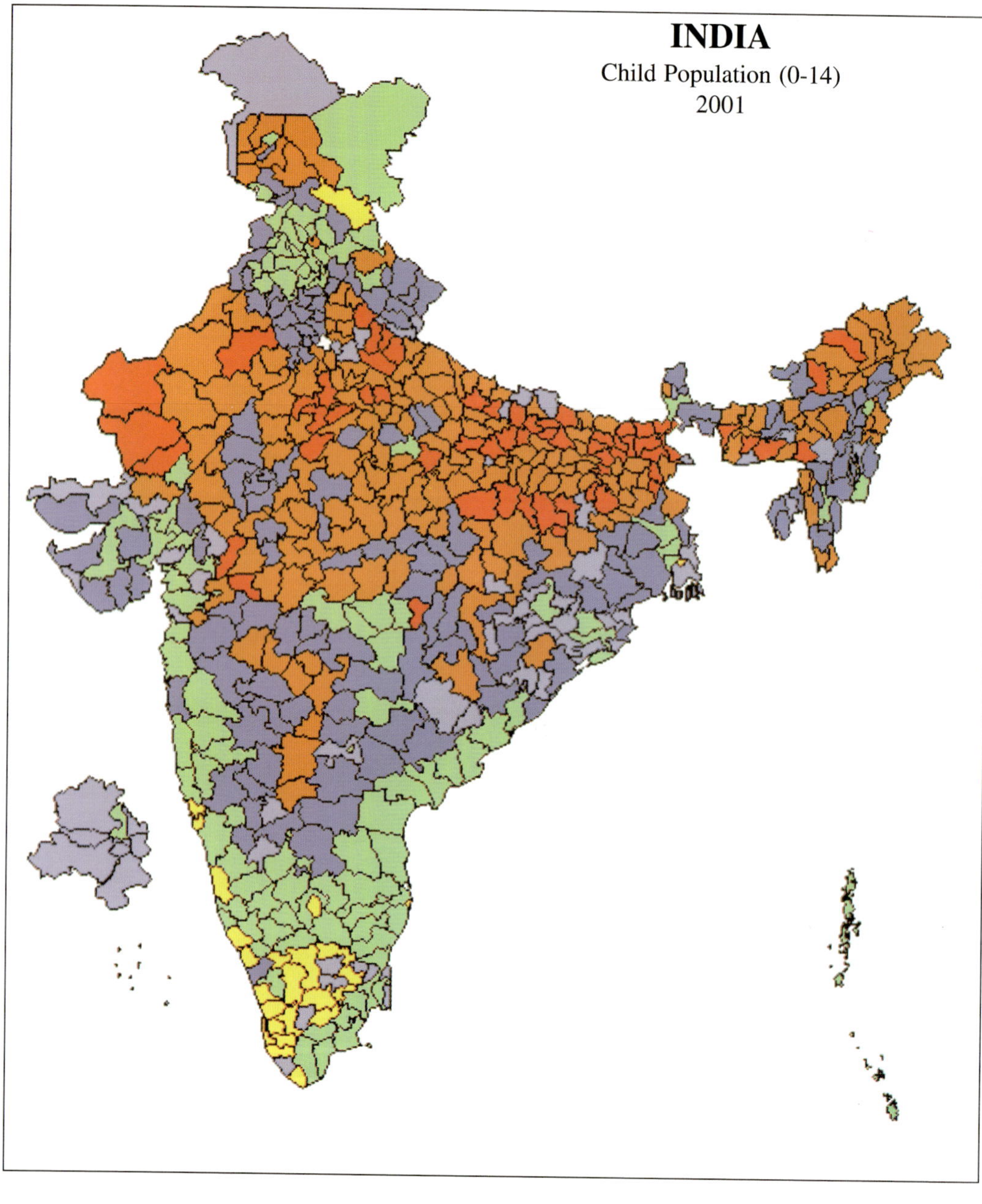

Fig 5.10: Choropleth map

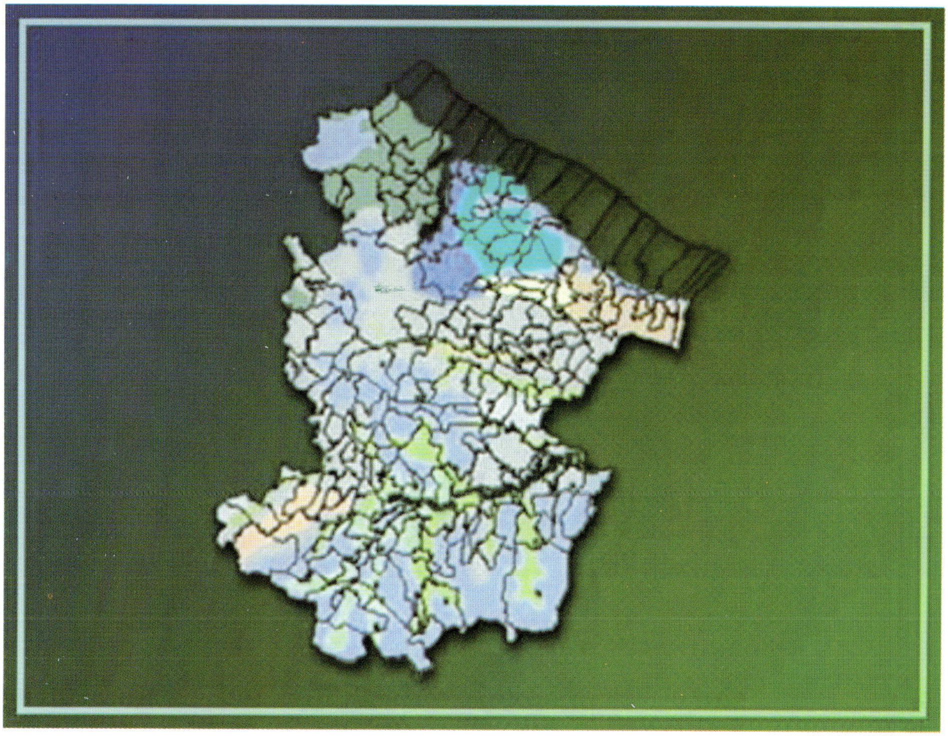

Fig 5.11: Soil map of Banjora Block (*Source*: NRDMS)

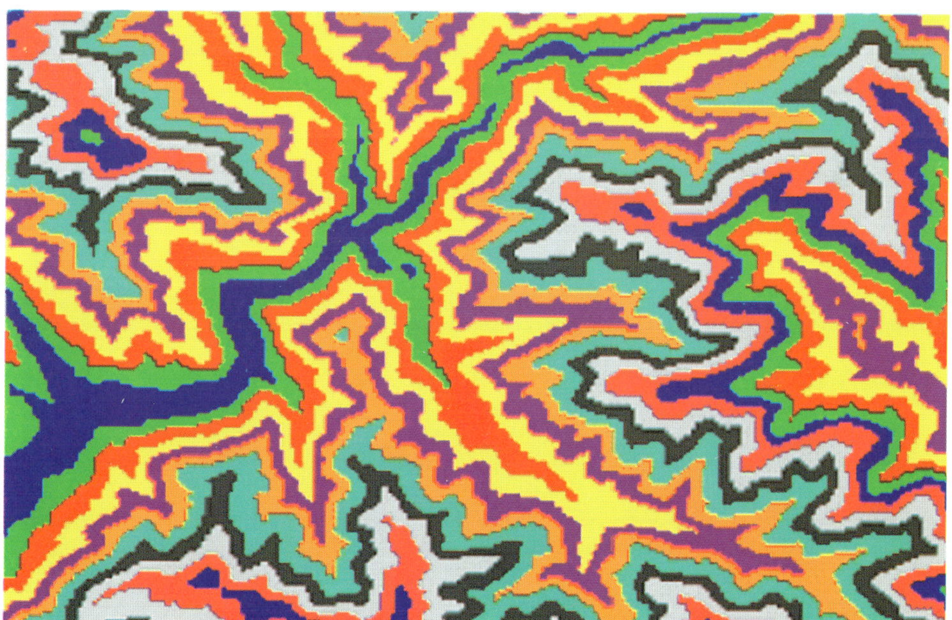

Fig 5.12: Isopleth maps (Color Coded Elevation Zones)

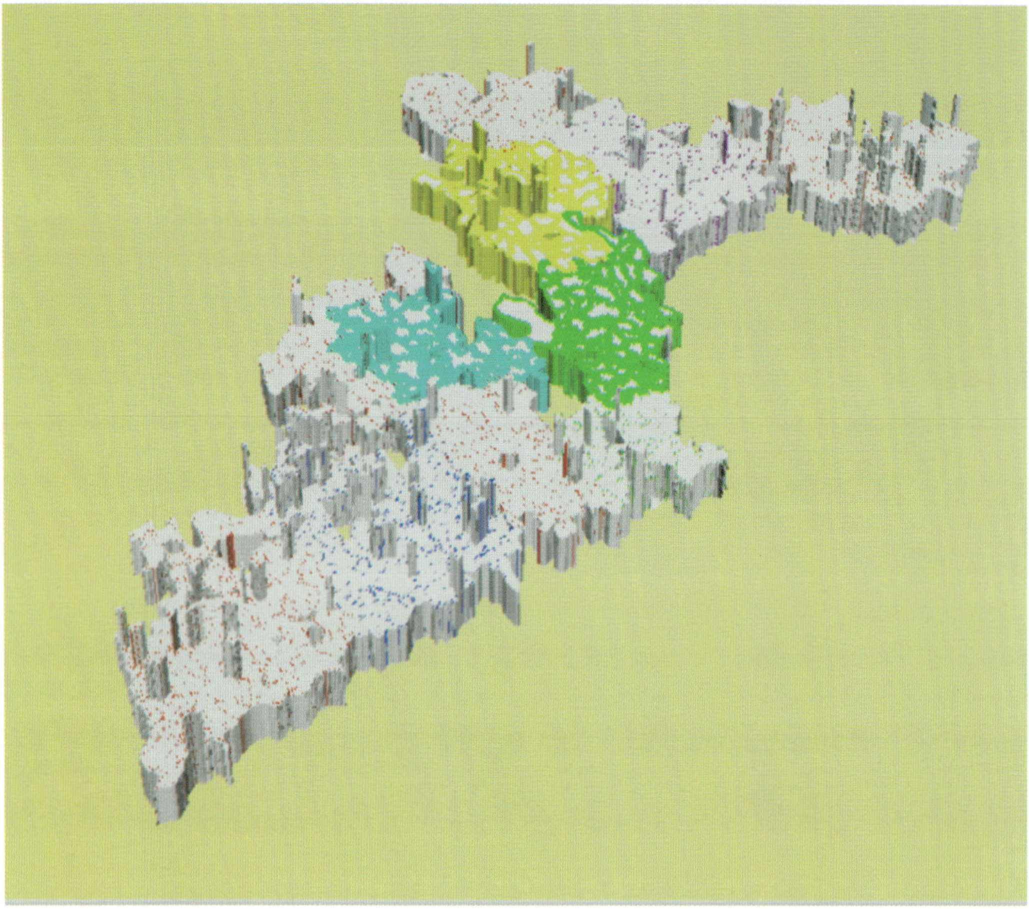

Fig 5.13: Digital elevation model : Nadia district—Educational institutions

- —Elements that are found on virtually all maps: distance or scale, direction, legend, sources of information and how processed.
- —Elements that are sensitive to context: title, projection, cartographer, and date of production.
- —Elements that are used selectively to assist effective communication: neat-lines, locator maps, inset maps, index maps.

With the help of Fig 5.14, some of the elements of map designing have been shown. Such elements are generally included in an Indian map.

Elements that are found on virtually all maps

Distance or scale: Distance or scale must always be indicated or implied, unless the audience is so familiar with the map area or distance of such little relative importance that it can be assumed by the audience. The scales on maps are linear scale. A scale is the ratio which a length on a map bears to the actual length on the ground. To say that a map is drawn to the scale of one cm to one km means that one cm on the map represents one km on the ground. Scale is shown in three ways (Fig 5.15) :

(a) The statement of scale or verbal
(b) The representative fractions or numeric scale
(c) The plain or linear scale also known as graphic scale.

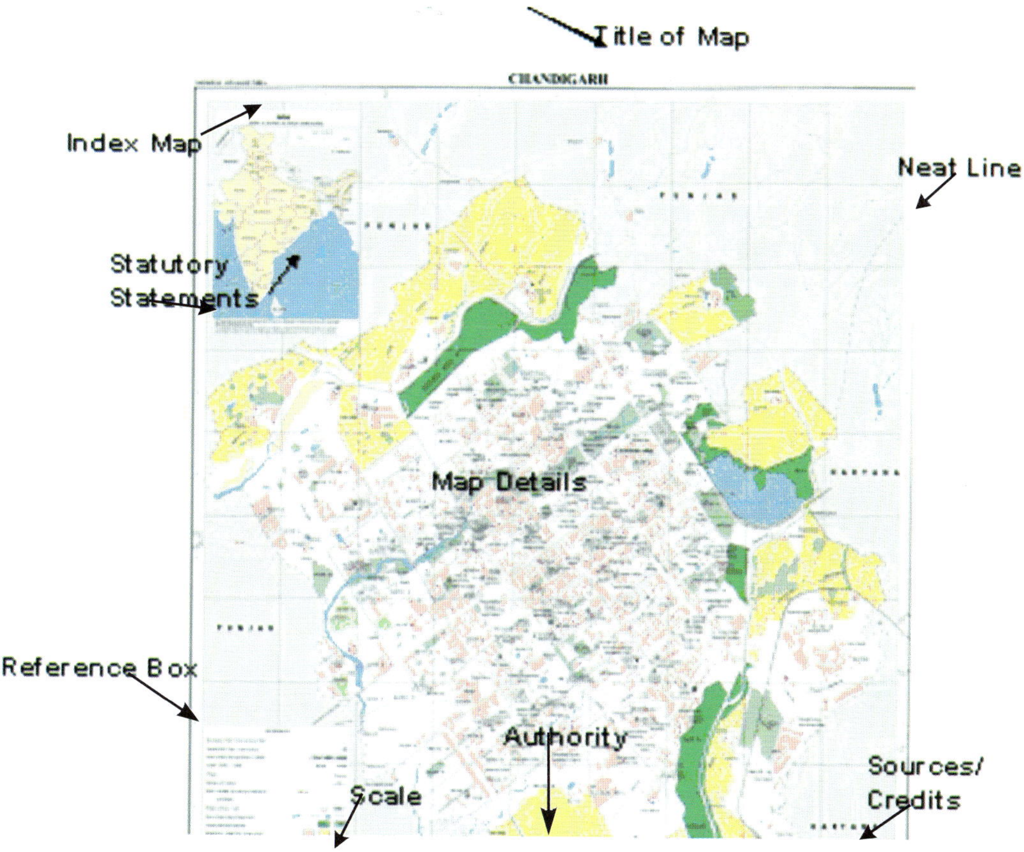

Fig 5.14: Basic elements of a map

Verbal, numeric, and graphic means of representing Scale

2 cm equal 1 km 1 : 50,000

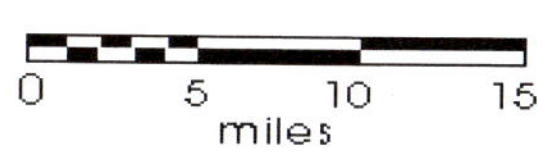

SLC 4/96

Fig 5.15: Types of scales

Verbal Scale: In this case it is noted that that one cm represents so many km or one km represented by so many cm. It is written in the map in the following manner.

1 inch to 1 mile or 1 cm to 1 km.

Scale may be either large or small

Representative Fraction (RF) Scale: In this case the proportion existing between length on the map and actual length on the ground is indicated by a fraction whose numerator is always 1. Such a fractions called as "Representative fraction" and is usually abbreviated as R.F. If a map has as R.F. 1/25,000 it indicates that any length on the Map represents 25,000 times that length on the ground, whether it be cm, km etc. This scale is independent of any units and is used in all countries.

Plain or Graphic Scale: It consists of line drawn at the bottom of the map conveniently divided and sub-divided so that distances on the map can easily be read from it by using a divider. In using computer systems, the graphic form of representing scale is often preferred. With computers, maps are often drafted at different scales than they are printed. In using verbal or numeric scales, the cartographer must be certain that the map is printed at precisely the scale indicated. If a graphic scale is inserted in a digital map, it will always maintain its relative size with respect to the digital map no matter how it is printed.

Map Scale: Small vs. Large : The definition or even understanding of *small* and *large* scale maps vary from country to country. In larger countries, some of the smaller scale maps are considered as large scale maps as well. Following criteria is to be taken into account for understanding the significance of scales of the maps and implications of reduction and enlargement :

- Small scale refers to the RF ratio. A 1:250,000 scale is small compared to a 1:2,000.
- The ratio is small and the amount of reduction is large, producing a map of a large area.
- Large scale means less reduction and a map covering a small area.

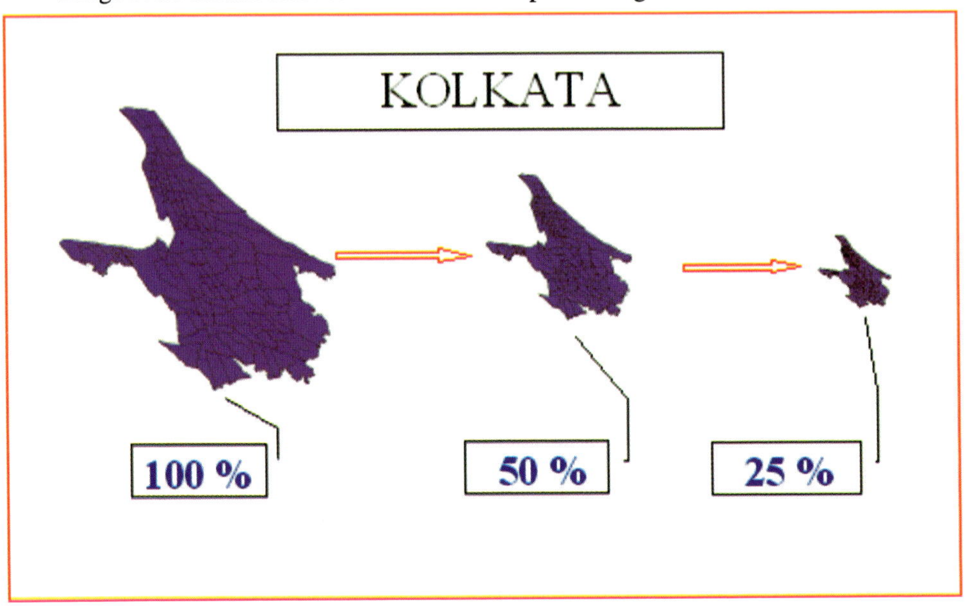

Fig 5.16: Reduction and Enlargement

In Fig 5.16, the implication of reduction of the map of Kolkata City has been shown. In a reduced map, there will be generalization and loss of information which may not necessarily be recovered by enlargement in the reverse fashion. However, powerful GIS software has reduced such problems a large extent. Further, it is to be remembered that the scale varies significantly across the area of some maps. One has to make necessary adjustments in order to circumvent this problem.

Direction and Bearings : The question of what is north can be an issue on some maps. On the earth, true north (the direction to the North Pole) differs from magnetic north, and the magnetic north pole moves due to changing geophysical conditions of the earth's crust and core. Many reference maps indicate both. Most maps we compose are oriented to true north, even though compass readings in the field are angled to the magnetic pole. Adjustments for these compass deviations are made routinely (Fig 5.17).

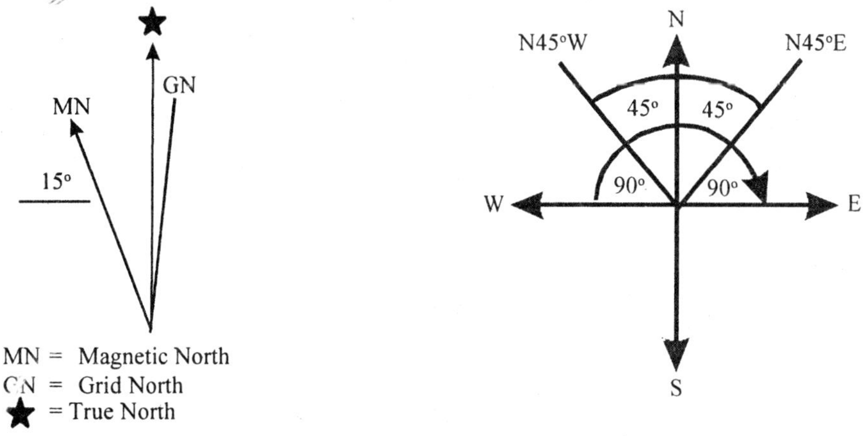

Fig 5.17: North directions

The bearing of an object is the angle which a line from the position of an observer to the object makes with the north and south line measured from it, in a clockwise direction (Fig 5.18) :

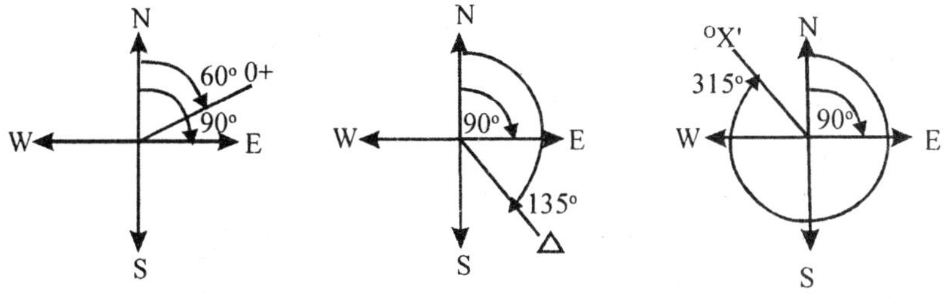

Fig 5.18: Bearings

Legend: The legend lists the symbols used on a map and what they depict. These symbols should appear in the legend exactly as they are found in the body of the map and be described clearly and fully.

Sources of information and how processed: Unless it is absolutely clear from the context in which a map appears, readers will need to know about the sources from which the map was derived. Often the age, accuracy, and reliability of sources is critical to the interpretation of a map and should be noted. Sometimes, it is also important to indicate how the data was processed, grouped, generalized, or categorized. In India, the basic source of all map related information is Survey of India. Recently several changes have been made by them in order to provide digital data to the users (Appendix I and II). For thematic maps, the *National Atlas of India* and other publications of the National Atlas and Thematic Mapping Organisation (NATMO) is the basic source of information.

Elements that are Sensitive to Context

Title: The title of a map is usually one of its most essential features. As such, it should receive very careful attention so as to match the needs of the theme and audience. The content of the title should also be measured against other lettering applied to the map, for example in the legend or annotations.

Projection: The projection used to create a map influences the representation of area, distance, direction, and shape. It should be noted when these characteristics are of prime importance to the interpretation of the map. Some widely used locational reference systems such as the U.S. State Plane Coordinate system or Polyconic and Universal Transverse Mercator (UTM) system in India are based on predefined projective geometries that are implicit in the use of the coordinate systems themselves (Fig 5.19 and also see Fig 5.24).

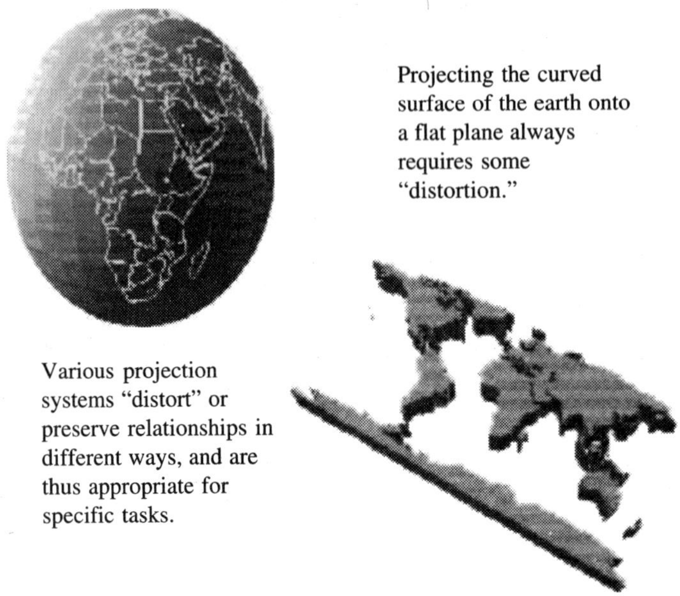

Fig 5.19: Projection and distortion

Cartographer: The authority lying behind the composition of a map can be of prime importance in some situations. In India, Surveyor General of India is authority for topographical maps while Director, National Atlas and Thematic Mapping Organisation is authority for thematic maps produced by them.

Date of Production: The meaning and value of some maps, such as those relating to temporal information, are time sensitive. An out-of-date road atlas or city map can cause tremendous frustration. However, there may some gap between the date of survey or data collection and the actual production of maps. This may be due to processing of field information and completion of the formalities which may vary from country to country, institution to institution.

Elements that are used Selectively to assist Effective Communication

Neatlines: Neatlines or clipping lines are used to frame a map and to indicate exactly where the area of a map begins and ends. The outer neatline of a map, i.e. its border, helps to frame the entire map composition to draw the reader's attention to the various elements of information.

Index maps, Location maps : Some maps portray areas whose locations may be unfamiliar to readers. In such cases, the cartographer adds a "helper" or index map that places the body of the map within a larger geographical context with which the reader can be expected to be familiar.

Inset maps: Sometimes observations and data are so densely clustered in small sections of a larger map that the cartographer must provide the reader with additional close-up, "zoomed-in" maps of these smaller areas. These close-up detailed maps are called *insets.*

Map Projection

As mentioned earlier, the surface of the earth is projected in the form of sphere-like ellipsoid for getting the longitudes and latitudes. Map projections are attempts to portray the surface of the earth or a portion of the earth on a flat surface with the help of such graticules. There are several projections or numerous ways to depict the three dimensional surface to two dimensional surface. However, map projection has to be purpose-specific. If the main purpose is served, other properties can be considered in a best possible way. Map projections fall into four general classes (Fig 5.20).

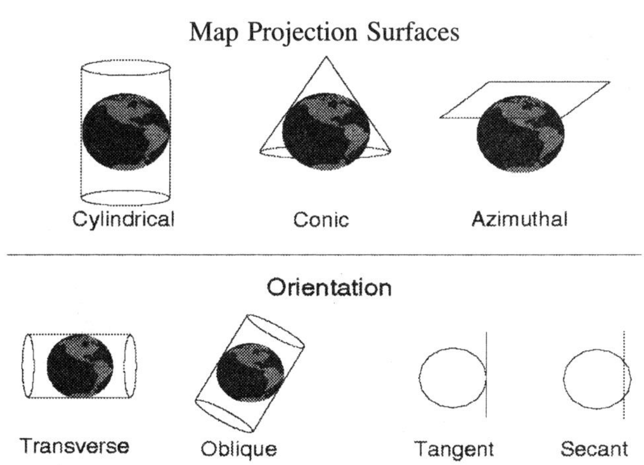

Fig 5.20: Types of projection

Cylindrical projections result from projecting a spherical surface onto a cylinder. When the cylinder is tangent to the sphere contact is along a great circle (the circle formed on the surface of the earth by a plane passing through the centre of the earth). In the secant case, the cylinder touches the sphere along two lines, both small circles (a circle formed on the surface of the earth by a plane not passing through the centre of the earth). When the cylinder upon which the sphere is projected is at right angles to the poles, the cylinder and resulting projection are transverse (Fig 5.21). When the cylinder is at some other, non-orthogonal, angle with respect to the poles, the cylinder and resulting projection is oblique.

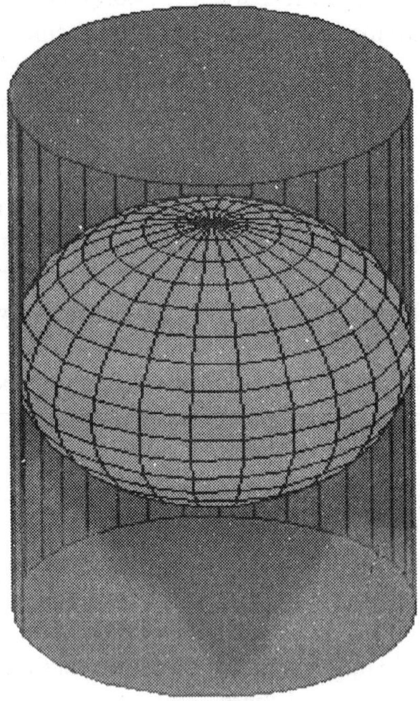

Fig 5.21: Cylindrical projection

Conical projections result from projecting a spherical surface onto a cone. When the cone is tangent to the sphere contact is along a small circle. In the secant case, the cone touches the sphere along two lines, one a great circle, the other a small circle (Fig 5.22).

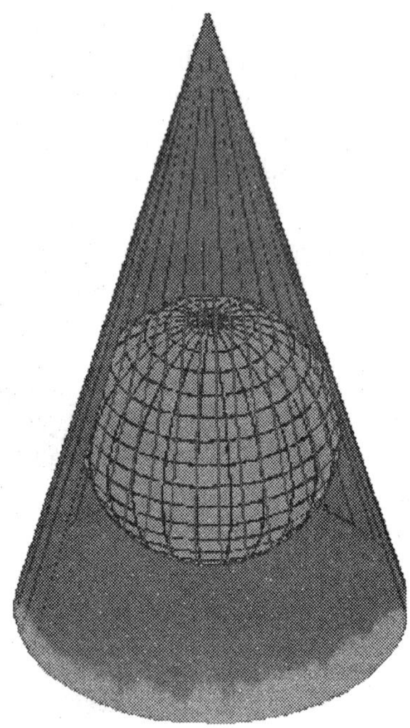

Fig 5.22: Conical projection

Azimuthal projections result from projecting a spherical surface onto a plane. When the plane is tangent to the sphere contact is at a single point on the surface of the earth. In the secant case, the plane touches the sphere along a small circle if the plane does not pass through the center of the earth, when it will touch along a great circle (Fig. 5.23).

Miscellaneous projections include unprojected ones such as rectangular latitude and longitude grids and other examples of that do not fall into the cylindrical, conic, or azimuthal categories.

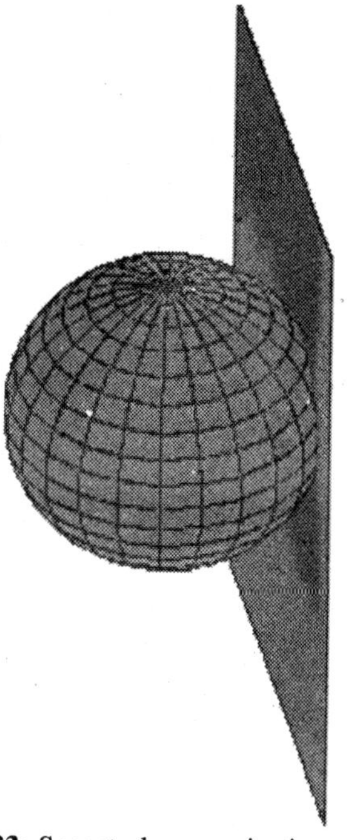

Fig. 5.23: Secant planar projection

Some Important Projections

Cylindrical Projections
Cylindrical Equal Area
Behrmann Cylindrical Equal-Area
Gall's Stereographic Cylindrical
Peters
Mercator
Miller Cylindrical
Oblique Mercator
Transverse Mercator
Universal Transverse Mercator (UTM) projection

Pseudocylindrical Projections

Pseudocylindrical projections resemble cylindrical projections, with straight and parallel latitude lines and equally spaced meridians, but the other meridians are curves.

Mollweide
Eckert Projections
Eckert IV Equal Area
Eckert VI Equal Area
Robinson
Sinusoidal Equal Area
Conic Projections
Albers Equal Area Conic
Equidistant Conic

Lambert Conformal Conic
Polyconic
Azimuthal Projections
Azimuthal Equidistant
Lambert Azimuthal Equal Area
Orthographic
Stereographic
Miscellaneous Projections
Unprojected Maps
Texas State-Wide Projection
Space Oblique Mercator

Basic Problems in a Projection

In order to represent the geoid (actual shape of earth—putative sea level) simplified into ellipsoid (a surface created by rotation of best fit ellipse to geoid] or spheroid on a flat map various forms of distortion are inevitable, but it is required to choose type and amount. No map projection transformation can maintain scale everywhere. Angles, areas, distances and directions will be altered in the planar representation of the ellipsoidal earth. The distortions created during the map projection transformation may be analyzed using a measure of distortion. Comparison of Mercator (cylindrical) and Mollweide (hyperelliptical) projections has been shown in Fig 5.24.

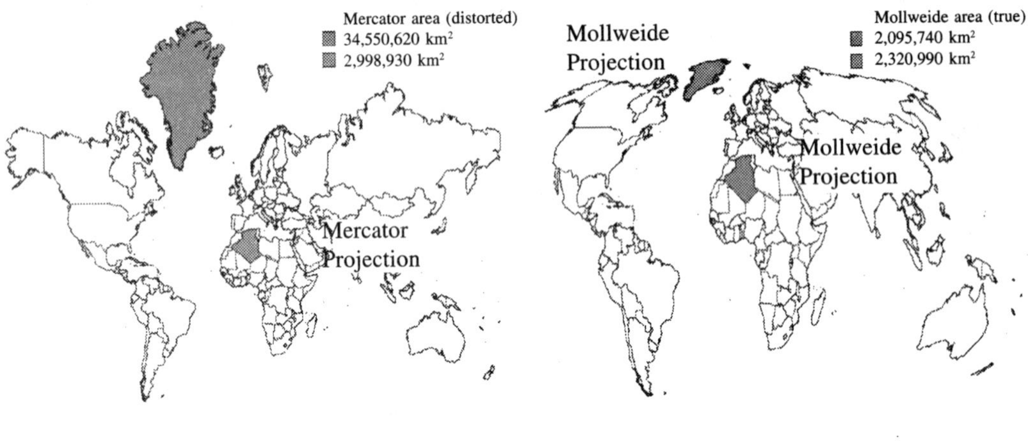

Fig 5.24: Transformation and distortion

Nicolas Auguste Tissot was one of many contributors during the 19th century of the application of sound mathematical principals to map projection science. During this period there was more than an eight fold increase in the number of publications relating to map projections. Some distortions of conformality, distance, direction, scale, and area always result from this process. Some projections minimize distortions in some of these properties at the expense of maximizing errors in others. Some projections are attempts to only moderately distort all of these properties.

Conformality : When the scale of a map at any point on the map is the same in any direction, the projection is conformal. Meridians (lines of longitude) and parallels (lines of latitude) intersect at right angles. Shape is preserved locally on conformal maps.

Distance : A map is equidistant when it portrays distances from the center of the projection to any other place on the map.

Direction : A map preserves direction when azimuths (angles from a point on a line to another point) are portrayed correctly in all directions.

Scale : Scale is the relationship between a distance portrayed on a map and the same distance on the earth.

Area : When a map portrays areas over the entire map so that all mapped areas have the same proportional relationship to the areas on the earth that they represent, the map is an equal-area map.

Distortion on different Developable Surfaces

Projections "map" the 3D solid onto a plane. Geometric construction methods had to use geometric tricks to do this—three primary developable surfaces (and then some more options).

Planar

—Orthographic (perspective view from parallel lines (infinite distance))
—Stereographic (from opposite side of earth) preserves shape
—Gnomic (from center of earth) great circles are straight lines (seismic/radio waves)

Cylindrical

—Rectangular (each degree is the same...)
—Mercator preserves loxodromes (straight compass bearings)
—Miller (modified Mercator to include poles)
—Peters/Gall's equivalent—preserves area (makes stringy shapes)

Conic (one or two standard parallels)

Albers Preserves area (in limited areas, shape not badly distorted)
Lambert's Conformal Conic: Preserves shapes (in limited use, area not badly distorted)

Mixtures of developable surfaces:
Goode's Homolosine (Mercator plus Sinusodial)

Mathematical Projections (not on a surface)
Sinusodial preserves area for large portions of globe, radical shape distortions.
Hyperelliptical General Projection, many possible results.

Mollweide (special case of Hyperelliptical)
Each projection can be in a given orientation (or "aspect")

Properties to preserve

Area preserved—equal area (Equivalent)
Preserve angles—Conformal
Distances preserved from a single point—Equidistant
Preserve compass directions (Loxodromes)
Preserve great circles as straight lines (gnomonic projection)

Like geodesy, map projection is also well defined field of specialization. With the availability of powerful GIS software, the field has got a boost. The readers are advised to consult books on map projections and the manuals of GIS software for details.

Indian Map Projection

Indian topographical maps have been based on *Polyconic projection.* Due to colonial legacy, the neighbouring countries are also using the same projection. With the introduction of the new National Map Policy in 2005, the Universal Transverse Mercator (UTM) projection has also been adopted by Survey of India (Appendix I). The Open Map Series (OSM) is being developed on UTM projection. Hence, in this book the Polyconic Projection and the UTM projections has been discussed in some details.

Polyconic Projection

In the Survey of India, Auxiliary Tables (1972), the details of the Polyconic projection at 1:10,000, 1:25,000, 3:100,000, 1:50,000, 3:200,000 and International Polyconic projection at 1:1,000,000 scales have been worked out. This publication also includes calculation for Lambert Conical Orthomorphic project at various scales. The calculations include length (in straight lines) for parallels, meridians and diagonals for each graticule squares for each map-scale. The calculations for Polyconic projection are the basis for the preparations of the Indian topographical sheets (Fig 5.25).

Polyconic projection belongs to the family of conical projections. It was probably developed by Hassler in 1820. It is called polyconic because several cones are involved in developing this projection. Here all parallels are without any distortion, i.e. the scale is correct along all the parallels. Further, scale is also correct along the central meridian. The parallels are arcs of circles but not concentric. Hence, it is neither conformal nor equal area. Further, the central meridian and equator are straight lines. It has rolling fit with adjacent topographical sheets in east-west direction. However, serious errors are introduced if the adjoining sheets are joined. Obviously, there are problems in preparing seamless data in a rectangular coordinate system because each sheet is projected separately. Hence, one has to take sufficient precautions if a large area is taken for GIS application.

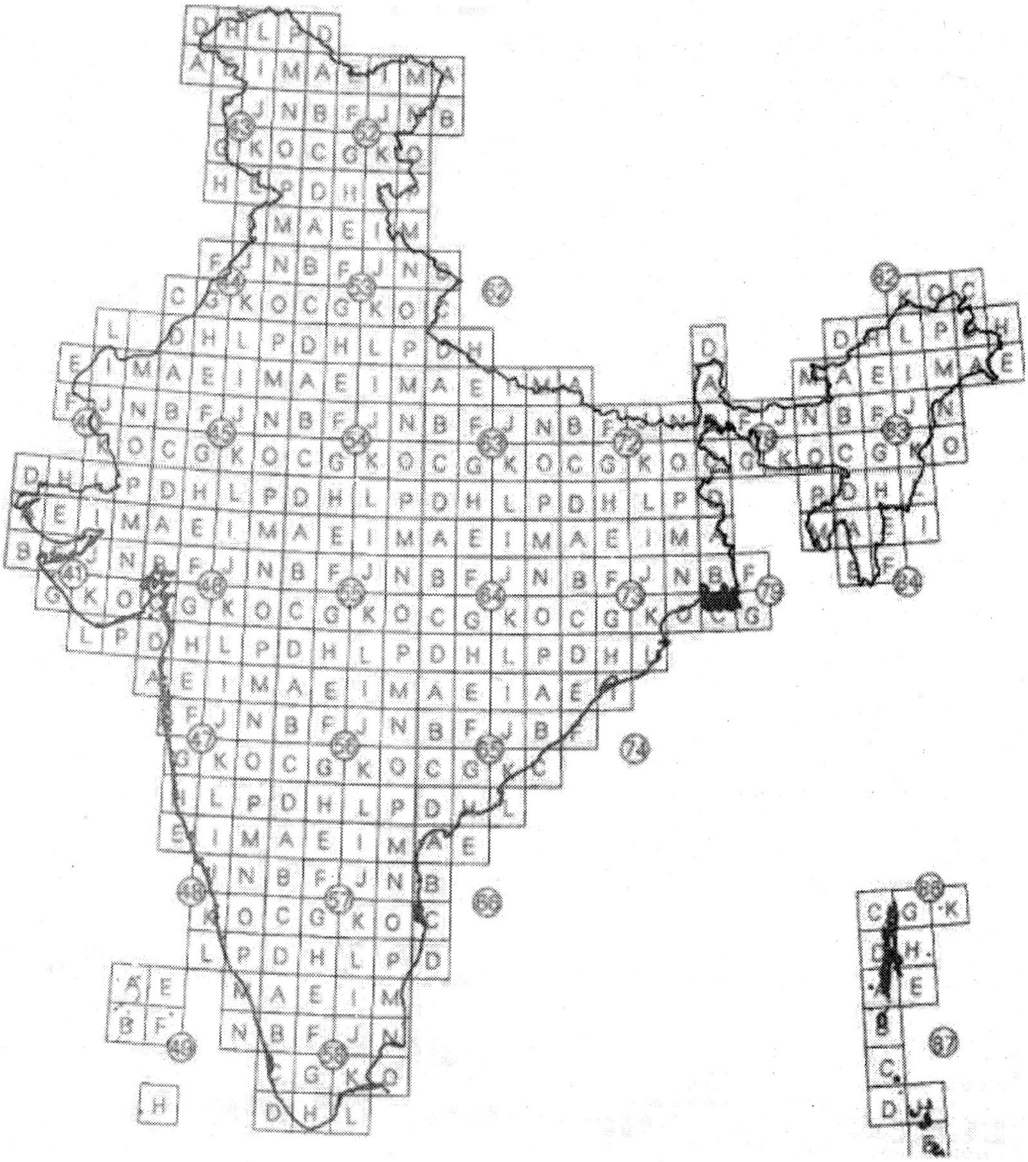

Fig 5.25: Coverage of Indian topographical sheets on conical projection

Universal Transverse Mercator Projection

This projection is some times referred as UTM grid system as well. Based on the experience during the Second World War, it was decided that all the North Atlantic Treaty Organization or NATO countries should bring their military mapping on to one system. For this purpose, Transverse Metcator projection was chosen as it was conformal projection suitable for topographical mapping. Further, in 1952, United Nations Cartographic Committee recommended that all topographical maps should adopt UTM grid system.

This grid system does not cover the whole world. It covers an area lying between latitudes 80°S and 84°N. The Arctic and Antarctic regions are covered by the Universal Polar Stereographic or UPS projection.

In order to bring the distortions within acceptable limits, longitudinal portion are mapped on to the projection surface which are called as 'gores' having 6 degrees longitude in width within 80°S to 84°N. Further, the 360 degrees of longitude is divided into 60° of these gores, beginning at 180° west longitude and working eastwards these gores are numbered from 1 to 60. They are known as UTM grid zone numbers (Fig 5.26).

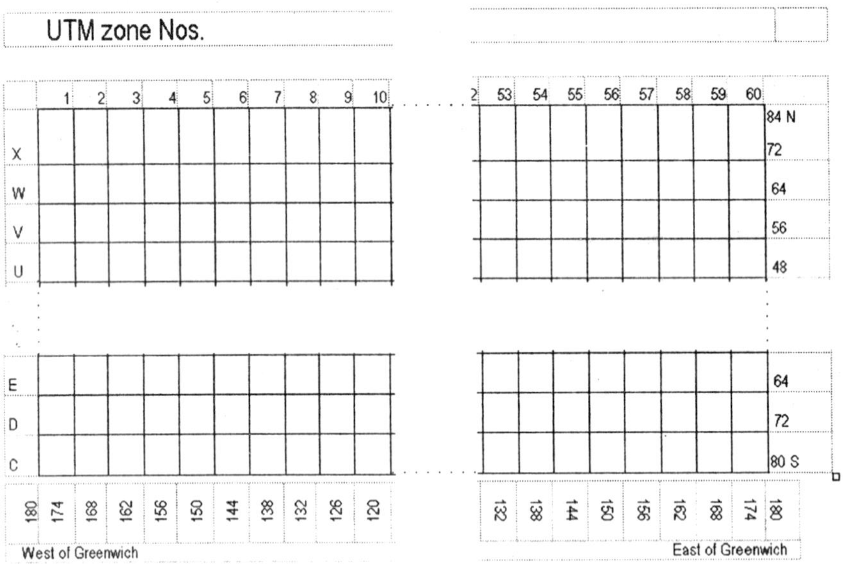

Fig 5.26: UTM Sheet numbering system

For locating any point on the earth surface, the UTM grid is further divided in to 20 latitudinal belt starting from 'C' in the southern extreme to 'X' in the northern extreme. The letters 'I' and 'O' are omitted because they can be confused with 1 or 0. Each belt is having a width of 8° latitude, except the northern most belt which is of 12° latitude. The concerned area of the earth is covered by 1200 UTM grid zones, and each such zone is identified by the number of the longitudinal zone followed by the letter of the latitudinal belt, as mentioned above. In addition, there are following four elements in a UTM grid reference :

1. UTM grid zone
2. A pair of letters identifying a 100,000 metre grid square
3. An eastings value in metres
4. A northings value in metres

Since the central meridian is a vertical straight line and equator is a horizontal straight line, both crossing at right angles, it is an ideal case for rectangular grids. The assumed northings (Y) is 0 metre is for the northern hemisphere and 10,000,000 metres for the southern hemisphere and the assumed easting (X) is 500,000 metres in both the cases. Since the maps are based on rectangular gird, it is easy for computer application and GIS as well.

Maps of the National Atlas and Thematic Mapping Organisation (NATMO) are at 1:1M, 1:2M, 1:6M, 1:12M and 1:14M scales. The projection of the most of the maps is conical equal area with two standard parallel.

National Map Policy

Considering the growing requirements of maps, geospatial data and GIS, Survey of India, the premier mapping institution in the country, has adopted a new vision which is as follows :

Survey of India takes a leadership role in providing customer focused, cost effective and timely geospatial data information and intelligence for meeting the needs of security, sustainable national development and new information markets.

The objective is to meet the requirements of socio-economic developmental activities, conservation of natural resources, planning for disaster mitigation and infrastructure development and the like. Accordingly, in May 2005, two map series have been accepted by the government. First one is the Defence Series Maps or DSMs to cater the requirements of the defence and national security. This series will be on Everest or WGS-84 datums and on polyconic or UTM projections with heights, contours and full content without dilution of accuracy. This series, in analogue and digital forms for the entire country will be classified and its use will be restricted. The methods of sheet numbering for both the series are shown in Fig 5.27.

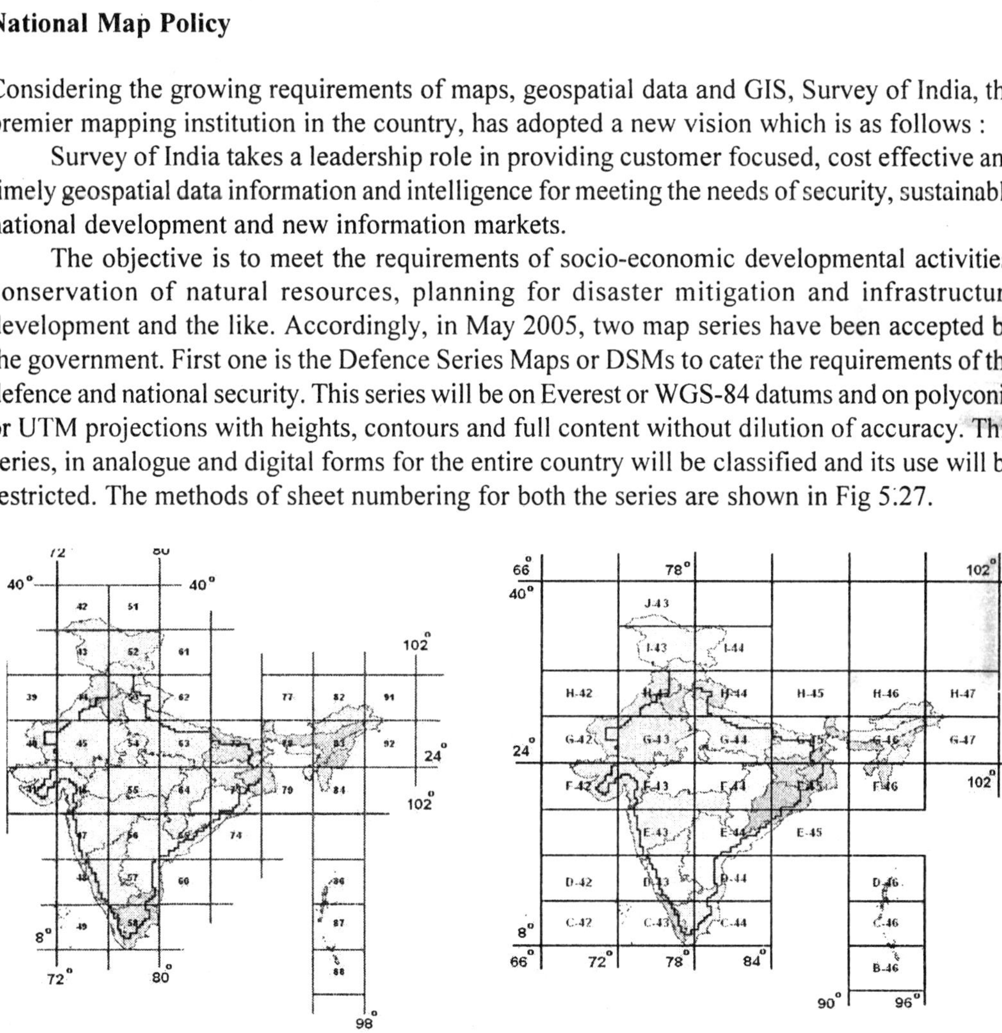

Fig 5.27: DSM and OSM sheet numberings

On the other hand, the Open Map Series or OSMs is basically to meet the requirements of the developmental activities of the country. This series will have different sheet numbers and will be on WGS-84 datum and UTM projection (Fig 2.28). This series either on hard copy or digital forms wili become unrestricted after obtaining one-time clearance of the Ministry of Defence. The content of the OSMs will be slightly reduced as indicated in Appendix I. But the requirements of the developmental activities will be taken care of. Further, in the Guidelines for Implementing National Map Policy, the details of the availability of data were announced just a year later, i.e. in May 2006. The readers advised should go through the Appendix I and further details are available in the Survey of India web site (www.surveyofindia.gov.in).

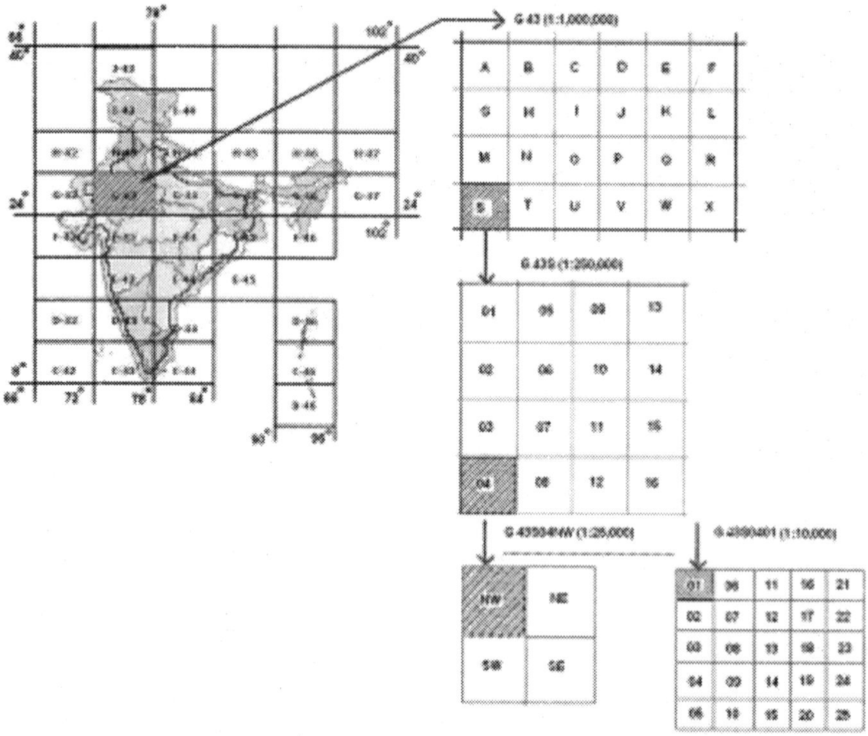

Fig 5.28: Details of OSM sheet numbering system

Coordinate System

Coordinate systems are used for georeferencing of local and global systems for use in precise positioning, navigation, and geographic information systems for the location of points in space (Fig 5.29). There are many different coordinate systems, based on a variety of geodetic datums, units, projections, and reference systems in use today. As an example, this overview often uses the position of one of the thousands of geodetic control points in the United States, the star in the hand of the Goddess of Liberty atop the Capitol building in Austin, Texas.

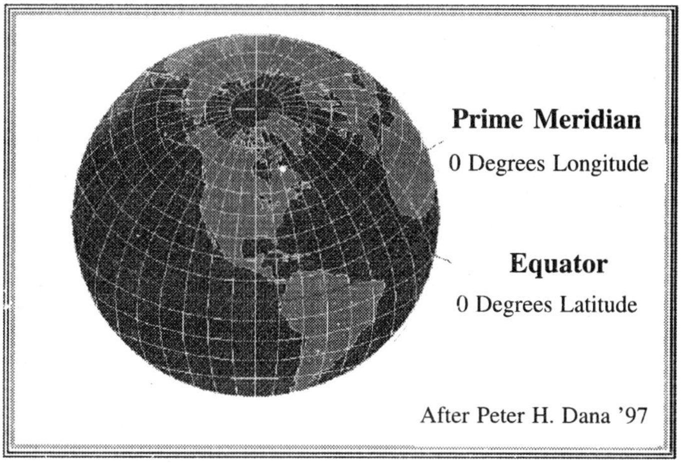

Fig 5.29: Longitudes and latitudes

Basic Coordinate Systems

There are many basic coordinate systems familiar to students of geometry and trigonometry (Fig 5.30). These systems can represent points in two-dimensional or three-dimensional space. René Descartes (1596-1650) introduced systems of coordinates based on orthogonal (right angle) coordinates. These two and three-dimensional systems used in analytic geometry are often referred to as Cartesian systems. Similar systems based on angles from baselines are often referred to as polar systems.

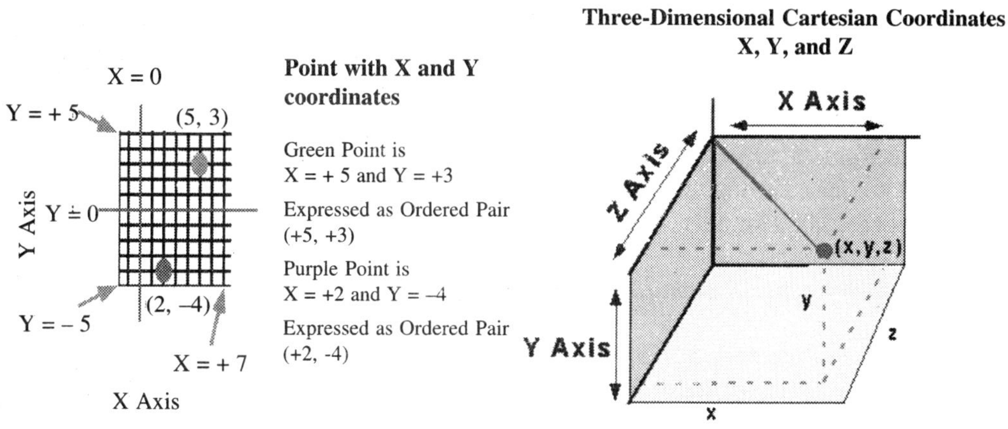

Fig 5.30: Coordinate systems

Plane Coordinate Systems : Two-dimensional coordinate systems are defined with respect to a single plane.

Three-Dimensional Systems : Three-dimensional coordinate systems can be defined with respect to two orthogonal planes.

Global Systems

Latitude, Longitude, Height

The most commonly used coordinate system today is the latitude, longitude, and height system. The prime meridian and the equator are the reference planes used to define latitude and longitude. The geodetic latitude (there are many other defined latitudes) of a point is the angle from the equatorial plane to the vertical direction of a line normal to the reference ellipsoid. The geodetic longitude of a point is the angle between a reference plane and a plane passing through the point, both planes being perpendicular to the equatorial plane. The geodetic height at a point is the distance from the reference ellipsoid to the point in a direction normal to the ellipsoid.

ECEF X, Y, Z

Earth Centered, Earth Fixed Cartesian coordinates are also used to define three dimensional positions. Earth Centered, Earth-Fixed, X, Y, and Z, Cartesian coordinates (XYZ) define three dimensional positions with respect to the center of mass of the reference ellipsoid. The Z-axis points toward the North Pole. The X-axis is defined by the intersection of the plane define by the prime meridian and the equatorial plane. The Y-axis completes a right handed orthogonal system by a plane 90° east of the X-axis and its intersection with the equator.

Universal Transverse Mercator (UTM)

Universal Transverse Mercator (UTM) coordinates define two dimensional, horizontal, positions. UTM zone numbers designate 6° longitudinal strips extending from 80° south latitude to 84° north latitude. UTM zone characters designate 8° zones extending north and south from the equator. There are special UTM zones between 0 degrees and 36° longitude above 72° latitude and a special zone 32 between 56° and 64° north latitude. Each zone has a central meridian. Zone 14, for example, has a central meridian of 99° degrees west longitude. The zone extends from 96° to 102° west longitude. Eastings are measured from the central meridian (with a 500 km false easting to insure positive coordinates). Northings are measured from the equator (with a 10,000 km false northing for positions south of the equator).

Military Grid Reference System (MGRS)

The Military Grid Reference System (MGRS) is an extension of the UTM system. UTM zone number and zone character are used to identify an area 6° in east-west extent and 8° in north-south extent. UTM zone number and designator are followed by 100 km square easting and northing identifiers. The system uses a set of alphabetic characters for the 100 km grid squares.

World Geographic Reference System (GEOREF)

The World Geographic Reference System is used for aircraft navigation. GEOREF is based on latitude and longitude. The globe is divided into twelve bands of latitude and twenty-four zones of longitude, each 15° in extent. These 15° areas are further divided into one degree units identified by 15 characters. Two numeric characters designate the integer number of minutes of longitude east of the one degree quadrangle boundary longitude. Two additional numeric characters designate the number of minutes of latitude north of the one degree quadrangle boundary latitude.

Local Systems

Universal Polar Stereographic (UPS)

The Universal Polar Stereographic projection (UPS) is defined above 84 degrees north latitude and south of 80° south latitude. The eastings and northings are computed using a polar aspect stereographic projection. Zones are computed using a different character set for south and north polar regions.

National Grid Systems

Many nations have defined grid systems based on coordinates that cover their territory. Australia, Belgium, Great Britain, Finland, Ireland, Italy, The Netherlands, New Zealand, and Sweden are a examples of nations that have defined a National Grid System.

State Plane Coordinates

In the United States, the State Plane System was developed in the 1930s and was based on the North American Datum 1927 (NAD-27). NAD-27 coordinates are based on the foot. While the NAD-27 State Plane System has been superseded by the NAD-83 System, maps in NAD-27 coordinates (in feet) are still in use. The State Plane System 1983 is based on the North American Datum 1983 (NAD-83). NAD-83 coordinates are based on the metre. State plane systems were developed in order to provide local reference systems that were tied to a national datum. Some smaller states use a single state plane zone. Larger states are divided into several zones. State plane zone boundaries often follow county boundaries.

Lambert Conformal Conic projections are used for rectangular zones with a larger east-west than north-south extent. Transverse Mercator projections are used to define zones with a larger north-south extent. One State Plane zone in Alaska uses an oblique Mercator projection for a thin diagonal area.

Public Land Rectangular Surveys

Public Land Rectangular Surveys have been used since the 1790s to identify public lands in the United States. The system is based on principal meridians and baselines.

Townships, approximately six miles square, are numbered with reference to baseline and principal meridian. Ranges are the distances and directions from baseline and meridian expressed in numbers of townships. Every four townships a new baseline is established so that orthogonal meridians can remain north oriented.

Metes and Bounds

Metes and Bounds identify the boundaries of land parcels by describing lengths and directions of lines. Lines are described with respect to natural or artificial monuments and baselines defined by these monuments. The metes and bounds survey is based on a point of beginning, an established monument. Line lengths are measured along a horizontal level plane. Directions are bearing angles measured with respect to a previous line in the survey.

Miscellaneous Systems

Postal Codes

Postal codes such as the United States ZIP code can be used to identify areas. Three digit codes identify large areas.

Maidenhead Grid Squares

The Maidenhead Grid Square system was designed to facilitate the designation of geographical positions for use within the amateur radio community. The Maidenhead grid identifies "Fields" consisting of an area twenty degrees of longitude by ten degrees of latitude with two alphabetic characters. An additional set of two numeric digits locates a specific two-degrees of longitude by one-degree of latitude "grid square" area within the Field. Two additional alphabetic characters can be used to refer to a 5.0 minutes of longitude by 2.5 minutes of latitude "sub-square" within the grid square. In each case the longitude character precedes the latitude designator.

- Variations and extensions to the Maidenhead system are in use. Some Global Positioning System (GPS) receivers display positions in an extended Maidenhead system that appends one or two additional sets of numeric and alphabetic pairs, increasing the precision with which a location can be specified.
- Some amateur radio operators use other terms for the Maidenhead system such as World Wide Locator (WWL) squares or QTH locator squares.
- Ham operators are use these grid designators to communicate transmitter positions to each other. Several utility programs are available to convert between latitude and longitude and the Maidenhead Grid Square system. Some of these also allow computation of distance and azimuth between stations.

AT&T V and H Coordinate System

The AT&T V and H (Vertical and Horizontal) coordinate system was devised in 1957 by Jay K. Donald for the easy computation of distances between telephone switching centers. The system is based on the Donald Elliptic Projection, a two-point equidistant projection covering the land masses of the continental United States and Canada. The system is based on units of the square-root of one-tenth of a mile. Once the coordinates of switching sites are known, distances between sites can be simply found by calculating the square root of the sum of the squares of the differences in the vertical and horizontal coordinates divided by ten. Designed for simple distance calculations that could be accomplished in the field with a slide rule, the system is still found imbedded in some telephone rate computation software.

Navigation System Coordinates

Navigation systems can define locations by referencing measurements of electronic signals. Loran-C time-differences can identify positions with an accuracy of one-quarter of a mile. Omega phase-differences can identify positions with an accuracy of 1-5 km. VOR-DME (Very high frequency Omni Range—Distance Measuring) measurements from an aircraft can identify locations with an accuracy of 0.5-3 km. Navigational buoys, and other aids to navigation can be used as visual reference points, bearings to visual references can identify locations with varying accuracies.

Registration and Conflation

Registration and Conflation are two related procedures in which two or more geographic datasets are combined, compared, or merged. *Registration* is the means by which one or more of the datasets are mathematically manipulated to a common coordinate system. Inherent in any digital spatial data is a mathematical structure that locates where the data is relative to itself. For a point coverage these may be X and Y coordinates that may or may not be geo-referenced. Suppose the two point coverages above are UFO sightings on the surface of the earth. Assume the first coverage merely contains the relative locations of the sightings with respect to one another on an arbitrary coordinate system. Suppose the second coverage is actually geo-referenced in the sense that the coordinates of the 12 points are the actual longitude and latitude of the sightings. A brief inspection of the relationship should produce the simple transformation equations between one coverage and the other (Fig 5.31):

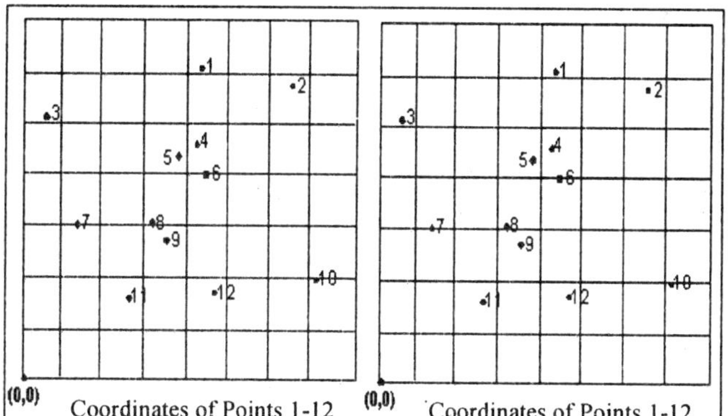

Coordinates of Points 1-12

Point1: X = 4.4 Y = 6.1
Point2: X = 6.6 Y = 5.8
Point3: X = .6, Y = 5.1
Point4: X = 4.3 Y = 4.6
Point5: X = 3.8 Y = 4.3
Point6: X = 4, Y = 4.5
Point7: X = 1.4, Y = 3
Point8: X = 3.2, Y = 3
Point9: X = 3.6, Y = 2.7
Point 10: X = 7.2, Y = 1.9
Point 11: X = 2.7, Y = 1.6
Point 12: X = 4.8, Y = 1.7

Coordinates of Points 1-12

Point1: X = 44, Y = 31
Point2: X = 66, Y = 28
Point3: X = 6, Y = 21
Point4: X = 43, Y = 16
Point5: X = 38, Y = 13
Point6: X = 40, Y = 15
Point 7: X = 14, Y = 0
Point 8: X = 32, Y = 0
Point 9: X = 36, Y = -3
Point 10: X = 72, Y = -11
Point 11: X = 27, Y = -14
Point 12: X = 48, Y = -13

X -> X : 10 * X ; Y -> Y : (10 * Y) 30

Fig 5.31: Transformation of coordinates

Registration is an issue because these transformation equations are rarely this simple to obtain; however, the identification of these transformation functions is the crucial element of the registration procedure. These transformation equations can be applied to point coverages (as in the above example), line coverages, polygon coverages, and images. In the case of line and polygon coverages the transformation equations are applied to the points that define the lines and/or polygons and the topology of the coverage is used to rebuild the coverage in the new transformed space.

In the case of images another issue comes into play because images are not really point data in the strict sense. The pixels of an image have a value that is representative of an area the size of a pixel. Determining the values of the pixels in the transformed image (from reprojection or rubbersheeting) raises the issue of resampling. The registration procedure usually involves the stretching, bending, or warping of the original space to match a geo-referenced space. The means by which the transformation equations are obtained for rubbersheeting is basically an ordinary least squares regression on the coordinates of "Ground Control Points" that are selected from each image.

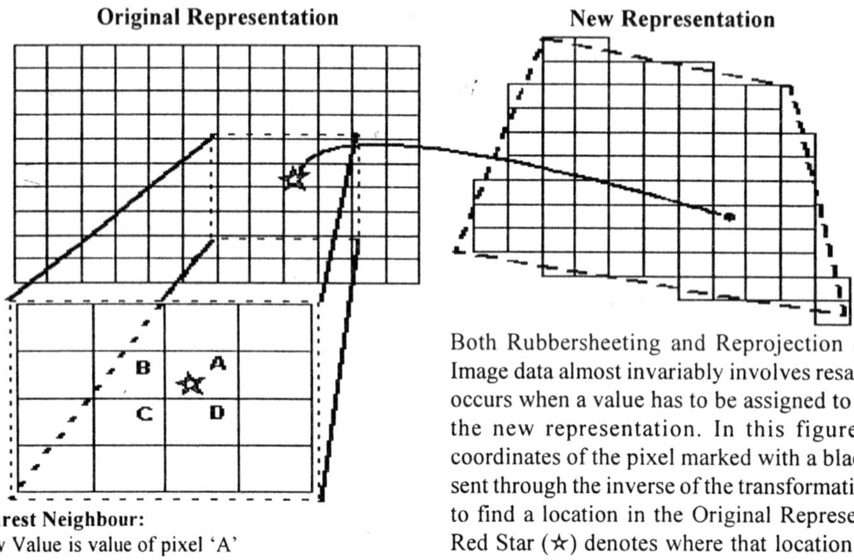

Both Rubbersheeting and Reprojection of Raster or Image data almost invariably involves resampling. This occurs when a value has to be assigned to the pixels in the new representation. In this figure the center coordinates of the pixel marked with a black dot (•) are sent through the inverse of the transformation equations to find a location in the Original Representation. The Red Star (☆) denotes where that location would be in the original image. The problem is finding an appropriate value for the pixel in the new representation. This is an interpolation task usually referred to as **Resampling.** The three most commonly used resampling techniques are: Nearest Neighbour, Bilinear Interpolation and Cubic Convolution. (See example at left)

Nearest Neighbour:
New Value is value of pixel 'A'

Bilinear Interpolation:
New Value is inverse distance based weighted average of Pixels A, B, C and D

Cubic Convolution:
New Value is inverse distance based weighted average of all sixteen pixels shown.

Fig 5.32: Transformation of image

Conflation is a less well defined term that is associated with the many issues and Conflation is not a well defined term. Here it is treated as many of the issues associated with merging, combining, and comparing geographic datasets. The previous discussion of registration describes some of the aspects what is an early step of the conflation process. However, other issues to consider are: data structure compatibility, accuracy compatibility, scale of measurement compatibility (e.g. nominal, ordinal, interval, and ratio data), and the various logical, mathematical, and statistical means by which two or more datasets can be combined or compared. The number of ways different kinds of datasets can be combined or compared clearly becomes quite large (Fig 5.32). Several examples of various types are listed here:

Co-Registering Datasets

Registering an image to a geo-referenced vector dataset
Potential purpose: Compare satellite image to census data
Registering one image to another image
Potential purpose: Create new images based on map algebra, measure cross-correlation between the two images, etc.

Combining Datasets

Intersecting two vector datasets
Potential purposes: Site selection based on combined attributes
Joining tabular data to spatial data
Potential purposes: Data compression, ease of spatial query, etc.

Spatial Scale Resampling / Registration

Aggregating a fine scale image to another coarser scaled image
Potential purpose: Allow for map algebra and statistical comparison
Note: There are many decisions to be made here: e.g. Type of aggregation
Dissolving a high resolution polygon coverage to a lower resolution
Potential purpose: Generalization, Check for scale dependence, etc

Data Model/Structure Conversions

Converting an image to polygons
Potential purpose: create a vegetation map from classified satellite data
Convert a vector dataset to a raster image
Potential purpose: allow for comparison of nighttime satellite image to census data.

These are only a few of many examples of what can be loosely defined as conflation. Our take on the big picture with respect to conflation is that the merging of spatial data allows for manipulation, analysis, and comparison within and between co-spatially extensive data. It also allows for the creation of datasets derived from diverse conflated data.

Global Positioning System

Global positioning system or GPS has given a new perspective to the surveying and mapping activities world over. This space based technology is entering in day to day life of even of common person. Nevertheless, it has boosted a new multi-dimensional sub-discipline now known as *positioning technology*. This technology has been found useful, apart from surveying and mapping, for telecommunication, disaster management, navigation, defence and so on. The development of GPS was initiated in 1973 by the US Department of Defence to support positioning requirements of the US armed forces and its allies. Accordingly, at an altitude of about 20,200 km, a constellation of 24 GPS satellites, located in six orbital planes inclines at about 63 degree to the equator ensuring at least four satellites are visible at any point on the earth (Fig 5.33). GPS uses signals from these rotating satellites to determine location on the earth surface. Further, GPS positioning is based on satellite ranging which implies that the position on the earth is by measuring the distance from a group of satellites in space acting as precise reference points. The basic principle of GPS is also based on triangulation from the satellites. In order to form triangulations, GPS measures distance using the travel time of a radio message from satellite to ground based antennas and receivers. Mathematically, four measurements are required to determine the exact position and therefore at least four satellites are required to determine a location on earth on WGS-84 datum. The graticule values can be converted to other datums and local coordinate systems using converting algorithms. The basic features are as follows :

—It receives satellite signals
—Provides latitude, longitude and height of a given point
—Able to store data during field operations
—Stored data can be taken on the computer or PC to generate maps or develop GIS
—It can work in all weather conditions, day and night and through out the year
—It is easy to carry and use and can even float on water
—The accuracies can be achieved up to sub metre level in differential mode.

GPS receivers are designed for different applications based on the accuracy required. Broadly they can be classified into (a) standalone GPS, and (b) differential GPS or DGPS which can further be divided into (i) mapping and GIS GPS, and (ii) survey grade or geodetic GPS. Each type of GPS have its own advantages including accuracy requirements, applications and cost. Standalone GPS provides accuracy of 10-15m while mapping or GPS can give accuracy of 2-5m. Sub-metre accuracies can be achieved in differential mode. Further, survey grade GPS have accuracy of 5mm or more. An example of typical GPS is shown in Fig 3.34.

Earlier GPS use to deliver two types of signals : (a) Precision, and (b) Selective Availability. The latter one has been switched off since May 2000 though it was able to provide location anywhere in the world with in 100m. This facility was popularly used all over.

As mentioned earlier, GPS maintained by the US Department of Defence has been in operation since 1995. Ministry of Defence of the Russian Federation has mooted the development of Global Satellite Navigation System or GLONASS. The objective here is also to continuously provide coordinates and time information. Further, China is also developing its Beidou

Navigation System and two satellites have been launched in 2000 which is a mix of medium earth satellites and geo-stationary satellites. In addition, European Union has proposed 30 medium earth satellites in civil domain under Galileo programme by 2008. The medium orbit planes will be inclined at 56°. The altitude of the satellites will be 23,616 km. However, the full constellation may not be achieved until 2011. Nevertheless, India has agreed to join the Galileo programme. All the above satellites are planned to become core systems of the Global Navigation Satellite System or GNSS.

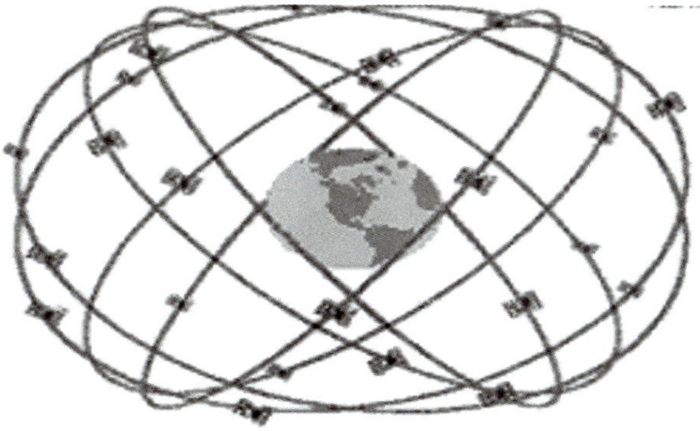

GPS Nominal Constellation
24 Satellites in 6 Orbital Planes
4 Satellites in each Plane
20,200 km Altitudes, 55 Degree inclination

Fig 5.33: GPS constellation

Fig 5.34: GPS on tripod

GPS has 24 satellites rotating around the earth. GLONASS during 1982-2007 launched 79 satellites and 12 are in orbit. The combination of both the systems is included in Tables 5.1 and 5.2. Further, combined view of GPS and GLONASS is shown in Fig 5.35.

Table 5.1: GPS and GLONASS—Constellations

	GPS	GLONASS
Number of satellites	24 (29)	24 (14)
Number of orb. Planes	6	3
Orbital inclination	55°	64.8°
Orbital Radius	26,560 km	25,510 km
Orbital Period	11^h58^m	11^h15^m

Table 5.2 : GPS and GLONASS—Reference frames

	GPS	GLONASS
Time	UTC (USNO)	UTC (SU)
Leap sec corr.	No	Yes
Datum	WGS-84	PZ 90
Semi-major axis	6378137.0 m	6378136.0 m
Inv. flattening	298.257223563	298.257839303

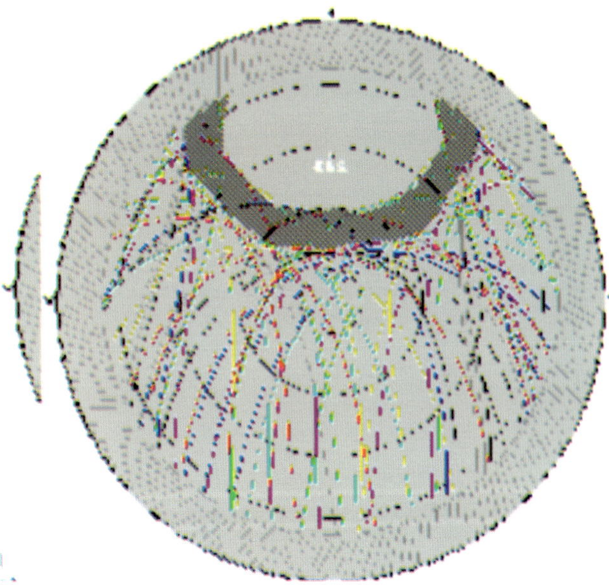

Orbital inclination
GPSS : 55°
GLONASS □ 64.8°

Fig 5.35: View of GPS and GLONASS combined

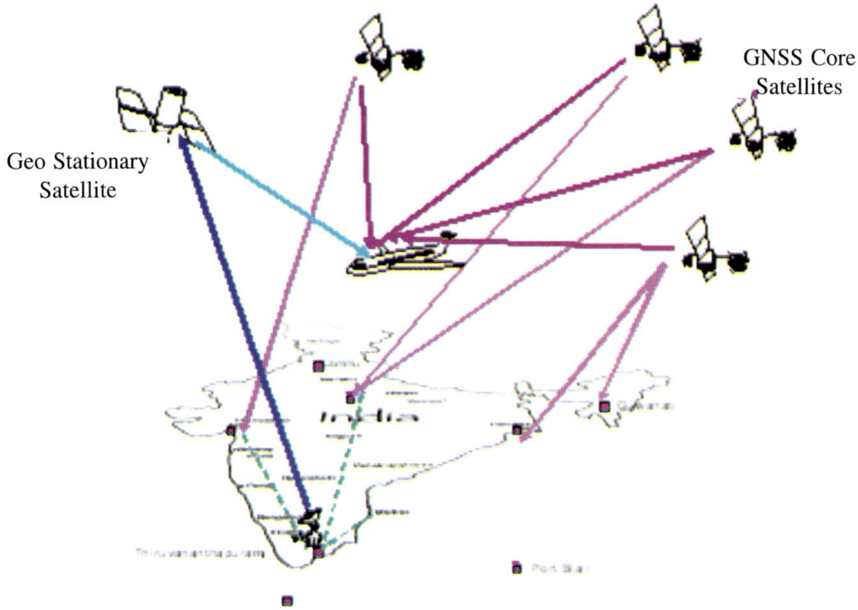

Fig 5.36 : GAGAN system

India also plans to have its own satellite based GPS system to serve the concerned region covering India. A cluster of satellites have been planned to form a hub in air navigation and to provide other services. This programme is known as GAGAN which stands for GPS and Geo-Augmented Navigation. The Union Ministry of Civil Aviation with the support of the Indian Space Research Organisation (ISRO) has mooted this programme. The objective here is to fill gap between the European and Japanese space augmentation systems and thereby provide seamless navigation of aircrafts from east to west or *vice-versa*. It would also take help of the geo-stationary satellites and the GNSS core satellites mentioned above (Fig 5.36). The objective is to develop a low cost satellite based navigation system having seven geo-stationary satellites and always visible and covering the region.

There are several other applications of GPS particularly as a valuable input in the GIS related activities. Three representative examples are given here in brief. First one deals with aerial information while the second one with linear information and third one concerns with neighbourhood mapping.

(a) GPS for Mapping Parcel of Land

In order to develop GIS or to update a GIS base with a certain level of accuracy, differential GPS or DGPS can be conveniently used. The differential segments of GPS operation in the field is shown in Fig 5.37. There are space, control and user segments. From Fig 5.38 it may be gathered that with the help of known point which can be used as base station, the roving GPS can take reading of the desired points in order to form a polygons or land parcels. The difference of the known coordinates and the observed coordinates at the base station can be introduced at other points and fairly accurate map can be prepared which becomes a valuable input for GIS development.

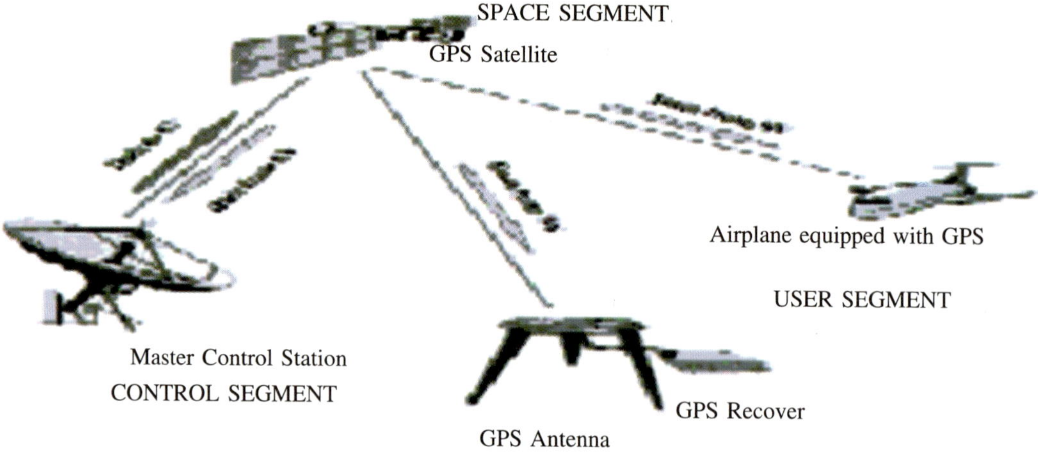

Fig 5.37: Segments of GPS

There are real time solutions as well. Currently real time GPS processing software is embedded in the GPS receiver—not residing on Window PC. Such a combination is known as 'firmware'. The real time firmware processing technique is commonly known as real time kinematic or RTK which is extensively used for mapping and for developing dynamic GIS applications. RTK method use differential GPS solution. A fixed GPS base transmit correction factor in real time to the remotely located GPS receivers. The firmware in the remote GPS receives the broadcast correction factors from the base station and applies to compute real time solution. With the movement of the surveyor from one location to another, the map denotes his position in real time. The powerful real time mapping or GIS software helps in graphical interfacing.

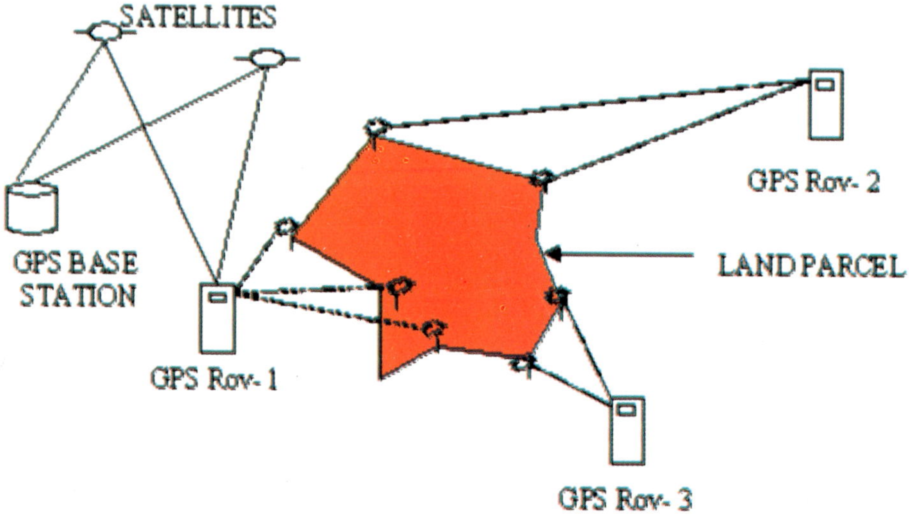

Fig 5.38: GPS operation

(b) Updation of Route Map

Survey of India now provides topographical data in digital form from their state geospatial data centres or from National Geospatial Data Centre located in Dehra Dun. The information contained in the topographical sheets is available in the form of different layers. The desired layers can be stitched and a digital base or a GIS of the concerned area can be developed. On this base, the GPS operation can be carried out for plotting new or additional information. In Fig 5.39, Survey of India prepared the Bangalore-Mysore route map. Based on similar methodology and using palm tops the topographical maps are being rapidly verified to produce the new Open Map Series or OSM.

Fig 5.39: Application of GPS for route map (*Source*: Survey of India)

(c) Neighbourhood Mapping

Under the aegis of the Department of Science and Technology, Government of India, a neighbourhood mapping by the school going children was launched. Here the objective was create awareness about the neighbourhood by preparing maps thereby taking mapping science to young minds. The students were provided basic training and instruments, including GPS, for mapping their surrounds. It was launched initially in Uttaranchal with the help of Centre for Spatial Data and Media Studies or CSDMS (www.csdms.org). The mapping process was as follows :

- —Informal discussions on settlements types, significance of maps and elements in preparing maps
- —Students were asked to develop mental maps of the concerned area
- —Discussions on shortcoming and strengths

—Redrawing on paper
—Discussions on the usages of GPS and PDA
—Surveying with basic measurements, sense of scale and direction, GPS and PDA
—Final drawing on hard copy (Fig 5.40).

The neighbourhood mapping programme is to be extended to other Indian states as well.

Fig 5.40: Neighbourhood mapping

(d) Assessment of Topographical Changes in Andaman and Nicobar Islands

Survey of India has a long association with Andaman and Nicobar islands. The maps were prepared considering the islands as separate entity. Lately, with the introduction of the GPS, efforts were made to connect these islands among themselves as well as with the mainland. In order to assess the changes in the landmass, several sets of observations were taken with the help of GPS. Differences in heights were also calculated. In order to further assess the changes, the pre- and post-*tsunami* coordinates of the remaining 12 stations in the islands were calculated with broadcast ephemeredes and with respect to GT stations (Table 5.3).

Table 5.3: Comparison of coordinates between pre- and post-*tsunami* observations

S.NO.	*STATION NAME*		*DIFFERENCE (POST-PRE)*	
			Latitude	*Longitude*
1.	AERIAL BAY	H.S.	-0.094	-0.096
2.	EAST ISLANDS	S.	-0.062	-0.074
3.	LONG ISLANDS	S.	-0.016	0.001
4.	UDAIGARH	S.	-0.050	-0.037
5.	GOVINDGARH	S.	-0.007	0.006
6.	PORTBLAIR	H.S.	-0.043	-0.067
7.	PASSAGE ISLANDS	S.	-0.124	-0.077
8.	RAMAKRISHNAPUR	S.	-0.117	-0.097
9.	TARASSA	S.	-0.077	-0.148
10.	KARDIP	S.	-0.034	-0.086
11.	MEROE	S.	-0.067	-0.198
12.	CAMPBELL BALL	S.	-0.051	-0.175

Fig 5.41 GPS operation in Andaman and Nicobar Islands

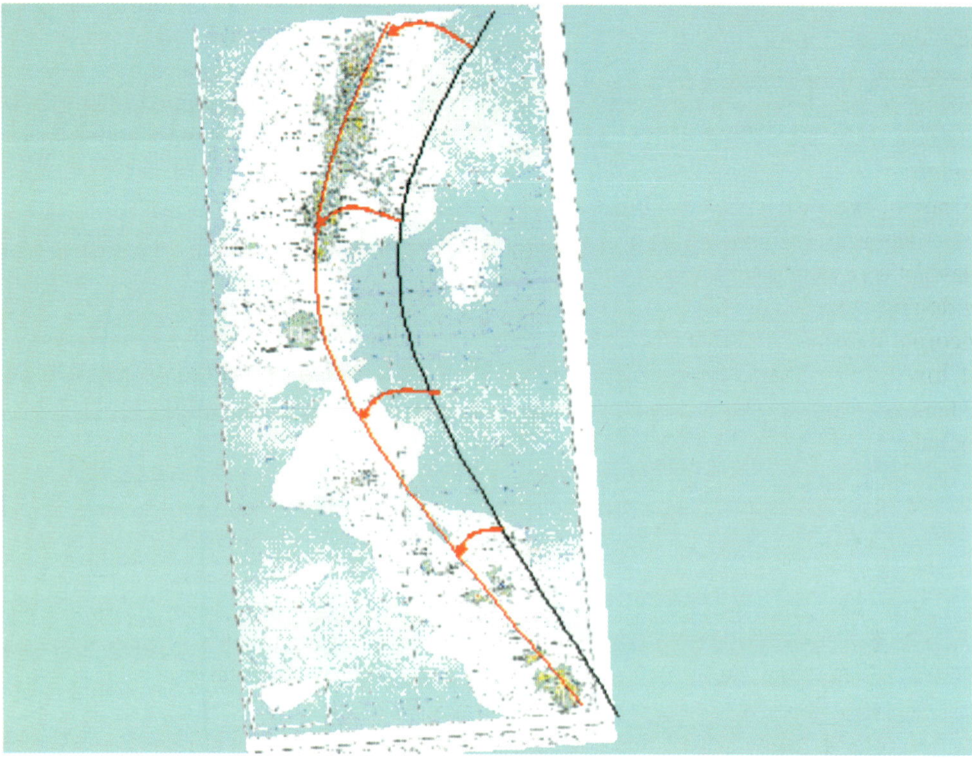

Fig 5.42: Apparent changes in topography in Andaman and Nicobar Islands

Preliminary investigations of the GPS results from April 2004 and January 2005 epochs of observations reveal that there is a small change in position of Port Blair station (Fig 5.41). GPS observations were carried out at this station immediately after *tsunami.* A vertical subsidence of approximately 1.2m was observed. However, there has been a history of changes of landmass in all the three directions in Andaman and Nicobar Islands as per the records now available. Hence, the changes brought by *tsunami* in December 2004 are not that alarming. The trend of changes in topography is shown in Fig 5.42. Such changes are now considered to be precursor for earthquakes followed by *tsunamis.* Nevertheless, the changes brought by the *tsunami* were good enough for making a case for fresh topographical mapping because there have been changes in the mean sea level; there has been subsidence and emergence at different places; coordinate and control points have changed; and values of baseline have also changed.

Geodesy and GPS have developed as a separate field of specialization. Apart from the examples mentioned above, they have a wide range of applications. The applications include crust movement, earth system studies, earthquakes, landslides, missile trajectory, dam and building deformation studies and the like. Since the details of this field do not fall in the scope of this book, the readers are advised consult relevant literature on the subject.

Conclusion

The technology of the geoinformatics is now becoming user friendly. The sophistication of new data gathering technologies and geographic information system (GIS) technologies is increasing in developing spatial data applications. In the present scenario, we may identify various technologies where its products and services are directly representing its application in sustainable development. These are like navigation and positioning, high resolution sensing, image analysis, GIS, data visualization, database management, geospatial data and its infrastructure and user application and solutions.

The principal navigation and positioning technologies in use today are the GPS, Inertial Navigation Systems (INS), radio-navigation systems such as differential GPS, beacons, satellite wide-area systems and local radio links. Recent developments in GPS technologies include the incorporation of the Russian Global Navigation Satellite System (GLONASS) and the networks of low Earth orbit satellites. Applications of navigation and positioning technologies are extensive, involving almost anything that requires positioning information. Application areas include:

- *Surveying*: geographic location and data analysis including high-resolution asset location mapping
- *GIS*: geographic location, data analysis and presentation for asset management
- *Vehicle navigation*: installed in anything from commercial passenger cars to tanks
- *Emergency location systems*: typically installed in cars, in call-911 and in telecommunications devices made to pass on precise location data
- *Airborne crop spraying*: field positioning devices
- *Marine navigation systems*: from commercial fish finders to redundant systems on oil tankers
- *Marine docking systems*: centimeter level, short range systems for ferries and cargo vessels
- *Aircraft navigation systems*: from simple, en-route systems in light aircraft to full landing systems
- *Aircraft CFIT (controlled flight into terrain) avoidance systems*: which match aircraft position to three-dimensional maps and warn crews of dangerous approaches to terrain
- *Aerial survey systems*: measuring and recording data from the air
- *Hand-held personal navigators*: from recreational use to search-and-rescue operations
- *Wide area navigation systems*: land-based, fixed data gathering and analysis systems to improve accuracy and integrity of the basic GPS system
- *Robotic vehicles*: mining ore extractors, helicopter surveillance, dangerous area surveillance, security surveillance
- *Weapons training pods*: attack aircraft weapons training systems

REFERENCES

Agarwal, N.K. (2002), GPS—Security and restriction, *Proceedings of the Asian GPS Conference 2002,* CSDMS, New Delhi, pp. 1-3.

Agarwal, N.K. (2004), *Essentials of GPS,* Spatial Networks Pvt. Ltd., Hyderabad.

Bugayevskiy, Lev M. and John P. Snyder (1995), *Map Projections: A Reference Manual.* London: Taylor and Francis.

By, Rolf A. (ed) (2001), *Principles of Geographic Information Systems : An Introductory Textbook*, International Institute for Aerospace Survey and Earth Sciences, Enschede

Chauniyal, Devi Dutt (2004), *Sudur Samwaydan aavm Bhologik Suchana Pranali*, Sarada Pushtak Bhavan, Allahabad (in Hindi).

Defense Mapping Agency (1977), *The American Practical Navigator Publication* No. 9, Defense Mapping Agency Hydrographic Center.

Ferndndez, R. Norberto, D. Fabidn Lozano-Garcia, Greg Deeds & Cris J.Johannsen (1991), Accuracy assessment of map coordinates retrieval, *Photogrammetric Engineering and Remote Sensing*, vol. 57, No. 11, pp. 1447-52.

Guha, P.K. (2003), *Remote Sensing for the Beginner*, East-West Press, New Delhi.

Kulkarni, Madhav N. (2005), Enter the GPS era, *Geospatial Today*, vol. 4, issue 4, pp. 42-6.

Laurila, Simo H. (1976), *Electronic Surveying and Navigation*, New York: John Wiley & Sons

Monkhouse, F.J. & H.R. Wilkson (1951), *Maps and Diagrams*, University Paperbacks, London.

Muehrcke, P.C and Juliana O. Muehrcke (1992), *Map Use*, Madison, WI: JP Publications.

Muehrcke, Phillip C. and Juliana O. Muehrcke (1998), *Map Use: Reading-Analysis-Interpretation*, 4th ed. Madison, WI: JP Publications.

Nag, P. (2005), Water and land in Andaman and Nicobar Islands, *Indian Journal of Landscape Systems and Ecological Studies*, vol. 28, No. 2, pp. 13-20, also published in A.L. Singh (ed)(2006), *Use and Misuse of Land and Water Resources*, B.R. Publishing Corporation, Delhi, pp. 38-51.

Nag, P. *et al* (1978), Perspective mapping of a broad range data: A case study of India, *Geographical Review of India*. Vol. 40, pp..42-6.

Nainwal, B.P. (u.d.), *Lecture Notes on Physical Geodesy*, Survey Training Institute, Survey of India, Hyderabad.

Nainwal, B.P. (2003), *Notes on Satellite Geodesy and Global Positioning System*, Survey Training Institute, Survey of India, Hyderabad.

Nainwal, B.P. (2003), *Lecture Notes on Map Projection*, Survey Training Institute, Survey of India, Hyderabad.

National Imagery and Mapping Agency (1997), *World Geodetic System 1984 (WGS-84) - Its Definition and Relationships with Local Geodetic Systems*, 3rd Edition. Washington, DC: National Imagery and Mapping Agency.

Padhi, R.C. (2005), Precision mapping, *Graticule 2005*, CSDMS, October 2005, pp. 26-32.

Ramachandran, R. (2000), Public access to Indian geographic data, *Current Science*, vol. 79, No. 4.25.

Robinson, A. H., Sale, R. D., Morrison, J. L., Muehrcke, P. C. (1984), *Elements of Cartography*, Fifth Edition. New York: John Wiley & Sons.

Sanyal, A.K. (1991), *Map Projections*, Survey Training Institute, Survey of India, Hyderabad,

Snyder, John P. (1987), *Map Projections: A Working Manual*. USGS Professional Paper 1395. Washington, DC: United States Government Printing Office.

Snyder, John P. (1987), *Map Projections : A Working Manual*. Washington, DC: US Govt. Printing Office.

Snyder, J.P., and R.M. Voxland, (1989), "An album of map projections", *USGS Professional Paper 1453*, 249 p.

Survey of India (1972), *Auxiliary Tables*, Part I : Graticules of Maps, Dehra Dun.

Survey of India (2003), *Souvenir of the National Seminar on GTS to GPS : Geodesy on the Move*, 9-12 February, Dehra Dun.

Survey of India (u.d.), *Proceedings of the National Seminar on GTS to GPS : Geodesy on the Move*, Dehra Dun.

Survey of India (2006), *National Map Policy*, Department of Science & Technology, Dehra Dun.

Thomas, P. D. (1970), *Spheroidal Geodesics, Reference Systems and Local Geometry*, Washington, DC: U. S. Naval Oceanographic Office.

U. S. Air Force and Navy. (1983), *Air Navigation*, AFM 51-40. Washington, DC: Department of the Air Force.

U. S. Army (1967), TM 5-241-1 *Grids and Grid References. Washington*, DC: Department of the Army.

6

GIS for Decision Support System

The concept of Decision Support System (DSS) is based on the seminal work by Simon and associates in 1950s and 1960s (Simon 1960). In the 1970's the concept of decision support systems (DSS) began to develop in the Information Systems (IS) community, notably with the work undertaken at the Massachusetts Institute of Technology (Gorry and Scott-Morton, 1971, Little, 1971). By the early 1980's there were many books and papers published in the DSS field and DSS had become a recognized part of IS (Sprague, 1980; Bonczek *et al.*, 1981). DSS had evolved out of the business data processing tradition and usually dealt with the financial and operating data associated with business use. The volumes of data involved with such systems were relatively small compared with those in the geographic domain. As computer systems became more powerful, some DSS type applications evolved that used map display or employed spatial information. A good example is the Geodata Analysis and Display System (GADS) which was used for routing applications (Grace, 1977). Nevertheless, the technology it used had limited graphics and inadequate processing power to exploit the full potential of spatial applications.

While these developments in DSS were taking place in the IS community in the 1970s, a largely separate trend of development took place in GIS, with developments largely concentrated on geographic data processing applications (Nagy and Wagle, 1979). Spatial applications had placed heavy demands on the technology, and this slowed the progression from data processing to decision support applications. However, over time improving computer performance led to increasing interest in spatial *what-if* analysis and modelling applications. The idea of a Spatial Decision Support System (SDSS) evolved in the mid-1980's (Armstrong *et al.*, 1986), and by the end of the decade SDSS was included in an authoritative review of the GIS field (Densham, 1991). This trend was evident in the launch of research initiative on SDSS in 1990 by the US National Center for Geographic Information and Analysis (Goodchild and Densham, 1993).

Consequently, by the early 1990's, SDSS had achieved a recognized place in the GIS community and was identified by Muller (1993) as a growth area in the application of GIS technology. The delay in the recognition of SDSS, compared to other DSS in other domains reflects the greater demands of spatial processing on IT. Nevertheless, despite these developments SDSS does not occupy a central place in the GIS field; and many introductory GIS textbooks do not mention SDSS at all (Clarke, 1997; Bernhardsen, 1999). This may reflect a feeling among many in the geographic disciplines that SDSS applications involve a diversity of techniques from different fields largely outside the geography domain. SDSS is an interactive, computer-based system designed to *support* a user or group of users in achieving a higher *effectiveness* of decision-making while solving a *semi-structured spatial decision problem.* Three terms i.e. semi-structure spatial problems, effectiveness, and decision support capture the essence of the SDSS concept.

Less attention was paid to SDSS within the DSS research community, until the mid-1990's when some work in this area began to appear (Wilson, 1994). One of the first papers in an IS related publication illustrated the effectiveness of SDSS technology (Crossland *et al.*, 1995). Recently the benefits of SDSS for both inexperienced and experienced decision-makers (Mennecke *et al.*, 2000) were discussed in *MIS Quarterly*. The field has now grown to the point that it is made up of many threads with different, but related names, such as collaborative SDSS, group SDSS, environmental DSS, spatial knowledge based, expert systems and the like. The main characteristics of spatial decision problems include:

- a large number of decision alternatives,
- the outcomes or consequences of the decision alternatives are spatially variable,
- each alternative is evaluated on the basis of multiple criteria,
- some of the criteria may be qualitative while others may be quantitative,
- there are typically more then one decision maker (or interest group) involved in the decision-making process,
- the decision makers have different preferences with respect to the relative importance of evaluation criteria and decision consequences,
- the decisions are often surrounded by uncertainty.

Basic Decision-Making Process and GIS

Simon (1960) suggests that any decision-making process can be structured into three major phases (Fig. 6.1):

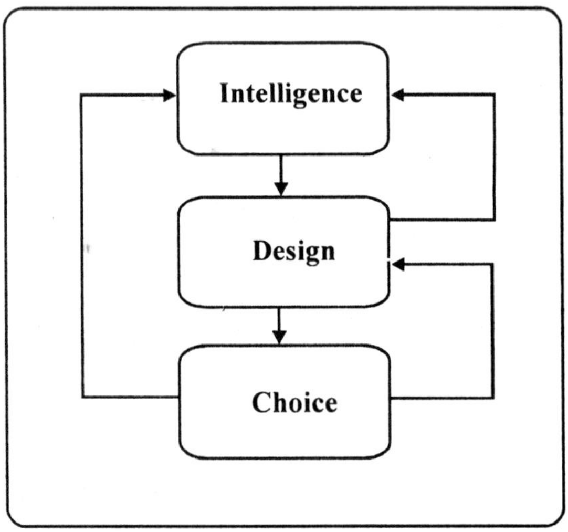

Fig. 6.1: Phases of decision-making

- *intelligence*—is there a problem or an opportunity for change?
- *design*—what are the decision alternatives?
- *choice*—which alternative is best?

It is important to find out how and to what extent GIS technology provides support required in each of the three phases of decision-making:

In the intelligence phase, which involves searching or scanning the environment for conditions calling for decisions, requires an exploratory analysis of the decision situation. GIS can play a vital role at the initial stage of spatial decision-making by helping in coordinating decision situation analysis through its ability to integrate and explore data and information from a wide range of sources. GIS can also effectively present information in a comprehensive form to the decision makers.

The design phase involves inventing, developing, and analyzing a set of possible decision alternatives for the problem identified in the intelligence phase. In this stage a formal model is typically used to support a decision maker in generating the set of alternatives. While an increasing number of GIS systems are described as systems for supporting the process of designing and evaluating spatial decision alternatives, but most commercially available GIS lack the kinds of spatial analysis and modelling required by decision makers. The capabilities of GIS for generating a set of alternative decisions are mainly based on the spatial relationship principles of connectivity, contiguity, proximity and the overlay methods. In current GIS environments, models for generating decision alternatives operate in the background, detached from users' insights and qualifications.

The choice phase involves selecting a particular decision alternative from those available. In this stage each alternative is evaluated and analyzed in relation to others in terms of a pre-specified decision rule. The decision rules are used to rank the alternatives under consideration and the ranking depends upon the decision maker's preferences with respect to the importance of the evaluation criteria. The critical for use of GIS in the choice phase is the capability of incorporating the decision maker's preferences into the decision-making process and in general, GIS systems do not provide a mechanism for flexible incorporation of the decision maker's preferences into the decision-making process.

It can be observed that GIS systems have limited capabilities of supporting the design and choice phases of the decision-making process. The systems provide a very static modelling environment and thus reduce their scope as decision support tools—especially in the context of problems involving collaborative decision-making.

Problems Semi-structured Decision

Any decision problem falls on a continuum that ranges from completely *structured* to *unstructured* decisions (Simon, 1960). The structured decisions occur when the decision-making problem can be structured either by the decision maker or on the basis of relevant theory. The structured decisions are programmable and can be solved by computers. On the other hand, the unstructured decisions occur when the decision maker is unable to structure the problem and the problem cannot be structured on the basis of a relevant theory. These decisions are non-programmable and must be solved by the decision makers without any assistance from a computer. Most real-world spatial decision problems, if not all, can be found somewhere between these two extreme cases of completely structured and unstructured decisions; these decisions are called *semi-structured* ones. The SDSS concept has major application this area. The structured (programmed) part of the problem may be amenable to automated solution by the use of a computer, while the unstructured (non-programmed) aspects are tackled by decision makers (Fig. 6.2).

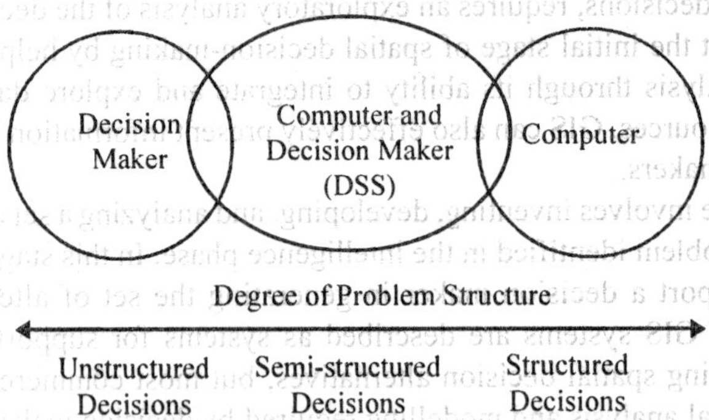

Fig. 6.2: Degree of decision problem structure.

The main aim of the system is to improve the effectiveness, rather than efficiency of the decision-making process. The higher effectiveness is achieved by incorporating the decision maker judgments and computer-based programmes into the decision-making process. To be effective the system must be easy to use. The system also helps the users to explore the decision problem in an interactive and recursive fashion in all phases of the decision-making process.

Principles of SDSS

The technology for a DSS must consist of three sets of capabilities in the areas of *dialog, data,* and *modeling* (the DDM paradigm) (Sprague and Watson, 1996). A well design SDSS should have balance among the three capabilities (Fig. 6.3).

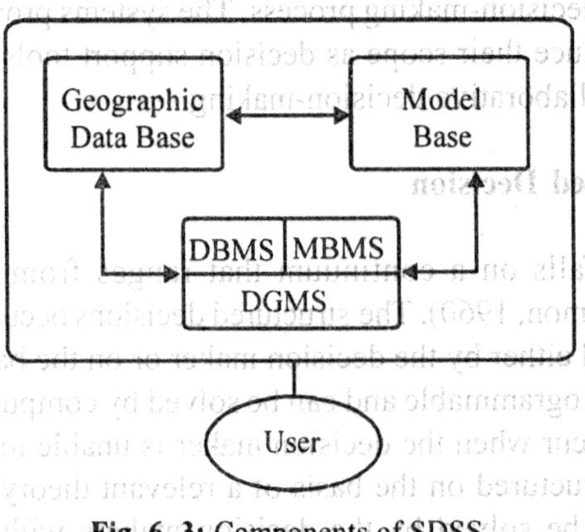

Fig. 6. 3: Components of SDSS

- the *Data Base Management System* (DBMS) contains the functions to manage the *geographic data base;*
- the *Model Base Management System* (MBMS) contains the functions to manage the *model base;*
- the *Dialog Generation and Management System* (DGMS) manages the interface between the user and the rest of the system.

Technologies for developing SDSS

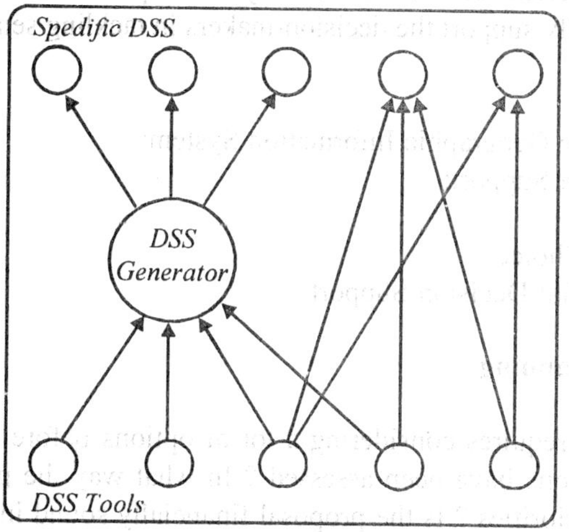

Fig. 6.4: Three levels of DSS technology (*Source*: Sprague and Watson, 1996)

DSS tools facilitate the development of either a DSS generator or a specific DSS (Fig. 6.4).

Following are examples of some programming languages:

- The procedural programming languages and code libraries (e.g., Arc Macro Language (AML) scripting tool of ARC/INFO, Avenue—ArcView GIS software's built-in object-oriented scripting language, TransCAD—Caliper Script macro language, MapInfo—MapBasic);
- visual programming language (e.g. STELLA II, Cantata and Khoros);
- inter-application communication software (e.g. dynamic data exchange (DDE), object linking (OLE), open database connectivity (ODBC));
- simulation languages and software (e.g. SIMULINK, SIMULA);
- application programming interfaces (API) (e.g. the IBM's geoManager API, Java Advanced Imaging API, TransCAD's API);
- applets (e.g. GISApplet, Microsoft Visual J++),
- visual interfaces, graphics and color subroutines (e.g. graphical user interfaces—GUI).

DSS generator is a package of related hardware and software which provides a set of capabilities to quickly and easily build a specific SDSS; examples include:

- GISystems (e.g. ARC/INFO, ArcView, ARCNetwork, Spatial Analyst, MapObjects LT, GRASS, IDRISI, MapInfo, TransCAD);
- database packages (e.g. dBase, Access, Paradox);
- decision analysis and optimization software (e.g. LINDO, EXPERT CHOICE, LOGICAL DECISION);
- statistical and geostatistical software (e.g. S-PLUS, SPSS, SAS);
- simulation (e.g. Spatial Modelling Environment)

Specific DSS are systems devoted to the analysis of a particular set of decision problems; the systems which actually support the decision-makers in tackling semi-structured problems; examples include:

- Active Response Geographic Information System;
- IDRISI Decision Support;
- GeoMed;
- Spatial Group Choice;
- winR+GIS Spatial Decision Support.

Decision Support in Planning

Any planning initiative requires considering a lot of options before taking a final decision. Whether the pros and cons have been assessed ? In what way the proposed action is most viable ? What are the priorities ? Is the proposal financially sound in immediate and in long terms ? What are the alternatives ? Finally, whether such action is politically acceptable ? Issues similar to above are likely to crop up whenever a planning exercise is taken up. Answers to such questions are essential. A flow diagram (Fig. 6.5) is included here to show the role of GIS in such complicated decision-making process. Further, decision-making process in heterogeneous multi user environment is shown in Fig. 6.6.

DSS vs SDSS

A SDSS is different from a DSS in the fact that it is used to support decision processes where the spatial aspect of a problem plays a decisive role. Despite ongoing discussion, there is no agreement on a single definition. Some compare this to describing a car where there seems to be general agreement on what it is, but not on what are the necessary attributes for giving it such a label. This explains why most authors simply end up with a list of characteristics of what a SDSS might comprise. For this study we prefer the statement by Densham (1991) that SDSS are "explicitly designed to provide the user with a decision-making environment that enables the analysis of geographical information to be carried out in a flexible manner". The lack of widely supported and unambiguous definitions of DSS and SDSS do not seem to hamper the development of such systems. On the contrary: there are innumerable systems under development for supporting a variety of decision questions. Most popular application areas are, at least in The Netherlands, coastal zone management, river management, water resource management and land-use planning.

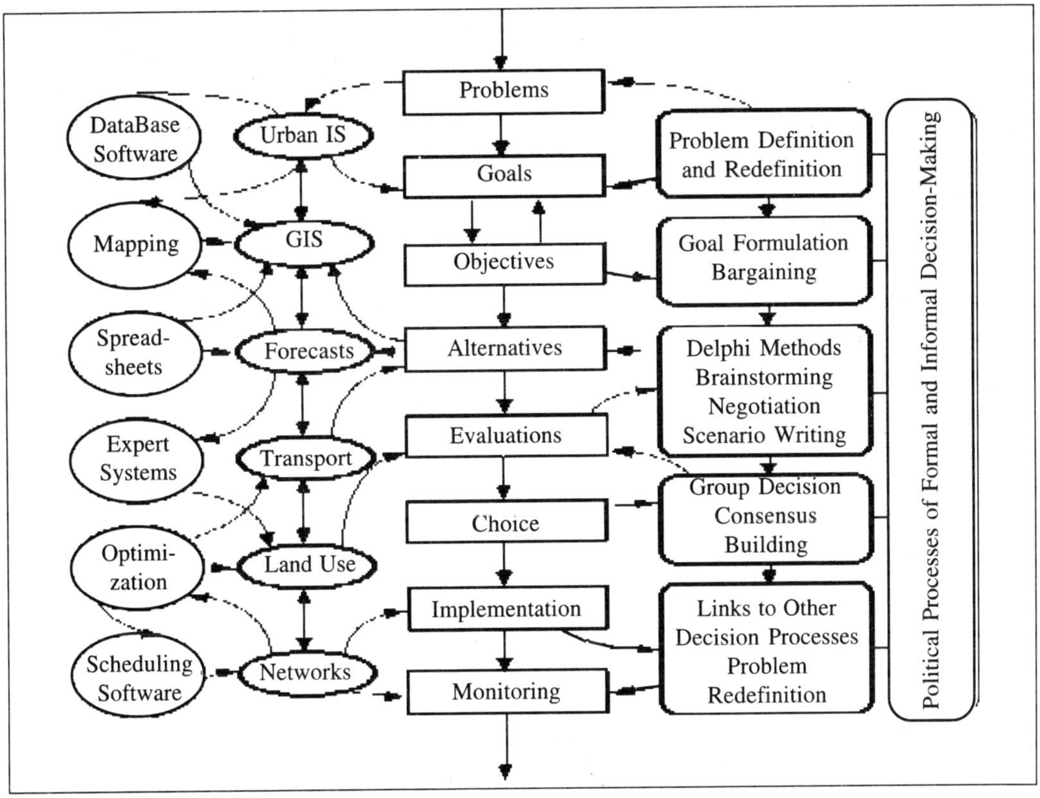

Fig. 6.5: Process of decision-making

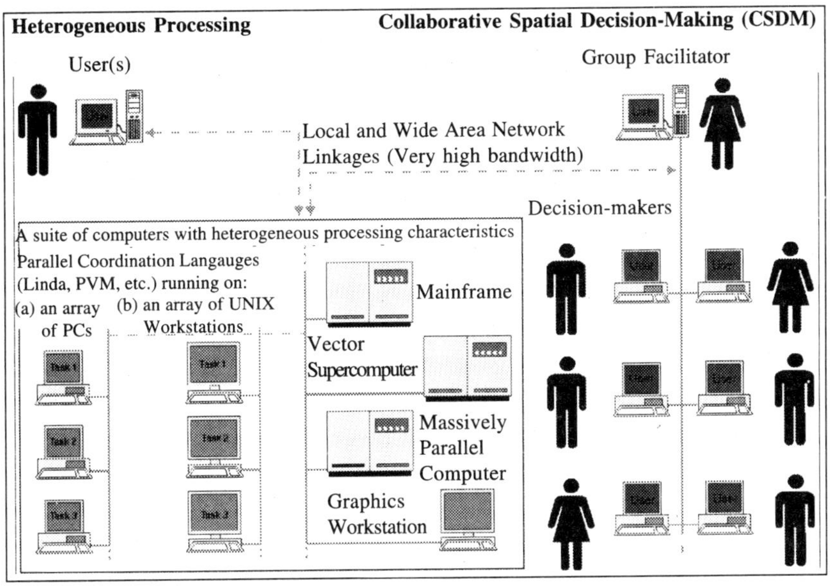

Fig. 6.6: Heterogeneous processing environments of SDSS

Despite the high effort and large amounts of money spent on developing DSS, there are many decisions that are never or hardly used (Adelman, 1992). There are many reasons for this lack of success of decision support systems. One reason can be that the users find the system too detailed, time consuming and costly to use (Ubbels and Verhallen, 2000). Other reasons are related the general complexity of the systems, while still other reasons are related to the uncertainty of the model output and on the appropriateness for solving the decision question. It is stated that the limited involvement of users in the development phase can lead to unsuccessful DSS. In addition, the need for training in the use of each particular DSS has been emphasized. Only then can a DSS be used effectively, the results free of errors and useful for decision-making. Four categories of factors which constrain the use of analytical tools in decision-making has been described.

These are too little, too much and wrong information, time and resource limitations, lack of access, and the hurdles of communication and trust. Some of the reasons for unsuccessful DSS mentioned earlier are strong indications that users are not always able to take systems into use as intended or expected by developers. In other words, the functionality of the systems is not in all cases optimal. Functionality is a very wide term and can include almost any aspect of a Spatial Decision Support System (SDSS).

There are probably just as many reasons for poorly functional SDSS and, therefore, unsuccessful SDSS as there are systems built. There are several suggestions to be found in the literature describing the characteristics of a DSS on which to base a comparison or an evaluation of the systems. Ubbels and Verhallen (2000) evaluate a set of DSS according to their suitability for collaborative planning processes using the characteristics user-friendliness, prerequisite for collaborative processes, transparency, flexibility, and the way the effects of possible actions are estimated. Using these characteristics the authors determine the suitability of tools for specific user groups and decision-making phases. Crossland, Wynne and Perkins (1995) investigated the impact on the decision makers performance when using SDSS in a decision making process. The performance was measured by the accuracy of the solutions and the time it took to come to the solutions.

It was concluded that SDSS users have a shorter solution time and fewer errors compared to those not using SDSS. Adelman (1992) presents three methods for evaluating a DSS in technical, empirical, and subjective evaluation terms. These methods of evaluating a DSS are intended for implementation during the development phase of the DSS, to keep the development on track. However, the methods can also be used when evaluating a system after completion. The technical evaluation method focuses on system characteristics and includes, among other criteria, the adequacy of the selected analytical methods, software development costs, software tests and verification, and adequacy of the knowledge base. The empirical evaluation method on the other hand focuses on obtaining objective measures of the system performance. An important question in this type of evaluation is whether decision-makers perform better with a system than without. The subjective evaluation method focuses on assessing the system from the perspective of the potential user in order to identify whether the end-users find the system useful or not.

This study focuses on the functionality of SDSS. Because the emphasis is on characteristics of the system this approach can be labelled as a technical evaluation. It is very difficult to perform an empirical or subjective evaluation because of the inevitable dependency on end-users. As stated earlier many SDSS are not taken into use and if they are, documentation of end-users experiences is rare. Five examples of Spatial Decision Support Systems (SDSS) for coastal zone management are used to test the research questions. These examples can be seen as representative for the state of the art in The Netherlands in this field.

1. Nature Development and Valuation (NDV Module)—which supports the design, development and valuation of nature on the Dutch coast. The potential for nature development is thereby quantified, analyzed and presented. The application is the land reclamation project Maasvlakte 2, located South of Rotterdam (Ruijgrok, de Vries, Baptist, Meulen, and Zaadnoordijk, 1999).
2. Ecotope evaluation DSS (ECOPEIL)—which supports questions concerning the expected ecotopes and the consequences of water level changes on the naturalness and biodiversity in the IJsselmeer area (Jans, Platteeuw, Tosserams, and Schiereck, 2000).
3. Shoreline DSS (Shoman)—which supports suitability assessment of locations for large-scale nature development projects on the Dutch coast. The application is the development of a wet dune valley in the Province of North—Holland (Heuvel, Katwijk, Triest, and Goedhart, 1993).
4. Wadden Sea DSS (WadBOS)—which supports the analysis of current policy and management issues in the Wadden Sea. It permits assessment of the e.ects of policy decisions and the feasibility of management measures (Uljee and Englen, 2000).
5. Eco-morphological DSS (EMM)—which supports the analysis and evaluation of morphological and ecological effects of different dredging and deposition interventions. The application area is the Western Scheldt estuary (Baptist, de Vries, and Wang, 1998).

One of the main objectives for taking a SDSS into use is that it must be possible to test alternative solutions. In creating alternatives, users can incorporate ideas and knowledge of the decision problem and possible solutions can be tested in terms of their feasibility. It is important that definition of alternatives can be managed by users. When making an alternative gets too complex for a user, the user can easily feel overwhelmed by the task and it is less likely that the SDSS will be a success. The complexity of making an alternative depends, among others, on how many parameters need to be changed and how easy it is to make changes to the parameters.

An example of a situation in which some users can feel overwhelmed by the task of making alternatives can be found in the eco-morphological DSS (Baptist *et al.*, 1998). Here only two decision variables play a role, which implies that it should be an easy task to make an alternative. However, this is not the case. The decision variables are the amount of dredged material and the amount of deposited material, both in m3. These are entered using two separate maps, one for deposition and one for dredging.

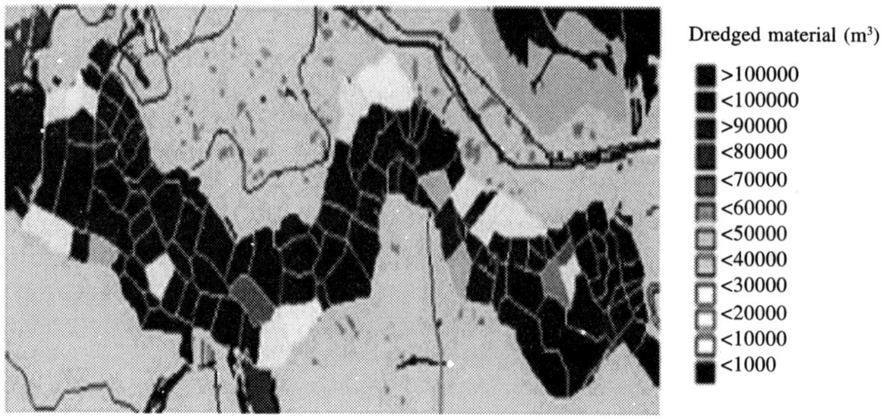

Fig. 6.7: Standard input in the eco-morphological DSS for decision variable dredging for one time step (*Source*: Baptists *et al.*, 1998)

Fig. 6.7 shows an example of a dredging input map. Input data must be provided for 112 possible locations for each time step. Not only does the user, therefore, have to think of the amount of material either to dredge or deposit when making an alternative, but also the spatial pattern of the input data has to be taken into consideration. The time step of the eco-morphological DSS is thirty days and because the DSS calculates long-term morphological and ecological effects on the estuary, putting in data for a time period of twenty years is not inconceivable. In addition to the challenging task of putting in data for one time step, the user has to repeat the input task a large number of times, both for the dredging and deposition decision variable. Not only is then the task very time consuming but also users easily lose the overview of the spatial patterns of the input data that very likely change in time. Generating alternatives therefore becomes a very complicated task.

In the four other SDSS, generating an alternative is different from the eco-morphological DSS and is more directed towards entering and changing parameters. In the Nature Development and Valuation DSS users need to design a land reclamation area and add management practices and functions of the designed area. A similar option is also available in the Ecotope Evaluation DSS where users can change the elevation of an area and add the management practice of this area. In the Shoreline DSS users must delineate an area which represents the form of a wet dune slack. The elevation of this new area must then also be filled in. In the Wadden Sea DSS, contrary to the Shoreline DSS, users are expected to fill in many, highly diverse, parameters. The emphasis on putting in data using maps is much less apparent, with the only input of data using maps being when users wish to open or close areas for functions like fishery or recreation.

Acceptance of a DSS to a large extent dependent of its ease of use. An important element in this respect is how well a user is guided though the system, for example whether the next step is self-explanatory or not. The organisation and number of windows at any one time is also important as is navigation through the system and its facilities. An example of a screen that is self-explanatory is shown in Fig. 6.8 and is taken from the Nature Development DSS (Ruijgrok *et al.,* 1999).

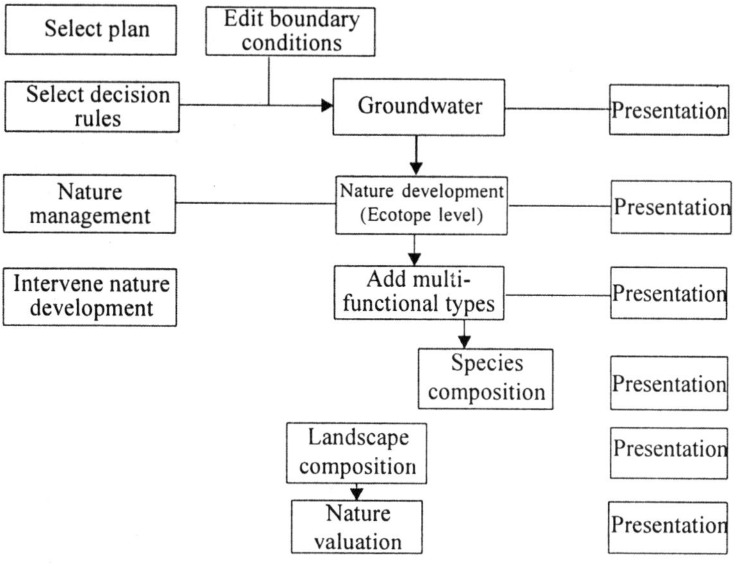

Fig. 6.8: Nature Development DSS (*Source*: Ruijgrok *et al*, 1999)

The boxes represent different tasks. The tasks that already have been performed can be shown in green while those still not done in yellow. Red or any other colour can be used to indicate tasks that for some reason cannot be done at this moment and lilac shows the currently running task. In this way, users keep a clear overview while performing the different tasks in the SDSS.

The same type of guidance is found in the eco-morphological DSS. However, the developers of the Wadden Sea DSS have chosen a very different approach, where the path through the system consists of many levels. Because of this it is easy for users to lose overview of which level they are on and what task to perform next. Another disadvantage is that the number of windows open at any given moment can be very large. Systems like the Shoreline DSS and the Ecotope Evaluation DSS also possess

Conclusion

Currently, most SDSS are tailored for specific problem domains. Users often will not recognize that SDSS in quite different application domains have similar architectures and combine similar elements. The applications use the attribute language of the domain, even though the spatial data inputs, manipulations, and outcomes may be identical. SDSS represent the integration of ideas, methods, and techniques from many of the social, behavioural, and management sciences. Their usefulness in the policy and planning sciences is based on the degree to which they successfully incorporate the findings and approaches of these sciences. Their focus on the spatial and the geographic data lead to a very important link with the evolving field of geographic information science. The rapid development of national spatial data infrastructures provides a rich and expanding database to which attribute information can be attached.

Spatial Decision Support System (SDSS), at their inception, were a derivative science that built upon the successful development of models of behavioural, social, and political processes. However, with the expanding use of SDSS in a wide variety of decision situations, theories and models of these processes will have to account for the use of SDSS. When normal decision-making patterns involve SDSS use, they will become an important element in social science theories and models to explain decisions in these areas. This evolution from the use of social and behavioural models as exogenous units in SDSS to their incorporation as endogenous units is an important development in the social sciences. This development will mark the more explicit incorporation of geographic variables in social science theory and method.

REFERENCES

Adelman, L. (1992). *Evaluating Decision Support and Expert Systems*. Wiley Series in System Engineering. New York: Wiley.

Alter, S.L., (1977), Why is man-computer interaction important for decision support systems?, *Interfaces*, Vol.7, No.2, pp.109-15.

Alter, S.L. (1977), A taxonomy of decision support systems", *Sloan Management Review*, Vol.19, No.1, pp. 39-56.

Alter, S. (1980), *Decision Support Systems: Current Practice and Continuing Challenges*, Addison-Wesley, Reading, USA.

Armstrong M. P. (1994), Requirements for the development of GIS-based group decision support systems, *Journal of the American Society for Information Science*, vol. 45(9); pp. 669-77.

Armstrong, M. P., Densham, P. J. and Rushton, G. (1986), Architecture for a microcomputer based spatial decision support system, *Second International Symposium on Spatial Data Handling*, Vol. 120-31, International Geographical Union,.

Armstrong, M.P., De, S., Densham, P.J., Lolonis, P., Rushton, G., Tewari, V.K., (1990), A knowledge-based approach for supporting locational decision-making. *Environment and Planning B: Planning and Design*, Vol. 17, No. 341-64.

Baptist, M. J., de Vries, M. B., Wang, Z. B. (1998), *Ecomorphological Decision Support System for the Western Scheldt (software and manual)*. Delft, The Netherlands: WL/Delft Hydraulics.

Bernhardsen, T. (1999), *Geographic Information Systems: An Introduction*, Wiley.

Bhargava, H., and D. J. Power (2001), *Decision Support Systems and Web Technologies: A Status Report*, Proceedings of the 2001 Americas Conference on Information Systems, Boston, MA, August 3-5.

Bonczek, R. H., Holsapple, C. W. and Whinston, A. B. (1981), *Foundations of Decision Support Systems*, Academic Press, Orlando.

Carver, S. J. (1991), Integrating multi-criteria evaluation with geographical information systems, *International Journal of Geographical Information Systems*, Vol. 5(3); pp. 321-39.

Church, R. L., S. R. Loban and K. Lombard (1992) An interface for exploring spatial alternatives for a corridor location problem, *Computers and Geosciences*, Vol. 8(9); pp. 1095-105.

Crossland, M. D., Wynne, B. E. and Perkins, W. C. (1995), Spatial Decision Support Systems: An overview of technology and a test of efficacy, *Decision Support Systems*, Vol. 14, No.3, 219-35.

Clarke, C., Clarke, M., (1995), The development and benefits of customized spatial decision support systems. in P. Longley and Clarke, G. (eds.), *GIS for Business and Service Planning*. Geoinformation International, Cambridge, pp. 227-54.

Clarke, K. C. (1997), *Getting Started with Geographic Information Systems*, Prentice Hall, Upper Saddle River, NJ.

Davis, G. (1974), *Management Information Systems: Conceptual Foundations, Structure, and Development*, McGraw-Hill, New York.

Dhar, V. and Stein, R. (1997), *Intelligent Decision Support Methods: The Science of Knowledge*. Prentice-Hall, Upper Saddle River, NJ.

Dickson, G. W., M. S. Poole and G. DeSanctis (1992), An overview of the GDSS research project and the SAMM system", in R.P. Bosttrom, R. T. Watson and S. T. Kinney, *Computer Augmented Teamwork: A Guided Tour*, New York: Van Nostrand Reinhold, pp. 63-79.

Densham, P. J. (1991), Spatial decision support systems, in D. J. Maguire, M. S. Goodchild and D. W. Rhind (eds), *Geographical Information Systems: Principles and Applications*, London: Longman, pp. 403-12.

Densham, P. J. and M. F. Goodchild (1989), Spatial decision support systems: A research agenda, in *Proceedings GIS/LIS'89*, Orlando, FL., pp. 707-76.

Eastman, J. R., P. A. K. Kyem, J. Toledano and W. Jin (1993), *GIS and Decision Making*, UNITAR, Geneva.

Ferguson, R. L. and C. H. Jones (1969), "A Computer Aided Decision System", *Management Science*, pp. B550-B561.

Gerrity, T. P., Jr. (1971), The design of man-machine decision systems, *Sloan Management Review*, Vol. 12, No. 2, pp. 59-75.

Grace, B. F. (1977), Training Users of a Decision Support System, *IBM Research Report RJ1790*, IBM Thomas J. Watson Research Laboratory, May 31, 1976.

Gray, P. (1981), The SMU decision room project, *Transactions of the Ist International Conference on Decision Support Systems*, Atlanta, pp. 122-9.

Gray, P.(1987), Guide to IFPS (Interactive Financial Planning System), McGraw-Hill Book Company, New York, 1987.

Houdeshel, G. and H. Watson (1987), The management information and decision support (MIDS) system at Lockheed-Georgia", *MIS Quarterly*, Vol. 11, No. 1.

Huber, G. P. (1982), Group decision support systems as aids in the use of structured group management techniques, *Transactions of the Second International Conference on Decision Support Systems*; pp. 96-103.

Goodchild, M. and Densham, P. (1993), *Initiative 6: Spatial Decision Support Systems (1990-1992)*, National Center for Geographic Information and Analysis, Santa Barbara.

Gorry, A. and Scott-Morton, M. (1971), A Framework for Information Systems, *Sloan Management Review*, Vol. 13, Fall 1971, 56-79.

Grace, B. F. (1977), Training Users of a prototype DSS, *Data Base*, 8, 3, 30-36.

Little, J. D. C. (1971) Models and managers: the concept of a decision calculus, *Management Science*, Vol. 16, No. 8, pp. 466-85.

Keen, P. G. W. and M. S. Scott-Morton (1978), *Decision Support Systems: An Organizational Perspective*, Addison-Wesley, Reading, MA.

Heuvel, T. van, Katwijk, V. F. van, Triest, H. van, and Goedhart, K. (1993), *Shoreline DSS (Shoman) (software and manual)*, The Netherlands: National Institute for Coastal and Marine Management.

Inmon, W. H. (1990), *Using Oracle to Build Decision Support Systems*, QED Press.

Jans, L., Platteeuw, M. Tosserams, M., and Schiereck, M. (2000), *ECOPEIL decision support system (software and manual)*, The Netherlands: Institute for Inland Water Management and Wastewater Treatment

Jankowski, P., T. L. Nyerges, A. Smith, T. J. Moore and E. Horvath (1997), Spatial group choice: a SDSS tool for collaborative spatial decision-making, *International Journal of Geographical Information Systems*, Vol. 11(6), pp. 566-602.

Keen, P. G. W. and M. S. Scott Morton (1978), *Decision Support Systems: An Organizational Perspective*, Addison-Wesley, Inc., Reading, MA.

Kimball, R., W. Thornthwaite, L. Reeves and M. Ross (1998), *The Data Warehouse Lifecycle Toolkit*, John Wiley and Sons, New York.

Klein, M. and L. B. Methlie (1995), *Knowledge-based Decision Support Systems with Applications in Business*, John Wiley and Sons, Chichester.

Little, J. D. C. (1971), Models and managers: The concept of a decision calculus". *Management Science*, Vol. 16, No. 8, pp. B466-85.

Malczewski, J and W. Ogryczak (1996), The multiple criteria location problem: 2, Preference-based techniques and interactive decision support, *Environment and Planning A*, Vol. 28(1); pp. 69-98.

Mennecke, B. E., Crossland, M. D. and Killingsworth, B. L. (2000), Is a map more than a picture? The role of SDSS technology, subject characteristics, and problem complexity on map reading and problem solving, *MIS Quarterly*, 24, 4, 601-4, 625-9.

Muller, J.C. (1993), Latest developments in GIS/LIS, *International Journal of Geographical Information Systems*, Vol. 7, No. 4, pp. 293-303.

McCosh, A. M and M.S. Scott Morton (1978), *Management Decision Support Systems,* Macmillan, London.

Nagy, G. and Wagle, S. (1979), Geographic data processing, *Computing Surveys*, Vol. 11, No. 2, pp. 139-81.

NCGIA (1990), Research initiative 6: Spatial decision support systems, *NCGIA Technical Paper 90-5*. UC, Santa Barbara.

NCGIA (1995), Report from the specialist meeting on collaborative spatial decision making, *Initiative* 17, National Center for Geographic Information Analysis, UC Santa Barbara.

Powell, R. (2001), DM review: A 10 year journey, *DM Review*, URL www.dmreview.com.

Power, D. J. (2000), *A Brief History of Spreadsheets*, DSSResources.COM, World Wide Web, http://dssresources.com/history/sshistory.html, version 3.3, 03/11/2000.

Power, D. J. (2000), Web-based and model-driven decision support systems: Concepts and issues, *Proceedings of the 2000 Americas Conference on Information Systems*, Long Beach, California, 10-13 August.

Power, D.J. and G. L. Rose (1976), Improving decision-making behaviour using the Hewlett Packard 2000/ Access System. *Proceedings of the American Institute for Decision Sciences*, pp. 47-49.

Rockart, J. F. (1979), Chief executives define their own data needs, *Harvard Business Review*, Vol. 67, No. 2 (March-April), pp 81-93.

Ruijgrok, E. C. M., M.B. de Vries, M.J. Baptist, F.V.D. Meulen, and W.J. Zaadnoordijk, (1999), *Nature Development and Valuation Decision Support System (software and manual)*, The Netherlands: WL/ Delft Hydraulics.

Simon, H. A. (1960), *The New Science of Management Decision,* New York: Harper and Row.

Sprague, R. (1980) A Framework for the development of Decision Support Systems, *MIS Quarterly*, Vol. 4, No. 1.

Sprague, R.H. (1989), A framework for the development of decision support systems. In: Sprague, R.H., Watson, H.J. (eds.), *Decision Support Systems: Putting Theory Into Practice*, Prentice-Hall, London, pp. 9-35.

Sprague Jr., R.H. and J.E.D. Carlson (1982), *Building Effective Decision Support Systems*. Prentice-Hall, New Jersey.

Scott Morton, M. S. (1967), Computer-Driven Visual Display Devices—Their Impact on the Management Decision-Making Process, Doctoral Dissertation, Harvard Business School, 1967.

Scott Morton, M. S. and J. A. Stephens (1968), *The impact of interactive visual display systems on the management planning process,* IFIP Congress, pp. 1178-84.

Scott Morton, M. S. and A. M. McCosh (1968), Terminal costing for better decisions, *Harvard Business Review*, Vol. 46, No. 3.

Scott Morton, M. S. (1971), *Management Decision Systems : Computer-based support for Decision Making*. Boston, Division of Research, Graduate School of Business Administration, Harvard University.

Sharda, R., S. Barr and J. McDonnell (1988), Decision support systems effectiveness: A review and an empirical test, *Management Science*, Vol. 34, No. 2, pp. 139-59.

Sprague, R, H., Jr. (1980), A framework for the development of decision support systems, *Management Information Systems Quarterly*, Vol. 4, No. 4, pp. 1-26.

Sprague, R. H., Jr. and E. D. Carlson (1982), *Building Effective Decision Support Systems*. Prentice-Hall, Inc., Englewood Cliffs, N.J.

Sprague, R. H., Jr. and H. J. Watson (1979), Bit by bit: Toward decision support systems, *California Management Review*, Vol. XXII, No. 1, pp. 60-8.

Sprague Jr. R. H. and H. J. Watson (1996), *Decision Support for Management*, Upper Saddle River, N.J.: Prentice Hall.

Turoff, M. and S. R. Hiltz (1982), Computer support for group versus individual decisions, *IEEE Trans. Communications*, COM-30, Vol. 1, pp. 82-90.

systems. It causes damage to lives, natural resources and environment as well as the loss of economy and health. The impact of floods has been increased due to a number of factors, with rising sea levels and increased development on flood plain. However, information is required for the development of warning system

With the increase of population and industrial growth, more and more settlements and development are encroaching in flood plains. Hence, floods in Bangladesh have been causing the most severe economic damage among all kinds of natural disasters around the world. In order to formulate flood management and control policies, a thorough understanding of flood problems and their solution is required. This can be done through data collection and analysis along with flood impact assessment. A study was attempted to develop an integrated approach of a new model for warning of floods rather than flood mapping for flood risk assessment in flood prone countries for regular monitoring of damages.

Hydrodynamic model MIKE 11-FF (NAM, HD and FF) was successfully integrated with GIS in the Arcview GIS environment. MIKE 11 FF has been designed to predict the variation in discharges and water levels in a river system as a result of catchments rainfall and inflow and outflow through boundaries in the river system. GIS has been used intensively to create watershed models from digital elevation (DEM) data to trace flow paths to get a complete surface for identifying the actual flood flow, flood affected properties, agricultural areas, population and infrastructure.

Flood inundation maps were created by overlaying land use features and infrastructures to delineate flooded areas using dynamic spatial modelling. GIS was used for the analysis to determine hazard zones in the maps, which serve as risk zone identifiers for the general population. This information leads to locate safer location to evacuate the affected population utilizing the optimum route. Complete statistics will also be released for the predicted time stamps in advance by 72 hours, 48 hours and 24 hours. Finally, the information of population, land use and infrastructure are likely to be affected was to be made available in a user-friendly interface for timely decision for flood preparedness. Mostly, the northern part of the study area is always in the flood affected region. Flood inundation maps were prepared using simulation results. As the simulation results at every hour for the next 72 hours are available, so flood inundation maps can be prepared at any of those times using the forecasting water levels. Water levels in the flood phase maps (F0-F4) were classified into five classes—dark blue showing higher water depth.

There is considerable crop damage almost every year of moderate flood. Overlaying land use map over the flood map, the area of affected crops was found out. The area was then reclassified according to the types of crops. Possible percentage of affected crops in the next 72 hours was worked out.

Information of predicted populations and infrastructures affected were also established using overlay operation. Those maps indicated the accessible route to evacuate people during emergency. It is important for the warning system, to develop and find out the efficient route to provide direction to the vulnerable people to move out during emergency. Before doing this, the need was to come up with a plan that will enable to choose the best possible way to minimize travel time to reach to the destination in an effective manner. Arcview Network Analyst helped in using networks more efficiently. It assists in finding out the quickest way to get the safest location and to determine the best route to reach. Overlaying this information with population data, it is easy to understand the situation, from where most of the people getting service and where to be provided for additional facilities.

Warning is crucial in emergency management. Adequate flood warning is a crucial element in emergency management, and operations of all of the systems around flood forecasting that actually deals with. Being able to give information on what's going on, and provide an adequate warning time to all of the public and emergency responders is necessary. The aim was to develop a user friendly system to support flood warning in advance at 72 hrs, 48hrs and 24hrs. Maps of the flood affected land use features, such as roads; agricultural lands, population and households can be released for above warning time stamps. In addition, complete statistics can also be available about the affected region. The information of population, land use, infrastructure etc to be affected will be available in term of user-friendly queries for timely decision for flood preparedness.

Perhaps floods cannot be avoided, but by implementing effective flood prevention schemes, one can reduce damages from severity, if sufficient information for flood forecasting is given timely. Dynamic use of GIS integrated with hydrodynamic model provides useful measures towards disaster preparedness and planning. This dynamic model can be automated fully to be accomplished with the existing flood warning system to provide warning for the people without the knowledge of GIS to identify the possible inundated areas to take initiative to evacuate people to the safe places in time. It can help to promote public awareness in disaster management activities as a part of focusing the dissemination of forecast at the grassroot level. Therefore, new flood control and management schemes based on dynamic model can be implemented for important disaster management aspects like prevention and mitigation, preparedness, response and recovery and also planning for operational activities, immediately before, during and after the floods.

(http://www.gisdevelopment.net/application/natural_hazards/floods and http://www.dhigulf.com/upload/developmentofafloodforecastingsysteminbangladesh.pdf)

In developed countries the problems are also similar, but with the help of technology, management of flood hazards is better. In UK, one of the wettest years on record, Simmons Aerofilms acquired vertical aerial photography of some of the worst hit areas, including the River Medway in Kent, which burst its banks at Yalding. On its own, such photography is a valuable resource for many applications, especially those concerned with hydrological monitoring. Its value is further enhanced when orthorectified and combined with related geographic data sets in a GIS.

(http://www.simmonsaerofilms.com/links/index and http://www.simmonsaerofilms.com/links/index).

GIS Needs for City

Calgary is situated on the eastern slopes of the Rocky Mountains in the southwestern portion of the province of Alberta, Canada. Calgary has a population of over 840,000 encompassing

280 square miles. It is the sixth largest city in Canada with the second highest number of head offices located within a city. The populace is well educated with over 58 per cent having a post secondary education and 61 per cent having internet in their homes (the highest total in Canada). It is a growing, vibrant city with over 9,000 new housing starts and 15,056 building permits totaling $2 billion issued in 1999.

The concept of a unified city with a population approaching one million, the above average technical knowledge and growth averaging 25,000 per year has all contributed to the necessity for The city of Calgary to take a second look at the value of a GIS to meet its business needs.

The city started to investigate the feasibility of using GIS to meet its business needs in 1989 and began preparing an RFP in Nov. 1990. A consultant was hired in the spring of 1990 to determine the city's spatial data requirements and they concluded that the city was spending $28 million per year processing location-related data. They sold the concept of a GIS on the basis of small gains (1 to 2%) of the $28 million being spent annually. This led to the Calgary Spatial Information Management (CalSIM) project under which the GIS project fell. The CalSIM project lasted from 1991–1996 with a $15.6 million budget. The web (intranet) application developed was based on a web delivery strategy that;

- Applied the ESRI out-of-the-box functionality
- Deployed increasing functionality in each phase
- Added more data as it became available
- Supported reusable functionality for other similar data
- Had a short development time.

The intranet web application was delivered in five phases :

- Phase 1 : Lists of maps to select and display,
- Phase 2 : Search by address, parcel ID and name,
- Phase 3 : List of addresses within distance and mailing lists,
- Phase 4 : Parcel info based on complex queries
- Phase 5 : Thematic maps and statistical information.

The *Quick Hit* applications developed were based on a strategy that :

- Built upon the GIS out-of-the-box functionality (ArcView applications)
- Used replicated city data as soon as it was available
- That supported a wide corporate interest based on the survey
- Had reusable functions for other similar tacks
- Had a short development time.

The *web application* and *quick hit* applications using the above strategies demonstrated to the corporation that GIS could and should be considered as part of its business strategies. It was learnt during this process that the time estimates were suspect. As it was further learnt, the ESRI tool and gained insight into its functionality, these estimates needed to be revised. When the three-year business plan was presented to the senior management, every effort was made to ensure that they understood this and as it turned out they were very supportive.

The city was fortunate our spatial data was stored in Oracle and that it did not have to deal with a number of legacy systems, including no data maintained from an ESRI environment. The city's roll out strategy therefore focused on the use of ArcView and its extensions to access and query of spatial data and display the results. Replication of spatial data became an immediate strategy with the understanding that future maintenance of spatial data would take advantage of the ArcInfo 8 technology.

Maintaining spatial data using ArcInfo 7 technology was not promoted and has been limited to only a few special circumstances. The ability to replicate data has gone smoothly because of good source data and the excellent functionality of the ESRI tools. This provided the city an opportunity to leave maintenance in its current environment (MicroStation). This is important to the users as it provides them the freedom to learn the new tool and determine where their spatial data maintenance should be performed (it is believed in a number of the cases they will choose ArcInfo 8 as their maintenance environment).

An organization (GIS Centre of Excellence) to support and sustain GIS technology in the corporation is now in place. The roles for the business unit are to coordinate and facilitate GIS integration into the business units (a) to ensure GIS investments are managed, (b) to coordinate and manage data standards and data sharing, and (c) to develop strategic policies and assume a leadership role in the GIS technology within the corporation.

(www.esri.com)

Identify Additional Customers for a New Service

British Telecommunications (BT), owner and operator of the United Kingdom's dominant telephone network, sells local, long-distance, international, wireless and internet services as well as equipment. Competitive pressures in its markets continue to push BT to develop new products and services that satisfy customer's communications needs and to compete effectively with available alternatives. BT recently launched in the United Kingdom a new Asymmetric Digital Subscriber Line (ADSL) service that allows ordinary home and business phone lines to be converted into high-bandwidth digital links. ADSL represents an important opportunity for BT to compete head-on with the cable companies and other high-speed data service providers for a share of the growing market for broadband services. The problems were :

- BT wanted to find a way to profile and pinpoint the "ideal" high-bandwidth customer
- BT needed to prioritize ADSL deployment in locations with greatest potential demand
- BT also needed a tool to perform cost/benefit analysis essential for a cost-effective nationwide roll out of the service

GIS technology of mapping, data visualization and analysis software combined with data of postal, demographic and road has produce a powerful ADSL market analysis and network planning tool for BT. The benefits were :

- BT has been able to pinpoint more than 400 high potential exchanges covering six million households and businesses for its initial ADSL roll out
- The system has enabled a ranking of all exchanges offices for ADSL upgrade
- System is helping BT plan the most cost-effective nationwide ADSL roll out

GIS technology enabled BT to plan an extremely focused and highly targeted ADSL launch for London and surrounding areas like Cambridge, Coventry and Manchester, where demographic factors most closely correspond with the profile of ADSL subscribers and significant potential for high bandwidth services has been identified. Deployment of capability into the appropriate areas to maximize the early adoption of new high bandwidth services into the UK is a critical factor in the business planning for implementing ADSL on the national network. With such an enormous programme to be undertaken over a period of time, profiling and pinpointing ideal customers with GIS is helping BT achieve the most effective roll out of ADSL. This ADSL market analysis and network planning solution is the latest addition to BT's portfolio of GIS-based applications. The company is one of the largest users of GIS technology in the world with more than 1,000 software licenses for a wide variety of key applications in network planning, operations and sales and marketing.

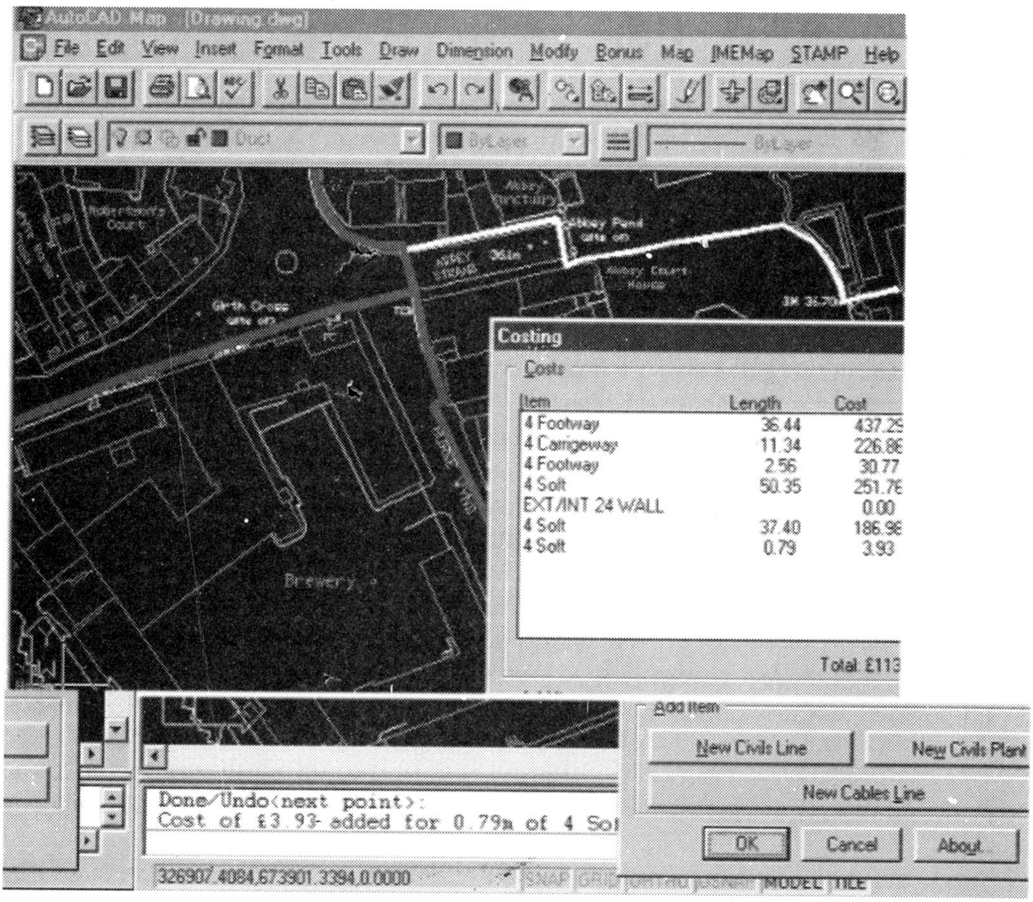

Fig. 7.5: Print screen of BT's web site

Further, in order to improve customer service, BT retail equipped 17,000 field service engineers with a comprehensive MapInfo location-based solution. Aimed at reducing time spent locating customer premises and pinpointing faults, the solution will result in a predicted savings of $33.4 million over the next five years. When BT's field engineers relied on paper-based maps and diagrams, they lost valuable time searching for information needed to respond to customers' needs. Then, BT replaced field terminals with Panasonic laptops for the engineering work force, and built a MapInfo solution enabling sharing of maps and diagrams. The solution, built using MapInfo MapX® and incorporating MapInfo®, MapMarker® and MapInfo®. DriveTime TM provides simple yet tremendously beneficial access to fundamental information. MapInfo MapX, a mapping engine, allows for highly visual display of location-based data. MapMarker enables engineers to locate the customer on a detailed map. DriveTime delivers the quickest route to the customer or fault.

With the solution, engineers first locate a property and then progressively zoom in to view road networks and local plant and network details; the location of underground ducts; and, finally the individual cables within them. Engineers can see the proximity of the network to the customer, search by customer's address and postal code and plot the shortest route to get there, all saving further time and effort. "With this new technology, we knew that we could provide engineers with accurate information on where the customer was, as well as providing detailed maps of the ducts and cables supplying our business and residential customers," said BT's programme manager for customer service. The solution provides engineers with access to extensive plant diagram electronic records—held on 32 databases throughout the United Kingdom—along with local and national road maps for each engineer's territory.

(www.gis.esri.com/library/usrconf/telecomos/papers/enterprise-gis.pdf).

GIS and Railway

Belgium's TUC RAIL company was formed with one aim in mind: to design and project manage the country's ambitious and economically-important Thalys high-speed (TGV) rail link to France, Holland and Germany. The company (a venture between SNCB, the national rail carrier, the Tractbel engineering concern and others) employed software from Star Informatic from the outset for technical design and cartographic purposes. The country's dense urban environment presents a particular challenge, with licenses needed at both local and regional level before construction can proceed. The ability to outline proposals in a clear and concise manner is vital in obtaining the necessary permissions.

To aid in this process, TUC Rail employed STAR software to generate DTMs for the two routes. These models were populated with buildings, trees, roads and other detail and rendered as full 3-D images, examples of which are pictured. This imagery gave politicians, officials and the public a clear idea of the impact of the lines in each locality and helped speed the process of obtaining licenses. In the Republic of Ireland, public acceptance was critical to proposals that sought to upgrade the N8 (Dublin to Cork) primary road to dual carriageway standard.

A joint venture company (Parkman, Carl Bro, Punch, or PCP) was commissioned by Cork County Council to undertake the project, following which its specialist subsidiary, Parkman LandAspects, went to work preparing public consultation material. A key element in the package was a series of 3-D virtual worlds, created in Erdas Imagine Virtual GIS v8.5, that allowed 'fly throughs' of the various route options. Primarily used as a viewing and awareness tool, 3-D visualization allowed members of the public to assess the level of impact on their own localities.

It also helped Parkman LandAspects respond to individual queries and concerns over the effects of the scheme. A detailed virtual world was generated using orthorectified aerial photography covering 211 sq km (commissioned from Simmons Aerofilms) and 3D DXF models of the proposed road designed using Infrasoft's Mx software and AutoCAD. A 1.5m resolution pixel-based Digital Elevation Model (DEM) was created that combined both existing and proposed elevation data. This yielded an 'as built' terrain model over which, high-resolution (1,000 dpi) orthorectified photography was draped.

This basic virtual world was enhanced by incorporating extruded vector layers (to represent buildings) and tree images (to replicate woodland), before adding 3D DXF models of the proposed road scheme. To these were added AutoCAD-generated bridge structures, street lights, traffic signs etc. Several flight paths were recorded as movies (MPEG and AVI format) and these, together with the Erdas software, proved invaluable in supporting 'live' presentations during the public consultation exercise (www.autodesk.com).

GIS in Armed Forces

GIS in an Armed Forces organisation is not a matter of choice of a certain software product or a matter or a choice of a specific platform, but is a systematic way of working with a number of new tools to handle the geographic information as one of the basic factors in the assessment and decision process in a headquarters (HQ), and it does not matter if one talks about a battalion HQ or an armed forces HQ. This implies that the use of GIS, necessarily, will affect both HQ organisations and their way of working. This awareness has grown in the Swedish Armed Forces (SAF) since the end of the 80s. This has also led in many way of trying to consider all aspects on the development and implementation of GIS which SAF wants to try out.

In 1989, there were more brigades in the Army, more combat vessels in the Navy, more fighter squadrons in the Air Force than today. Sweden had a rather static planning resulting in training of their units again and again in the same geographic area (the area in case of a war) and through this training the security forces got an extensive geographic knowledge—about the area where they were supposed to execute combat, but nowhere else. At the same time digital geographic data began to be available and the software packages as well to reasonable prices. A number of computer enthusiasts started to develop simple GIS applications.

In 1990 the Armed Forces and the National Land Survey signed an agreement concerning a shared GIS-centre, with people from the Land Survey organisation as well as with, initially, one officer from the Armed Forces. At the same time they tried to get an idea about what was going on in several places and they tried to get all these development activities to face a common objective. Further, in 1991 they had our first tests on GIS-applications during a divisional exercise in the northern part of Sweden. They also had the first agreements with the civilian producers of digital geographic data about availability for the Armed Forces. In order to establish some sort of umbrella organisations for the GIS efforts, the Armed Forces GIS Steering Committee was established with following mandate :

- To give entitlement to start new development activities
- To co-ordinate GIS development efforts
- To allocate the resources at the GIS-centre

- To allocate other resources for GIS development as SW, data sets, test facilities etc.
- To encourage the initiation of new GIS efforts

In 1993 they started developing the present GIS strategy which was signed by our Supreme Commander in 1995. Thus it has continued, more and more rapidly.

One of the most important tools for establishing a consistent GIS development and implementation in a large organisation, independent of which type of organisation it is, is the presence of a strong GIS strategy. It is not enough to have the strategy developed, signed, distributed and read. A strategy always consists of a number of outer frames for what is accepted and what is not accepted. This is often considered limitations for the developer but is absolutely necessary in order to achieve the desired co-ordination with the overall context for a C3I-system. If that strategy, with its limitations, is to be complied with, it has to offer something very beneficial for the individual who has been given the task to develop a GIS application. The GIS-strategy offers :

- GIS consulting resources
- geographic data
- software recommendations
- hardware recommendations
- development equipment
- application test facilities
- project manager training

Further, in addition to above, it offers a couple of other facilities without costs for the responsible project manager or his commanding officer. These assets are funded centrally in the Armed Forces. In this way the strategy will become close to loved.

One core issue in the strategy is the mandatory development process, with a limited local prototype phase initially, where the main responsibility is with the project manager, who always has to be an end user of the application under development. And this emphasis on the end user will be a central part of the entire development process until the implementation of the application. After the local prototype phase, where real trial and error is allowed we have a coordinated project phase aiming to the implementation. During this phase attempts can be made to widen the project organisation, involve reference groups and external test groups and start preparing the training of future users. Another core issue is the careful separation of common GIS tools as (a) projection handling, (b) positioning, and (c) best map functions etc. which are required by every user, and also for activity related applications which are requested by representatives for a specific branch, service or function. Such activity related applications can be (a) artillery sites, (b) radio link communication establishment, (c) bridging and (d) low-flying paths for attack helicopters and so on.

The responsibility for development of those common functions is a matter for a central project organisation. On the other hand is the responsibility for the development of activity-related application for a particular issue for those who are in the middle of a specific activity, with both feet in the mud. Those persons are the only individuals able to describe the specific activity careful and systematic enough to become subject to an activity-modelling process subsequently followed by a data modelling process. And here it is vital to be extremely careful. Otherwise one can not get the desired results.

(www.gis.esri.com/library/usrconf/europroc99/proceed/papers/paper904/p904.html).

False Alarms

Linking GIS capabilities with an automated software system such as CryWolf for the tracking and processing of false alarm calls has resulted in a more efficient use of time and resources and a reduction in false alarms. The Charles county, Maryland (111,050), False Alarm Reduction Unit (FARU), in collaboration with Public Safety Corporation (PSC), has served as the initial site for the operation and evaluation of the CryWolf software. Alarm data is imported from computer-aided dispatch and used to send letters and prepare reports. Data and statistics can also be graphically presented on local county maps using MapObjects programming. Additionally, graphs and charts can be prepared and exported to other programs for use in reports to management.

The number of municipalities adopting alarm legislation is rapidly increasing due to the growing concern over police officer safety and the costs related to the reduction in manpower and resources due to false alarms. Unfounded alarm calls reduce law enforcement effectiveness and undermine their ability to respond to more critical calls. For residential and business property owners who are concerned about crime, alarm systems offer an added sense of safety and security. They can be an effective crime deterrent, but if they are unreliable and result in false alarms, they pose a major problem for everyone.

Due to the requirements for new legislation and the resulting delays in trying to create a quick fix in-house, there is a growing need for a sophisticated software system to assist in the tracking and processing of alarms. Public Safety Corporation (PSC) has created a GIS-based application that brings the power of geo-spatial analysis to the false alarm arena. Utilizing ESRI's MapObjects v2.0 allows CryWolf to be translated into a comprehensive programme planned, designed, and created specifically to track false alarms. Unlike current software solutions, CryWolf is a very user-friendly system that maps alarm locations, allowing for the visual interpretation of alarm patterns. Agencies can customize multiple programme features, such as clearance codes, letter formatting, and custom letterhead design and utilize a variety of data collection modules for alarm companies, alarm registrations and alarm call entry. CryWolf is designed to use enhanced TIGER data 'shape' files for the current jurisdiction, and offers the mapping capabilities that geo-spatial users expect. Truly customized software for tracking false alarms can cost up to several hundred thousands dollars. The more elaborate the requirements are and the more customization requested by the user, the more costly the software will become. CryWolf is designed as a generic false alarm application with a variety of customization features. This allows the software to accommodate the needs of many jurisdictions without having to resort to custom code.

Linking GIS capabilities with an automated software system such as CryWolf™ for the tracking and processing of false alarm calls has resulted in a more efficient use of time and resources and a reduction in false alarms. During the initial testing and verification phase, almost 7,000 false alarm calls have been processed through the programme. Test notifications, and invoices have been mailed to hundreds of county residents to date. The receipt of these letters has already impacted the county by accelerating the number of registrations being received each day. Director of FARU remarked that with the CryWolf application we have been able to accomplish more in the past four months than what we were able to do in the preceding sixteen months!

The reduction in false alarms is essential to address the rising costs to tax payers, raise additional funds to be allocated towards community education and crime prevention, increase response time and presence in the community and free up personnel from manual tasks by providing the billing mechanism to generate fees, which will more than pay for the initial investment in the CryWolf application. Furthermore, alarm users will gain a better understanding of the impact of repeated false alarms, with more alarm businesses and owners inspecting and updating their systems to prevent further false alarms. (Ref : Using Maps to Track the Incidence of False Alarms Track: Law Enforcement and Criminal Justice by Edward Chicca, Denise Shaw; also http://gis.esri/library/userconf).

Summary

The above examples do indicate the potentialities to apply GIS in different spheres of law enforcement, criminal justice, environmental and resources management, defence and public access . But these case studies are generally from western world where the governments and welfare departments have sufficient funds to use GIS related technologies The conditions in the developing countries are entirely different. The day-to-day problems have over taken the long term preparations such as development of GIS data base, training of personnel, and creation of GIS infrastructure. Nevertheless, 9/11 types of emergency, terrorism and high-tech crimes have compelled the governments all over to give due attention to GIS. The decisions and actions are now taking into consideration of the information generated from spatial technologies such as air photos, GIS, remote sensing and GPS.

8

Spatial Data Infrastructures

In last few years a huge volume of spatial information has been available, stored and managed in the form of GIS for various applications which has opened a new vista for business opportunities. International Geo-sphere Biosphere research, digital earth, digital regions and several global and national projects related to digital information have given a boost to a new type of applications. These applications or projects are based on geo-spatial data and knowledge about such databases. The management of such data, information or knowledge depends on efficient and user-friendly GIS. The GIS technology now permits us to collect basic common sense knowledge linked with geo-spatial data, and also allows this knowledge to be closely associated with such data. The association of knowledge with spatial data makes the option of having feature-oriented objects in the GIS environment. The new applications look towards the database which can produce intelligent levels. Such a combination of data and knowledge has opened types of applications not only in the developed world but also in developing world. Hence, there has been a greater urgency to understand the spatial or geospatial data infrastructures and their various facets.

Geospatial data Infrastructure plays a crucial role in the development of the economy and promotion of industrialization at national and regional level. While being an expedient for economy, infrastructural support a highly complex and heterogeneous life system, having varied expectations and accountability depending upon its socio-economic structure. Infrastructure therefore, cannot be merely visualized as physical infrastructure. These have to be planned as a life support system for the man as well as his environment with milestones indicating infringement limits and related course correction. The Spatial Data Infrastructure (SDI) concept continues to evolve as it becomes a core infrastructure supporting economic development, environmental management and social stability in developed and developing countries alike (Williamson, Rajabifard and Feeney, 2000 and 2003). Spatial data has been considered as an essential perquisite for any GIS initiative (*Geospatial Today,* 2004). SDI hierarchy can be organized hierarchically starting from a local to global data infrastructure (Fig. 8.1).

We live in information age which has become essential to solve problems of national development. Further, spatial information has proved to be an importance element underpinning decision-making for many disciplines because it deals with location. In the sophisticated digital environment, it has expanded to include databases, satellite positioning, communication networks including internet, web technology and wireless applications. It draws experiences from the geographic information systems, computer science, land administration, geography, surveying and mapping, legal and public administration disciplines. SDI initiatives in different

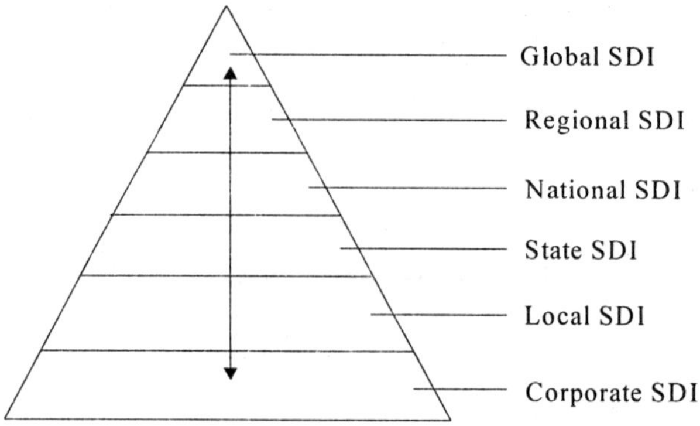

Fig. 8.1: SDI Hierarchy, After Williamson, Rajabifard and Feeney (2003)

countries have a common objective, i.e. to create an environment or framework where all stakeholders can cop-operate and interact with each other with relevant technology in order to meet objectives at different political and administrative levels. SDIs are meant to share data by avoiding duplication linked with generation, maintenance and integration of data and development of innovative applications related to business and national development. It appears that spatial information industry has evolved and will continue to grow. SDIs can become instrumental in having higher savings of all sorts. The expected structure of SDI is shown in Fig. 8.2.

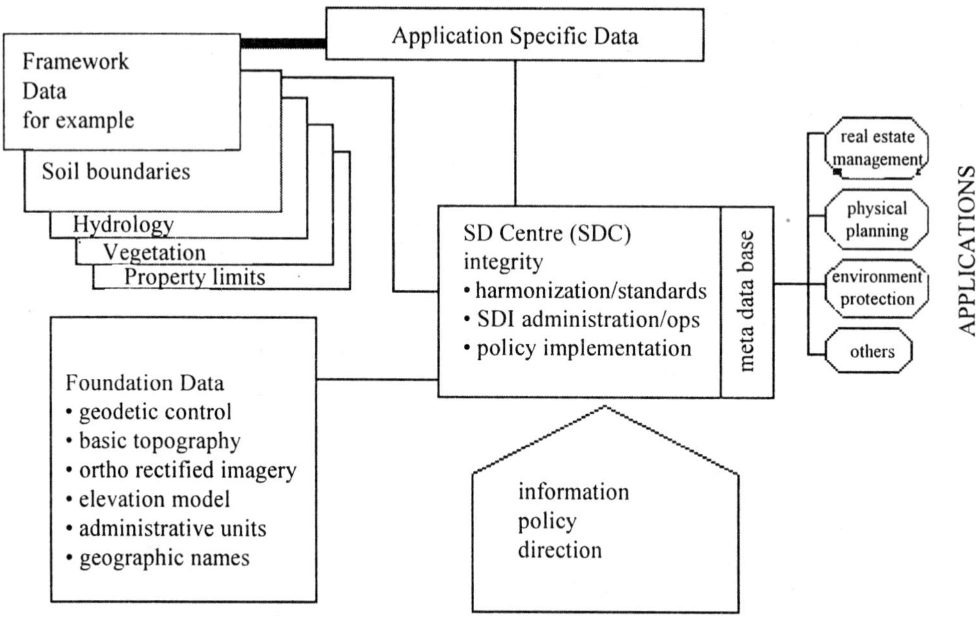

Fig. 8.2: Structure of SDI

At the global and international levels, SDI initiatives have been taken through different forums. The most important are being the Global Spatial Data Infrastructures (GSDI) conferences. UN sponsored Permanent Committee on GIS Infrastructures for Asia and the Pacific (PCGIAP), Permanent Committee for Spatial Data Infrastructure for Africa, Digital Earth Conferences, United Nations Committee on Development Information (CODI), International Cartographic Association, International Steering Committee for Global Mapping, International Society of Photogrammetry and Remote Sensing and the like are contributing to the understanding of the SDI concept. At the national level, the Executive order from the President of the United States Office of 1994 was a great boost. On one hand, complimentary initiatives followed in the US; while on the other, similar orders or initiatives were taken in other countries. The examples of the latter are Australia, Germany and Canada.

The Indian Situation

Technological advancements in the fields of computers, satellite sensors, global positioning systems, geographical information systems and digital photogrammetry combined with the increased user aspirations in a way to obtain, analyse and apply the spatial data, have forced even the developing nations to think of evolving mechanisms to make both spatial and non-spatial data accessible to users to a set standards. India has possibly the longest known tradition of systematically collected scientific spatial data. Survey of India having been doing so for the past 235 years. Survey of India (SOI) being the leader with 235 years of collection of topographical data followed by Indian Meteorological Department (IMD) and Geological Survey of India (GSI) each over 150 years. Similarly, other Departments such as Central Ground Water Board (CGWB), Forest Survey of India (FSI), National Bureau of Soil Survey and Land Use Planning (NBSSLUP) etc., have collected wealth of data over long periods of time (Fig. 8.3). In the past three decades, Department of Space (DOS) has mastered the space and

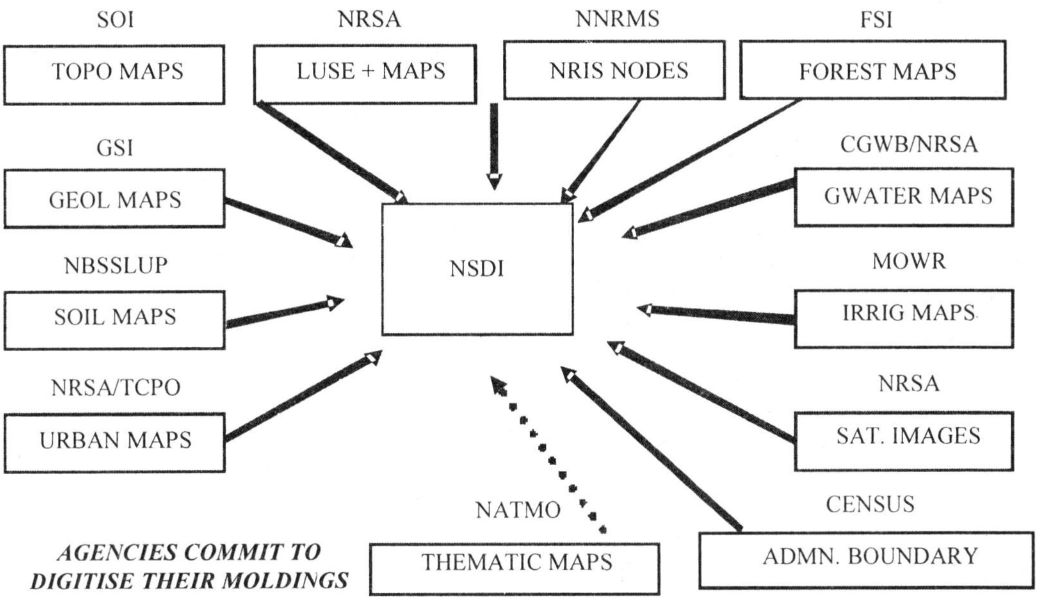

Fig. 8.3: Indian NSDI

produced a variety of data from various satellites. However, this data could be utilized for limited developmental activities. The basic reasons could be classified under three categories viz. (a) technical, (b) institutional and attitudinal, and (c) policy issues. Technical problems are mainly platform oriented. Fortunately, in India, most of the spatial data is based on Survey of India topographic maps on scale 1:50,000 or 1:250,000. However, the analogue data when converted into digital format is obtained in the machine formats of the platform on which it is generated. In spite of the best conversion software available, there is loss of data when converted from one format to another. SOI has taken the lead in preparing a national standard exchange format for Digital Vector Data (DVD). However, it does not cater to all requirements.

The institutions who have generated the data are highly possessive of the data generated and do not want to part with that. Even if they are willing to share, in the absence of metadata, it has become really difficult to know what is available. Lastly, the present policies of the government do not permit digital data of topographical nature being utilized freely and there is a specific ban on putting the data on network.

Considering these issues, a Task Force was constituted by Department of Science and Technology (DST) in November 2000 with Surveyor General of India as the Chairman. Members were drawn from various departments, organisations, users and NGOs providing spatial data. The Task Force has met several times and has set up Working Groups for metadata, data exchange formats, network protocols, communication strategy etc. Three national/international Workshops were held and the vision of NSDI was decided. Meanwhile, Department of Science and Technology obtained all approvals for bringing out a second series of maps which could be made available without any restrictions. The task force has brought all the data providers to a common platform and to think alike. Thus, there is a better understanding of each other view points and a consensus has emerged to evolve a National Spatial Data Infrastructure (NSDI). The data providers are engaged in the task of conversion of analogue data into digital domain and also to generate the metadata. The metadata server is ready and will be launched soon.

What is NSDI ?

India, has over the past years, produced a rich "base" of map information through systematic topographic surveys, geological surveys, soil surveys, cadastral surveys, various natural resources inventory programmes and the use of the remote sensing images. Further, with the availability of precision, high-resolution satellite images, data enabling the organisation of GIS, combined with the Global Positioning System (GPS), the accuracy and information content of these spatial datasets or maps is extremely high.

India has a National Spatial Data Infrastructure (NSDI) strategy and the *NSDI Strategy and Action Plan* report has been adopted in 2001, through the coordinated efforts of the NSDI Task Force. The NSDI will be an important element for supporting economic and sustainable growth in the country. With the recognition that spatial information is a national resource and citizens, society, private enterprise and government have a right to access it, appropriately. Only through common conventions and technical agreements, standards, metadata definitions, network and access protocols will it be easily possible for the NSDI to come into existence. While government must provide the lead, private enterprise, NGOs and academia have a major role to play in making the NSDI reality. NSDI will require for coming together of various *groups* and harmonizing their efforts in making this national endeavour a success. The NSDI model for India has been proposed (Fig. 8.4).

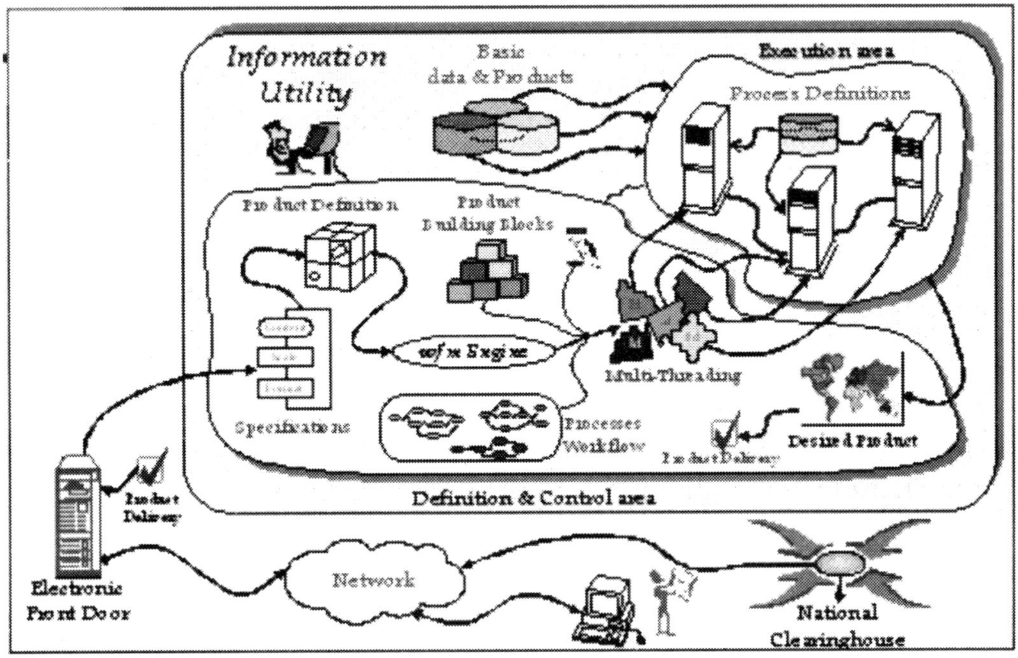

Fig. 8.4: Example of process model

Since the adoption of the NSDI Strategy in 2001, the NSDI Task Force has been working on developing a NSDI implementation mechanism through drafting an executive mechanism and also drafting a possible NSDI bill. The NSDI vision is:

> " National Infrastructure for the availability of and access to organized spatial data, use of the infrastructure at community, local, state, regional and national levels for sustained economic growth."

It has been decided that the National Spatial Data Infrastructure (NSDI) be established under a legislation that will set out essential operational features of the NSDI including its composition and administrative structure, decision-making powers, and the broad policy issues on which the NSDI will be entitled to make rules and regulations. The stated objective of the NSDI is the creation of a national infrastructure for the generation, availability of and access to organized spatial data to be used at community, local, state, regional and national levels to meet the requirements of spatial data for sustainable development. It is proposed that the NSDI be established as a statutory body, empowered to carry out various functions necessary in order to achieve the broad objectives stated above. The functions of this overarching framework will be as follows:

(a) Function as a repository for spatial data currently available with, as well as all future data created and maintained by, participating government agencies;

(b) Co-ordinate, control and manage the receipt of multiple sources of spatial data from participating government agencies and supervise the appropriate format conversion of this data to ensure uniformity of information and inter-operability of data;
(c) Develop the NSDI metadata of NSDI spatial information;
(d) Develop a physical infrastructure compromising a network of servers on which an intranet for the storage, exchange, control and access to NSDI data operates;
(e) Determine the requirement for particular categories of spatial data and require the creation or collection of spatial data, by any of the participating agencies, to fill such requirement;
(f) Co-ordinate and authorize the regular revision of existing spatial data in order to reflect changes in spatial features;
(g) Establish an appropriate standard (the NSDI standard) defining the content, scheme, design, process, quality, protocols and other standards necessary in order to ensure standardization of the data contained in the NSDI repository as well as appropriate formats for its exchange and interoperability;
(h) Function as a repository for spatial data meeting the NSDI standard, that have been generated by industry and private enterprise;
(i) Develop and co-ordinate appropriate quality control measures to ensure the integrity, standardization, accuracy and quality of NSDI data, including but not limited to the establishment of appropriate certification and accreditation procedures;
(j) Create mechanisms, procedures, tolls and applications (including appropriate search and access protocols and an NSDI user interface) to facilitate public access to NSDI spatial data;
(k) Develop common solutions for the discovery, access and use of spatial data in response to the needs of diverse user groups;
(l) Offer consultancy services on specialized subjects including but not limited to geodetic, geophysical, topographical surveys and digital database to government and non-government organisations;
(m) Co-ordinate multi-disciplinary mapping activities in collaboration with governmental and non-governmental organisations; generate a digital database of unrestricted Open Series Maps (OSM) as per National Map Policy; and
(n) Increase the awareness and understanding of the vision, concept and the NSDI.

In the event of establishing NSDI on the approval of the Government, the Union Department of Science and Technology will create the proposed body through the administrative act of appropriate order constituting the body. Similar action was taken by the US by Executive Order 12906, "Coordinating Geographic Data Acquisition and Access: The National Spatial Data Infrastructure" which was signed in April 1999. (http://www.fgdc.gov/fgdc.html).

Major Constraints for the Spatial Database

Many planning department and other national organizations such as Survey of India, National Remote Sensing Agency and National Atlas and Thematic Mapping Organisation have a long history of using topographical maps and remotely sensed imageries for preparing base maps

Ubbels, A. A., Verhallen, A. J. M. (2000), *Suitability of decision support tools for collaborative planning processes in water resource management*, Report No. 9.067, The Netherlands: Institute for Water Management and Wastewater Treatment.

Uljee, I., and Engelen, G. (2000), *Wadden Sea decision support system (software and manual)*, Research Institute for Knowledge Sytstems., Maastricht.

Wilson, R. D. (1994), GIS and decision support systems, *Journal of Systems Management*, Vol. 45, No. 11, pp. 36-40.

Web References

- Batty M. and P. J. Densham (1996), Decision support, GIS, and urban planning, http://www.geog.ucl.ac.uk/~pdensham/s_t_paper.html
- *Journal of Geographic Information and Decision Analysis* (JGIDA), http://www.geog.uwo.ca/gimda/journal/journal.htm
- Malczewsk, J. (1997) The internet resources for geoinformation-based decision analysis, http://www.geog.uwo.ca/gimda/intres.htm
- Larry, D. (1992), SDSS for location planning, or the seat of the pants is out, The Geographer's Craft Project, Department of Geography, University of Texas at Austin http://www.utexas.edu/depts/grg/gcraft/notes/gisapps/sdss.html
- NCGIA Initiative 6: Spatial decision support systems (SDSS), http://www.ncgia.ucsb.edu/research/initiatives.html#i6
- NCGIA Initiative 17: Collaborative spatial decision-Making, http://www.ncgia.ucsb.edu/research/i17/I-17_home.html
- Power, D.J. (1997) DSS research resources, http://dss.cba.uni.edu/DSShome.html
- Watkins, Jr. D. W. and D. C. McKinney (1996) Recent developments associated with decision support systems in water resources, http://earth.agu.org/revgeophys/watkin00/watkin00.html

7

GIS Applications

In today's competitive world, a successful utility must take maximum advantage of each of its considerable resources, from people to equipment to information. Integrating geographic and other corporate data to make maximum use of this valuable asset has become absolutely vital. Utilities require a spatially enabled technology infrastructure on which their operational, customer, and decision support systems can leverage the information they have about their assets, employees, customers, and their relationships. In addition to solutions that are scalable, supporting work for individuals, work groups, and the entire enterprise, utilities need access to their information from the office, the field, or anywhere the user may be.

Geography plays an important role in sustainable development and humanitarian affairs. Planning and implementation of complex development goals and humanitarian response efforts often require spatial analysis and map creation capabilities. Data for these efforts is often in a variety of formats (maps, databases, spreadsheets, images, and so forth) from multiple agencies and multiple sources. The ability to access and process information quickly while displaying it in a spatial and visual medium allows organizations to improve decision-making and promote better organizational integration and knowledge management. Geographic Information System (GIS) software uses geography and computer-generated maps as an interface for integrating and accessing massive amounts of location-based information. GIS allows international agency staff to integrate data for planning and implementation purposes. Analysis of geographic features with a GIS allows the analyst to view new patterns, trends, and relationships that were not clearly evident without visualization of the data. GIS can also be used to get critical information about a humanitarian crisis to appropriate response agencies in a coordinated and efficient manner. Once in the field, the coordination can continue as new data can be added and disseminated via wireless applications and internet or intranet connectivity.

AM/FM/GIS implementation

As part of their AM/FM/GIS implementation project, Allegheny Power directed the work of a GeoResearch Inc. to perform electric distribution system field inventory and mapping services along 24,500 circuit miles in a 12,000 square mile service area. Over 500,000 distribution utility poles needed to be mapped in a 13-month period, representing one of the largest and fastest GPS/GIS field inventory and data collection efforts ever undertaken by an electric utility. The project required the rapid staffing and effective management of over 40 field data collectors—a challenging task.

This project was to involve the management and planning issues involved in completing such a large-scale, fast track field data collection project on schedule. Project planning considerations, recruiting and training, management of a large field staff, field data post processing issues, and QA procedures was to be addressed. The Electric Distribution mapping was part of an initiative at Allegheny Power to build a geospatial inventory model of their electrical facilities for their AM/FM system. This initiative also includes paper-to-vector conversion and attribution of other division's data. Once the database is complete, Allegheny Power will have a connectivity model of their network for load analysis, design, and outage management processes. The following statistics of this project are provided. Allegheny Power's System includes :

—23,000 distribution circuit miles
—520,000 distribution poles
—13 month deadline
—95 per cent- 100'/40 attribute accuracy
—3-5 meter positional accuracy

GeoResearch

—5 field offices
—35 data collectors at one time (58 total collected poles)
—5 supervisors
—8 GIS/data processors

The field methodology was simple enough, they would walk to each pole once, starting from the substations and follow each circuit outward until it ended its electrical flow. They would carry a GPS receiver integrated with a personal computer running a DOS software which allowed the collection of the required attributes, and tagged the attributes to the latitude and longitude of each pole.

This project was completed on-time, though not without considerable effort, overtime and costs. As with most large scale projects, many mistakes occurred, particularly in the first couple months, most of those mistakes resulted in lost productivity (poles per day), and we were constantly scrambling to adjust our staff numbers to keep pace with delivery schedules, and to replace staff lost to turnover. One of the largest mistakes from the beginning was overestimating the number of poles the collectors could map each day. The rugged terrain in percentage of the collection area prevented reaching the estimated productivity goals on most days. However, in the relatively flat, rural area of Mercersburg, PA, the collectors averaged well over the target on most days. In addition to the overestimating of daily productivity, it was difficult to overcome the productivity losses created from numerous equipment failures throughout the length of the project (especially at the beginning). Additionally, since the company was constantly adding new staff, they almost always had some number of new collectors "in training" which required personnel resources that would otherwise have been utilized as supervisors, auditors, GIS processors, and collectors.

The company feels that this job could have been performed much more efficiently and profitably in two years rather than in one year. It takes a few months after training to really develop a smooth and efficient working environment, and it takes time to absorb the significant start-up costs. The technology for this type of project is here, and it works very well. The key to success is good management starting from the estimating and proposal stage through recruiting, training, planning and production.

(www.gisdevelopment.net/proceesings/gita/1993/mobile/mobile001pf.htm)

Relief Operations in Africa

Several organisations have created web based and freestanding GIS to support humanitarian relief projects for the Horn of Africa. The United Nations, Humanitarian Assistance Information Center, and U.S. Department of State are working in concert to produce a geospatial data system that will support humanitarian relief efforts for natural disasters and political upheaval. Various data sources have been searched and pertinent information retrieved. What is lacking are land use maps and some form of physical imagery to "see" what is happening on the ground. A global mapping project sponsored by NASA and NIMA has provided Landsat imagery and land use maps from that imagery for the continent of Africa. LANDSAT imagery and a 13-class land use map created from that imagery are delivered through an ESRI ArcIMS site for the Horn of Africa. This presentation outlines the development and implementation of the remote sensing in support of humanitarian relief in eastern Africa (www.esri.com).

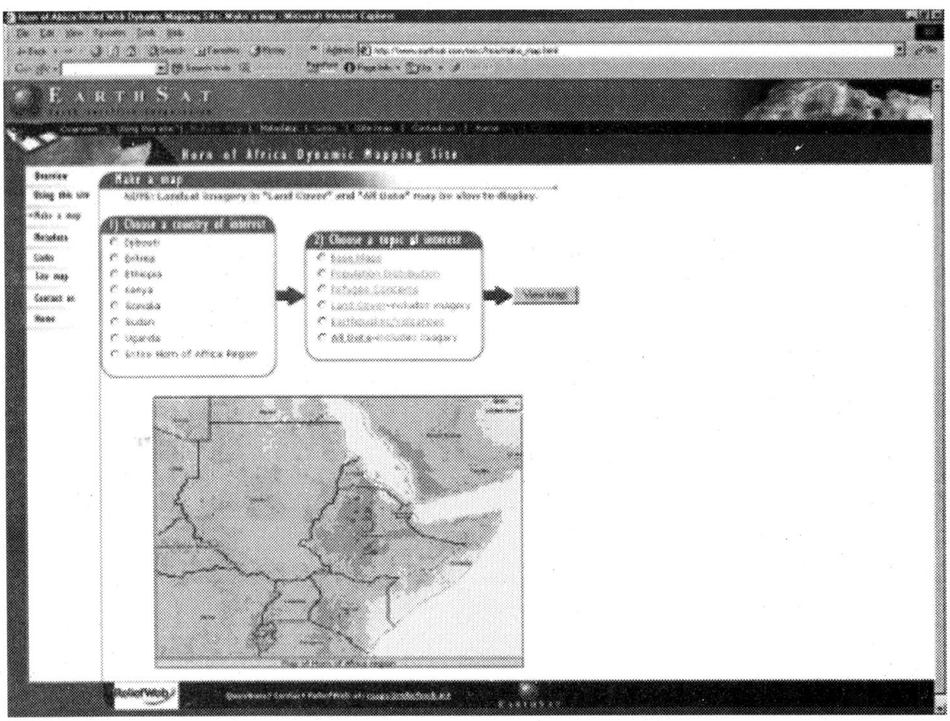

Fig. 7.1: Orientation of the Seven Countries in the Horn of Africa

In January 2001, the US Department of State (DOS) made a request of NIMA for aid in generating humanitarian planning maps for countries in the Horn of Africa: Djibouti, Eritrea, Ethiopia, Kenya, Somalia, Sudan, and Uganda (Figure 7.1). The maps were to assist relief workers in the planning of humanitarian efforts to relieve suffering associated with natural and man-made disasters. The data were to be accessible through an on-line mapping system utilizing ESRI's ArcIMS™ software. Earth Satellite Corporation of Rockville, Maryland was retained to create the web site and populate the mapping system with raster and vector data sets.

GIS for Libraries, Museums and Public Access

Making information accessible and easy to understand is the goal of libraries and museums everywhere. A GIS organizes and presents information in ways that spark greater understanding of places and events around the globe and close to home. It helps in administration, exhibition and displays, public access, research and teaching. Libraries and museums provide a gateway to all kinds of information. GIS manages census data, zoning and tax assessment maps, digital aerial photographs, and satellite images, providing public access to information that previously was difficult to use. GIS also provides tools for exploring information through maps and images, helping inquisitive minds to see things in a new way. Museum exhibits about topics such as natural history and geology come alive with dynamic maps to illustrate processes and events.

Regarding public access of information, there is an example of Monmouth Battlefield State Park. In 1770s, seven well-established farms occupied an area of hilly farmland and hedgerows of 2,000 acres of Manalapan and Freehold townships in New Jersey. This bustling farming community was occupying a future battlefield. Monmouth Battlefield State Park preserves this splendid 18th century landscape, which includes miles of hiking and riding trails, picnic areas, a restored revolutionary war farmhouse, and a visitor's centre. On June 28, 1778, one of the largest and hottest battles of the American revolution took place in Monmouth county, New Jersey. As the British Army marched from Philadelphia to New York City, General Washington and the Continental Line (Army) attacked the rear guard of the British Army.

Even with a short review of the battle, the many intimate details that the public could be and is interested in learning more about are seen. This project utilized ArcView GIS to develop an interactive interface that the public can utilize at the park to gather the volumes of information collected by the state park historian. Integrating the volumes of information with pictures, maps, and explanations allows the visitor to experience the fullest the battlefield has to offer.

This presentation was to review the goals of the project, the choices made, the lessons learned, and the plans for the future. A demonstration of the interface and the information used was to be the highlight of the presentation.

With information technology and electronic communications making inroads everywhere, the media is making use of GIS, particularly when it is dealing with aerial of geographic description of an event. GIS combines layers of information about a place to give a better understanding of that place. GIS software, such as ArcView, provides the data model and tools needed to store, analyze, and display information about places. Reporters can map everything from census data to crime statistics to traffic accidents quickly and accurately. With wireless and internet access, data can be retrieved at the scene or before the reporter even leaves the office. GIS provides journalists with more effective tools to report the news and put it in context for readers, and also manages the networks that facilitate media subscriptions and distribution.

There are several examples where GIS has been successfully used for academic purposes. The 3-D visualisations of the beautiful Parco del Corno alle Scale in Italy's Northern Apennines were generated by Faenza-based GEOgrafica for a multimedia teaching CD ROM. This aimed to show geomorphological features of the area, vegetation distribution and trail paths—all data being derived from the National Park's ArcView-based GIS or from GPS trail readings. The software employed to generate the imagery was 3D Nature's Visual Nature Studio (VNS) and in which vector isolines were gridded to produce 20m DEMs of the area. ArcView shapefiles were imported and gridded in a single-pass operation. The entire gridding process was extremely rapid, taking just five minutes to tile an area of 30x30 km into 64 DEMs of around 20m grid size. CORINE Landcover coverage was used to attach ecosystems to the terrain. Some fifteen different eco-classes were prepared in VNS to fit the CORINE legend; then, while importing the CORINE shapefiles, they were automatically attached, through attributes, to the imported and georeferenced vectors. Further, each ecosystem is a combination of ground, grasses and trees that fit the description of CORINE landcover and includes real images of plants, 3-D building objects and simulation algorithms of natural distribution. In this way, it was possible to simulate a higher resolution than that of CORINE itself. Furthermore, TIFF with TFW scanned maps were used to drape the ground and display a more "technical" view of the terrain. With the help of multiple cameras, several views of the area, from zenithal to birds-eye, were rendered. This was accomplished on a simple PC with 512MB RAM and running Windows 2000. These renders were inserted in the multimedia CD ROM and also used at visitor information points in the park. The public appreciated the views, especially for their "virtual reality" look, while the park management was delighted with this fusion of 3-D photorealistic visualisation with validated and georeferenced data.

(www.gis.esri.com/library/usrconf/educ02/abstract/a5127.html and www.geoplace/ge/2001/0111/0111vs_2.asp).

GIS in Real Estate

The real estate industry has always known geography matters since location is what GIS is all about. ESRI is actively engaged in implementing GIS-based solutions in all segments of the real estate industry, from map-based contact management to sophisticated investment analysis in large real estate investment. GIS allows the real estate practitioner to integrate a wide variety of data into one common format, a map. The presentation of a wide variety of data affecting the desirability and value of a property on one or two maps can give a far more accurate picture of the property's suitability as a first time residence, acquisition for a portfolio, or site for a retail outlet than any number of generalized market studies, photographs, and marketing text.

In case of RealtiCrop a real estate company, the first six months were spent laying out groundwork to make GIS. After some time the growth and flexibility were factored into the design of the organization so that very little lead time is needed to open up offices in new areas when good opportunities are identified. The company makes extensive use of the internet utilizing frame relay and T1 connections to transmit information immediately to regional offices. The firm has operated primarily in the southeastern portion of the United States.

RealtiCorp is built on four interlocking components: (a) Research and Investment, (b) Sales and Marketing (c) Technology Departments, and (d) regional offices. Sales and marketing staff works with past and potential clients to identify their requirements and impending needs. The Technology Department constantly develops new ways to utilize information technology to spot and manage opportunities. The Research and Investment Department uses ArcView GIS and ArcView Spatial Analyst to analyze data and determine areas of high suitability that can be referred to regional offices for further development. The company uses GIS to get know their customers better than the customers know themselves and for decision support. These robust applications analyze risk and returns on commercial real estate investments. It has chosen ArcView GIS because the company was looking for an enterprise wide solution that offered flexibility in terms of customization of the interface in using data in various formats from public and private sources. The lower start-up costs associated with a desktop application and the ability to make ArcView GIS work with Word were also attractive to the company. Analysts use ArcView GIS to track expanding MSUs and research their specific locational requirements. This information is overlaid with the company's base GIS to locate areas of high suitability.

A network of regional offices provides local savvy and locates promising deals on either a to-order basis or by seizing an opportunity they discover. Regional offices validate property characteristics for specific requirements and monitor the progress of ongoing deals. Lotus Notes ties the company's departments together. As areas of high suitability are located, that information gets forwarded to broker partners using Lotus Notes. These local representatives are given information on promising areas and the specific site requirements for clients. Brokers visit properties to qualify the physical characteristics of the property. RealtiCorp plans to bring the power of GIS out of the back room and make it more accessible to brokers through developing more intuitive applications and is exploring the use of MapObjects to supply low-tech users with real estate information.

(www.gis.esri.com/news/arcusers/1088/regis..html).

GIS for Environmental Management

Geography is the key to helping government and commercial entities monitor the environment more efficiently and cost-effectively. It also is the key to helping organizations comply with the multilateral environmental regulations that are the result of growing global markets. The digital version of such expertise, in the form of GIS, has been useful for site remediation, natural resource management, waste management, groundwater modelling, environmental impact assessment, policy assessment, environmental compliance and permit tracking, water quality and vegetation mapping. Nevertheless, since the inception of the technology, GIS has been an integral component in environmental management. A GIS links geographic and attribute data, such as lab and field data, which allows you to do predictive modelling for planning and compliance purposes. GIS technology reduces the cost of doing environmental business while giving the organisation that uses it a competitive advantage in an ever-evolving marketplace.

An example of the application of AutoCAD Map in arboreal activities may be mentioned here. Arbortech Consulting Ltd. offers consulting services to assess, diagnose and manage clients' trees and shrubs. These services include preparing management plans and specifying pruning, fertilizing, and monitoring trees for insects or diseases, as well as providing specifications for planting, transplanting, and removing trees. The company is often called on to consult with clients on issues such as how existing trees and landscaping might affect building, land development, and related infrastructure installation and renovation.

Relying heavily on AutoCAD® and AutoCAD Map® software, Arbortech collected survey data about the trees and landscape, while associated companies collect information regarding storm, water, sanitary system data and the like. The resulting GIS is to help to take important decisions regarding proposed land use and community development. With approximately 1,000 acres of land and water resources and more than 8,000 trees to inventory, the team had quite a challenge. Moreover, the team needed to collect and store the data in a GIS format so that the future contractors, land developers, and university staff could use and build on.

As surveyors from Murray and Associates collected field data about storm, water, and sanitation sites, Arbortech created a Microsoft Excel database containing the size, species, location, and condition of all trees on the property and surrounding residential areas. This effort included using AutoLISP® routines for plotting to scale the radii of tree drip lines. Each tree's drip line was represented by a circle, so one could easily visualize the radius of each tree. The data collected by both companies was then output to Excel and dBASE files for compilation.

As long time AutoCAD users, Murray and Associates turned to AutoCAD Map software to create the final maps of the area. AutoCAD Map simplifies the process of creating and maintaining maps and makes it easy to share that data with others. The team depended on AutoCAD Map software's ability to integrate data from a variety of sources. AutoCAD Map gave the power to present a large amount of complicated data in a way that anyone could easily understand. Obviously, getting ideas across in layman's terms is crucial. The specific applications were (a) to select a specific piping and view the database to find its age, material, diameter, depths, and so on; (b) the mapping and inventory of trees for planning future development; and (c) preparation of plant operations schedule and budget treatments for pest control and to keep track of hazardous conditions like sick or dying trees.

Another example of GIS application is for the development and use of an Environmental Information System (EIS) for power station atmospheric emissions. ArcInfo has been used to develop an EIS for atmospheric data. The system combines the output from computer models of atmospheric stack and cooling tower plume dispersion and deposition with monitored air quality data, and third-party datasets (such as SSSI boundaries), to provide a range of information about power station stack emissions, their dispersion and their potential environmental effects. This multi-user application includes custom grid and contour visualization, automatic context-sensitive legend generation and complete "loading" and "saving" of all displays. It uses AML, C, AWK and UNIX shell scripts for OpenWindows or Motif environments.

National Power (NP) is the UK's largest electricity generator holding about 25 per cent of the market and an increasing investment in electricity generation worldwide. Effective management of the environment is vital to the commercial success of National Power. As well as managing the existing assets with the environment in mind, it is necessary to develop the business strategy to take account of the likely direction of future legislation and technology—and the views of everyone with an interest in the business. The national power environmental approach is guided by five principles:

(a) To integrate environmental factors into business decisions,
(b) To monitor compliance with environmental regulations and to perform better than they require, where appropriate,
(c) To improve environmental performance continuously,
(d) To review regularly at Board level, and to make public, the company's environmental performance, and
(e) To establish a reputation for effective environmental management.

The above principles were considered important to understand its impacts on the environment, manage them cost-effectively, and to communicate its performance and the progress.

The NP environmental performance is backed by a team of environmental specialists who provide the underlying scientific and technical expertise. Its Environment Unit uses modelled and measured information about the environmental effects of power station operation both for development of environmental strategy and for input to site-specific environmental effects evaluations and impact assessments. Historically these modelling studies were performed using separate systems and integrated only at a later stage. The continuing drive for increased productivity motivates the search for opportunities to integrate existing systems within a single framework. This integration has several clear benefits: (a) it reduces the number of separate software systems that need to be supported, (b) it provides many opportunities for synergies between the combined areas (sometimes unforeseen), (c) the ability to contextualize model output using other datasets adds considerably to effectiveness of use and ease of interpretation, and (d) it provides a unified approach to data management. The integrated Environmental Information System (EIS) used ArcInfo initially for atmospheric environmental studies.

There is a similar case in France where GIS was used for environmental concerns for power transmission. During the sixties, this country needed electricity everywhere. There was no understanding of the best way to transmit electricity from one point to another. Decisions were being based on economical considerations—the straight line was quite often the best one. At the end of the 1970s, when the Very High Voltage (in French Tres Haute Tension) network was being reinforced, the context changed : the global needs had been satisfied and ecological consciousness (following the events of May 1968 and other demonstrations like Woodstock) was rising. It was at that time that the French power establishment, EDF, began to take account of the environmental concerns. In 1982 it signed an agreement with state authorities to show its good will. Accordingly, in 1990 its strategic plan included environmental measures. EDF also concluded an agreement with the authorities to take steps to integrate the transmission network components into their surroundings. Later, EDF asked its regional divisions to plan the enhancement of the regional network considering the environmental constraints.

At the end of 1993, some years after the EDF's decision to take account of the environmental concern, Energie Aquitaine (an EDF Generation and Transmission division) decided to build Pro-SIGEA, a dedicated GIS to deal with its region's environmental data. It was planned to show interferences from the power transmission network on maps and to form these interferences into a hierarchy. In addition, the system might improve communication between different organisations dealing with long term planning problems. In order to achieve the goals, ArcView with Access was chosen as the main query tool. Now technical EDF staff use Pro-SIGEA for various purposes within their areas of operation. Today, such a system is strongly dependent on the data input into the computer. Though it is not difficult to form a hierarchy of classifications based on juridical aspects, it is not the same for the landscaped and ecological criteria, partly as a consequence of input data. Hence, the main conclusion is that the task of modelling interferences between the environment and human activities is difficult precisely because they interact together in space and time. Nevertheless, Pro-SIGEA can be considered as a first step towards that goal.

The database and ArcView based tool that form Pro-SIGEA, the EDF version of SIGEA, are useful. Energie Aquitaine, a regional division of EDF Generation and Transmission, uses it as an aid in the decision-making process in its technical field. The environmental database and its use through SIGEA is now the responsibility of Aquitaine DIREN (Directions Régionales de l'ENvironnement). An agreement has been concluded between EDF and Aquitaine DIREN to protect it and its use by third-party members. SIGEA which allows the exchange of data between EDF and DIREN facilitates communication between these two entities.

The results that can be issued from Pro-SIGEA (as for SIGEA) depend strongly on the data input into the computer from both a quantity and quality point of view. Fortunately, data is usually updated quickly. For the transmission network geo-databases the updating is permanently undertaken by EDF. Further, if the object of forming a hierarchy of environmental constraints has been achieved for the legal aspects, the results are not so good for the ecological and the landscaped aspects. This can be explained by the difficulty of measuring the constraints attached to those criteria and also by some limitations of the software and of the environmental database. Either the precision of data location is not sufficient or knowledge of the land's cover is missing making it difficult to obtain correct results.

Another interesting case related to environment and power is regarding oil spill trajectory and fate model using GIS. It is widely understood that the oil spill is a great threat to environment in general and mankind in particular. It is a challenge for petroleum industry as well. The conventional processes are much tedious, less accurate and time consuming. The solution lies in using GIS along with all the features of conventional processes with decision support system, analysis and visualization capabilities, developing spill predictive model, providing strategies and cleanup plan, charting optimum shipping roots and to support contingency plan for quick action.

The study area was Rawa oil field located along the east coast of India. The methodology adopted for this purpose is as follows :

- After pre/post-monsoon surveys, available data like physical and chemical parameters are processed using Geomedia Professional.
- Identification of sensitive areas like mangrove, reserved forest, fishponds, paddy fields, terrestrial and aquatic ecosystems.
- Analysis in this fragile environment, by GIS to predict the extent and location of a spread
- Use of mathematical model to calculate amount of oil evaporated, spread etc.
- Probable spill behaviour under local conditions
- Creating maps to analyze different zones, for resultant directions and speeds of wind and current.

- To predict the extent of spread at specific time interval and also the spreading of spill towards the coast.
- Analysis of relationship of dispersion rate with wind speed.
- Calculation of the affected shoreline area.

The GIS system was considered for planning for the clean up and predicting how the oil will move, accessibility, preparedness in a situation of oil spill, quick analysis and follow up action. The application of GIS is meant to save valuable time which in turn can be used to save the nature and human life.

(www.gis.esri.com/library/usrconf/europroce96/PAPERS/PN18/PN18F.HTM; www.autodesk.co.jp/adsk/servlet/item; www.gis.esri.com/library/usrconf/europroc96/PAPERS/PN26/PN26F.HTM; and www.gis.esri.com/library/usrconf/europroc96/PAPERS/PN40/PN40F.HTM)

GIS for Banking and Insurance

USAA has developed a web based, graphical model that provides users flexibility in estimating the number of potential, active and remaining claims for any type of catastrophe. The model uses ESRI software applications with some custom development on the base GIS infrastructure. For this purpose, a CAT model is designed based on ArcIMS application to assist loss reserving and cat operations personnel in the viewing and analysis of homeowners, fire and renter's policy in force data, as well as other catastrophe-specific data. There are four options with which to view the policy in force information :

- *Quick view:* An interactive tool which provides tools for simple viewing and querying of the U.S. policy in force data.
- *Java view:* An interactive tool similar to the *Quick View*, but with additional analysis capabilities.
- *Pre-processed policy in force maps:* This option shows policy in force maps at the U.S., state, county and ZIP code levels.
- *Policy in force (PIF) reports and geocoding hit rates:* This option shows policy in force counts (both total PIF and geocoded PIF) at the state, county and ZIP code levels. The geocoding hit rate is the percentage of PIF that were successfully geocoded to the address level, however other levels of geocoding are available in the application.

When a major catastrophe occurs, senior management will rightfully expect accurate, timely (daily) information that they can use to make critical decisions. Sending the wrong number of adjusters with the wrong skill sets to the wrong places can result in poor service to the membership when they need us most, and can result in unnecessarily high expenses. Inaccurate estimates of loss and Additional Living Expenses (ALE) costs can lead to improper decision-making by senior management. Also, the push towards shorter financial closing schedules (eventually, to a virtual close) makes it imperative that robust catastrophe data be accessible on a daily basis to facilitate the accurate estimation of catastrophe liabilities. Waiting until delivery of correct month-end data to analyze catastrophes is not an option.

The benefits of such initiatives could be estimated at the time of crises. The case of the early benefits in San Diego County (Alpine) Fire can be mentioned here. Around 5:00 a.m. on Wednesday, January 3, 2001 a fire started about 30 miles east of San Diego. With the winds blowing at 65 mph toward the west the fire quickly got out control. It started threatening the moderately populated city of Alpine. By mid-morning, evacuations began to take place. Looking at a 3-hour visible satellite loop the smoke from the fire could be seen blowing across San Diego and then out over the Pacific.

By late morning the WESTRO (USAA Sacramento office) requested that Cat Operations plot a map showing the southwest portion of San Diego county. They asked that the map include the homeowner and fire policies in force (PIF), by risk location, so that they could assess the exposure. This was a standard request, but given that the fire was out of control and moving quickly WESTRO had need for immediate information. They expressed concern about how quickly that need could be met. Plotting a hard copy map in the home office generally takes about an hour, if the map is fairly simple. Then the map must be shipped over night to the region which means they wouldn't receive it until the next morning. With the new GIS web application in place, WESTRO did not have to wait. It only took about 5 minutes for them to pull up a web browser and plot their own map on the screen. This new application provides the same information (plus more) that is available in a hard copy plotted map and it is available immediately.

Fortunately, the fire did not continue to rage. By the next day, the winds were down to a mere 30 mph and then by Friday, the fire fighting teams were able to start controlling the fire. USAA only received 6-7 claims and only one of those actually had any fire damage. The rest were for smoke damage or Additional Living Expenses. But it could have easily been much worse.

(www.gis.esri.com/library/usrconf/europroc01/professional/papers/pap1105/p1105.htm)

GIS in Internet Mapping

Internet has evolved as a viable tool for public access to information. Governments, institutions, industries, corporations, municipalities and the like have found this web based technology to promote their agenda and business interests. Internet combined with GIS has created further wonders in making the spatial data more accessible to people at large. Web GIS or internet GIS, as they are popularly called, have wide range of map based solutions ranging from applications on finding directions to mapping biodiversity to creating the Geography Network. In case of India, the new National Map Policy announced in 2005, allows the users to publish maps on the web with or without GIS database. This liberalization has yet to make an impact in terms of internet mapping. However, internet has been accepted as a powerful tool for public access and transparency in the functioning of the government and its agencies because the information is easily available to all and reduces time for making inquiries through telephones or otherwise. Nevertheless, some examples from other countries are given in Table 7.1:

Table 7.1: Examples of internet GIS

Title	*Publisher and Place*	*Content summary*	*Purpose etc.*
Iowa Gap Analysis Stewardship Application	Iowa Gap Analysis ProgramAmes, Iowa	Data includes conservation and recreation boundaries from municipal, county, state and federal agencies.	For conservation managers, planners
Sistema Informativo Territoriale—Istituto Promozione Industriale (SLI)	Istituto Promozione Industriale	Location of a new industrial plant Using ArcView IMS and actually is only in Italian language	To provide public Access to the geographic content
Jefferson County, CO GIS	Jefferson County, CO GIS	Delivers address related property, political, and physical information to residents using HTML, Perl/CGI, SQL/Informix, and VB/MapObjects IMS components	To provide public access to the geographic content
Johnson County, KS—AIMS Interactive Maps	Johnson County, KS	Provides various interactive mapping solutions using property maps, aerial photos, planimetrics, and other pertinent county information	To provide public access to the geographic content
Ruhrtal à la Karte	Kommunalverband Ruhrgebiet (KVR)	Shows public points of interest, tourist sights, landmarks, and bicycle paths in German.	To plan bicycle tour
LOJIC Home Page—Interactive Maps Section	LOJIC	Information on polling sites and contact information and flood plain, property, zoning, sewers and much more.	Public access to the geographic content
Natural Resources, Physical Features	Louisiana Dept. of Natural Resources	Displaying and overlaying the mineral, energy, water, biological and other natural resources	For navigational, engineering
Real Estate Information System of Lucas County, Ohio	Lucas County Auditors Office	Parcel attributes, tax information, interactive digital base maps and aerial photos	To provide public access geographic content
Maine Office of GIS	TopoZone—USGS Topographic Maps; Maps a la carte, Inc. North Chelmsford, MA, USA	Interactive, seamless topographic map of the entire United State. Maps are available a 1:25,000, 1:50,000, 1:100,000 and 1:200,000 scales.	TopoZone is designed for all topographic map users, including engineers, planning professionals, outdoor recreation enthusiasts, genealogists, and others.
Meridian Live Internet Demos	Meridian Systems Oy	Online geocoding, automatic vehicle tracking demo, and MOSAIC land information service	Public access to the geographic content
Nashvill/Davidson County Online Property Maps	Metropolitan Nashville/Davidson County Planning	Viewing and querying of property, ownership, zoning, topographic, photographic and planimetric data.	Access to public property and ownership records. This is not a

(contd.)

(contd.)

Title	*Publisher and Place*	*Content summary*	*Purpose etc.*
	Department, Nashville, TN	Print maps and property records. Search by owner, address, subdivision etc.	legal document
Montana State Library Natural Resource Information System	State Library	To create detailed thematic maps based on user-selected criteria. "Topographic Map Finder" lets you browse USGS quadrangle images or zoom to the image of any feature or Township/Range/Section in the state.	Public access to the geographic content
On-line Browser for Topographic Maps of Montana	Montana State Library	For viewing MrSID-compressed USGS 1:24000, 1:100000, and 1:250000 Topographic Maps. Search tool for any Public Land Survey Section, Latitude/Longitude coordinate, UTM Coordinate, or any feature name on a Topographic Map.	Finding topographic maps of any place in Montana.
Zoning and Platting Cases On-line	Municipality of Anchorage	To have access to information on case descriptions, staff reports, zoning regulations and Municipal codes, hearing agendas, schedules, and meeting minutes and view many GIS data layers and print a map.	To give the public access to timely information and for public comment.
National Atlas of the United States Solution	National Atlas of the United States®	It delivers easy to use maps of America's natural and socio-economic landscapes. More than 20 Federal organizations have collaborated to produce reliable, integrated, and documented national maps. These are offered online in interactive mapping presentations that allow Americans to visualize and understand complex relationships between environments, places, and people.	This site is intended to promote geographic literacy and national self-awareness. The Atlas is designed to provide a reliable summary of national-scale geographical information. Partnered with ESRI to develop innovative and reliable products and services that characterize America through maps
The National Geographic MapMachine	National Geographic Society	The MapMachine is an interactive mapping site that enables users to access a wide variety of geographic information covering the world. It offers dynamic maps from many of the world's leading data providers, atlas maps from the National Geographic 7th Edition Atlas.. It is powered by ESRI Internet mapping	MapMachine is intended to support the education and conservation communities as well as others interested in geographic information about our world.

(contd.)

(contd.)

Title	*Publisher and Place*	*Content summary*	*Purpose etc.*
		technology and built on content from the Geography Network.	
National Land Survey of Finland	National Land Survey of Finland	These web sites in English provide you with information on the maps and services in real estate, topography and the environment for the needs of citizens, other customers and the community at large and information on the NLS and its international activities.	It also promotes the shared use of geographic information.
NGS Survey Control Map	National Oceanic and Atmospheric Administration	It provides a map of the US, which allows you to find, and display survey control points stored in the database of the National Geodetic Survey and access the geodetic data associated with the points.	Provides detailed information on the NGS CORS sites located throughout the nation.
National Atlas of Canada	Natural Resources Canada	Using a combination of high-quality small- and large-scale digital data, users can select custom maps and perform advanced spatial analysis. Users can zoom in on a particular area of a map, create, query, save and print customized maps, download base maps, and access educational resources.	This site is intended to provide public access to the geographic content described above.
Nelson and Wallery, Ltd./Nursing Home INFO	Nelson and Wallery	Find a nursing home based on location, religious affiliation, medical specialization, language skills, operator, and other database navigational aids. Create a map of your vicinity—then generate a list of nearby nursing homes from a list of over 18,000 nursing home in the U.S.	This site is intended to provide public access to the geographic content described above.
AfriView	SDS Mapping	An interactive mapping site of Africa and Madagascar based on a publicly funded watershed mapping exercise sponsored by the UN/FAO and WWF. By simulating Internet Map Serving (IMS) technology, AfriView allows "users": to dynamically browse, pan, and zoom through six layers of increasingly higher resolution map graphics of Africa.	It introduces users to NIMA and USGS that provide near global data coverage of the world. It is a preliminary map display interface that is part of a much larger and as yet complete data viewing tool entitled "NIMA in a Nutshell".

(contd.)

(contd.)

Title	*Publisher and Place*	*Content summary*	*Purpose etc.*
Regional Economic Development Information (REDI) System	San Diego Association of Governments	Assists industrial developers, economic development analysts, small business owners, community planners, and others to explore land development opportunities and visualize relationships among available employment land, existing land use, planned land use, average daily traffic volumes, and other geographic layers. A unique aspect of the REDI system is the ability to map on top of a digital ortho quad aerial photo base.	This site is intended to provide public access to the geographic content described above.
Inmarsat GPS Tracking and Display Solution	Sky Satellites System and ESRI France	This service can be used to display GPS position and status through Inmarsat satellite link.	This service can be used to display the position and status of a fleet of various vehicles (car, truck, boat, etc.).
eMAP	Systematics Technologies	Provides commercial services in Hebrew and English to many other Israeli sites and portals, including demographic data, generate driving directions between any two addresses or find all the gas stations, banks and health centers in Israel.	This site is intended to provide public access to the geographic content described above.
TBO.com	Hillsborough County Office, Tampa Police Department, and Pinellas County Office	The crime data is available at http://crimetracker.tbo.com/crimetracker/definitions.htm. Data fields include the date, location, and offense for each crime reported. Segmented the data is available into four separate Map Services: 2000 Crime Data for both Hillsborough and Pinellas Counties and 2001 Crime Data for both Hillsborough and Pinellas Counties. The street data was purchased from GDT.	Allows to find out about the reported crime data. Helps in taking decisions about personal safety, home protection, and relocation, and increases community awareness
Afghanistan GIS and satellite image visualization service	Telemorphic, Inc., Berkeley, CA	It is a free web browser-based interactive GIS and satellite image visualization service powered by Telemorphic's Maplicity and ESRI's ArcIMS technology. Telemorphic built the service entirely with public domain data including low-resolution digital elevation models, medium-resolution Landsat Thematic Mapper	To provide a basic understanding of Afghanistan, along with geospatial data such as satellie imagery, digital terrain models, and basic map information related to place names, rivers,

(contd.)

(contd.)

Title	*Publisher and Place*	*Content summary*	*Purpose etc.*
		(TM) satellite imagery, and small-scale roads, rivers, populated places, and political boundaries.	roads, and country boundaries.
Australian White Pages Directory	Telstra Corporation	It used SDE to build a White Pages directory that provides online access to 8,000,000 residential, business, and government listings. The directory is updated daily.	This site is intended to provide public access to the geographic information.
Thailand Environment Institute	Thailand Environment Institute	The institute provides Internet based access to various environmentally related spatial information maintained by them.	Provides public access to the geographic content.
Centre for GIS, Qatar	CGIS, Qatar	The Map Server application called "Explore Qatar" lets you locate street names and addresses, and landmarks such as hotels, post offices, banks, and schools. A spatial query feature is also available.	This site is intended to provide public access to the geographic content.
National Seismic Hazards Mapping Project	U.S. Geological Survey	USGS offers seismic hazards maps showing peak horizontal and horizontal spectral response accelerations on a national scale.	To provide public access to the geographic content.
Malaysia Internet Yellow Pages	Cybermall Sdn Bhd	Business and travel guide to Kuala Lumpur (Klang Valley), Malaysia. Powered by MO IMS, AV IMS and customized Java client	To provide public access to the geographic content described above
Flood Risk Mapping Solution	ESRI, Redlands, CA	Users can access FEMA Q3 Flood Data to help determine the relative flood risk of 1,200 counties across the US. The ESRI Flood Risk Maps are created using FEMA Q3 Flood Data and GDT Dynamap/2000 street data. The maps are intended as a tool to assist in screening property addresses or other locations to determine their flood prone status.	The Flood Risk Maps help identify a location falls in or near a Special Flood Hazard Area (SFHA). This flood area is the national standard on which the flood plain management and insurance requirements.
Base map for the Lithuania at Scale 1:50000, Vilnius map	Hnit-Baltic GeoInfoServisas	Digital Base map for the Lithuanian Territory at Scale 1:50000, was produced as National Basic Map in digital and paper format with the help of existing maps information, digital data bases together with the most recent panchromatic ortophoto material from SPOT satellite and with geometrically rectified multi spectral space images.	To be used for production of printing thematic maps. Ellipsoid GRS80, TM projection with the central meridian Lo =24°. Scale 1:50000. Coordinate system. ETRS-89 (ARC/INFO data format)

(contd.)

(contd.)

Title	*Publisher and Place*	*Content summary*	*Purpose etc.*
Korea-Japan 3D Tour	Infmedia Srl	The website is spectacular and interactive, for an impressive flight experience over the sites of the 2002 FIFA World Cup(TM). With 3DIMap technology, landscape meshes are dynamically built, and overlay images precisely positioned. On the Korea-Japan 3D Tour you can freely fly inside a virtual world with an extent of more than 3000 km!	To demonstrate the 3DIMap technology (3D Interactive Maps) as a powerful and spectacular way to deliver geographic content on the Web.

World Location-based Services

In 2002, Webraska (Brazil) was to open on-line access to geospatial platform for third party developers and content providers. Webraska, the leading provider of location-based services (LBS), telematics applications and enabling platforms, launched its full range of products with comprehensive map data coverage of Brazil. These products include the SmartZone Geospatial and Application Platforms, SmartZone LBS Applications and Webraska's wireless navigation applications for PDAs and in-car devices. "Location-based services depend on geospatial software platforms to provide vital functions such as address validation, route calculation and mapping," says CEO of Webraska Brazil. "The biggest challenge facing developers and wireless carriers in Brazil, has been the absence of geospatial platforms with comprehensive, accurate, navigation-enabled map coverage. The launch of the SmartZone Geospatial Platform in Brazil fills this gap." (www.apontador.com.br).

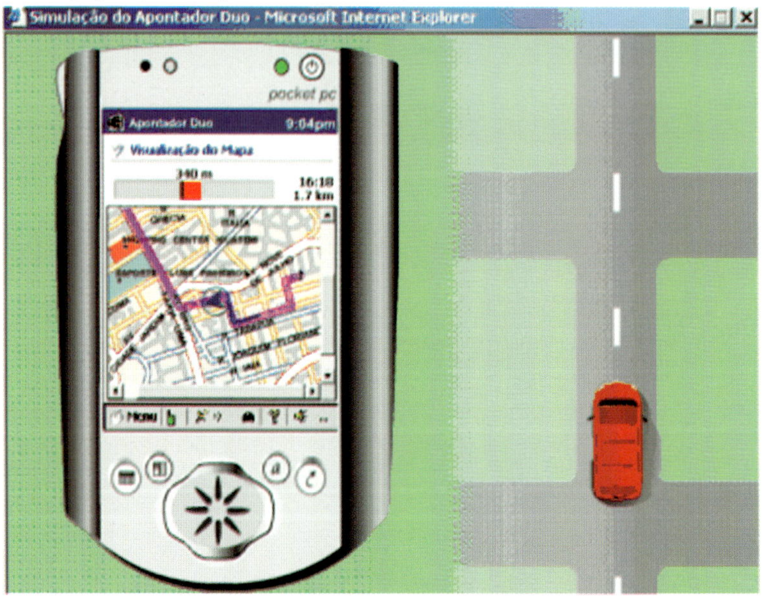

Fig 7.2: Location based services in Webraska, Brazil

In order to address the lack of high-quality digital map-data, Webraska Brazil has launched a nation-wide initiative to develop and maintain comprehensive navigation-enabled map data. Since December 2000, a team of field data collectors has been gathering detailed street-level information for all towns and cities with over 200,000 inhabitants, covering over 90 per cent of the population with internet access and 100 per cent of the wireless internet market. "Webraska has a proven track record in assisting customers and partners to qualify and integrate datasets of varying quality suppliers in Europe, North America, Australia, China, Asia Pacific and Latin America", says Marketing Director for the Geospatial Platform at Webraska. "This project demonstrates our commitment and our ability to address key issues of data quality outside our core markets."

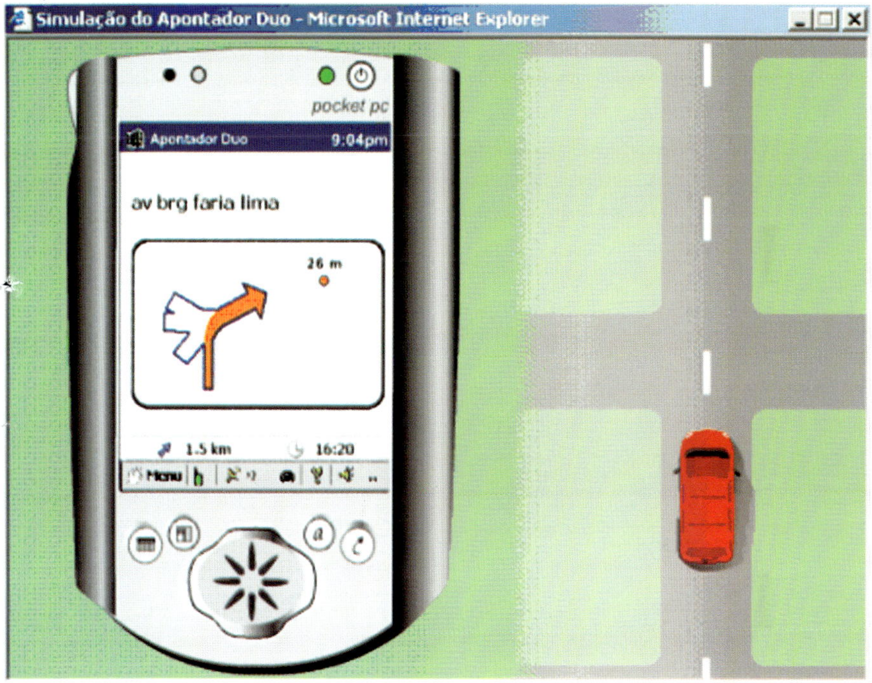

Fig 7.3: Help in navigation

For stimulating the creation of LBS applications, Webraska Brazil is also providing on-line access to a hosted SmartZone Geospatial platform via an open Application Programming Interface (API). This will enable local software developers to create their own consumer and corporate mobility applications, as well as to add new LBS functionalities to their existing solutions. "We"ve already signed the first access agreements with local software developers," said CEO. "These companies have already begun development and will play a vital role in helping Webraska to develop the market for location-based services in Brazil."

Revolutionary Safari Network

In July, 2002 NavCom Technology, Inc. announced the introduction of its Safari Network SR-7100 System. The Safari Network product line utilizes NavCom's wireless communications expertise to provide mobile users in harsh environments with local wireless networks. The SR-7100 System offers a revolutionary data communications solution for systems integrators looking to expand their customers' network communications systems to the remote mobile environment. "The SR-7100 is a private, dedicated system that seamlessly integrates fixed and mobile assets into a single flexible network," said NavCom Product Manager. "The rugged design of the SR-7100 components creates an always on' communications system for vehicles in harsh environments such as farm fields, construction sites, and mines."

The Safari Network SR-7100 System handles multiple sessions of varied data types simultaneously and has a 20 mile range that can be extended through its repeater capability. It builds on NavCom's communications systems expertise to provide the necessary features and capabilities to wirelessly meet most fleet management, data acquisition, machine control and local area networking application requirements. Systems integrators who are developing these applications benefit by providing greater savings and efficiencies to their customers. The innovative SR-7100 System, with its ability to simultaneously receive and transmit data in real-time, brings the sophistication of multi-service, multi-session communications networks to a new level by expanding it to the mobile wireless environment (www.direction.mag.com).

Dynamic Flood Warning System

Floods are the major disaster affecting many countries in the world year after year. It is an inevitable natural phenomenon occurring from time to time in all rivers and natural drainage

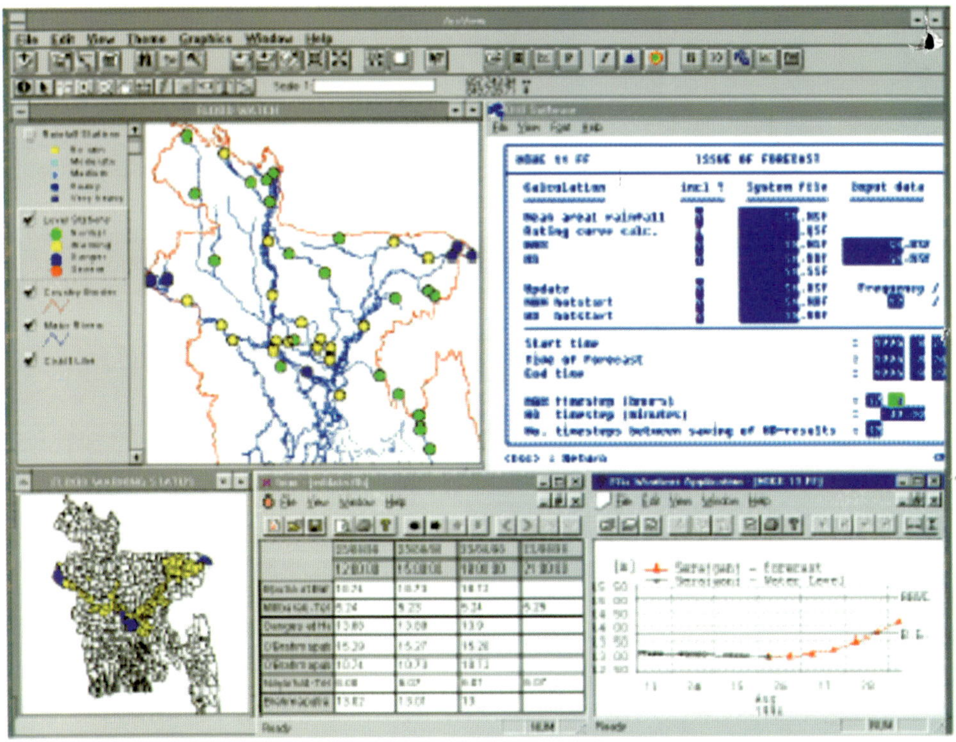

Fig 7.4: MIKE 11 Flood watch system

for spatial planning. The methodologies of designs in such projects are invariably worked out on the basis of requirements projected by the resource users. The multidisciplinary approach to planning is often effected by non-availability of reliable, systematic and accurate data to all those involved in spatial planning. A number of planning organizations have developed need based GIS to assist in analysing and making suitable projections. Non-availability of accurate and consistent spatial data base is, however, proving to be the biggest constraint. There also appears to be hesitation in standardization the database content on account of subjectivity in the requirement of data. Major constraints for standardizations of data are as follows:

(i) Geodetic framework and projections of topographical maps (base map) are often not in conformity with that of the cadastral maps.
(ii) The information on the base map is generally outdated and does not contain the details required by the planners.
(iii) The scale of map does not conform to the requirement of spatial planning at village/ town level.
(iv) Value added information on artifacts have to be collected from widely distributed sources and ground survey for items not depicted on map.
(v) Metadata, i.e. definitions and description of parameters are not standardized.
(vi) The mechanism for public participation depends on the local legal provisions and practices.
(vii) Lack of technology upgradation.
(viii) With no commitment available, the data producers are hesitant to invest in creating and maintaining the infrastructure for spatial database.
(ix) Lack of legal commitment for dissemination of data to resource users.

The Department of Science and Technology and the Department of Space have come together and taken an initiative to establish a platform, viz. National Spatial Data Infrastructure (NSDI) for coordination and integration of activities between various data producing agencies to enable standardization, collection, compilation of data on a variety of related themes in a sustainable manner. Such a platform would also play a crucial role in formulating policies on dissemination of data and regulations for commercialization/privatization in various theatres of information technologies.

Major Functions of NSDI

After a series of discussions with the members of the Task Force, the major functions of the NSDI have been identified as follows:

(i) To define policy concerning dissemination of spatial data to the government and private organisations.
(ii) To identify and fix the standards of data to form fundamental data sets for topographical, cadastral and thematic applications.
(iii) To lay down standards to which digital technology should conform and remain valid for a fixed period.

(iv) Identify sponsors and formulate guidelines for funding for sustaining the infrastructure.

(v) Initiate standardisation of geo-referencing systems, plan for generating transformation parameters to support transportation of data in other popular geo-referencing systems and projections commensurate with the user requirements.

(vi) To identify potential users, their resources and data requirements, economics of data transfer and legal cover etc., for setting up clearing houses.

The current model of the spatial data infrastructure in the country is shown in Fig. 8.5. The data users and the data providers have one to one association and such connections do not necessarily process the transaction through the NSDI gateway. Obviously, this model will change with the implementation of NSDI.

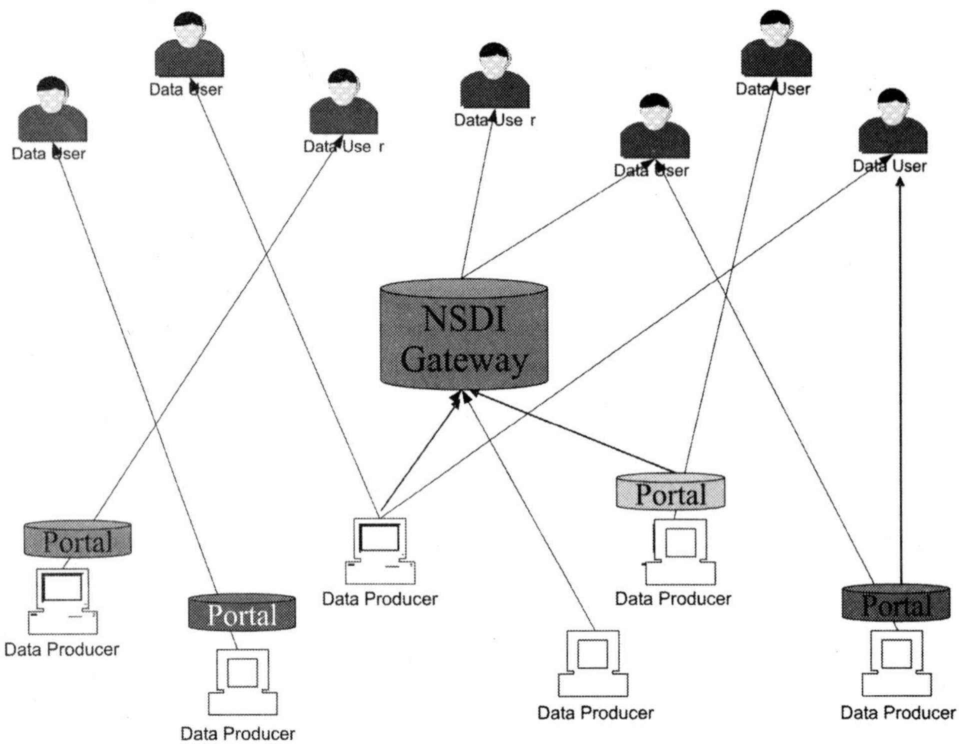

Fig. 8.5: Current model of NSDI

Expected Results from NSDI are:

(i) Single window clearance for all users of geospatial digital data.

(ii) Standardisation of geospatial digital databases including format, hardware, software, linkage, access and sharing and meta data.

(iii) Finalisation of geospatial accuracy requirements for different users.

(iv) More effective modelling for planning and development projects.

The future model of NSDI is indicated in Fig. 8.5 where portals of different data producing institutions will be connected with the NSDI portal. The queries can be directed to the NSDI portal as well.

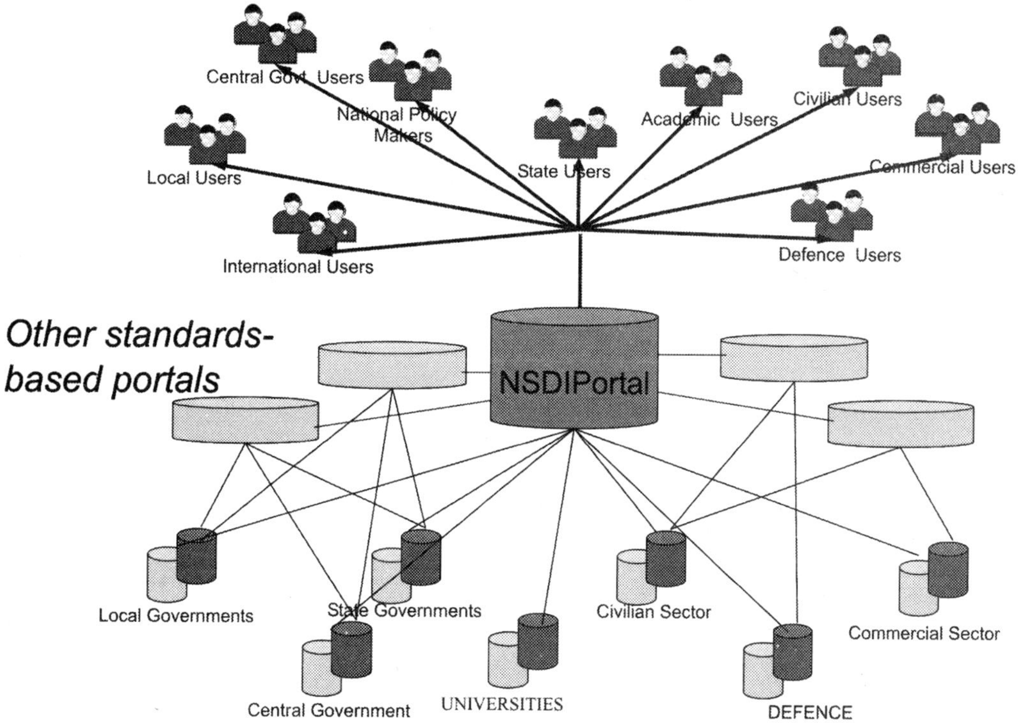

Fig. 8.6: Future model of NSDI

Major Issues Related to NSDI

The issues related to NSDI are being formulated through various sub-groups which are deliberating with user agencies to assess their requirement of data set. Some of the issues are as under:

(i) To lay down priorities and time frames for the development of infrastructure.
(ii) To lay down standards for metadata in the fundamental data sets comprising of topographical and thematic data.
(iii) To identify data generating organizations who can meet the standards of metadata and work on mechanism for transfer of standardized data to the fundamental data sets.
(iv) To set up subcommittees to evaluate and scrutinize the on data transfer format, exchange format and other technical standards for the infrastructure.
(v) To identify the custodians of fundamental data sets, methodologies for its upkeep, updation, operational linkages and evaluation *vis-à-vis* metadata standards.

(iv) Initiate standardization of geo-referencing systems and projection commensurate with the user requirements.
(vii) To identify potential users, their resources and data requirements.
(viii) To identify data requirements of users and lay down the contents of user data sets inconsonance with the government policies on desensitization of spatial data. Policies on networking for access to data for browsing and value addition are defined.

These are many and varied stakeholders to the NSDI and would include government at local, state, national levels (as both users and data collectors/owners); NGOs, educational and scientific organisations, private and public agencies. As of now, the major players as contributing to NSDI are: Department of Space, Survey of India (SOI), India Metereological Department (IMD), National Atlas and Thematic Mapping Organisation (NATMO), Geological Survey of India (GSI), National Informatic Centre (NIC), Census Department and the like.

Progress so far

- Various Central Government Departments, DST, DOS, MoEnF, Mines, Urban Development, Rural Development, NIC, Census Department, DAC, DARE, DOD etc. have initiated demonstration programmes towards NSDI Agency activities.
- The finalization of the text of a NSDI Bill—which will now be processed by DST—first as an Executive Order and then the NSDI activities as an inter-agency programme could start.
- Drafting of seven NSDI Standards documents—through inter-agency committees. The following documents have been initiated and are in different stages of finalization:
  —NSDI Content Standard—being coordinated through a committee and based on inputs from different NSDI agencies.
  —NSDI Metadata Standard—being coordinated by a committee. Design and development of the Metadata Server also being taken up.
  —NSDI Exchange (NSDE) Standard—being coordinated through a committee. The NRIS and SOI DVD formats are being merged into a NSDE Standard.
  —NSDI Search and Access Protocol—being coordinated through the committee for NSDI Metadata.
  —NSDI Networking Framework—being coordinated through a committee.
  —NSDI Quality Standard. Just initiated as a draft based on NRIS Standard.
  —NSDI Access Rules—being worked out by ISRO/DST.
- The task force has developed a NSDI web-site and a NSDI Metadata Server—which, when launched, will be able to provide information on spatial data availability to users. The NSDI Metadata Server is a catalogue of all spatial information of the NSDI Agencies, as per a NSDI Metadata Standard. The design of the clearinghouse search for spatial data based on internet has already been proposed (Fig. 8.7).

Major Issues

The major issues still to be considered—and very crucial to the implementation of NSDI are:

Use of NSDI

The utility of NSDI and its possible support to sustainable and economic growth in the country and for supporting developmental activities is well accepted by all. During 2nd NSDI Workshop held at Ooty in July 2002, the business potential of NSDI was discussed and it has emerged that the value of data and data products could be of the order of a few billion rupees. There is a need to evolve a business plan for NSDI as well as the organisations participating in NSDI.

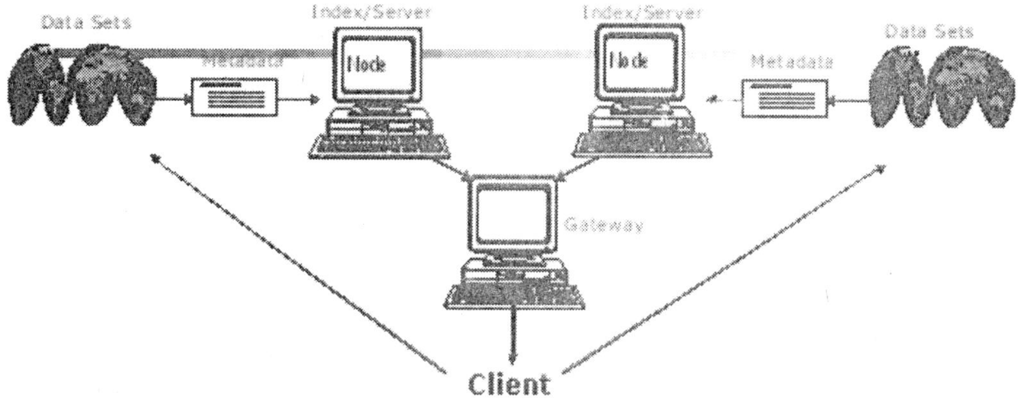

- Multiple catalogs searched on internet in parallel by gateway
- Results are collated for presentation to client
- A single session can search hundreds of collections and accesses current content!

Fig. 8.7: Clearing house search metaphor

Discussion and Debate on NSDI Standard Documents

The NSDI Standard documents prepared need to be discussed, debated and, if endorsed, adopted for implementation. Now they are available on the web.

www.nsdiindia.org/www.nsdindia.com/www.nsdiindia.net

Map Policy and a NSDI Policy

The Map Restriction Policy needs review and modification if NSDI has to be implemented. Actually the restriction on digital map data (topographical sheets and other maps; including datum and projection parameters restriction) is an important issue for NSDI. The other issue is making the data available on networks and for public domain. In terms Ooty Communique, DST has prepared a note for discussion on policy issues.

There is a need for a NSDI policy that takes of from a modified and tuned map policy and defines the policy scope of each NSDI agency (SOI for topographical data, GSI for geological data, FSI for forest data and so on). The NSDI policy must address issues like—access rules, costing, copyright and licences, value addition and so on. This issue needs to be debated in the NSDI workshop. In June 2006, NSDI has got the approval of the Government of India.

Institutionalization of NSDI

How should NSDI be implemented? As decided in Ooty, a draft government resolution is ready for discussion amongst ministries. Further, after three years of effort by the task force on NSDI, the following have emerged.

- A clear understanding on the next steps and the institutional framework—Government resolution with an ultimate aim to have the act.
- NSDI Task Force to continue its effort in the direction resolved at Ooty.
- NSDI Agency to commit specific agency programmes and resources for Tenth Plan. Major thrust is to convert analogue data into digital and prepare metadata.
- Each agency designs its own business plan.
- Private sector to gear up for providing value added services and products to users.

Recognizing the need of accurate, consistent and updated geographical information, it is inescapable that the planners and decision makers should adopt an integrated approach for holistic planning and aided with geo-informatics, arrive at rational solutions. Reliable and complete information on various themes will facilitate better understanding of complex issues on conflicting objectives and reduce the conflicts in decision making, especially where the conflict has its origin in disagreement on 'facts' and specially 'spatial awareness'.

Keeping these aspects in mind, two major initiatives, viz. introduction of new series of topographical maps and establishment of NSDI taken by the Department of Science and Technology will go a long way to claim g-readiness of India to provide geographical information to resource users and industry for sustainable development of the nation.

Geospatial information, including maps and images, are vital to support decision-making at various level and implementation of action plans. With the availability of space borne imagery, Global Positioning System (GPS) data and Geographic Information System (GIS) technology, users are now able to process maps—both individually and along with tabular data and crunch them together to provide a new perception—the spatial visualization of information.

Space based images has the characteristic to cover vast areas and have larger information quality as well as multi-spectral and repeat observations ability, thus, it is suitable for regional and global environment monitoring. The application of satellite images not only used for mapping but also to monitor temporal changes in the environment.

The science of surveying and mapping encompasses a broad range of disciplines including Surveying and Mapping, Remote Sensing (RS), GIS and GPS. The technology of digital image processing, the global positioning system and GIS are being used to integrate, processed and analyzed the spatial data for sustainable development of the natural resources. The digital data (e.g. high resolution images, scanned aerial photographs, scanned maps and digital orthophotos) are currently available for this purpose. GIS is powerful system for decision-making tool is now in the hand of cartographers. Beginning with a computerized topographic map as its base, a GIS overlays and integrate graphic and textual information from separate databases. The end result is a customized and reliable tool that can support decision making and problem solving and provides almost instantaneous answers to complex questions.

In view of the present day requirement, Survey of India (SOI) plans to provide digital topographical databases for entire country on 1:50,000 scale initially, in a years time. Topographical data/maps in digital and analogue form on WGS-84 (World Geodetic System-84) is made available for general public without any restriction (existing topographical maps of SOI are on Everest datum). The department uses the state-of-the-art technology in glossarial data acquisition, management and dissemination. Airborne Laser Terrain Mapping (ALTM), GPS, Electronic Distance Measuring (EDM) instruments, digital photogrammetry, modern printing technology and computer hardware/software are being used in map-making process. High-resolution satellite imagery, viz. SPOT, IKONOS data etc. being used for updating of topographical maps. All these data sets would be useful in developing Spatial Data Infrastructure that will better facilitate the availability and access to spatial data for all level including Government sectors, Commercial sectors, Academia and citizens in general. Survey Training Institute of Survey of India also provides short training course on the insight to NSDI which includes topics like standards, nodes and network, capacity building, role of service providers, legal framework and metadata. (www.soisti.gov.in).

User Applications and Solutions with more Flexible Solutions

While technologies such as GPS and GIS have matured during the 1990s, increasing attention has been focused on the development of user applications and solutions. This technology segment can be defined as software/hardware solution bundles developed specifically to solve a geospatial information user's problem. The critical challenge in this area is for the technology supplier to gain an in-depth understanding of the user's business environment so that the solution is optimized to address the user's key business issues. Built on top of the core technologies, user applications/solutions are developed using combinations of database management, object-oriented programming and systems integration tools and techniques. Clearly, geospatial data and GIS were considered to be extremely important technologies for addressing solutions in a number of applications.

India, over the years, has generated a rich base of information through systematic data collection in the form of topographical surveys, geological surveys, soil surveys, cadastral surveys, various national resource innovative programmes and the use of remote sensing images. Further with the availability of digital topographic databases, high-resolution imageries and sophisticated techniques of data collection using GPS and tidal station, the accuracy of information content of these spatial data sets or map is very high. With the growing awareness for Information Technology among users, an initiative was taken by Department of Science and Technology in the form of National Spatial Data Infrastructure (NSDI). NSDI is the mechanism through which need for access to reliable, timely and spatial data are met. It is a first step in the direction of bringing together data providing agencies and to obtain commitment from them to make special data available and accessible to the user community, which conforms to accepted standards, The NSDI framework aims at decentralized approach to achieve the following objectives:

—Develop and maintain standard of collection of spatial data.
—Develop common solution for discovery, access and use of spatial data in response to the needs of diverse and new user groups.

—Relationship among participant's organizations to support the continuing development of NSDI.
—Increase the awareness of the vision, concepts and benefits of the NSDI.

This framework when fully implemented will be utilized for various developmental activities including preparedness, response and recovery purposes during disasters. The availability and use of spatial data shall have impacts on every aspect of society and will be made available to people who need them, when they need them and in the format at they can make decision with minimal pre-processing.

Open Geospatial Consortium (OGC)

For data standards at the international level, Open Geospatial Consortium or OGC has been established. It is a non-profit making international voluntary organisation. The core mission is to develop standards that enable interoperability and seamless integration of geospatial information, processing software and geospatial services. Geospatial information and processing encompass GIS, remote sensing, surveying and mapping, navigation, location based services, access to geospatial databases, sensor webs and other IT geospatial technologies and information sources. Further, its facilitates a consensus process to define, test and approve standards that promote the integration and use of geospatial data and services. In the forum provided by the OGC, participants from different information communities collaborate to achieve the vision of a world in which everyone is likely to benefit from geospatial information. Representatives of about 310 industries, academic and government organisations work together to make geospatial information and services accessible and useful with all kinds of applications.

In order to ensure that the OGC standards are harmonized with broader information technology standards, the OGC works closely with other standards bodies such as the International Organisation for Standards or ISO, the European Committee for Standardization or CEN, the World Wide Web Consortium or W3C and the like. The importance of OGC standards has been felt because the more one knows about the interoperability, the more he or she saves in technology lifecycle. The interoperability standards enable to go to users to increase their plug and play options. Developers and integrators go to market faster with more flexible solutions. Organisations join OGC not only to create the standards that enable such benefits, but also to be a part of the world's premier network of such users and providers of geospatial software, service and data. As a result, the technology developers submit interoperability requirements to meet anticipated needs of the market place.

The development of the standards does not require the members to give up any intellectual property or trade secrets. The use of open standards to connect components, applications and content—allowing a white box view on the components' functionality and interrelationships without revealing implementation details—fulfils the industry requirement for protection of intellectual property as well as the user requirement for transparency. Such transparency supports both interoperability and the credibility of the enterprise, or federated solution. Nevertheless, OGC offers its specifications free of charges to all and adheres to rigorous process to ensure that the specifications remain frees of royalties for use, a position taken by the World Wide Web Consortium as well. It promotes free and open web. For further details, log into the http://www.opengeospatial.org.

Conclusion

The characteristics of the geospatial data set are changing. First and foremost, in order to meet users demands effectively, the capacity for the real-time collection, synthesis and access must exist; data currency is essential. The data should be scale less, seamless, without artificial boundaries, and linked to a time component that has become critical to many applications, for example, traffic flow management, routing and delivery, and tidal and marine traffic. Moreover, as technologies become more advanced, geospatial information will be both more readily available and in greater demand.

There will also be a growing trend toward the collection and integration of non-traditional data using secondary reference systems like voting, culture and housing patterns, gender, sales and industry. Furthermore, as technologies and applications become more globally used, geospatial data will spread to and originate from non-traditional sources such as the voluntary sector, health councils, communities and peoples. However, regardless of what data are collected by whom, unless they are easily and readily accessible, their value diminishes; hence the importance of an exceptional geospatial data infrastructure. Furthermore, a well-developed national information infrastructure, enabling the dissemination and sharing of valuable, geographically referenced information, and with an ever-increasing audience of businesses, entrepreneurs, students and researchers, and communities, is widely accepted as an essential asset for any country to maintain and to advance its social and economic well-being. As such, geospatial data and the infrastructure in which they are organized can be considered to be a technology in its own right within the rubric of this technology.

NSDI is a reality and not just a dream of few visionaries. NSDI will be playing key role in the developmental activities of the country which has the requisite expertise, data, networking and other infrastructure. NSDI offers tremendous opportunities are there for the wealth of geospatial data to be for the benefit of all. The big picture is as follows (Fig. 8.8).

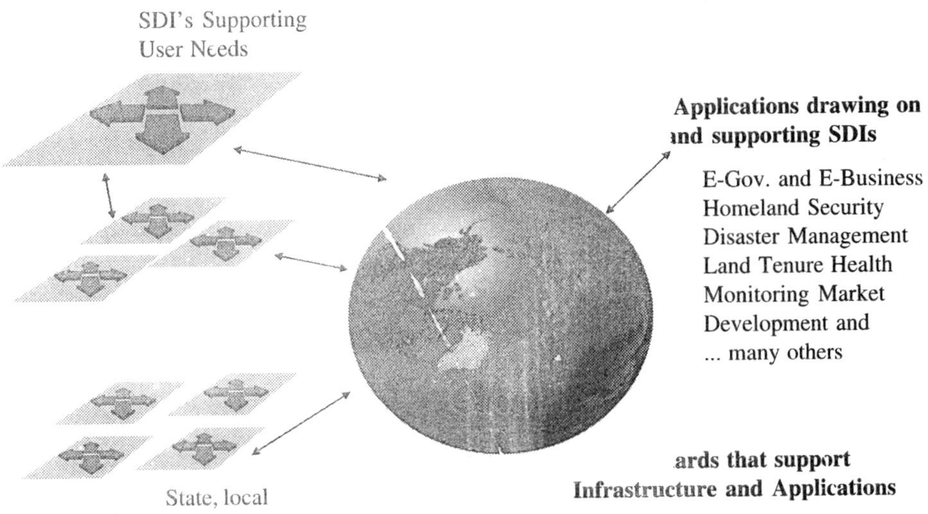

Fig. 8.8: The big picture of NSDI

Geospatial scientists must take advantage of the technology and the information available with it. The information will be available through net, web and other electronic media. The geographically referred information will be in the form of bits, bytes and pixels; and the volume of data in kilobytes, megabytes and gigabytes. This will also help in horizontal and vertical analysis of smallest areas such a building in a town or a parcel of land in the village. Further, the electronic flow of data will effect spatial applications in twenty first century by shifting (a) from so-called analogue theories to empirical models; (b) geographical enquiry from bottom up to bottom down; (c) developing indigenous models; and (d) revalidating established models, theorems and postulates.

REFERENCES

Adams, T. and S. Dworkin (1997), *WfMC Work or Handbook*, chapter Work of Interoperability Between Businesses, pages 211-21. John Wiley and Sons.

Alameh, N. (2001), *Scalable and Extensible Infrastructures for Distributing Interoperable Geographic Information Services on the Internet.* Ph.D. thesis, Massachusetts Institute of Technology (MIT), Massachussets, United States of America,.

Alfred Leick, Alfred (1995), *GPS Satellite Surveying*, John Wiley and Sons Inc.

Batty, M., *et al.* (1994a), Modelling inside GIS. Part 1: Model structures, exploratory spatial data analysis and aggregation, *International Journal of GIS*, Vol. 8, pp. 291-307.

Bernus, Peter and Laszlo Nemes (1996), *Modelling and Methodologies for Enterprise Integration.* In Proceedings of the International Federation for Information Processing (IFIP) Conference on Models and Methodologies for Enterprise Integration. Chapman and Hall.

Berry, J.K. (1987), Fundamental operations in computer-assisted map analysis, *International Journal of Geographical Information Systems*, Vol. 1, no. 2, pp. 119-136.

Bishr, Y., W. Kuhn, and M. Radwan (1999), *Interoperating Geographic Information Systems, Chapter Probing the semantic content of information communities: a first step toward semantic interoperability*, Kluwer Academic Publishers, pp. 211-21.

Blakemore, Michael (2004), Globalisation, infrastructure and agenda, *GIM International*, Vol. 18, No. 1, pp. 11-3.

Buehler, Kurt (ed) (2003), *OpenGIS Reference Model.* Version: 0.1.2, No. OGC 03-040, Open GIS Consortium Inc.

Burrough, P.A. and McDonnell, R.A. (1998), *Principles of Geographical Information Systems: Spatial Information Systems and Geostatistics.* Oxford University Press.

Coordinates (2005), NSDI in India:The reality behind the dream, *Coordinates*, December, pp. 34-38.

Dasgupta, A.R. (2003), Spatial data infrastructure: Concepts and efforts, *Geospatial Today*, Vol. 2, issue 4, pp. 16-20.

Forghani, Alan and Denise Gaughwin (2002), Spatial information technologies to aid archaeological site mapping, *GIS@Development*, Vol. VI, issue 6, pp. 31-35.

Gavin, Elizabeth (2003), Roadmap to SDI in Africa, *GIS@Development*, Vol. 7, issue 9, pp. 29-31.

Georgiadeou, Yola (2003), Back to future, *Geospatial Today*, Vol. 2, issue 4, pp.21-23.

Geospatial Today (2004), National SDI initiatives, *Geospatial Today*, Vol. 2, issue 5, pp. 25-46 (based on www.ngdf.org.uk).

Groot, Richard and John McLaughlin (2000), *Geospatial Data Infrastructure: Concepts, Cases and Good Practice.* Oxford University Press, Oxford.

Hutchinson, M.F. (1995). Interpolating mean rainfall using thin plate smoothing Spline. *International Journal of GIS*, Vol. 9, pp.385-404.

ISO/IEC (1996), *Information Technology—Open Distributed Processing—Reference Model: Foundations. Draft International Standard*, International Organization for Standardization, Geneva, Switzerland, No. 10746-2.

Konecny, Gottfried (2002), *Geoinformation: Remote Sensing, Photogrammetry and Geographic Information Systems*, Taylor and Francis, London.

Kottman, C. (1999), *Interoperating Geographic Information Systems,* (chapter on *The OpenGIS Consurtium and Progress Toward Interoperability in GIS*). Kluwer Academic Publishers.

Lieberman, Josua (ed) (2003), OpenGIS Web Services Architecture. Discussion paper, Version: 0.3, No. OGC 03-025, Open GIS Consor-tium Inc.

Luisa Liliana, Alvarez C. (2003), *Geoinformation virtual enterprises design and process management.* Masters thesis, International Institute for Geo-Information Science and Earth Observation ITC, Enschede, The Netherlands.

Masser, I. (1998), *Governments and Geographic Information*, Taylor and Francis, London.

Mitasova, *et al.* (1995), Modelling spatially and temporally distributed phenomena: new methods and tools for GRASS GIS, *International Journal of GIS*, Vol. 9, pp. 433-46.

Moellering, H., H.J.G.L. Aalders and A. Crane (eds)(2003), *World Spatial Metadata Standards*, International Cartogaraphic Association (forthcoming).

Mohammadi, Hossein, Abbas Rajabifard, Andrew Binns and Ian P. Williamson (2006), Bridging SDI design gaps, *Coordinates*, Vol. 2, issue 9, pp. 26-29.

Morales, Javier and Mostafa Radwan (2002), Extending geoinformation services: A virtual architecture for spatial data infrastructures, in *Proceedings of the Joint International Symposium on GeoSpatial Theory, Processing and Applications* ISPRS Commission IV, Ottawa, Canada.

Nag, P. (2002), Maponomics – Map products and their commercialisation, *GIS@Development*, Vol. 6, issue 2, pp. 19-23.

NSDI (2002), *National Spatial Data Infrastructure: Strategy and Action Plan*, Department of Science and Technology, New Delhi.

NSDI (2003), *Data Exchange Format – NSDE Version 1.01*, National Spatial Data Infrastructure, New Delhi.

OGC (2002), *OGC Web Services*, DVD NTSC, Open GIS Consortium, Inc.

OGC (2006), *OGC Web Services*, OWS-3 Demonstration, Open Geospatial Consortium, Inc. DVD Rom/ Video.

Percivall, George (ed)(2002), The Open GIS Abstract Specification Topic 12: Open GIS Service Architecture. Version 4.3, Open GIS Consortium, Inc.

Pulusani, Preetha (2003), Interoperability—Trend or reality ?, *GIS@Development*, Vol. 7, issue 9, pp. 16-19.

Rajabifard, A. and I.P. Williamson (2002), Spatial Data Infrastructures: An initiative to facilitate spatial data sharing, in R. Tateishi and D. Hasting (eds), *Global Environmental Databases—Present Situation and Future Directions*, Vol. 2, International Society for Photogrammetry and Remote Sensing (ISPRS-WG IV/8), GeoCarto International Centre, Hong Kong.

Rajabifard, A., M. Feeney and I.P. Williamson (2002), Future directions for SDI development, *International Journal of Applied Earth Observation and Geoinformation*, Vol. 4, pp. 11-22.

Rhind, D. (ed) (1997), *Framework for the World*, Geoinformation International, Cambridge.

Roy, P.S. and Samir Saran (2003), An interoperability model for Indian NSDI, *Geospatial Today*, Vol. 2, issue 4, pp. 36-8.

Satyanarayana, P. (1999), Open GIS consortium, *GIS India*, Vol. 8, April-June, pp. 36-37.

Sipkes, Jacques (2004), AfricaGIS 2003: How far have we come ?, *GIM International*, Vol. 18, No. 1, pp.55-59.

Siva Kumar, R. (2005), Standards and interoperability: For an Indian standard, *Geospatial Today*, Vol. 4, issue 2, pp. 54-56.

Williamson, Ian, Abbas Rajabifard and Mary-Ellen F. Feeney (eds)(2003), *Developing Spatial Data Infrastructures: From Concept to Reality*, Taylor and Francis, London and New York.

Williamson, Ian and Abbas Rajabifard (2004), Capacity building for SDI in geo information science, *Geospatial Today*, Vol. 3, issue 4, pp. 31-39.

9

Status of GIS in India

Since very beginning, cartography in India had been considered very secretive and protective and the map-making activities lie within the four walls of the obscure government offices in the country. Maps were considered to be confidential document and only to be consulted if there are no other option. However, a few industries did take up cartography and map printing, not always as their prime activity. Since independence and till the time of the advent of the computer-assisted cartography, the scenario remained almost same. The digital cartography got impetus with (a) introduction of the Geographical Information System (GIS) and (b) easy availability of remote sensing data, that too from the Indian remote sensing satellites.

In India, digital mapping and GIS related activities in the public sector and industry go in different directions. However, such activities are either sponsored by the government or are from abroad. The priorities, policies, nature of work and issues are entirely different. This was due to the so-called Map Restriction Policy which was in force till recently. A New Map Policy has been announced which is meant to promote, apart from other things, the GIS industry in the country (Appendix I). As per this policy, the new maps, known as Open Series Maps (OSM), will have WGS-84 datum, UTM projection, a different map numbering system and reduced content. This series in hard as well as soft copy will be made available to users without any restrictions. Hence, OSM is likely to boost GIS activities further in the country. Nevertheless, the priorities have shifted more towards detail maps and GIS applications, particularly for urban areas. Hence, digital data at a larger scale is to be generated.

Within the fold of government, there are institutions, which generate the basic data for mapping such as, Survey of India and the National Remote Sensing Agency under the Department of Space. Then there are institutions, which have major activity for preparing maps required for their specific use, particularly, mapping of natural resources. Their cartographic divisions have repository of relevant information. These institutions are Geological Survey of India and the Forest Survey of India. Then there are institutions like the office of the Registrar General and Census Commissioner which collect data on human resources and also prepare maps and atlases related to population. The case of NATMO is somewhat unique. In this case, the information is collected from every possible source and maps of relevant themes are prepared.

Survey of India (SOI) was established in 1767. It is the national survey and mapping organisation of the country. Survey of India has already accomplished various landmarks in the field of cartography including the mapping of the entire country on the scale of 1:50,000, provided high precision geodetic control survey etc. Besides its commitment of topographic mapping of the country, SOI provides infrastructure data for national development programmes, such as survey for road, rail, tunnel, alignment, irrigation, large scale river valley project, mining and excavation. With the advent of new era in computer and space technology and its rapid proliferation in every sphere of human activities, SOI has also stepped in and taped the immense potential of this modern technology. Initially the digital cartography was adopted as an R and D activity with the installation of Automap System in 1981. These R and D activity results were encouraging and it was decided to adopt the state-of-the-art technology at operational level. Thus during the years 1986-89, three centres, viz. Modern Cartographic Centre (MCC) a project with UNDP assistance and two digital mapping centres (DMCs) at Dehra Dun and Hyderabad were established in Survey of India. These centres of SOI were fully equipped with state-of-the-art technology of digital mapping and are fully operational to process the digital cartographic databases required for planning and developmental activities. The 1:250,000 and 1:50,000 series of maps have been digitized. At present, 1:25,000 series is being digitized. The availability of data at these three scales will provide much required basis for the promotion of GIS activities in the country. For this purpose, Survey of India has opened geospatial data centre for each state.

National Atlas and Thematic Mapping Organisation (NATMO) also under the Department of Science and Technology is a pioneer and nodal organisation for preparing atlases and thematic maps of the country. It was established in 1956. As per technological advancement in different field of cartography, NATMO has also adopted newer technologies in preparation of different maps on different themes, area and scales.

NATMO has a special division on Digital Mapping and Geographical Information System. This division is very efficiently preparing digital thematic maps. Digital Mapping System (DMS) has started the digitization of spatial, statistical and non-spatial databases available in the organisation. It has already generated data under the NRDMS programme for a part of Bankura district, West Bengal. Further, a map on *Infrastructure Profile of West Bengal* has been also produced by this division for the West Bengal Industrial Development Corporation. The division is also coordinated the scanning and digitization of SMART maps of Calcutta Municipal Corporation (CMC) with an industrial partner (IDS) and CMC, Kolkata. A project is also in progress for preparation of *Satellite Atlas of India* in collaboration of the Department of Space.

Geological Survey of India (GSI), one of the largest scientific organisations of the Government of India, was founded in 1851 and has been carrying out different geo-scientific activities of the entire land and recently also of offshore areas of India (covering EEZ) with aims and objectives enshrined in the charter of functions as laid down by the government. It has played a vital role for development of country's economy in the industries and mineral, power or energy, agriculture, irrigation and water management sectors and in environmental and natural hazard studies. Geological and other geo-scientific thematic maps on different scales are some of the major end products of GSI activities. Geological maps on 1:50,000 scale form the basic documents and are prepared primarily based on ground surveys. These are also supplemented at times by aerial photos, satellite imagery and various remote sensing techniques and aero-geophysical mapping. Huge amount of map data in analog forms have been collected through a long period of about 140 years by geo-scientific activities. Since geological data are collected mostly in analog format and not always quantifiable, conventional cartographic techniques are being followed for compilation of geological maps. However, it is intended to introduce digital cartography in phases so that the colossal information so far generated could be stored and retrieve efficiently thereby avoiding time gaps between the dates of surveys and the publication of the results. Airborne Mineral Surveys and Exploration (AMSE) wing of the GSI is already processing and interpreting the airborne geophysical data in the form of maps by computer assisted methods in the Geophysical Mapping Centre at Bangalore. Application of GIS is also being attempted for storing and suitable retrieval of various geo-scientific and mineral data.

The Census of India is one of the largest producers of maps in South Asia. This has been achieved with intimate collaboration among geographers, statisticians and demographers. Census of India has been a discontinuous affair upto 1861. The restricted programme in 1941 due to Second World War restrained the cartographic activities. In 1951 Census, one of the major contributions was excellent *tahsil* or *taluk* maps published in *District Census Handbooks.* In 1961, *Census Atlas Project* was initiated for the first time while urban land use maps in state level town directory were introduced in 1971. It was also proposed to bring out administrative atlases for states and union territories, which may serve as a handy guide to the administrators and planners. In 1991, in view of the scientific and technical advancement, the Registrar General has initiated another plan project designated as 'Computer Assisted Cartography and Data Dissemination' to map the towns at ward or enumeration block (EB) level. The main objective of this plan scheme is to study the cities or towns through various mapping and cartographic techniques, fieldwork and with the introduction of computer aided cartography which aims at the generation of selected city or town maps at enumerative base (EB) level and the analysis of growth of urban centres through decades from different view points caused by religious, administrative and industrial stand points. The scheme also envisages preparing and updating selected town maps on large scale and create digital database for dissemination of census results as well as for uses in subsequent censuses and surveys etc. Final base maps of cities or towns were to be prepared on suitable scale indicating the ward. The base maps were to be utilized to generate thematic maps, utility maps, land use maps of city or town at micro level through computer. Similarly, the enumerative base maps will be digitized for further updation and use by various survey for conducting census and other vital statistics programme. Work has been initiated in a moderate way on a PC platform at Registrar General India (RGI) headquarters using ARC/INFO GIS software. Similar facilities have been extended in different states to generate cartographic database at town or ward level for further work on utility maps at micro level for area specific planning. In addition to pre-census and post-census cartographic activities, the census organisation also caters to the cartographic need of different departments and ministries.

In 1972, the successful launching of Landsat by the US provided a new source of data which has a lot of flexibility and with special processing, such data can be tuned for a variety of applications. Though the remote sensing establishments based on US data and later French data (SPOT) were established in late 1970s and early eighties, the successful launching of IRS 1A in 1988 provided a boost and a different orientation to GIS-related activities in the country. Space based observation system provide timely and accurate information for monitoring and optimum management of earth resources. The main advantages of space-based remote sensing are: (a) synoptic view-wide area can be covered by a single imagery; (b) repetitivity—can collect the data of any area repeatedly; and (c) coverage of inaccessible areas like mountains, swampy areas, thick forests and deep sea. To day, the IRS system, comprising several Indian remote sensing satellites, is providing data for monitoring and managing our natural resources. The payloads on board the IRS satellites operate in different spectral bands and provide varying spatial and spectral resolutions.

The first operational remote sensing satellite, IRS-1A, was successfully launched on March 17, 1988. Even though designed for three years, IRS-1A served for almost eight years. An identical satellite, IRS-1B, launched in August 1991, is still in service. IRS-1B is placed in the sun-synchronous polar orbit of 904 km with equatorial crossing time of the descending node being at 10.25 a.m. The repetitivity cycle is 22 days. The payload on this satellite comprises two Linear Imaging Self-Scanning Sensors, LISS-I and LISS-II. While LISS-I provides a spatial resolution of 72.5m, the LISS-II sensors provide a spatial resolution of 36.25m. LISS-I has a swath of 148 km the two LISS cameras provide a composite swath of 145 km. The sensors operate in four spectral bands—0.45-0.52 micron; 0.52-0.59 micron; 0.62-0.68 micron and 0.77-0.86 micron. IRS-1C, launched in 1995, and IRS-1D was launched in 1997 incorporate sophisticated sensors with resolutions of around 23m in multi-spectral bands and 5.8m in the panchromatic band. They have stereo-viewing capability and on-board data recording facility (Roy, 2004). Cartosat-1 has been successfully launched in 2005. Its pixel size is 2.5m for panchromatic band. The data will be made available soon which should give a big boost to GIS activities. The Cartosat-2 with pixel size less than 1m is planned to be launched in a year or so.

Planning Commission recognized the importance of establishing National Natural Resources Management System (NNRMS) and within its umbrella, the Regional Remote Sensing Service Centres were opened for the satellite data processing for specific application of the concerned region in which they were located. Due to better launching facilities and higher resolution of data, the cartographic activities are likely to improve further.

An Integrated Mission for Sustainable Development (IMSD) was launched in 1992 which now covers 174 districts in the country. Under IMSD, locale-specific prescriptions are arrived at mainly using IRS data, merged with other collateral social economic data using GIS. The mission involves generation of thematic maps showing current land use or land cover, types of wastelands, forest cover or types, surface water resources, drainage pattern, potential ground water zones, land forms, geology (rock types, structural features, mineral occurrences), soil types, etc. Maps showing slope aspect is prepared using topographic contour information and the meteorological data is collected from existing data-bases. Using all these information map showing land capability, irrigability, soil erosion, run-off potential, priority watersheds needing immediate treatment, etc., are prepared. The action plans, for example, will include identification of sites for water harvesting through ponds and check dams, soil conservation through terracing and contour bunding, afforestation, agro-forestry and agro-horticulture, fuel wood and fodder development, sand-dune stabilization, mining and necessary conservation measures.

National Remote Sensing Agency (NRSA), an autonomous organisation under Department of Space was established in 1978. One of the primary responsibilities of NRSA is to provide remote sensing data not only from the Indian IRS series of satellites and from other satellites, such as Ikonoss or SPOT. The services provided by this agency include (a) generation and dissemination of standard and special satellite data products, (b) generation of large-scale line maps, (c) 3-D digital data generation for input GIS and generation of DTM or DEM, (d) long-term and short-term training courses in the field of remote sensing, GIS and allied areas at

IIRS, Dehara Dun and NRSA, Hyderabad, and (e) GIS as decision support system for natural resource management. All these activities have supported GIS activities in the country. Further, the world is being exposed in yet another aspect of the revolution of information technology the increasingly widespread acceptance and use of GIS. Owing to their strong analysis capabilities, GIS is playing an important role in the management of a country's natural resources. Efficient use of GIS is providing decision-makers, the information, for the proper utilization of these resources. Keeping this in view, NRSA has acquired several state-of-the-art GIS and applications are being developed using them.

Space Applications Centre (SAC) of the Indian Space Research Organisation (ISRO) at Ahmedabad is engaged in the research and development activities of remote sensing technology and GIS applications. The major activity of this centre are design and development of optical and infrared payloads for space and airborne platforms, design and development of microwave remote sensing payloads (both active and passive), image processing and data processing related software development and operationalization of applications and development of models for meteorological and oceanographic applications. SAC has been actively engaged in ISRO satellite programmes through contribution of payloads for IRS series of satellites, the processing of payload data for data products generation and their applications. SAC has been closely interacting with various user agencies and departments to developmental methodologies for remote sensing and GIS applications in the field of resources survey and environmental monitoring under the NRIS programme (Dasgupta, Rao and Gopalan, 2000). Some of the notable applications that have been carried out are the crop acreage and production inventory, coastal zone mapping, wetland mapping, monitoring of forests, geological applications for mineral targeting and the like.

Forest Survey of India (FSI) deals with forest mapping at national level by making use of GIS and remote sensing techniques. Its predecessor organisation, the Pre-Investment Survey of Forest Resources (PISFR), took up aerial photo interpretation and mapping for forestry purpose for the first time in India in the year 1965 with the help of FAO experts. In 1968, FAO collaboration ended successfully and by that time the Indian counterpart acquired enough expertise in this field to carry out work independently. From 1st June, 1981, PISFR was converted to Forest Survey of India and reorganized with effect from 1st July, 1986, under the Ministry of Environment and Forests, Government of India. FSI prepares thematic maps on 1:50,000 scale by interpretation of existing latest aerial photographs on ten years cycle for the whole country and forest vegetation maps on 1:250,000 scale by interpreting satellite imagery in two years cycle. Maps under special projects are also prepared. Thematic mapping and vegetation mapping activities are described as under.

The conventional method of map production is tedious and time consuming. The advent of computer technology for handling spatial data sets has brought in powerful techniques for analysing and archiving map data. It increases flexibility in map making, as the digital cartographic databases are capable of carrying much more information that could be legibly presented on any one map. Digital mapping combined with other data sources can be used for variety of end uses such as forest cover monitoring and mapping selection and location of plantation sites, environmental change detection, assessment of damage done to forest, estimation of area under shifting cultivation etc. Data in digital form facilitate linkage between all other data sources—which are both physical (forest cover, soil capability classes, elevation, slope), and socio-economic (demographic distribution, marketing facilities, production and demand etc.). Digital cartographic system and GIS have been installed in FSI at Dehra Dun. The FSI has also taken up special projects for preparation of land use maps in Bhutan and Nepal. Further, this institute has developed a GIS model and village database for Uttari Kosi sub-watershed (Almora) for watershed management and wasteland development at micro-level planning during 1993. At present FSI is preparing a GIS project using remote sensing and GIS techniques for Sariska Tiger Reserve and its environ.

Natural Resources Data Management System (NRDMS)—a multi-disciplinary programme of the Department of Science and Technology, Government of India aims at developing a scientific data based approach for operationalising the concept of decentralized planning. Launched in 1982, the programme is a major initiative of the government to promote research and development-cum-demonstration in the field of GIS technology and its application to the problems of natural resources management. During the Eighth Plan, the focus has been on preparation of spatial district resource profiles, upgradation of the low cost GRAM GIS package, area specific application studies and training of users, academic institutions. Centre of Studies in Resource Engineering (CSRE), Indian Institute of Technology and R&D laboratories provided the technical back up. Priority areas have been identified for research and development. Over the last few years the areas of research studied were (a) land system classification, (b) watershed modelling, (c) energy planning, (d) groundwater budgeting, (e) large scale resource mapping, and (f) landslide studies. NRDMS is now poised towards further consolidating these efforts by upgrading the GRAM package to suit the data processing requirements of local level planning. Concerted efforts are being made to develop kits and tutorials for the training of potential users so that the methodology strikes roots in the resource management, practices at different levels of the planning hierarchy. UNDP supported NRDMS in this endeavour.

National Informatics Centre (NIC) was set up in March 1975 by the Government of India to play a promotional role in creating computer awareness and for developing and implementing computer-based information systems for decision support in the ministries and departments of the Central Government and the corporate sector. NIC has set up a satellite-based nation-wide computer-communication network, called NICNET, with over 1,000 nodes connecting the national capital, the state capitals and district headquarters to one and another. The services provided by NIC includes development of application software packages, systems software, database systems, distributed databases, networking, electronic mail, Electronic Data Interchange (EDI), access to international databases, computer-aided design, GIS, modelling and training.

The spatial information and GIS are fast becoming an important prerequisite in most government departments. With the development and availability of more than 2,000 data bases on NICNET consisting data of almost every activity of the government, it was felt to encompass the spatial dimension of data into various databases. Applications ranging from demographic census to soil conservation, pollution control to location of a facility, all require one or other spatial information. After getting acquainted with the power and usefulness of computerized information, available through various databases on NICNET, there has been sufficient awareness and demand for spatial data component. Further the decreasing cost of hardware, versatile graphic peripherals and high speed network have made the prospects of developing spatial data bases at decentralized location a viable preposition.

NIC has developed an indigenous GIS software, GISNIC, with user-friendly interface and complex GIS functions built into an integrated package. GISNIC is designed to provide the users with a set of tools for retrieval, transformation and analysis of both spatial and non-spatial data which provides an environment to manipulate and manage locational and thematic data. The product comprises of modules DIGITIZE for digitization, MAPBASE for generating thematic maps with facility to query on attribute data base, PLOT for off-line generation of hard copy maps and SHOWMAP a callable routine for integrating maps with existing applications.

NIC has initiated the process of making many large spatial databases available on NICNET. GISNIC is being used for the development and retrieval of these databases. It is also being used as a presentation tool for preparing thematic maps for deriving the attribute information from the existing databases.

NIC has also set up the facilities for processing of remote sensing data with programmes to utilize data from IRS and other satellites. A major component of project under remote sensing programme envisages development of national level thematic database. Software design and development in this connection has been for processing raster data. Indigenous software on NICNET would provide a powerful tool for processing images, which would yield the thematic maps and strengthen GIS programme for spatial data analysis. NIC has been encouraging various pilot projects in the area of GIS and thematic mapping with a view to establish the procedures and technology. This technology can be horizontally transferred to various districts and states. NIC has been working in the area of natural resource informatics for sustainable development at grass root level.

In order to support the city administration NIC has also established a group, which looks after the mapping of utility lines. This group is fully equipped with digital photogrammetry equipment, GPS receivers and treated software. The group offers its services for establishment of utility mapping centres in ever-growing cities. The group at NIC is also involved in a facilities management project for entire National Capital Territory of Delhi. In this project an arbitrary grid has been established for Delhi. Same grid has been used for digital base mapping by NRSA. The base maps are prepared at 1:1,000 scale. These maps are then used for overlaying underground utility networks. The planned utilities include water, sewage, electricity and the road network. The objective is to support the city administration and the respective utility organisation. The project is also helping in the better management of city traffic and fire-fighting management. NIC has executed several projects on natural resources data management system for sustainable development at grassroots level through GIS based mapping.

Within the fold of the state governments, there are two major sources of cartographic information: (a) cadastral map available with the land records and (b) detailed ward-wise maps with the Corporations and Municipal bodies. These maps are at a standard scale but they are old and almost in all cases, serious efforts are to be made for its updation and validation. Such maps are also important because due emphasis has been given on infrastructural development in urban and rural areas of the country. *Panchayati Raj* system has been empowered with 73rd and 74th Amendments of the Constitution for taking up the developmental issues at the lowest level. Both types of maps have been used for GIS purposes as well.

Municipal GIS: Example of Kolkata

Several examples are now available where municipal GIS has been introduced in urban areas. Such cases are generally sponsored by the municipal bodies or by the government. Universities and the NGOs have taken up such projects as well. Further, in several cases, there have been joint efforts in this regard. The case of Kolkata Municipal Corporation (KMC) is a typical example of such initiative. About sixty per cent of its area is covered by Smart's maps which were prepared during the early part of the last century (1903-10). The example of Smart's map of a part of Calcutta, (now Kolkata), is included here (Fig. 9.1).

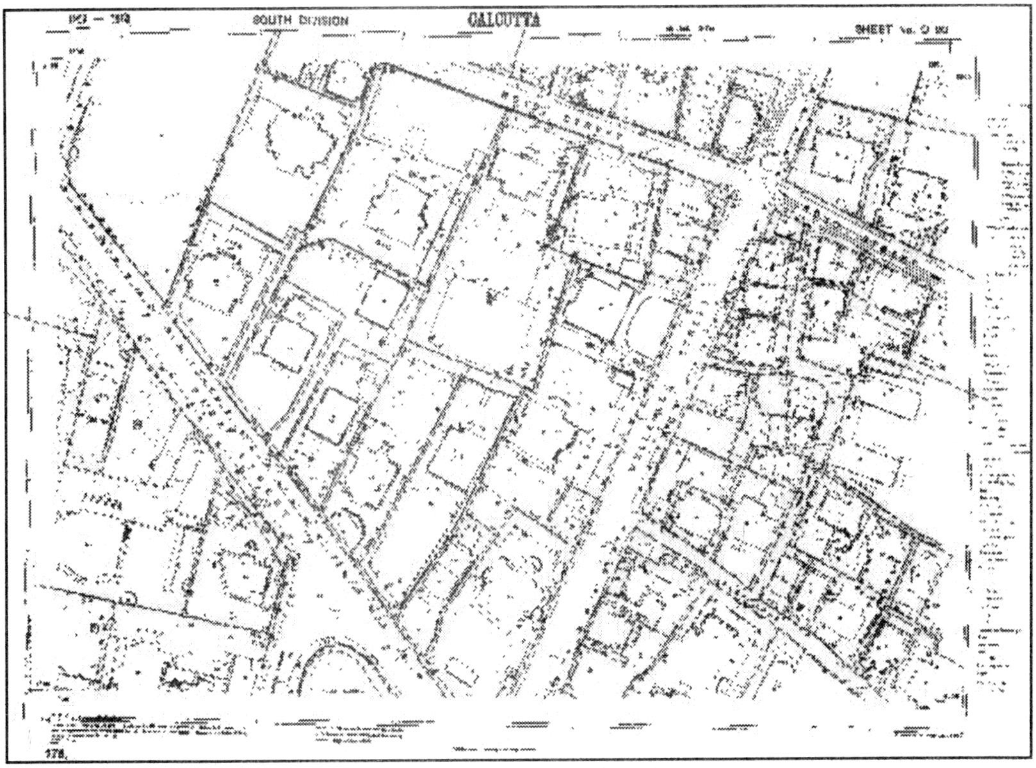

Fig. 9.1: Smarts map of a part of Kolkata (Ballygunge)

These maps contain some physical features and property boundaries and are considered as a legal document. Obviously, it has become necessary to update these maps along with other relevant information. Hence, a project was taken up by KMC in collaboration with the Kolkata based National Atlas and Thematic Mapping Organisation (NATMO). Integrated Data System (IDS) from the GIS industry was engaged in this project as well (Nag, 1999). The objectives were as follows:

(a) To scan the old Smarts maps with out making changes.
(b) To keep film positives/negatives of the original maps as a backup.

(c) To incorporate the latest changes on a separate file.
(d) To attach data base required for record and tax administration.

It was felt that the digital data would be useful for following activities:

(a) Property records
(b) Development and planning initiatives
(c) Revenue collection
(d) Development of GIS
(e) Decentralization of development and taxation processes.

Recently, Kolkata Municipal Corporation or KMC has taken the implementation of Municipal GIS with the funding of UK's Department for International Development (DFID) and technical support of Tata Consultancy Services and the National Atlas and Thematic Mapping Organisation (NATMO). KMC's vision is: "*KMC will be efficient, effective, equitable, citizen responsive, financially sustainable and transparent, delivering quality services to its citizens.*" The objective of the project is to develop a GIS based system to assist KMC in strategic planning and resource mobilization and utilization, streamlining municipal taxation (especially the property tax, but encompassing all the taxes and fee structures as envisaged under the Municipalities Act, 1980) and management functions, and planning and management of day-to-day operations in all sectors of activity. In addition GIS would provide a dynamic information base on poverty and poor settlements in the KMC to serve as a basic planning, monitoring and evaluation tool to improve governance of poverty alleviation programmes particularly the targeting and the delivery of services.

In KMC views computerization as a very important tool, which can be used to convert its vision into reality. Whether it is making KMC more efficient and effective in its operations or improving the financial health of KMC, or making KMC more responsive to citizens' needs and making its operation more transparent, the role of computerization cannot be overstated. With a view to resolve all the problems inherent in the current scenario, an integrated and comprehensive Municipal Administration System (MAS) is proposed to be developed by a separate group of consultants called MAS consultants. The proposed MAS would comprise of the following main components:

1. Municipal Transaction Processing System (MTPS)
2. Municipal Data Bank
3. Citizens Interface System
4. Municipal Geographic Information System (MGIS)
5. Municipal Management Information System (MMIS)

A suggested schematic diagram is given here (Fig. 9.2).

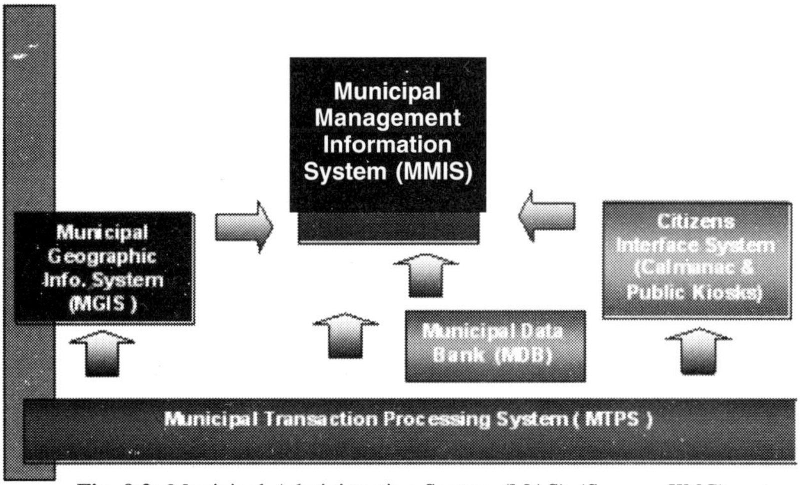

Fig. 9.2: Municipal Administration System (MAS) *(Source: KMC)*

The Municipal Transaction Processing System (MTPS) would provide the on-line transaction-processing interface, which would be used by operational users to key in all the transactions. This component would comprise of the various functional modules that would take care of the working of all the departments of KMC. The following functional modules are envisaged for TPS. This module would cover the following functions (Fig. 9.3):

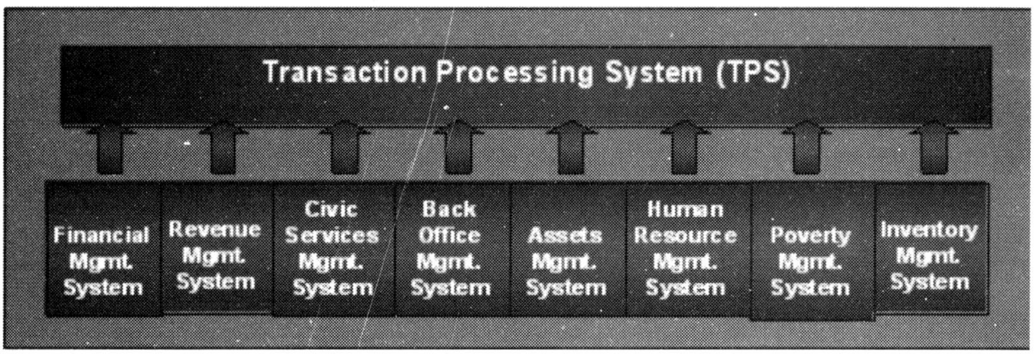

Fig. 9.3: Schematic diagram of TPS (*Source: KMC)*

- Finance
- Treasury
- Auditing
- Budgeting

Being the central module, this system would be linked to all the other management systems. Any financial transaction carried out in any department would automatically generate the required accounting entry in this module. Further, an important element in the process of strengthening planning and management capabilities is the development of an integrated Geographical Information System (GIS) for KMC. This will strengthen the strategic planning capacities and capacities and day-to-day management of core municipal functions by the local bodies. In this regard, it is proposed to implement a fully integrated GIS in KMC which has a population of roughly 4.6 million citizens and the city spread over an area of 187 sq. km.

This GIS would help the KMC in restructuring and rationalization of its revenue from various sources such as property tax, water supply, rental from municipal properties etc. and would provide a platform for effective planning, operation and maintenance of municipal utilities and infrastructure. It would also enable KMC to concentrate on improving access to municipal services in poorer areas of Kolkata. So the GIS would provide the different thematic layers superimposed on the base map of Kolkata. As in any GIS, the purpose here is to attach KMC's attributes data as different thematic layers to the spatial data. The air photograph based digital base map was prepared by the National Remote Sensing Agency (NRSA). Example of such map is as follows (Fig. 9.4).

The GIS proposed to be developed should have the following dynamic features:

(a) Interfacing the alphanumeric data from the Transaction Processing System of MAS, or any other deemed fit by KMC/Consultants.
(b) The updation of data (spatial or alphanumeric) should be dynamic in nature so that the GIS always produces up-to-date information of reality on ground.

The development of enterprise GIS map for this purpose has been shown in Fig. 9.5.

On the rural scene, all the states and union territories of the country have a directorate of land records. The directorate is on state level and there used to be subordinate offices known as superintendent of land records at district level and further as revenue inspectors or *kanoongo* at *tahsil* or *taluk* level. In these offices various kind of maps are prepared for different level. The directorate prepares and update the maps at state level that is when there use to be any change in administrative boundaries of state or districts or other units the map is redrawn and updated. Similarly, if any change at *tashil* or block level happens, then at the office of Superintendent of Land Records (SLR) the district and *tashil* level revenue maps are updated. At the *tahsil* office the cadastral or village maps are updated or redrawn. The cadastral map are updated or redrawn after the consolidations of landholding are made. This process is generally goes on periodical bases for one or the other village. Thus in the directorate of land records or department revenue, cartographic or map making is a permanent activity. From time to time the directorate of land records also publishes statistical maps and atlases based on available up-to-date statistics on the development of the states. Such maps are also published by the department of agriculture showing land use, crop distribution, crop production, soil quality, irrigation etc.

Now, many state governments' directorate of land records have introduced or adopted the technology of LIS or GIS and GPS for the management and maintenance of their land records and related matters. Government of Madhya Pradesh was the first state to adopt this modern technology. The other states like Andhra Pradesh and Karnataka, are following the same. It may be concluded that the map making and cartographic activity is widely done in almost in all government departments in one form or the other.

Fig. 9.4: Example of Air photographs based digital map of a part of Kolkata City by NRSA

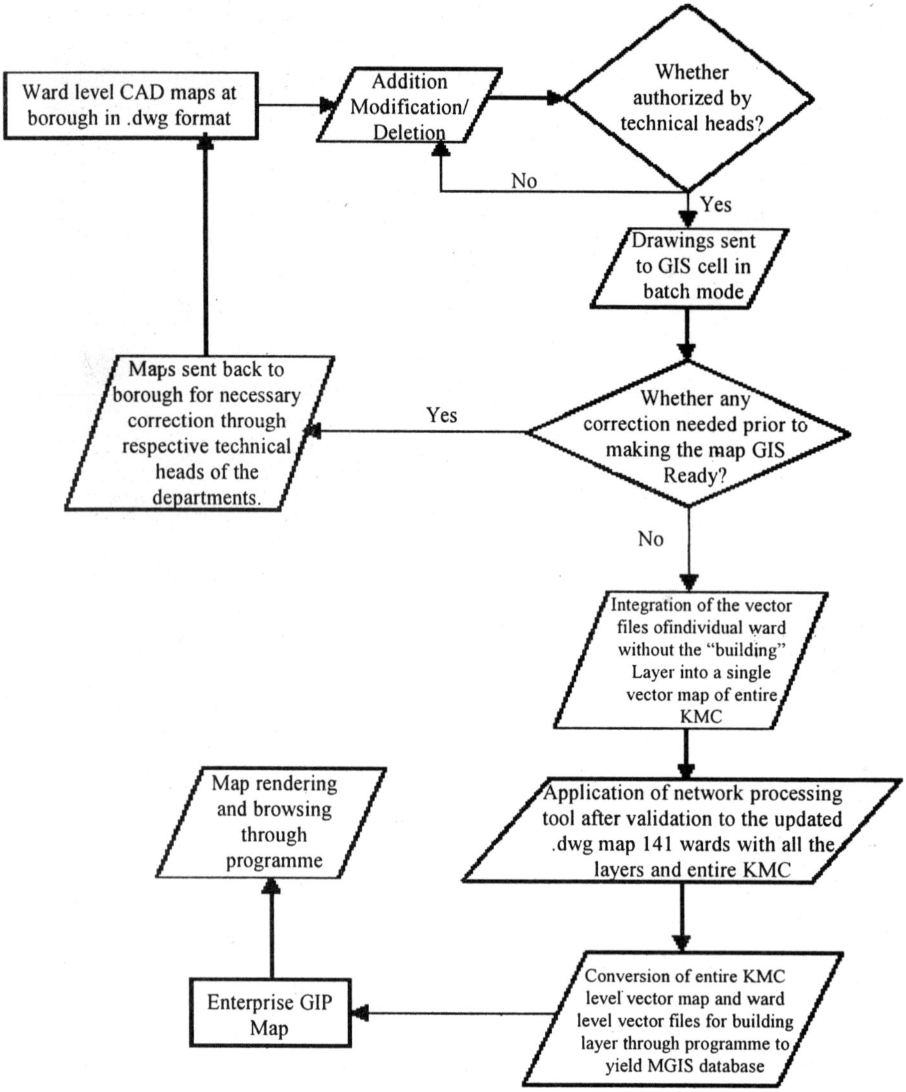

Fig. 9.5: Enterprise GIS map (*Source*: KMC)

GIS in Indian Industry

In the Indian Industry also the cartographic activities are fairly old as well. The making and production of maps for educational purposes and guide maps for tourists and pilgrimage were largely made by private publishing industries or companies. But now in the age of Information Technology (IT), industry is also involved in production of different kinds of maps and graphics. The GIS has revolutionised the field of cartography. In fact, GIS industry has taken over cartography outside the folds of the government. About hundred IT companies in the country are involved in GIS related activities. They are mostly involved in the application GIS packages in exploration, natural resource management and telecommunication, power, ports, roads, soil, water, forest, land, etc. However, it is difficult to find out the full strength of the industry since projects handled by them are hardly publicized. One has to go through their web sites, annual reports, company profiles, conference proceedings and through personal contacts in order to make some assessment. National Spatial Data Infrastructure is trying to involve them but not yet fully successful. There is no association or confederation of GIS industry in the country. Attempts made so far have not yielded any results. Based on the information available, the activities of some of the leading companies or industries in the area of GIS are as follows:

(a) Supplying GIS software
(b) Development of digital bases for GIS applications
(c) Undertaking GIS projects basically from government sponsored activities
(d) Undertaking digitization and GIS projects from abroad, primarily from the west
(e) Organising conferences, exhibitions and events related to GIS
(f) Bringing out GIS related journals

GIS related activities of some industry are described here. This list is not exhaustive but only indicative. Wherever, the Indian industry has taken up GIS projects, they are mostly government sponsored (Marutish, 2004). The examples are regarding enterprise system for electric utilities, command and control system for safety. Recently Surveys of Indian and GIS industries have been made by CSDMS and Spatial Network Pvt. Ltd. (www.spatialindia.com). They have been taking up projects on local problems as well. An example of Chennai based WTI Advanced Technologies (www.wtiatl.com) is given in Fig. 9.6.

Besides the above mentioned companies or industry there are many other industries who are engaged in preparation and production of maps. These are the text and other books publishers, wall and guide or tourist city maps producers. The newspaper publishing houses also do a lot of cartographic or map making jobs. Unpublished maps are available in large numbers in the theses submitted by the students for their different degrees in the educational institutes also.

As apparent form the above, the industries which are involved in cartographic work are of different sizes and specialization. There are purely cartographic institutions who have been dedicated to this work as their primary field of activity. Over time, they have changed their orientation due to digital methods and the quantum jump in technology. However, such activities in industry are of recent past. They got impetus due to the possible higher profitability in this area owing to the introduction of automated or computer assisted systems. Nevertheless, the units of this industry have not only kept the cartographic work as their sole objective. There is a set of back-up activities on which they can fall, in the event of lack of activities in cartography. Such back-up activities are related to sale of hardware, software, data product, specialized services, conversion activities and supply of qualified manpower. At the lower end, these companies do carry out data processing, DTP, project evaluation and related activities. Hence, their outfits can take shape as per the market forces. In fact, we could find out the units, which can raise highly train personnel and infrastructure as and when required. They are waiting for projects, which may exclusively or partially fall in the arena of cartography. Further, from the examples given above, some industries have been closed down, some have closed GIS activities while some have modified their vision and action.

Unfortunately, the origin of almost all the cartographic projects in the country is from government sources and this industry looks forward to Central and State Governments or their

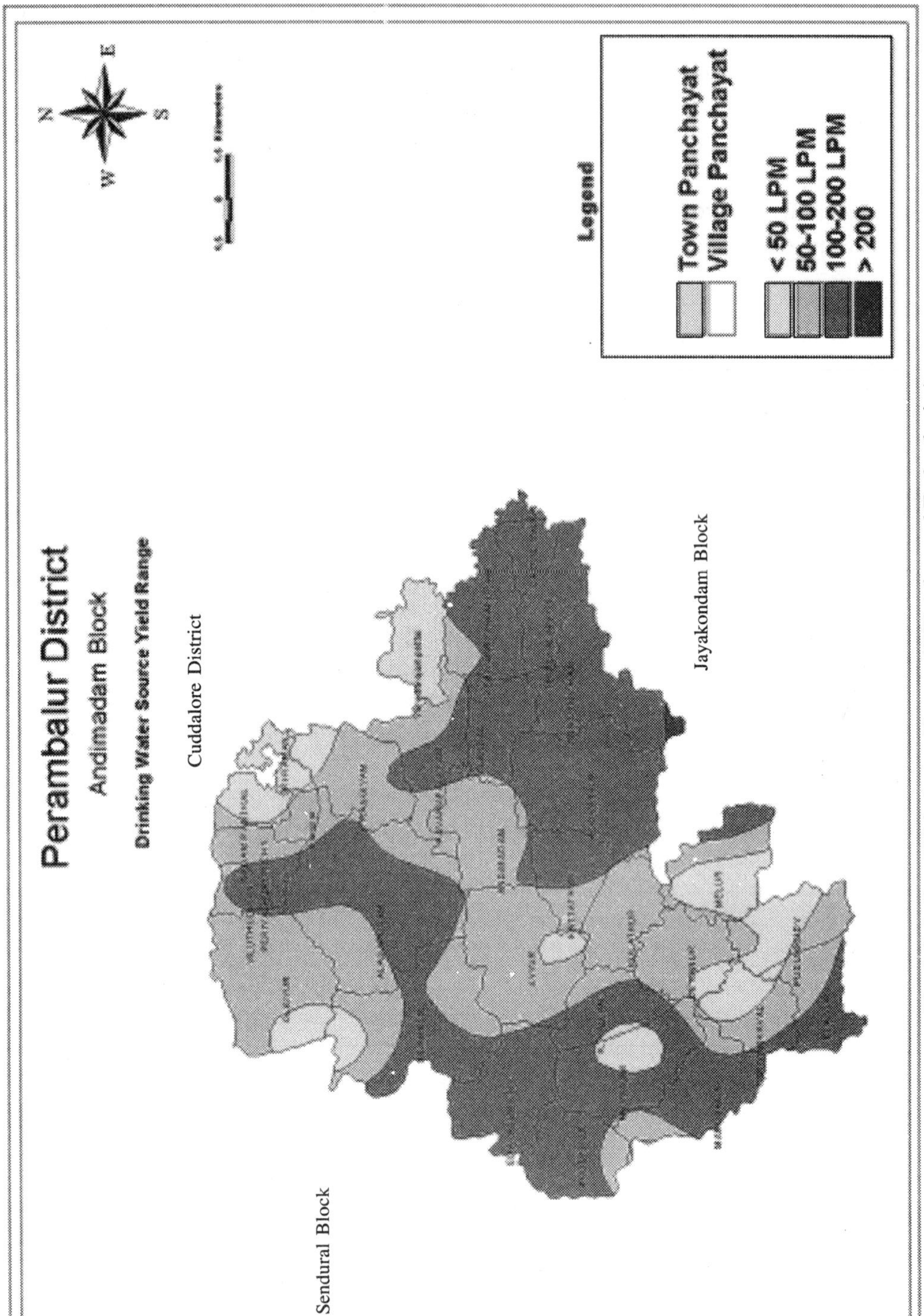

Fig. 9.6: Indian GIS industry (*Source:* WTI Advanced Technologies)

specialized agencies or to municipalities for different projects. There are some other companies in India who use GIS technology for their marketing and planning like HLL, ITC, Domino's Pizza, Tata Chemicals, Tata Power, Tata Teleservices, Lintas and the like. Since the cartographic services are specialized in nature, the comparison of the quality of the work at pre-project status is difficult to assess. Due to this reason, the current practices of awarding jobs through tender and contract system do not find acceptability among the captains of the industry. Their claim is regarding the professionalism, quality and the post-project technical support. Some of the government departments, like Survey of India and National Remote Sensing Agency (NRSA) did develop a method of short-listing the industries for out-sourcing. This system appears to be acceptable with the industry as well as the government. The identification of such industries also gives them recognition to take up jobs from other government departments as well.

With the liberalization of economy in the recent past, it has been seen that a lot of information in spatial format is required for NRI and multi-national investments. The information is to be processed so that the market strategies and location of industries can be worked out. Further, maps are also required for infrastructural development related to telecommunication, road development and urban development where the industry is directly involved. Hence, there is a healthy trend for more cartographic jobs to be done by one sector for another sector of the industry. But such development is likely to have a shorter life. Hence, the cartographic activities within the industry will have to find out new ways and options for its survival. Nevertheless, larger industry groups like Reliance and Infocom have initiated their own GIS projects for internal use. The same digital data is being used by other group of companies or sister concerns like Reliance Energy.

GIS Offshore Projects by Indian Industry

As mentioned earlier, GIS industry in India depends on the offshore projects for their survival. The bigger companies have their offices in major cities abroad. Projects are taken up by these outposts and transferred. The smaller GIS companies do get a share of such projects which are outsourced of partially off-loaded in India. In some cases, the smaller companies never come to know the original client in the countries in the US of Europe. Similar to the success of call centre type of out-sourcing, GIS projects are passed on to this country due to cheap technical manpower, knowledge of English and favourable time difference. But the socalled emerging GIS centres do not have the advantage of 24×7×365 situation. The projects come in phases and the flow is unpredictable. Nevertheless, Indian GIS industry is making its dent allover the world. It may not be too far when India becomes a hub for GIS activities on a global scale. Following Table 9.1 will indicate the pattern of offshore projects taken up by the GIS industry.

Table 9.1: Some examples of GIS projects undertaken by Indian industries from abroad

Project	*Country/Agency*	*Indian Company*
GIS based solutions and mapping	Tokyo Electric Power Corporation, Japan	Tata Infotech
GIS based solutions and mapping	Hokkaido Electric Power Corporation, Japan	Tata Infotech

(contd.)

(contd.)

Project	*Country/Agency*	*Indian Company*
A geographical management information system (GMIS) of Sri Lanka for the World Bank and provided various geographic and attribute data from a data hub to certain agencies of Government of Sri Lanka. The GMIS prototype are designed for specialise application like: • Land use and thematic mapping. • Utility network fault-call and proximity analysis.	World Bank	Tata Consultancy Services, New Delhi
• Environment Impact Assessment. • Monitoring of Environmental Quality. • Solid waste management.		
Developing a geographic database (GDB). Performing sector specific analysis and bringing out map and tabular outputs Regional Information System application development to help sector experts make *on-line* queries and *what-if* analysis.	The Japan International Co-operation Agency (JICA)-Sri Lanka.	Tata Infotech
Creation of data base for corridor in Madhya Pradesh, West Bengal, Orissa, Bihar, Northeast India, Assam and Himachal Pradesh.	Nynex, U.S.A and Reliance Telecommunication	IN-RIMT, Hyderabad
Creation of data base for Gujarat and Maharashtra.	AT&T, U.S.A. and Birla Communications	IN-RIMT, Hyderabad
Digitization of geophysical data.	Emron Oil and Gas, the largest oil companies in the world.	INDICOS, Mumbai
Digitization of several hundred land parcel maps and re-engineering of GIS software including digital land base files with the attribute information attached to them.	U.S.A. based company	INDICOS, Mumbai
Software re-engineering of GIS product to port it from BTOS platform to DOS and UNIX.	K and A Intl., U.S.A.	INDICOS, Mumbai
Digitization of maps and drawings for the C and P Bell Telephone, Florida Power, Seminote Electric, Con Edison, Burns and Roc. etc.	A Business Associate in U.S.A.	INDICOS, Mumbai
Digitised the drawings of railnet work and other engineering applications of New York City Transit Authority (NYCTA).	New York City Transit Authority (NYCTA)	INDICOS, Mumbai
Digitization of oil and gas exploration maps.	GEO-PROS, Venezuela.	INDICOS, Mumbai
Flood Rate Maps of USA digital conversion from paper maps with different flood levels, county boundary in AutoCAD with 0.0024′′ accuracy.	U.S.A. (prime contractor R.M.S India Pvt. Limited)	Caltech India, Kolkata
Digitization of drawings.	Imagine Technologies, USA.	DataVision, Pune
Digital Maps.	Map Info Corporation	RMSI, New Delhi
Digital Maps.	Motorola and ESRI.	RMSI, New Delhi

(contd.)

(contd.)

Project	*Country/Agency*	*Indian Company*
Digitization and GIS application.	M/s IMMCO INC., USA.	Pixel Infotek Pvt. Ltd., Bangalore
Digitization of Maps.	Canadian Government Highway Engineering Division.	DataVision, Pune
GIS based solutions and mapping.	Unisys, Canada.	Tata Infotech
CAD set up with GIS and digital mapping.	Voist Alpine Industrilagnabahu, Austria	RCE-I, New Delhi
GIS based application to maintain image, spatial and attribute data in a central ORACLE database.	British Telecommunication United Kingdom	Tata Infotech
Digital map needed for telecommunication net working planning.	Siemens, Germany	RMSI, New Delhi

With the above it is seen that Indian industries have done a number of GIS, digitization and mapping related projects for different overseas companies or governments. They have the potentialities to expand the GIS activities in future as well. Such activities can be carried out for the Indian projects. For this, the availability of base maps and date bases extremely essential. Indian industry would like to have access to such sources without any restriction. Since the present policy of government on map availability or production has already changed the GIS industry is likely to flourish with Indian projects. They may not have to look for offshore projects for their survival. However, they may be able to export GIS products and services in large volume and can work for more and more foreign countries or foreign companies in the field of GIS or cartography or mapping related work. The concerned industries are earning valuable foreign exchange by doing GIS related projects for foreign companies or countries.

Indian GIS Software

In India the standard GIS packages have been widely used. They are MapInfo (http://mapinfo.com), Idrisi (http://www.idrisi.clarku.edu), Geomedia ((http://www.intergraph.com), or ArcInfo (http://www.ersiindia.com) (GIS Development Team, 2000). But there is always demand for software by which both, GIS and remote sensing data, can be handled. Hence, a software package which can equally manage raster and vector is preferred. One of the important reason is the provision of in-built compatibilities for using any type of digital data at a reasonable cost. For example, GRASS (http://wgbis.ces.lisc.ernet.in/grass) is preferred because it is a open source GIS with capabilities such as raster analysis, vector analysis, image processing, map algebra, and other functionalities which are comparable to any commercially available software (Ramachandra *et al.*, 2004).

The first attempt to develop a GIS software in India was made by the Indian Space Research Organisation (ISRO). This was developed for wider application of the remote sensing data available from the Indian satellites in GIS activities. This software is known as Isrovision. On the other hand, in the Department of Science and Technology through the Natural Resources Data Management System (NRDMS) with the financial assistance of the United Nations Development Programme (UNDP) and technical assistance from the Centre for Studies in

Resources Engineering (CSRE) of the Indian Institute of Technology, Bombay developed a window based GIS software called Geo-References Area Management++ or GRAM++ software package which has the following features (www.nrdms.nic.in):

—Data preparation tool (DPT)
—GRAM Viewer
—Raster analysis kit
—Network analysis
—TIN Generation
—Statistics
—Digital image processing
—Mapping control tool
—GIS tutor

Modules like GRAMNET and GEO_SQL have been developed to facilitate resource network analysis and spatial querying. Further, a user friendly module VEC Viewer has also been developed to support viewing of geo-referenced layers with collateral attribute information with query facility. A GIS tutor around GRAM++ has also been developed (Venkatachalam *et al.*, 2001; Prasad, 2002).

GRAM++ is a comprehensive suite of GIS tools primarily useful in preparing and implementing development plans by district level line departments, Panchyati Raj institutions, communities and the Non-Government Organisations (NGOs). Obviously, its potentials for the development of business geographic are yet to be explored. Indian industry is taking some time to accept it in a big way. With training it is making way in the administration at district level. Universities and colleges are accepting it due to attractive cost. IIT Bombay (CSRE) is now marketing this software. Recently, Government of Madhya Pradesh in collaboration with the Department of Science and Technology has developed its own version of this software, known as GRAM Dristi (NRDMS, 2003).

Further, an initiative was also taken up by the industry to develop a low cost GIS package which can equally handle remote sensing data. The target group was universities and collages where GIS and remote sensing courses have been introduced in the courses but these educational institutions were not having sufficient funds for having commercially available software. JT Technologies, a Noida (near New Delhi) based company, specializing in the domain of development of software products and solutions in the field of GIS has recently released a comprehensive GIS software called JT Maps (http://www.jtmaps.com). This package has a variety of features which does address almost the entire spectrum of GIS activities. The software supports following features:

—Image processing which supports a range of satellite data formats
—Data creation
—Analysis: Spatial analysis, Network analysis
—SQL support
—External database connectivity
—Export-import compatible with industry standard data formats
—Projection transformation

There are also attempts to develop or customize the available GIS packages for specific applications or for specific institutional use. The case of the Maharashtra Remote Sensing Applications Centre (MRSAC) is an example where the data base development was considered to be cumbersome task when done through the basic GIS operations and commands. As a result MRSAC developed data base structuring tools (DBS-TOOLS) which is a menu driven, user friendly customized software package built over ARC-Info software using Arc Macro language (Tamhane and Despande, 2000). Further, GRDASS, a open source GIS with capabilities such as raster analysis, vector analysis, image processing, map algebra and other functionalities has been developed which is claimed to be comparable to any commercial software available in the market (Ramchandra, 2004). Similar attempts are being made in the GIS related institutions and industries. Unfortunately very little is known about such initiatives.

Indian Journals on GIS

With all the problems being faced by the GIS community in India, the publications of journals is doing fairly well in India. The existing journals connected with geography, remote sensing, aerial photography or science in general were giving due attention to GIS issues. But it was felt that there is more potentiality to commercially explore this field. Industries, NGOs, universities and government sponsored institutions appear to converge on this aspect of GIS. Organisation of conferences, exhibitions and events, publishing of journals, launching web sites and the like have lead to a new brand of professionalism, i.e. GIS journalism or Geojournalism. New journals are being brought out and at least in one case the quarterly has become a monthly journal. Issues to meet different regional requirements are also being brought out. As a result every activity related to GIS or mapping is closely watched and becomes a news item. The GIS related journals are as follows:

— GIS Development: www.GISdevelopment.net
— GIS India: gisindia@hd2.vsnl.net.in
— Geospatial Today: www.geospatialtoday.com
— Geography and You: www.geographyandyou.com
— ISG Newsletter: bsahai@wilnetonline.net
— Coordinates: www.mycoordinates.org
— Photonirvachak: Journal of the Indian Society of remote Sensing: www.isrsindia.org
— Manchitrian: Journal of National Atlas and Thematic Mapping Organisation: www.natmo.gov.in
— Manchitrian Taknik: Manchitran@hotmail.com (in Hindi)

GIS Training in India

Two decades earlier, when GIS was getting introduced in the country, no was sure in which discipline it should be included (Saraf, 2005). Hence, there was a degree of uncertainty about its status under the aegis of the University Grants Commission or All India Council for Technical Education. But with the success of the Indian satellite programme, remote sensing was introduced in a big way which included GIS. Hence, GIS got a foothold in the country and developed as a multi-discipline basically for assessment and mapping of natural resources.

The nature of GIS education has been influenced by such development. As a result, there is less emphasis on business GIS or GIS related activities in the industry. Obviously, the course contents are more inclined towards the development of GIS for natural resources using remote sensing data. However, GIS is being taught in the civil engineering, geography or earth sciences curriculum. Further, the emphasis or combinations with other specializations vary from social aspects to geodesy to surveying and mapping. GIS courses have wholly or partially been integrated with Bachelor of Engineering or Technology, M.Sc. or M.Tech. or even doctoral level research programmes (Kulkarni, 2005). In the Indian universities and colleges, the theoretical foundation has already been laid and trained persons are also available. It is up to the academics to take further initiatives in promoting GIS (Parihar, 2001).

A survey was carried out by CSDMS team in 2000 about the educational qualifications and training of the GIS users through out the world. The trends reveal that the large number of users in India is post graduates. They are mostly involved as GIS analysts, mostly with 7-10 years experience and employed in private sector (GIS Development Team, 2000). The GIS training institutes in India are as follows:

Indian Institute of Remote Sensing (IIRS), Dehra Dun
National Remote Sensing Agency, Hyderabad
Space Applications Centre, Ahmedabad
Regional Remote Sensing Service Centres: Kharagpur, Dehra Dun, Nagpur, Jodhpur, Bangalore
Centre for Studies in Resource Engineering, IIT Bombay, Mumbai
Institute of Remote Sensing, Anna University, Chennai
Geological Survey of India
Indian Institute of Technology, Kanpur
National Bureau of Soil Survey and Land Use Planning, Nagpur
Survey Training Institute, Survey of India, Hyderabad (www.soisti.gov.in)
National Atlas and Thematic Mapping Organisation, Kolkata
Birla Institute of Technology, Meshra, Ranchi
Jawaharlal Nehru Technological University, Hyderabad
Jadavpur University, Kolkata
The University of Calcutta, Kolkata
GIS Institute, CSDMS, Noida
Symbiosis Institute of Geoinformatics, Pune
Bhaskaracharya Institute for Space Applications and Geoinformatics, Gandhinagar
Visvesvarayya Technological University with Karnataka State Remote Sensing Centre, Bangalore

There is a need that the GIS course content which allows university programmes to remain abreast of market needs as well as providing students with a well-rounded education in the use of technology (Hall, 2005). "GIS is potentially much more than a technology. Education in and for GIS now needs to be much wider than has hitherto been normal, if we are to gain the full potential available. Good technology—and the skills to use it—form a necessary condition for success but not a sufficient condition" (Rhind, 2005; Roy, 2001; Venkatachalam *et al.,*

2001). GIS Education is not only limited to universities and specialized institutions, but it is being taught in schools as a tool to teach science (Phoenix, 2004).

Possible GIS Products and Applications in India

In order to have proper utilizations of GIS products, there is a need to identify potential market to commercialization of its data and products. The first step in commercialization process is to develop a consensus among the industry and users on what markets are critical to future growth. There could be ten types of uses, which are potential application of geospatial data. Obviously, the products and applications cannot be in isolation. There has been a convergence of technologies within GIS and related fields. As a result, different combinations of data, technologies, and delivery systems are being worked out. Now such integration of technologies is also identified under the realm of geomatics or geoinformatics. The possible GIS products and applications are:

Land and Natural Resources

The natural resources market for geospatial data is land, forestry, agriculture, fishing, mining etc. These are derived from the earth and its water. The principle of sustainability and bio-diversity protection are increasing demands for more and better data. High-resolution imagery, GPS and integration of GPS, GIS and remote sensing (RS) technologies can develop the required information. Attempts are made to model groundwater vulnerability using GIS technology, particularly pollution assessment applying Drastic model (Mohan and Gopi, 2004). The example of remote sensing based GIS application for preparing land use map a part of Dehra Dun town is shown below (Fig. 9.7).

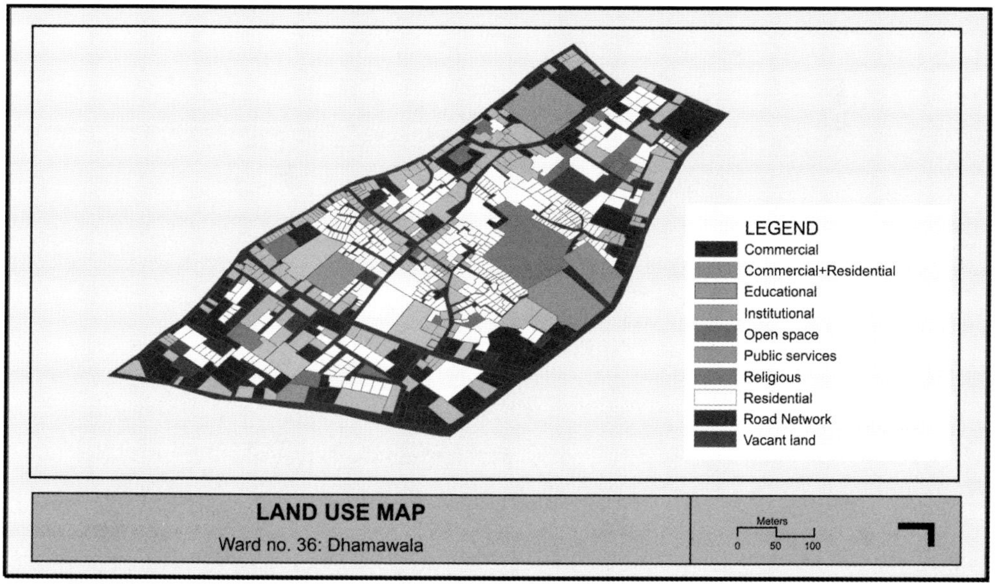

Fig. 9.7: Application of GIS and remote sensing for preparation of urban land use map of a part of Dehra Dun town (*Courtsey*: Suman Jain)

Environment

The environmental market for GIS goods and services is related to a wide range like pollution prevention, environmental conservation, protection, remediation and enhancement technologies, processes, products and services. Environmental reporting is increasing because of the growing recognition of the importance of environmental management (Ramaswamy, *et al.*, 2002). Two examples from the National Atlas and Thematic Mapping Organisation (NATMO) has been taken to demonstrate the possible GIS applications for (a) monitoring air pollution over major Indian cities, and (b) mapping of flood effected areas in Maldah district of West Bengal. The former example is for small scale GIS application (Fig. 9.8), while the latter one is for large scale operation (Fig. 9.9) in the field of environment.

The complexity of environmental issues demands the collection, analysis and presentation of massive quantities of data, and the mathematical and surface modelling of multifarious processes, usually over time. GIS technologies are ideally suited to this area (Patil, Ravan and Kaushal, 2003). Global Position System (GPS), for example, offers an excellent tool to monitor society's impact on the environment. It can be used to track the movement of everything from icebergs to polar bears, and offers exciting opportunities in the area of weather forecasting. Remote sensing technologies and GIS, which are ideally suited for handling such data sets, and many other technologies, are regularly being employed in many environmental applications.

Property and Urban GIS

In terms of GIS goods and services, the property market includes the building and maintenance of property infrastructures, which involve systems for survey, registry and assessment. Other activities include surveying and mapping in support of land development and providing information in support of property-related transactions such as conveyancing, mortgaging, and valuing and environmental assessing. The business of providing information in support of property transactions is also changing from its more traditional ways. Traditional conveyancing and mortgaging work is diminishing for a variety of market, economic and technical reasons; however, opportunities to provide new services such as environmental assessment are gradually increasing. Unless the industry can successfully introduce new business models, it can expect very little growth in this area. In general, it will become increasingly important to be able to work in an on-line world, conversant both with the latest developments for accessing and analysing property-related information and with the world of electronic commerce (Upadhyaya, 2002). Bhoomi Project in Karnataka state is a good example of some degree of success in providing access to the farmers about their land records (Chawla and Bhatanagar, 2003; Sipkes, 2003). More important than technology *per se*, however, will be an in-depth understanding of areas such as cost/benefit and risk management and the capacity to quickly incorporate them into new GIS business models (Umashankar, 2003). The case study of municipal GIS for property management, and related issues for Hyderabad city has been depicted (Fig. 9.10).

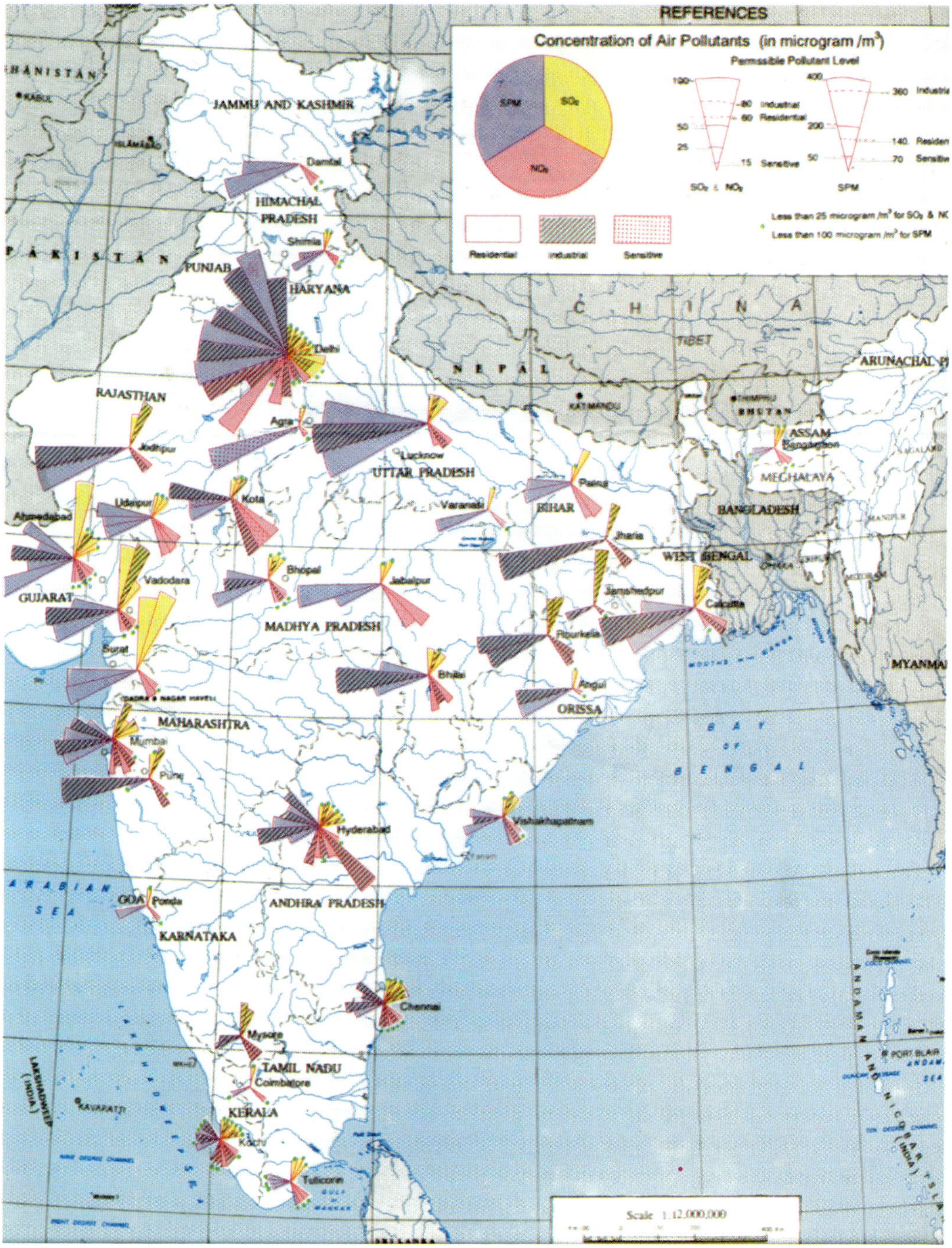

Fig. 9.8: Pollution over Indian cities (*Source*: NATMO)

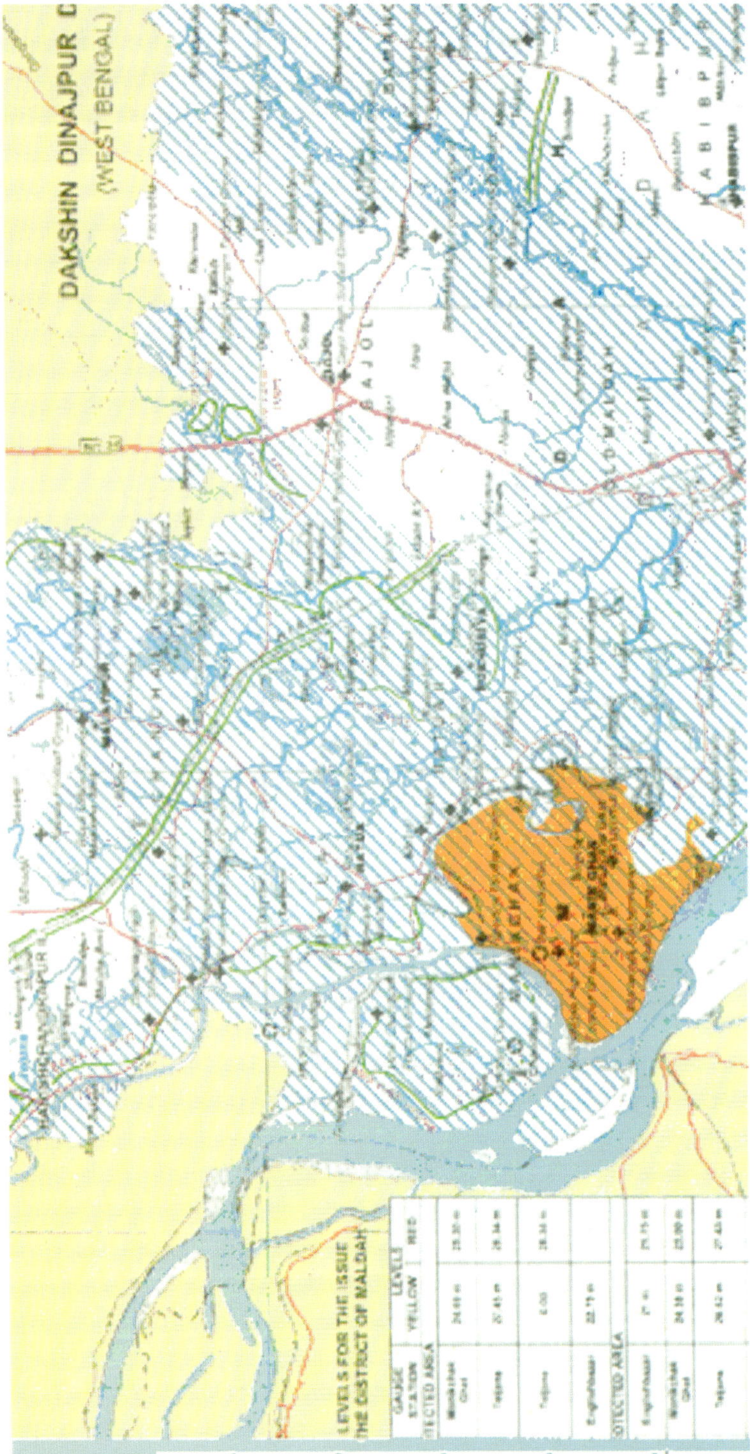

Fig. 9.9: Flood map of Maldah district, West Bengal (*Source:* NATMO)

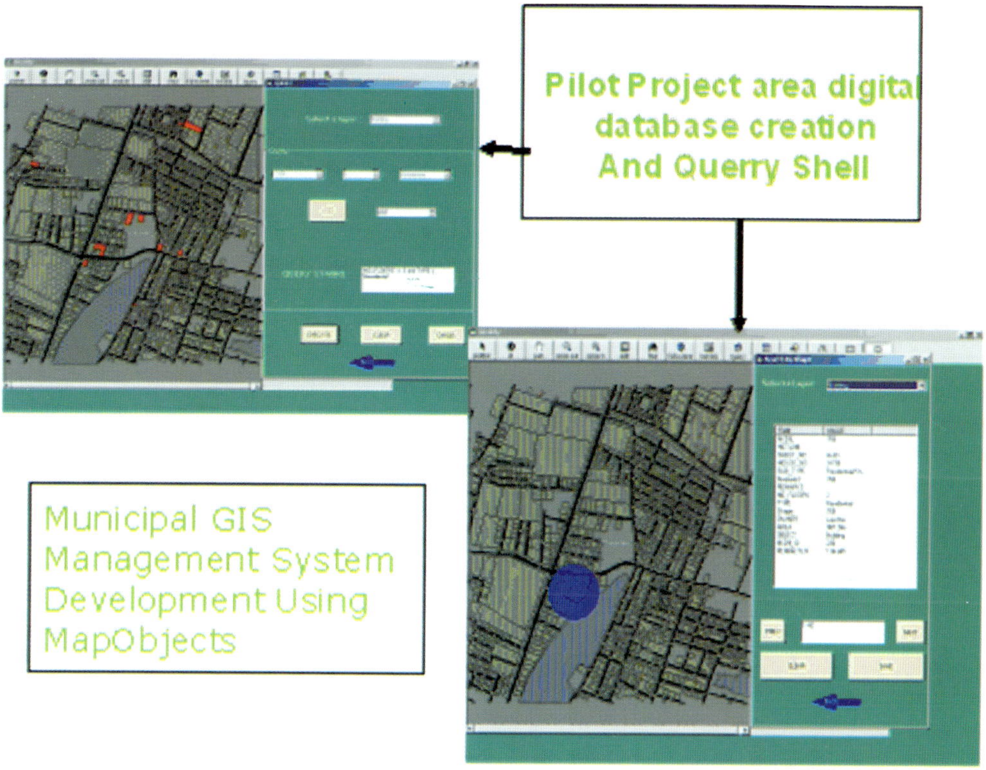

Fig. 9.10: Municipal system for property management, tax etc. (*Courtesy:* RSI)

The case study of Kolkata Municipal GIS has been discussed in this chapter earlier. National Urban Renewal Mission (NURM) is funding several major cities of the country for their development, including preparation of spatial data base for 137 cities. These data bases are to be used for development of GIS for the respective cities and towns. In addition, international funding agencies like DFID (UK) and JICA (Japan) are also keen in providing resources and expertise for several cities. Municipal GIS is also has the potentiality to be used for urban services such as electricity, transportation, fire, flood and drainage. Further, there is also scope for using this data base for business opportunities as well. Mobile, location based, retail, taxi and several services can be based on such GIS data base. In India there is a big scope for such businesses and services. In near future a good number of GIS trained man power will be required for managing such GIS activities as well.

Infrastructure

The infrastructure market for GIS goods and services includes engineering, construction, transportation and utilities industries. This sector supports both business and non-business activities in a country through the design, engineering, construction and maintenance of capital projects and the provision of ground, air, water and space transport services, telecommunications, and water and energy supplies. The use of GIS technologies in this area not only facilitates the efficient, cost-effective and environmentally responsible development and operation of these activities but also may preclude the need for the more costly expansion of the infrastructures. As such, the GIS industry will play an increasingly important role in the infrastructure market.

The construction sector is responsible for building the facilities and supportive infrastructure necessary for a nation to produce wealth and shelter its citizens. Some GIS-related activities include land planning and development, design, location selection, impact assessments, surveying, mapping and other associated municipal activities such as urban planning and development. Utilities, hydro grids, pipelines, communications systems and so on must monitor, maintain, accurately locate and coordinate extensive amounts of expensive, geographically dispersed capital stocks.

Global Position System (GPS) is another area that is having a major impact on the utilities market. The ability of GPS receivers to synchronize themselves to atomic clocks on satellites is proving to be a boon to the wireless communications industry.

There is currently in process a technology on electric power that will be of interest to the GIS industry. It will focus on four issues: asset optimization, intelligent power system delivery, end-use efficiency and convergence, and small-scale generation and renewable. An example of the use of GIS data base and GPS is included here for planning location of pylons for power transmission (Fig. 9.11). Here the options for alignments of transmission lines have been worked out on the GIS data base and also on the digital elevation model. Location of pylons, their coordinates and expected sag between the pylons have been calibrated as well.

In combination with advanced computing and communication technologies, GIS technologies are dramatically improving the efficiency and environmental cleanliness of all modes of transportation and commodity exchange. Intelligent Transportation Systems (ITS), for example, are revolutionizing the way transportation-related issues such as congestion; safety and logistical optimization are being handled. ITS can be applied in areas ranging from travel planning, adaptive traffic control, real-time traffic information, public transit operations, commercial vehicle operations, fleet management and emergency vehicle routing to, in the future, automated vehicle control systems.

Access to geospatial information will play an essential role in the development of these market opportunities. As we have seen, United States is currently the leader in the development and deployment of ITS and, as emerging industrial nations' transportation systems become more congested, their markets will become very important as well.

Health

As the GIS industry has evolved, there has been limited involvement between it and the health care sector. Where links have been made, they have been concerned with the location of facilities (i.e. clinics, ambulance stations and practitioners' offices) and the routing of emergency service vehicles. The growth of GIS in this area has been hampered by limitations on the data, technology constraints, lack of training and education, and the structure of the health care sector. These constraints are interrelated and, taken together, have resulted in limited use of geospatial data in the health care sector. However, both the GIS industry and the health care sector are undergoing major changes.

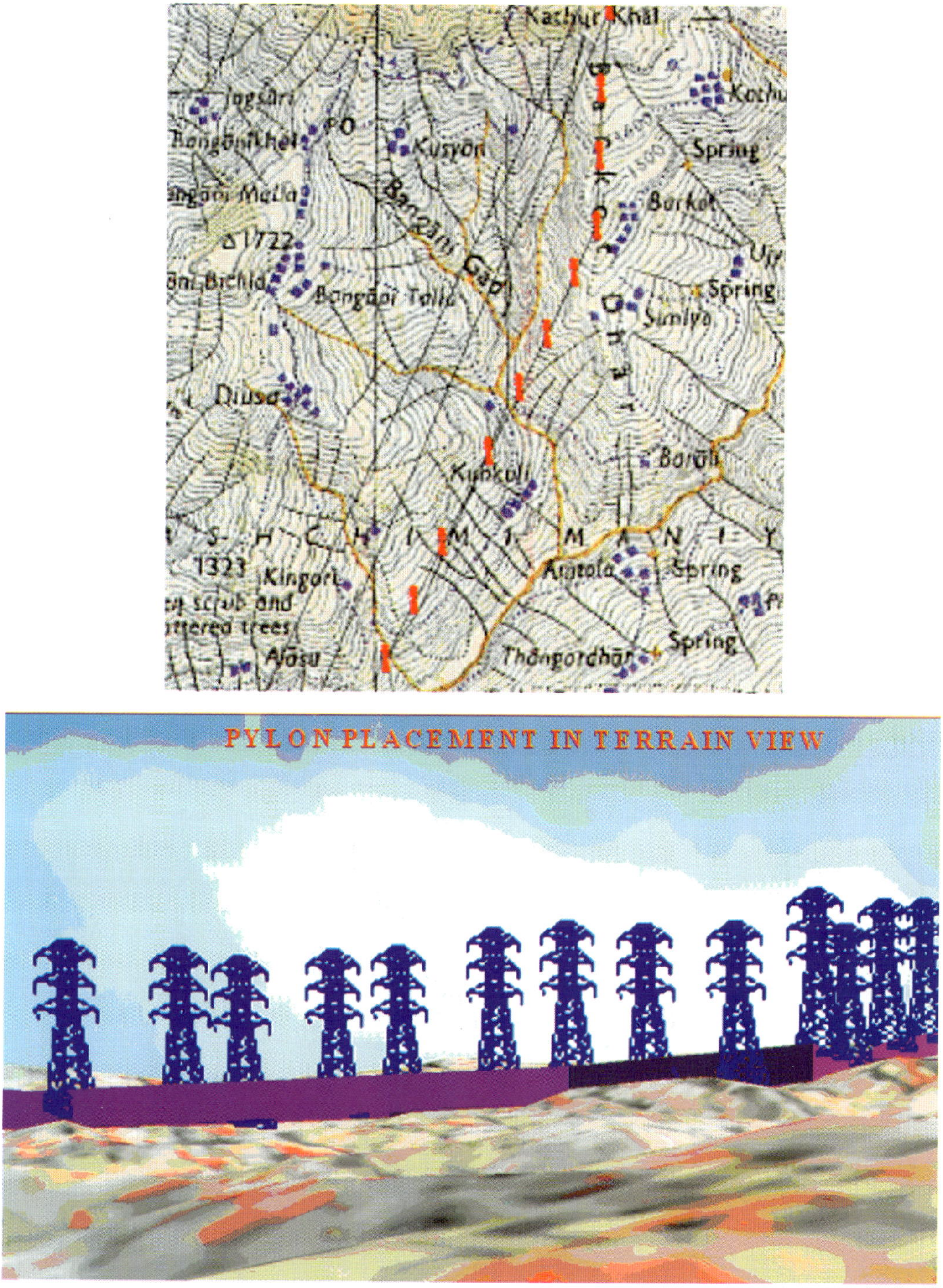

Fig. 9.11: Placement of pylons on GIS data base (*Courtesy*: M.V. Bhatt)

With GIS expertise residing in highly specialized units, there have been few opportunities for interaction between these experts and experts in the health care sector. As a result, awareness and appreciation of the potential usefulness of geospatial data in the health care sector have been limited. Furthermore, even where there was an awareness of the technology, it was sometimes discounted because of the presumed high costs associated with its use.

Geospatial data provide the health care sector with a new sub discipline for collecting, storing, retrieving and analyzing health care data. An example of Uttaranchal has been included here (Fig. 9.12). NATMO has not only prepared such a data base for a particular state, but an whole Atlas of Health and Diseases of India (Nag, 2000). To take advantage of this sub-discipline, the geospatial and health data must be coordinated across the organizations, institutions and individual providers in the community. A coordinated data infrastructure supports:

- Human resources planning
- Facilities planning and optimized service delivery
- Epidemiological studies
- Informed decision-making by users and providers
- Proactive and preventive health care
- Emergency preparedness
- Health imaging techniques development.

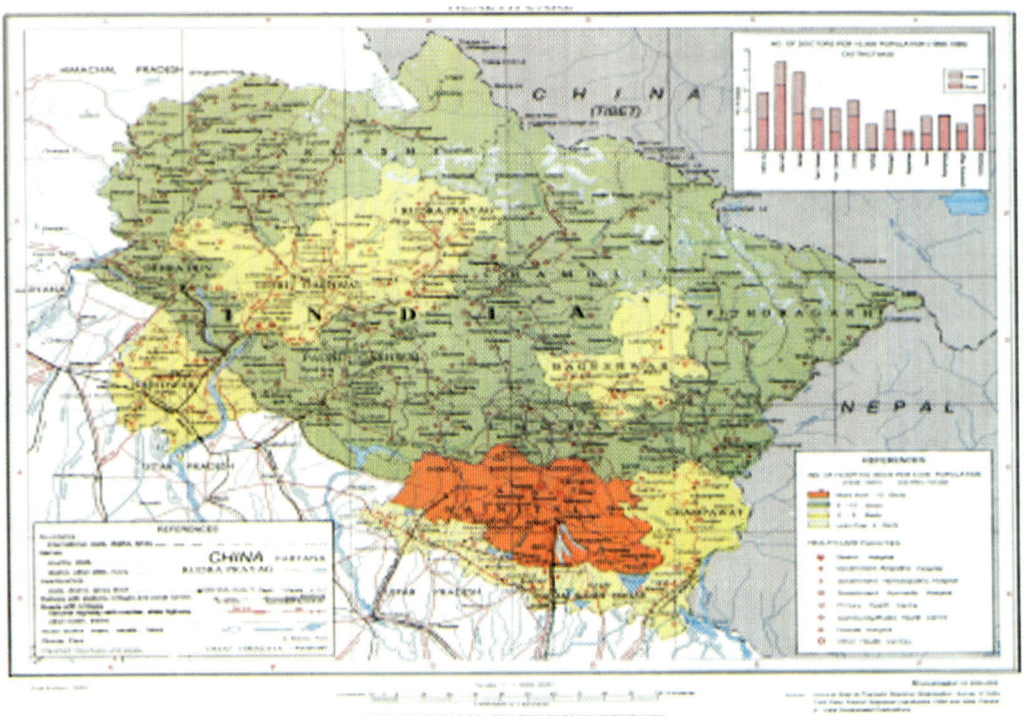

Fig. 9.12: Uttaranchal: Health Service (*Source*: NATMO)

Application of GIS in social, health and welfare activities are generally government sponsored. It is considered as an extension of thematic mapping. NATMO has been contributing a lot in this regard (Massey, 2002). However, there is a lot of scope to develop GIS for the social sector.

Emergency Response and Disaster Relief

Both natural and human disasters such as floods, tornadoes, storms, oil spills and the release of poisonous gases require timely, accurate and supporting geographic information for relief operations, logistics of supply, national search and rescue capabilities, prediction modelling and environmental monitoring associated with these disasters (Sharma, 1999). Considering the topography, village boundaries and information regarding the floods, a GIS data base has been developed for identifying effected villages (Fig. 9.13). This will help in drawing immediate attention for relief and rescue.

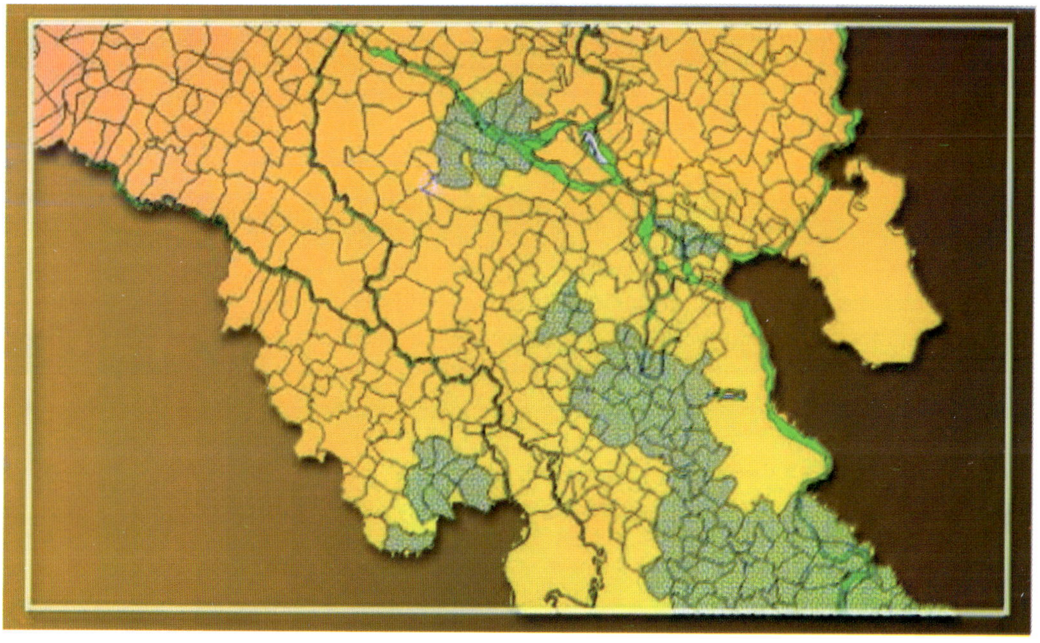

Fig. 9.13: Low lying villages effected by floods (*Source*: NRDMS)

Current and future GIS technologies used for emergency response and disaster relief efforts include integrated information systems, precision and mobile GPS, sophisticated planning and prediction modelling, wireless dissemination and multi-purpose operational support systems. As future disasters emerge and as relief efforts are mobilized, GIS technologies will be required more and more to accelerate the planning and response activities. Technologies that provide ease of access and integration will be important, as will online access to information sources through integrated infrastructure shared between the private and public sectors. Timely access and rapid dissemination are key variables to aid relief efforts. Information distribution must also support differentiated views of supporting geographic information, along with the tool kits to use wireless and remote communications to disseminate critical information to users in the field and in planning and coordination centers (Sarkar and Kanungo, 2001).

Military, Defence and Security

The military GIS market offers some of the greatest potential for new applications and integrated GIS technology and systems. The military GIS market has traditionally been one of the larger consumers of GIS technology, information and related services, although most production activity so far has been conducted in-house. As we move towards the millennium, the military market remains large and strong, with an emphasis on maintaining global capabilities that enable the mobilization of operations over land, sea and air (Singhal, 2000).

Overall, the military GIS market spans a range of different requirements. Military GIS systems and technology requirements are diffuse, vary greatly and tend to be user driven. The nature of the market has moved from customized large-scale developments to commercial-off-the-shelf (COTS) products. Similarly, the pace of technological development and changing operational environments have resulted in a change in requirements toward flexible components, interoperability, standardized tool kits and data, and integrated systems development.

Future market opportunities for GIS technologies in this sector are numerous. Key areas include advanced communications and secure networking to meet demands for on-line, just-in-time delivery of information and decision support capabilities. Development of experimental, soon-to-be-operational mobile warrior and integrated systems for in-field application continues to dominate research and early pilot phases. These and other technology areas such as advanced guidance, high-resolution imagery, operational picture and precise location determination remain critical areas for GIS technologies, often within larger operational command and control systems.

As ever, military systems require well-supported access to geographic information. These requirements fuel an ever-expanding demand for innovative geospatial information discovery, access, integration and management tools to support the provision of responsive, accurate and timely global geospatial information systems. Delivery of this information is moving from internally produced, discrete hard copy products toward interoperable, seamless, on-line delivery of services via national and global infrastructures, in essence, mobilizing technology to support the delivery, storage and accessibility of information from within and outside military sources.

Business Geographic

The business market includes applications in commercial sectors such as shipping, banking, retail sales, food services, real estate and insurance, and also in vehicle navigation and routing. There has been a clearly identified shift in the mapping or GIS business opportunities away from traditional surveying and mapping data capture to the outsourcing of data services. Large organizations with legacy databases are now asking the surveying and mapping providers to offer "value-added" services—where their legacy non-graphic databases are joined with their graphics, delivered by a data capture firm, to produce intelligent business analysis of their day-to-day operations. This trend will continue, and many organizations will find themselves moving from the existing traditional data capture model to a new and more lucrative business GIS model. In addition, the traditional business geographic market now has expanded to encompass this new and more lucrative market. An example of Hyderabad city is depicted here (Fig. 9.14). The GIS for this purpose was jointly developed by Survey of India and M/s Spinfo, Bangalore under the MapQue Project. Similar GIS data base for Bangalore is also available.

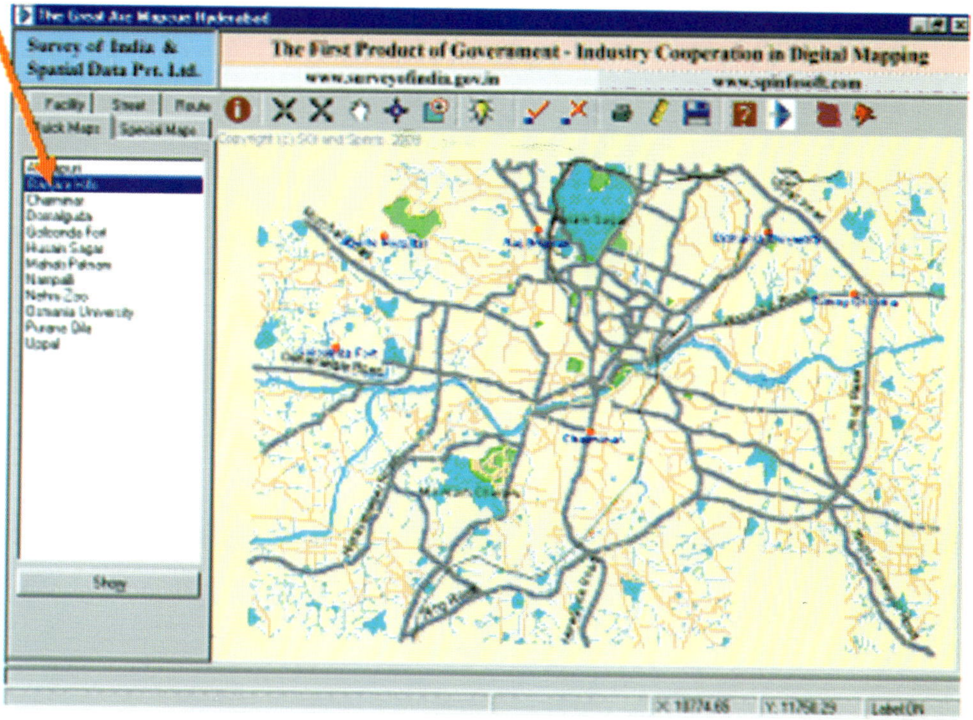

Fig. 9.14: Business geographic for Hyderabad City

The traditional business geographic market addresses the needs of the commercial business sector for business demographic and geographic market trend analysis, while the new business GIS market encompasses non-traditional GIS analysis in market segments such as health and welfare. Medical and social market analyses are but two of the new business market areas being addressed by GIS.

The market today includes any organization with large data sets that wishes to share those data on a corporate intranet. In addition, the market is also the family household, and their desktop personal computer, which offers both a marketing and sales opportunity for that large corporate database through the use of the Internet. Companies familiar with handling databases, both graphic and non-graphic, will have excellent opportunities to service these new market areas.

Currently, managing and manipulating digital map and non-graphic databases are now providing services for organizations not familiar with the graphic presentation of data in a geographic context. These new business opportunities are offering challenging and rewarding experiences for companies previously confined to the traditional mapping or GIS business model. The expression "a picture is worth a thousand words" is proving that the value of presenting previously proprietary non-graphic data in a public forum on a graphic backdrop is both effective and exciting. The analysis now available using the previously "graphically static" databases is very impressive.

Entirely new approaches are being used to distribute and display these data sets by marrying the legacy databases, graphic and non-graphic, with the companies, which produced them. The technology is intranet or internet software. The products necessary to "explode" this market include access to digital mapping, national, provincial, regional and local data sets, interoperable software and the associated non-graphic databases.

The convergence of technologies and applications in the mapping sciences has led to geographic information processing being used on the development of a much wider range of products. As mentioned, traditional GIS has concentrated on applications in environment, natural resources, infrastructure, cadastre and defence. More recently, business applications, vehicle navigation systems and health applications have increased in importance. Education and entertainment, especially the latter, are the most recent and promising markets for GIS. The entertainment segment of the "edutainment" market is probably larger but much less focused than the educational segment.

Consumer Services

This market currently consists of individuals whose primary use of geographical information is for such purposes as travel planning and navigation in both land and water vehicles and for outdoor recreational activities such as hiking, camping and hunting. These kinds of demands have traditionally been met through the use of paper map and atlas products and less frequently through the use of aerial photography or satellite image products. Advances in two key technologies, GPS and GIS, combined with other technological developments in data storage and communications (especially the internet) and computer visualization, make possible a broad range of new GIS applications. For example, consumers now have access to mass-produced, hand-held GPS receivers that are being used for a variety of recreational activities. Basic desktop mapping and GIS functionality are being bundled with off-the-shelf office management software, providing consumers with the capability to manage and use all kinds of information in a geographic context.

M/s Eicher is involved in providing GIS based services in several cities. This company has also taken up joint projects with the Survey of India. Example of GIS application in telecom and cellular services has been included above (Fig. 9.15). From their GIS data base, M/s Eicher has produced details maps of cities like Delhi, Mumbai and Chennai.

Within the next five years, other applications of GIS technology for the consumer market will include Yellow Pages searches by geographic parameters, and travel and play in three-dimensional virtual reality based on geography. Given the scope of daily activities that are location-dependent, applications for the integration of geography with other types of information are limited only by the imagination. Constraints on the rate of development of the society or the consumer market segment include the limited amount of up-to-date geographical information publicly available, restrictions on its access and the small size of the present-day consumer market.

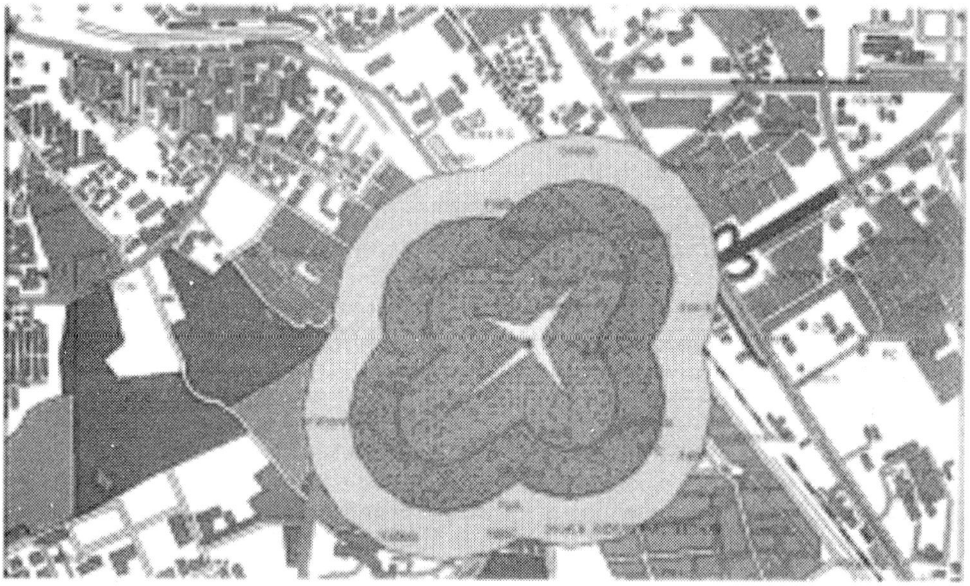

Fig. 9.15: Telcom and cellular services (*Source*: Eicher)

Note: Cellular Services Industry: Figure depicts usage of base digital map data for range analysis of cellular (mobile phone) coverage by superimposing clutter data as a layer.

Summary

Like elsewhere in the world, GIS in India too has its mating dance with the information technology mainstream, as the two players circle around each other, cautious about surrendering autonomy but conscious of the certainty of the conclusion (Goodchild, 2003). In the federal and state governments, a lot of GIS activities are being sponsored and promoted by their ministries and departments of information technology. Further, we would like to have the benefit of the IT mainstream without losing the special nature of GIS. With a very strong presence of IT community in the country, the threat of losing uniqueness is greater. Perhaps the best way out is to develop GIS in India as a very special application area of information technology.

REFERENCES

Agarwal, Pragya (2000), GIS in cultural resource management in historic urban centres, *GIS Development,* Vol. 4, Issue 11, pp. 32-34.

Ahuja, J.S. (2000), Legal, political, technical and commercial issues of land information system, Pre-conference *Proceedings of the National Conference on Land Information System,* 18-19 October, New Delhi, pp. 72-75.

Ahmad, Jamil (2003), Step-by-step approach to build a municipal GIS, *GIS Development*, Vol. 7, Issue 2, pp. 42-46.

Baijal, R., M.K. Arora and S.K. Singh (2002), A GIS assisted knowledge-based approach for military operations, *GIS Development*, Vol. 6, Issue 10, pp. 21-24.

Chawla, Rajiv and Subash Bhatnagar (2003), Bhoomi: A boom for farmers, *Geospatial Today*, Vol.1, Issue 5, pp. 13-17.

Cooper, Paul (2003), GIS gets cool!, *GIM International*, Vol. 17, No. 7, pp. 28-33.

Dangermound, Jack (2005), GIS: Helping manage our world, *GIS Development*, Vol. 9, Issue 12, pp. 26-27.

Dasgupta, A.R., Mukund Rao and A.K.S. Gopalan (2000), The National (Natural) Resources Information System, *GIS Development*, Vol. 4, Issue 12, pp. 18-23 and 47.

Dent, Borden (1999), *Cartography: Thematic Map Design*, WCB McGraw Hill.

Dorling, Daniel and David Fairbairn (1997), *Mapping: Ways of Representing the World*, Addison-Wesley.

Georqiadou, Yola (2003), Is Survey of India ready for the NSDI ?, *GIS Development*, Vol. 7, Issue 2, pp. 38-39.

Gibbons, Scott (1999), Mirzapur: A CIS that works, *GIS Development*, January-February, pp. 39-41.

Goel, R.K. (u.d.), National (Natural) Resources Information System (NRIS) – Design standards, unpublished paper from Space Applications Centre, Indian Space Research Organisation, Ahmedabad.

Goodchild, Michael F. (2003), GIS – Finding the mainstream, *GIS Development*, Vol. 7, Issue 1, pp. 8-15.

Gopal Rao, M. (2000), LIS in India – Perspective and Retrospective, Pre-conference. *Proceedings of the National Conference on Land Information System*, 18-19 October, New Delhi, pp. 1-6.

GIS Development Team (2000), GIS survey: An analysis, *GIS Development*, Vol. 4, Issue 10, pp. 25-28.

GIS Development Team (2000), GIS software at a glance, *GIS Development*, Vol. 4, Issue 11, pp. 20-21.

Gupta, V.K. (2001), Monitoring crime, *GIS Development*, Vol. 5, Issue 4, pp. 23-25.

Hadley, Clare and John Hammond (2003), Survey of India – A view from abroad, *GIS Development*, Vol. 7, Issue 2, pp. 29-31.

Halari, Rakesh (2000), Computerization of urban land records – Case study for city survey land records of Ahmedabad district – Gujarat state, Pre-conference *Proceedings of the National Conference on Land Information System*, 18-19 October, New Delhi, pp. 46-49.

Hall, G. Brent (2005), Curriculum development in university-level GIS education, *GIS Development*, Middle East, Vol. 1, Issue 1, pp. 18-21.

Harmsen, Karl, Yogesh Kant and Shilpi Singh (2004), Assessment of the use of RS and GIS in India, *GIS Development*, Vol. 8, Issue 4, pp. 36-41.

Jain, Suman (2005), Vulnerability identification of low income settlements, *GIS Development*, Vol. 9, issue 10, pp. 26-28.

Joao, Elsa Maria (1998), *Causes and Consequences of Map Generalisation*, Taylor and Francis.

JT Map (2004), *JT Maps 2005 User Manual*, Jlet Technologies Pvt. Ltd. *(info@JTMaps.com)*.

Khare, Manoj, Ashok Kaushal, Sandeep K. Srivastava and Krishna Kant (2005), Disaster management: Get real, pick sites for flood forecast, *Geospatial Today*, September-October, pp. 49-51.

Krishna, Bal (2003), Breaking inertia: Survey of India embraces a new era, *GIS Development*, Vol. 7, Issue 2, pp. 40-42.

Kulkarni, Madhav N. (2005), GIS-GPS academics in India: Present and future, *GIS Development*, Vol. 9, Issue 1, pp. 32-33.

Lemmens, Mathias (2003), Geospatial intelligence, *GIM International*, Vol. 17, No. 7, pp. 50-51.

Longley, Paul and Graham Clarke (2000), *GIS for Business and Service Planning*, Bell and Bain, Glasgow.

Massey, Jack (2002), A tribute to social GIS in India, *Geospatial Today*, Vol. 1, Issue 4, pp. 75-77.

Marutish, V. (2004), An enterprise GIS system for electric utilities: new approach to profitability, *Geospatial Today*, Vol. 3, issue 3, pp. 30-33.

Mishra, Prabhakar (2000), Cadastral Surveys in India – A critique, Pre-conference *Proceedings of the National Conference on Land Information System*, 18-19 October, New Delhi, pp. 15-17.

Mishra, Prabhakar (2004), Thoughts on LIS in India: Policies and reorganization, *GIS Development*, Vol. 8, Issue 4, pp. 20-24.

Mohan, S. and Kethari Gopi (2004), Modelling groundwater vulnerability using GIS, *Geospatial Today*, Vol. 3, issue 1, pp. 46-56.

Munshi, M.K. (2004), Integrated command and control system: dail your way to safety, *Geospatial Today*, Vol. 3, issue 3, pp. 34-36.

Nandan, Vaishali (2001), ...And Gorakhpur follows the suit, *GIS Development*, Vol. 5, Issue 5, pp. 29-32.

Nag, P. (1987), A proposed base for a geographical information system for India, *International Journal for Geographical Information System*, Vol. 1, No. 2, pp. 181-87.

Nag, P. (1999), Getting Smarter with Smart maps, *GIS Development*, January-February, pp. 20-24.

Nag, P. (2000), Health and Disease Atlas of India, *GIS Development*, Vol. IV, No.5, pp. 26-27.

Nori, Chandrasekhar (1999), Towards effective utility management – A GIS perspective, *GIS India*, Vol. 8, April-June, pp. 41-44.

NRDMS (2003), *Natural Resources Data Management System: Vision Document*, Department of Science and Technology, New Delhi.

Pandey, Piyush Kumar and Ruma Shukla (2005), Web GIS based asset life cycle management, *GIS Development*, Vol. 9, Issue 10, pp. 42-43.

Parihar, Seema M. (2001), GIS in Indian colleges, *GIS Development*, Vol. 5, issue 11, pp. 36-41.

Parker, V.N. and D. Sampat Kumar (1999), A facelift to Mumbai: *Courtesy, GIS Development*, January-February, pp. 16-19.

Patil, Uday, Shirish Ravan and Ashok Kaushal (2003), GIS based air pollution surface modelling, *GIS Development*, Vol. 7, Issue 8, pp. 45-47.

Phoenix, Michael (2004), Global GIS education, *GIM International*, Vol. 18, No. 2, pp. 30-33.

Prasad, Manish (2002), Product development, *GIS Development*, Vol. 6, issue 8, p. 19.

Raghavendran, S. (2003), GIS in water resource management, *GIM International*, Vol. 17, No. 7, pp. 41-43.

Raja Rao, T. and P. Raj Sekhar (2000), LIS-Issues, technologies and standardization, Pre-conference *Proceedings of the National Conference on Land Information System,* 18-19 October, New Delhi, pp. 62-65.

Ramaswamy, S.M., C.J. Kumanan and K. Palanivel (2002), GIS based solutions for waste disposals, *GIS Development*, Vol. 6, Issue 12, pp. 40-42.

Ramchandra, T.V., Uttam Kumar, Vaishnav and S.N. Pradad (2004), Geographic resources decision support system – an open source GIS, *Geospatial Today*, Vol. 3, issue 3, pp. 52-59.

Rhind, David (2005), GIS education for a changed world, *GIS Development Middle East*, Vol. 1, Issue 1, pp. 14-17.

Robinson, Arthur, Joel Morrison, Philip Muehreke, A. Jon Kimerling and Stephen Guptill (1995), *Elements of Cartography*, John Wiley and Sons.

Roy, Aniruddha (2001), Web mapping solution, *GIS Development*, Vol. 5, Issue 5, pp. 33-36.

Roy, P.S. (2004), A qualitative journey into the technique, *Geospatial Today*, Vol. 3, issue 3, pp. 19-27.

Roy, P.S. and P.K. Joshi (2000), Making of GIS: 2000, *GIS India*, Vol. 9. No. 3, pp. 9-11.

Roy, P.S. (2001), Remote sensing training and education needs in India, *GIS Development*, Vol. 5, Issue 1, pp. 20-23.

Saraf, Arun K. (2005), GIS education in India: Issues and challenges, *GIS Development*, Vol. 9, Issue 1, pp. 30-31.

Sarkar, S. and D.P. Kanungo (2001), Landslide identification and mapping using remote sensing and GIS, *GIS Development*, Vol. 5, Issue 10, pp. 42-44.

Satyanarayana, P. (1999), Open GIS Consortium (OGC), *GIS India*, Vol. 8, April-June, pp. 36-37.

Sharma, Vinod (1999), Use of GIS related technologies for managing disasters in India – An overview, *GIS Development*, Vol. 3, Issue 3, pp. 26-30.

Singh, Bhoop (2005), Landslide mitigation, a rugged tract, *Geospatial Today*, September-October, pp. 30-34.

Singhal, Alka (2000), In search of military GIS, *GIS Development*, Vol. 4, Issue 10, pp. 29-31.

Slocum, Terry (1999), *Thematic Cartography and Visualization*, Pretince Hall.

Sipkes, Jacques (2003, The Bhoomi Project: Computerisation of land records, *GIM International*, Vol. 17, No. 6, pp. 51-53.

Sreeshaila, G. and Satish Selvalumar (1999), The garden city blooms with GIS projects, *GIS Development*, January-February, pp. 33-38.

Subudhi, A.P., B.S. Sokhi and P.S. Roy (2000), Land information systems for municipal planning possibilities, problems and perspectives, Pre-conference *Proceedings of the National Conference on Land Information System,* 18-19 October, New Delhi, pp. 35-40.

Tamhane, Rajen M. and Ajay S. Despande (2000), Customised GIS software package, *GIS Development*, Vol. 4, Issue 11, pp. 18-19 and 22.

Tarafdar, Ayon (2002), Community GIS vs GIS community, *GIS Development*, Vol. 6, Issue 9, pp. 31-33.

Tarafdar, Ayon (2005), Understanding the Indian map policy, *GIS Development,* Vol. 9, issue 7, pp. 18-26.

TIFAC (1999), *Report of the Special Core Group on Government-Industry in Cartography*, Technology Information, Forecasting and Assessment Council, Department of Science and Technology, New Delhi.

Umashankar, C. (2003), Land record systems revisited, *GIS Development*, Vol. 7, Issue 10, pp. 30-36.

Upadhyaya, Alok (2002), State-wide digital database and GIS of cadastral maps, *GIS Development*, Vol. 6, Issue 12, pp. 28-29.

Uttarwar, P.S. (2000), Transparency in public dealing departments through GIS, *GIS Development*, Vol. 4, Issue 11, p. 28-31.

Ventatachalam, P., B. Krishna Mohan, J.K. Suri, Aarthi T. Chandrasekar and Vikas Mishra (2001), Teaching GIS principles through multimedia based GIS tutor, *GIS Development*, Vol. 5, Issue 1, pp. 24-27.

Verbyla, David (2002), *Practical GIS Analysis*, Taylor and Francis, London.

10

Conclusion

Cartography has now become a repository of enormous quantity spatial digital data acquired, processed, stored, retrieved and displayed. The instant communication to the user, in the form, mode, speed and variety desired by user is possible with electronic media. The technology and the information age have transformed cartography from the art, science and technology of making maps in to an "information transfer process". Cartography can be visualized as an information science performing a service to a wide range of other disciplines. Disciplines that are linked in spatial relationships are history, geography, geology, sociology, meteorology, astronomy, ecology, demography etc. There are many users and owners of different data involved in the building of Geographical Information System (GIS). It is the cartography who has always coordinating all sources of inputs and responsible for the concept and design of the product. It is also entrusted with the design and management of GIS and other sub-sets of the information system such as Land Information System (LIS) cadastral information system etc. Digital Cartographic databases have increased the capability and flexibility, which allow investigation of potential methods for information display to meet a wide variety of requirements. Perspective visual simulation is one of the most effective display methods to create natural appearing image, allowing associating geographic data with the real world. There are many applications of digital data such as tool for planners, Digital Elevation Model (DEM) which could be used to generate inter-visibility maps, 3D oblique view of terrain is useful for the military planner in assessing approaches and relative size and shape of key terrain features. DEM is especially useful in air-borne operation planning by providing a pilot's eyeview of the terrain that will be encountered in an actual mission. Spatial data in an information system and representation technology of cartography continue to make the scope of applications almost limit less, far outstretched from the old tasks of map making and data handling.

In the past, it sometimes took years to provide map product containing the new information. The delays in the cartographic procedures made such new information to obsolete ones. The map making tools have undergone revolutionary advancement such as high resolution imaging, Global Position System (GPS), Airborne Laser Terrain Mapping (ALTM) etc. which has the capability of rapid collection of data and has reduced the response time, which is becoming increasingly critical as a user requirement. In this context the aspect of 'Time' has become an additional dimension. A typical data that is collected with time as an element is that of meteorology. There are many other types of data that lend themselves for arrangement with time as the main element, like population censuses. Such data sets with time enable modelling of the processes to provide forecasts of future patterns. This is a field of "dynamic cartography", which will have a greater role with the technological advancement in the field of electronic hardware and software.

Cartography and GIS

Cartography and GIS are strongly related to each other, yet they remain fairly independent in their central objectives. Cartography is a much older field than GIS and some of its areas of interest are not well suited for GIS like visual impacts of different mapping techniques, map theory etc. In other areas cartography is strongly supported by GIS like data storage and manipulation, spatial inter-relational studies, experimentation with different mapping techniques. Therefore, GIS operations do not have to lead to the production of maps as final products. Although, it helps in producing more user input-friendly spatial products. Updating and changing of data and maps using GIS tools is easier than using manual method. GIS in their nature are strongly connected with fields like cartography, geography, computer science, data bases, computer graphic, land shape architecture, spatial statistic, management planning and other.

Cartography with GIS could be used in various fields such as, national topographic mapping, navigation, property and administrative boundary, geology, geography, surveying, other social and natural sciences, defense, emergency services, marketing and other commercial, population census, public administration and others.

GIS has come long way since manual GIS systems evolved from the discipline of cartography, where architects or site designers needed to visually compare the building plan with the site survey. With more elaborate urban planning in the 19th and early 20th centuries, *map overlays* became popular. These were translucent paper (and later, plastic) sheets upon which maps were drawn or printed. For example, a map of site drainage could be overlaid on a topographic map of the site, or on a street plan. This could help the urban planner or site designer determine whether or not to locate houses or other buildings at a given place (e.g., close to a creek or river floodplain). The same overlay process can be used to superimpose structural data on an aerial photograph. This technique was often used by site designers or architects when presenting designs to a prospective client.

In the 1960s, computers available to domestic government agencies and university laboratories had few graphics capabilities, generally operated in batch mode, and were quite expensive and unreliable by today's standards. Although little cost analysis information is available, it appears as though the automated methods were not yet competitive with manual map production, due largely to the high cost of procuring and maintaining computer hardware.

A more fortunate circumstance occurred with the establishment of the Environmental Science Research Institute by R. Dangermond in 1969, who eventually produced the ARC/INFO package that has become something of a standard in GIS. Intergraph, ComputerVision, and Synercom were also major players in the 1970s, and most of these approached GIS from the CAD/CAM area. But by 1976, there were at least 285 computer software packages that handled spatial data and were developed outside USGS, which number rose to over 500 in 1980. Due to lack of contact between software developers, much duplication resulted. Large states in the US developed their own GIS software, some using CGIS, others adapting SYMAP, and still others generating proprietary, in-house products. Tomlinson describes the 1970s as a period of "lateral diffusion" rather than innovation. In the defence community, the CIA developed a world data bank that subsequently was made available in the public domain. No comprehensive history exists of these local approaches, but the Bureau of the Census and USGS have made representative progress, namely linking cover from the USGS 1:100,000 topographic maps with tracts for the 1990 Census, which has motivated further GIS developments in the US.

In the 1990s, GIS has matured somewhat, with research directed away from the basic issues of map production and encoding, which have been solved after a fashion. The current issues of importance are more related to engineering concerns, such as map accuracy, precise co-registration of maps and imagery, and the vexing problem of combining datasets of diverse formats. In the latter case, much data has accumulated from earlier vector-format GIS, which is difficult to accurately integrate with current GIS systems.

For some time now GIS has been expected to turn a corner and enter into the mainstream of computing, and may be society. Chances are there that this will happen through a more 'location aware' society supported by services built on positioning, distributed geospatial databases and connected with ubiquitous telecommunication coverage. Will geographic information or GI science lead the way, or will GI technology drive developments ? Due to the potential of 'location' to serve as the critical link connecting the virtual and real worlds it is argued that the geospatial domain plays a critical role in establishing the business foundations of any information society.

GIS is not a pre-compiled map. Rather, it enables users to construct maps dynamically by displaying the selective information in the way in which they choose to view it. GIS software helps in developing the maps in chosen formats for better visualization of the geo-referenced data. Obviously GIS makes it possible to link or integrate data which is otherwise difficult. Since all most all the information vital to the government and industry are generally in one way or other location-based, GIS allows better understanding and clear visualization of the action and decisions. This aspect of GIS has a vital role in applications which we have seen in the earlier chapters.

There is definitely better business opportunity with the help of GIS. Linking location to information is beneficial for decision-making in business. As we have seen that GIS has helped the industry or business houses in choosing a site, targeting a market segment, planning a distribution network or to meet an emergency situation. Where are the customers or potential customers? In which neighbourhoods or zip code areas or pin code areas are important for a particular business? Anyone trying to take a right decision, the intelligent digital maps made in GIS environment is required. The location-based data allows government and the business managers to better understand and visualize the impacts of their decisions. For an emergency situation, not only the government but also the industry benefits from the intelligent use of GIS capabilities.

GI for Business

When we discuss current trends and a vision where we expect to be standing several years from now, we first have to establish a clear frame of reference. GIS as software architectures links technology and methods to application and business domains. Geographic Information (GI) science researches the conceptual background and methodology frameworks implemented in GI systems and guiding applications. GI manifested in geospatial data collections and data streams represents the real world, thus describing application domains and making them accessible for digital workflows. GI technology reaches beyond the software in GI systems to include sensor systems and dedicated interaction devices from handhelds to vision aids. For a discussion on trends and visions, we will have to pursue an integrated perspective with a focus on what we will be able to do many years from now. The science behind that, its translation into technology and systems, and the availability of multiple and detailed digital representations of past, current and future realities are being integrated towards future capabilities. All GI combined will serve the purpose of providing an explicitly 'spatial user interface' for our world across all its different scales, facets, themes and user perspectives.

Before attempting to look into future perspectives, we first need to review recent trends as an indicator of where substantial change is currently occurring. Of course, any such listing is highly subjective, but most experts will likely agree on the majority of items mentioned here:

- More or less isolated desktop systems have become part of distributed architectures. Recent and upcoming releases of GI software clearly demonstrate that they work best against a backdrop of services feeding into the workflow of the locally-based analyst.
- Developments initiated through the Open Geospatial Consortium (OGC) have crystallized the interoperability concept and contributed to a unified web of services evolving into our GI infrastructures.
- The past 'data bottleneck' where projects were stifled by the enormous efforts needed to supply data has by and large given way to a 'flood of data'. Data mining and 'intelligent' extraction of relevant information are critical to effective operations.
- Part of this 'flood' is provided by a multitude of sensors above and at the earth's surface. These sensors supply geo-referenced streams of measurements, effectively tying the virtual world of geographic models into the real world around our lives.
- Largely due to these sensors and ubiquitous connectivity, capabilities for near real-time monitoring have evolved. For a large part of GIS' past, we were essentially processing historical data, only recently we are moving towards 'current affairs'.
- This development has been made feasible by the now pervasive presence of positioning devices. Our societies are increasingly mobile in every respect: individual, environmental and business 'objects' change position frequently, without tracking our model of the real world would not be current.

Framework for GIS

All too often we are focusing on 'what can be done with technology' and 'how we can leverage progress in GI science'. Preferably we should look into 'use cases' and analyse what needs in society are being and need to be supported by progress in the GI fields from technology to science. GIS is still burdened (like many ICT fields) by the 'automation trap' of computers being fast and good at clearly defined repetitive tasks. Therefore, long established jobs are more effectively done, which led to file cabinets moving to databases, typing to word processing, drafting to CAD and of course mapping to GIS. The quest for the 'map looking like the one we've always had' is still on in many people's minds.

For real progress, GIS clearly needs to focus not on doing a better job at established tasks, but rather to see how tasks can best be organized to reach our societal, environmental and business objectives. Analysing how to reach these objectives, breaking down workflows into tasks and then researching any missing concepts or methods through GI science will lead to newly defined 'jobs' for GIS instead! So, what are GIS good at now? Documentation through static representations, change detection and visualization probably rank among the most frequent applications. Spatial analysis helps create new insights by generating information from data through advanced methods. Dynamic process models and scenario techniques support spatial decision support applications. All of these will be needed and provided in the future, but is there more to come?

Certainly, we will be moving from analysing and presenting manifest data sets towards working with streams of mostly georeferenced data. Without the 'where?' any data on 'how much?', 'what?', 'how?', 'when?' are missing much of their value, hence sensors deliver a fixed or moving location integrated into their reports. Through data networks, these reports are delivered right away. This means that instead of analysing past patterns, we are looking at the current picture. Insight into the past gives way to real time applications. Instead of learning from what did not work well to do it better the next time around, we can react right away. GIS, therefore, is moving towards real time control. Cartography develops into a human computer interface discipline for spatial views supporting interaction with current processes or for deep, recursive analysis of spatial data. In essence, the long feedback loop is shortened substantially: what used to be an elaborate cycle of data acquisition, analysis, visualization and decision making is for many applications being compressed into real time monitoring and control.

GIS Trends

Observing the above outlined current trends and changes to business logics involving geospatial information, we now might want to assess where we expect to stand several years from now. Distinguishing technology developments from GI science quests will be helpful while acknowledging that both areas are of course tightly linked in a symmetric stimulus—response relationship. First, looking at expected technological progress, the main drivers clearly are broad, multiple use developments demonstrating that GIS is increasingly becoming an integral part of mainstream ICT:

- Sensors, based on the sensor-web framework are penetrating our world widely. This means that the state of our environment, business assets and society's infrastructure is thoroughly monitored and available for diagnostic and controlling action.
- Positioning technologies are being integrated into sensor frameworks, and positioning/ navigation applications will become independent from specific technologies like GPS, GSM, Wifi or others. Transparently roaming across positioning technologies across a ubiquitous positioning infrastructure is a key enabler for novel applications.
- Due to pervasive networking and 'the network being the system' (a) local (geo-) processing capabilities are much less critical for GIS applications than access to the network.
- Interaction with digital information is becoming much less abstract. Working directly with spatial views (Google and Microsoft Virtual Earth currently leading the way) ties the 'online domain' directly into daily individual experience and perception. This likely will turn out to be a major driver towards the development of spatial data infrastructure services.

- Finally, access and interface devices will be defined by workflows, not by GIS. We need a big desktop screen to look at a map? Well, do we actually need to look at a map or can we relate the required information otherwise? A majority of geospatial applications will not be tied to map-like user interfaces, but rather provide 'location intelligence' in the background.

Interacting with technology trends like these, GI science will set priorities continuing from current research as well as explore new areas. Again, there are of course many legitimate perspectives assessing future emphasis which are:

- First and foremost, although this might come as a surprise to many, privacy and preservation of civil rights clearly top the list. Location is such a powerful key for integrating and accessing data collections that this issue has a huge potential for a backlash against leveraging spatial information to benefit society and our lives. How to protect individuals' locations and tracks, how to design applications without generating a 'big brother' effect will have to be high on our research agenda.
- As a related theme, geospatial enabling will be accepted as a means towards more public participation, supporting the empowerment of citizens. Citizens watch (instead of being watched!) and participate in planning and decisions affecting their lives and interests. Geospatial applications as a local communication platform require a high level of spatial literacy, this has to be confronted as an educational task.
- With increasing flow of data, methodology research is being focused on approaches to spatial pattern recognition in a wide sense. Filtering data streams to extract meaningful information is needed to environmental monitoring, security, infrastructure operations and transportation applications.
- Visual interaction (and other human sensory interfacing) ties the real world into our virtual models. Designing spatially enabled human-computer interaction (HCI) is a key theme supporting penetration of geospatial applications throughout personal and professional applications, providing assistance for mundane tasks as well as 'performance support'.
- towards 'intelligent' infrastructures integrates many currently leading topics in GI science. Infrastructures serve a purpose, they should do so efficiently and effectively and without causing undue harm. Research in semantic web, including cognitive aspects, emphasizing sustainability and minimum impact all come together in this meta-topic with a long overlooked strong GI science dimension.

Education Perspective

Education towards professional qualifications has frequently been named as a key factor towards extending the reach of the geospatial industry, with the idea that qualified specialists ranging from IT-focused via methodology-centered analyst to application experts. While this observation is certainly true and needs to be supported through the development of suitable curricula and implementation of programmes across traditional boundaries between disciplines, another aspect might be even more decisive. Traditionally, one aspect of schooling is the establishment of literacy as a foundation for critical thinking, creativity and participation. The spatial angle has been covered by the building of map reading skills supporting education towards spatial awareness and understanding many facets of our world. Well, map reading skills are changing substantially and are applied in the computing domain. 'Maps' now move, change perspective, can be actively navigated and are much more an orientation device and spatial user interface.

Most surveys and visionary perspectives claim that the geospatial industry is moving from a technology-oriented niche into the mainstream of society. GI only has limited growth potential in the business-to-business sectors of corporate decision support, workflow optimization and day-to-day operations. Rather, a customer-focused evolution towards spatially enabled and locally centred services is considered as a major growth opportunity. If we want customers to use a search engine's spatial portal as their default web page, bookmark their community's map interface, work out an itinerary from a spatial perspective and shop, eat and interact with friends based on ICT-facilitated proximity, we need support from general education.

Spatial literacy is not about teaching GIS in schools. Rather, creating spatial awareness, navigational ICT skills and an understanding that all or most of today's problems have a significant spatial aspect to them need to become objectives throughout schools curricula. Only spatially aware citizens with the requisite ICT skills will be able to critically participate and contribute to an information society.

Will our lives be different?

Of course, the answer will be yes and no, depending on what we consider the defining aspects. Will our future lives be different due to progress in geoinformatics? For all of us, and many others, yes. In what ways will life be different? This will depend largely on lifestyles, professions and interests, and of course, our geographical and societal positions on the globe. Geospatial information will more and more affect us on different scales. The 'global village' and the much quoted links between local action and global thinking will become less philosophical and more practical. Beyond the obvious assistance with travel and transportation, geography affects everything around us. The environment, sustainability and infrastructures are all inherently geographical issues. As they are being represented through geoinformation to enable direct, location aware interaction, digital geospatial interfaces essentially become our dashboards as we navigate life.

On a practical level, the trend towards networked services means that local availability of computing and processing power becomes much less critical than the availability of bandwidth. If ample bandwidth is provided differences in education, opportunities and participation will decrease. Bandwidth as an indicator for access to information and interaction, therefore, is the key differentiating factor in future geographies.

Towards the Geo-Information Society

The European Union's 'Information Society' concept claims that the spatial aspect is pertinent to a majority of information aspects and it also provides interesting perspectives to toy with the term *Geo-Information Society.* We believe that, in a way, we are moving into this direction. Let us briefly look at information from an abstract meta-perspective: information objects or features can be located in an 'information cube' which has defined geographical research since the early days of the (then) 'quantitative revolution'. Objects have attributes characterizing their position in subject, temporal and spatial domains. A cube with the dimensions of subject (what?), time (when?) and place (where?) can therefore be used to place information objects (Fig. 10.1)

This view provides support for the argument that GIS is not a distinct area along the lines of the 'spatial-is-special' phrase, but geography is rather a view, often a very important view on our world through the lens of information representations. In simple terms, the spatial dimension is a key component of the 'context' of objects, and which affects our actions. Providing full context for decisions and actions is essential, as we know from our many 'profiles' we use to make software and services more focused on our preferences and intentions. Context is a key enabler of semantic-rich interaction (semantic web, see reference), and context is rarely complete without location.

So, the concept of a geo-information society as a more context-sensitive, less abstract concept might not be so far-fetched after all! Technologically speaking, it is built on three pillars: (Fig. 10.2) geospatial data as digital representations of much of our world; positioning services putting people, assets and other 'objects' into this context; and mobile telecommunication connecting everything and providing access for users.

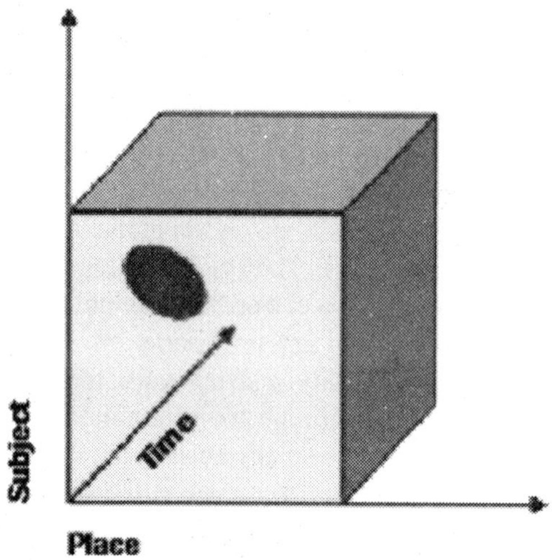

Fig. 10.1: Attributes to place information objects

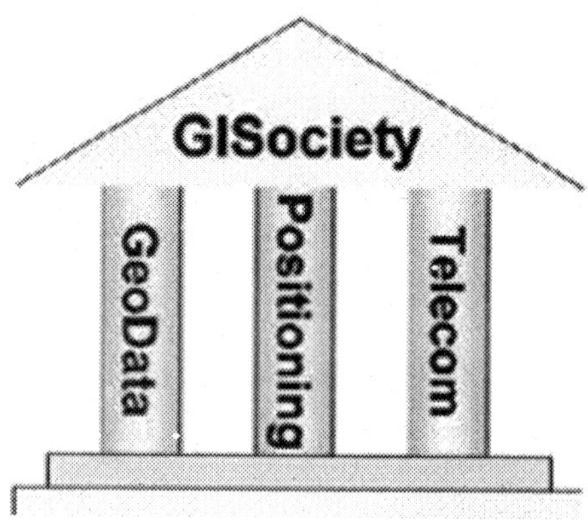

Fig. 10.2: The Geo-Information Society

GI Science and the Information Economy

In 1995 and 1997, the magazine *Economist* declared the 'death of distance' caused by information infrastructures making location largely irrelevant. Telecommunication, telecommuting and globalization appeared as clear indicators. Interpreting Tobler's 'First Law of Geography' (stating 'Everything is related to everything else, but near things are more related than distant things') this should mean that geography had to be declared defunct as well. Further, in March 2003, though, the same publication titled an article 'The revenge of geography', starting with geography', starting with the observation: "In the early days of the internet boom, there was much talk of the death of distance. The emergence of a global digital network, it seemed, would put an end to mundane physical or geographical constraints. There was some truth in this, but ..." Clearly, the authors had recognized that location provides the best general means of connecting virtual and real worlds. This connect in turn is required to provide business models for the information economy—Geography and GI Science are well placed to provide some of them!

Just as there are 'macro' and 'micro' aspects to economy, we recognize this as a valuable distinction in geography and now GI science. Up to now we have mostly worked in the frame of macro-geography. Now due to positioning services, detailed geospatial data, and powerful communication GIS moves more towards micro-geography. Actually, there would be no macro-economy without the micro-level of businesses. We wonder what it means that GIS has come the other way. Due to the availability of better and finer data and also the technology for processing the same, the concept of "digital geography" is getting currency. It is becoming increasingly possible to have a bottom-up approach in developing geographical (also spatial) analysis, models and postulated. A new set of indigenous concepts in geography is likely to evolve with the help of computer processed huge data sets. Earlier, models and concepts will be subject to revision, modification and revalidation.

Similar to above, in the business and industry, a fresh approach is likely to evolve for providing better services, functionally and economically, to the clients and people at large. In order to achieve this goal, GIS cannot be effective in isolation. It has to integrate with on one hand with other mapping technology like remote sensing of GPS; while on the other, with other emerging technologies such as information and communications technologies. In short, due to the expanding horizon of the digital technologies, integration is the core issue in any business solution involving GIS as well. In the previous chapters we have seen the GIS is perhaps a small part of a business solution, but effective and essential. The developing countries do become a virtual laboratory for the application of GIS as in these places the vast population reside in less organized way. For any attempt for economic or social development, application of GIS cannot be ignored.

Increasingly, state and local economic development activities are coming to rely as much on GIS technology as are environmental and natural resource concerns. These problems would be challenging enough, but they are complicated by the inefficiencies of traditional government planning processes. As authorities admit, "traditional planning is fractured, takes decades to implement, and often spawns lawsuits between developers and environmentalists." With population growing at skyrocketing rates, these were conditions different counties could no longer tolerate. Increasingly, national, state, regional and municipal, social and economic development activities are coming to rely on GIS technology. In order to have an economically viable approach, GIS capabilities are also being deployed in an effort to improve health, enhance emergency preparedness, and save lives. GIS has also been applied to reduce the incidence and cost of major diseases like diabetes, cardiovascular disease, and cancer, improve health outcomes through prevention and earlier detection, and increase the health of residents through fitness and nutrition education programmes. There are several instances where the government officials have used GIS technologies to accurately gauge the size of the affected population, the optimum sites for distributing medicines, the impact of weather conditions at the time of the event, the relative capacities of nearby hospitals, clinics, and pharmacies in the area, and the travel patterns of affected individuals in a given time before and after the event. Further, as governments have deployed GIS capabilities more widely, many have discovered that greater usage of GIS data may be the quickest route to more efficient government. The civic authorities can browse through a variety of general information, like census data, housing profiles, home values, and so on. San Francisco GIS Manager observes a few of the ways in which the city and county have already implemented GIS capabilities:

—The city assessor and real estate staff rely on GIS information to improve their ability to manage city property.
—Police officers can review Internet-based geographical data directly from their squad cars in order to help speed their response to emergencies.
—The police department later this year will begin using GIS-based crime data as the basis for reassigning task groups to neighbourhoods with the highest crime rates.
—A new application called SFViewer enables users to input an address or block number in order to view parcel information, aerial photographs, or street maps.
—As in Dallas, city residents can enter their address into the city's web site in order to locate a range of neighbourhood-related data, from the name of their legislative representative to times when their streets will be cleaned.
—And businesses can use an online application called SF Prospector to help them to determine the best location for a new business, based on office rental vacancies, neighbourhood demographics, and so on.

The key to the success of these efforts, GIS manager believes, has been an unprecedented degree of executive level support, including the Mayor of the City. This approach has enabled several departments to employ GIS successfully with minimal investment. Further, most government agencies deploy GIS technologies to manage such geospatial data as land parcels, roads, and facilities. By linking this GIS data with the supporting documents, such as engineering drawings, scanned deeds, or road construction contracts, employees are given quick and easy access to the data required to make immediate and informed decisions. All that changed with the adoption of a fully integrated electronic GIS system. For example, finding related information all of the documents pertaining to a certain project will involve in locating someone who knows which file cabinet to search. What typically happened therefore was that a request for information would be sent to the records section and it would take days for staff to collect the required information and respond to the request. Residents and other members of the public looking to the municipality for legal contracts, payment of taxes, financial records, engineering documents, geographical maps, satellite maps, and land parcel ownership records and correspondence hence often found it difficult if not impossible to obtain access to records that corresponded with information in the GIS.

As we have seen, with the newly integrated GIS system, documents can be associated with projects though text-based searches as well as geographically based searches. By linking geographic documents with corresponding map features in this way, staff will spend less time searching for records, thus increasing their productivity. Instead of spending time looking for records that might not be found, the staff can now email requests directly to the appropriate parties and receive a prompt reply, sometimes in a matter of minutes. This whole activity can be out-sourced as well. There is a potential business opportunity in such activities as well. Obviously, integrated GIS solution has become a very powerful strategic enterprise tool for managing costs, complying with government regulations, and improving customer service as well.

There are other instances also which require immediate response, such as firefighters and emergency medical services where the personnel work around the clock to put out fires and tc respond to the health emergencies of individual citizens; or engaged in daily operations to identify, track, and mitigate life-threatening diseases; or respond to accidents, keep roads safe and analyse accident patterns to develop strategies that reduce injury and death. All of these operations have two things in common: (a) they are responsible for saving lives each day, and (b) they have to depend upon the information that has a spatial or geographic context, such as digital aerial photography, accurate street and parcel maps, topographic maps, critical infrastructure maps, and maps of hazardous materials. These information sources when combined with geospatial and other complementary technologies as GIS, computer-aided dispatch systems, routing software, the GGPS, automated vehicle location, remote sensing, and others contribute to the foundation for a modern public safety information infrastructure. During major emergencies such as terrorist attacks, flood, fire, earthquakes, or hurricanes, the data can immediately be used to support the efforts of first responders. They also can support hazard mitigation operations such as the tracking of potential terrorists and environmental monitoring to prevent emergencies from happening in the first place.

Certain life saving operations, like *tsunami* or earthquakes, extend beyond the borders of local jurisdictions including state and regional boundaries, however international and national emergency management operations must use these same data so that support and response efforts can be quickly and easily integrated with efforts at national and global levels. It is, therefore, essential that spatial data be built to comprehensive, consistent, and nationally and internationally agreed upon standards. Global Spatial Data Infrastructure (GSDI), Permanent Committee for GIS Infrastructure for Asia and Pacific (PCGIAP) and similar regional bodies, International Steering Committee for Global Mapping (ICGM: www.iscgm.org) and the National Spatial Data Infrastructure (NSDI) and similar national bodies have been working on joint initiatives for developing common data bases. Spatial data must be created through national, regional and global initiatives to ensure that they are available to all who require their use. In case of India, following the approval of the Cabinet, the state governments are also taking steps to form the state level Spatial Data Infrastructure (SDI). This process will further filter down to district or even at sub-district levels.

Creation and deployment of the public safety components of the NSDI or state SDI will have many additional benefits. Local and state governments can use the same data to provide a base for countless non-emergency operations and applications as well, such as services, formulation of business strategies, e-government initiatives, economic development and planning, waste removal, street cleaning, code enforcement, environmental protection, construction, inspections, capital construction and the like. These applications of spatial data are known to increase workforce productivity, streamline business processes, reduction in costs and improve services delivered to the public. To take advantage of the technology, a GIS based on reliable data should already be available with every government agency deployed in extending the services to thc people. The public safety components of the NSDI need to be developed immediately particularly for a country like India. Necessary steps should be taken well before the measures are required to respond to a catastrophe.

Finally......

Geographic Information Systems (GIS) deal with the storage, management, retrieval, conversion, analysis, modelling and display of spatially related data in a systematic way. GIS software is potentially applicable in nearly any situation that calls for decisions involving a spatial component. In other words, GIS can be a useful tool in situations as diverse as area to determining the environmental impact. It is impossible to list all the possible applications of GIS, as they are practically limitless. As computer power grows, enabling the use of more complex models and the more efficient incorporation of a time component in analysis, the more this will be true. Advances in computer software and hardware, increased familiarity with the power and applicability of GIS and its related technologies, and a greater breadth of accessible data have driven the growth in demand for GIS and will continue to do so in the future. Throughout the knowledge-based economy, both software and hardware have become much more powerful and simple to use while becoming less expensive. Clients in all markets are now comfortable with computers and have access to the necessary equipment to run even complex GIS packages. Furthermore, not only has a greater range of geospatial data become more readily available but also non-traditional data such as those found in the social sciences.

In a developing country like India, having huge population base with democracy, public access to information and freedom of expression, GIS has several additional function to play. It is consider being a tool for getting in touch with the government, lodging protests, and having interface with people. In addition, there is a wide scope for business enterprise as well. GIS has a definite role to play in the following sectors:

(i) Highways survey and mapping.
(ii) Natural disaster management.
(iii) Telecommunication services.
(iv) Power projects.
(v) Ports projects.
(vi) Urban land use planning.
(vii) Rural land management or agricultural land management.
(viii) Waste land reclamation projects.
(ix) River basins or watersheds management.
(x) Imagination and Canal developments.
(xi) Consolidation of land holdings and land records.
(xii) Digitization of maps on various scales.
(xiii) Updating and digitization of wards and services maps of different cities and towns for their betterment both in terms of economy or finance and environments.
(xiv) Mapping of places of tourist interest or pilgrimages both in hard copy and soft copy forms.
(xv) National natural resource management.
(xvi) Monitoring of environment.
(xvii) Terrain and height information or positions for frequency requirement for the purpose of All India Radio, Door Darshan and telecommunication.
(xviii) For better and fool-proof security with up-to-date and accurate maps in the form of digitized or special software both for military and civilian purposes.

(xix) Development of accurate navigational maps both at ocean or water and land for better traffic-flow and least risk of accident.

(xx) Production of good quality maps and atlases for the use of students and others.

(xxi) Maps and GIS for government officials and the common people to understand the real positions and facts in their day-to-day jobs or life.

Although a number of national GIS initiatives are underway, many of the most interesting developments in the use of GIS technology are taking place at the state and local level. The use of geographic information over the past few years has increased dramatically, with a focus particularly on public safety and economic development. In some countries in the west, an extensive crime-mapping system is being implemented that will provide real-time information to police authorities regarding criminal activities. In terms of economic development, the city authorities in order to attract investments have sponsored interactive mapping web site that allows the public to review and analyse economic and demographic data for business site selection. The sense of achievement lies in getting support from the leadership to enterprise GIS effort, particularly freeing agencies from having to individually procure and develop data and systems on their own. This approach has enabled several departments to employ GIS successfully with minimal investment. The success is due to the collaborative efforts of other GIS professionals as well. Without support and initiative from all concerned, the GIS programme would be much more difficult to implement. Nevertheless, the main constraint has been the inability, at times, to get relevant agencies to collaborate toward a collective goal of using, sharing, and disseminating standardized geospatial data, the issue which is very significant for the implementation of NSDI in India as well. From spatial data technology perspective, implementing of a protocol that preserves data ownership, while still emphasizing the sharing and use of common GIS datasets, is a challenge. The solution is a hybrid approach where the agencies still retained ownership of their information, but in which enterprise GIS servers are able to dynamically connect to their own servers among themselves or with NSDI hub for data access and retrieval. The key to success is collaboration.

While the technology hurdles are considerable, organizational and political concerns have tended to be much more challenging. Resolving issues such as data distribution to the public, or privacy concerns related to public data, have proven to be more complex than implementing the technology. Some solutions have been worked for India by establishing NDSI, officially in June 2006. It has committees at two levels, first the National Spatial Data Committee (NSDC) for polices, strategies and programmes; and the second the NSDI Executive Committee for implementation and executive functions.

Funding and marketing have been the challenge. From a business perspective, policy makers, legislators, and administrators do not buy what they do not understand. The potential application of geography, digital geospatial data, and the ability to manipulate it in GIS environment are to be marketed well enough to obtain the necessary funding from government officials. The success lies in public/private partnership aimed at modernizing the land records and other development and public service related data in a standardized manner. Champions of GIS face the same challenges as those seeking to implement other forms of IT, i.e. how to demonstrate value to the business side of government institutions. This issue has become more vital due to government's emphasis on revenue generation. Each institution should have a revenue model as well. However, such initiatives have to take into account of (a) heavy initial costs, (b) data integration expenses, and (c) long payback periods. It will soon become necessary to quantify the financial benefits of GIS implementation in terms of public good, operational efficiency and also political returns. The latter involves public relation, public safety, economic development and increased participation in government sponsored activities. Obviously, GIS is an integrating technology; its value is realized when multiple data sources reveal spatial relationships between geography and a collection of information on economic, environmental, health, social and infrastructure. GIS implementations often begin, not at an enterprise level, but from organizational initiatives that have recognized the value of GIS as a part of their mission. GIS startup carries substantial costs for thee procurement of data, digital imaging, map digitization, data quality assurance, data quality control and software and hardware purchase and implementation. Solid project management and supportive governance are essential to a successful GIS implementation.

It is seen that the Indian industry is fully capable and able to take the challenge of the demand for mapping or cartographic work. The concerned Indian industry is fully prepared to take cartographic work in collaboration with the government departments or agencies. They are doing a number of cartographic or mapping jobs for and of the government. They are also doing Geographical Information System (GIS) or Land Information System (LIS), digitization of maps or documents etc. for different organisations and agencies. It is also found that there is a great demand for thematic maps on large and small scales on 1:1,000, 1:10,000, 1:50,000 and 1:1,000,000 on different themes for projects in the field of telecommunication, business, irrigation, power, transport and tourism, resource management, town and urban planning, etc. It has been assessed that there is a huge potential for export of cartographic products and services, especially in the field of GIS or land information system, scanning and digitization, remote sensing technology and developing different products, services and applications not only for India but other countries as well. Many Indian industries have done several mapping and digitization work for foreign companies or countries and are doing for many others. Many Indian industries are ready to collaborate in the field of cartographic or GIS with global partnership. Further, India has become a hub for in sourcing IT projects from abroad. Mini-Bangalores have cropped not only in the fringes of the bigger Indian cities, but the medium size cities have not been spared as well. With the successful experience of IT, trained man-power available with fairly good knowledge of English and capacity to pick up other foreign languages, India is likely to become hub for global GIS or geospatial industry as well. GIS or geoinformatic institutes for teaching, training and research are coming up in every city. The time is not too far when India will become a favourite destination for GIS industry as well. Geotechnology is the new challenge, a new determination and a new dimension to the world of science.

REFERENCES

Ramchandra, T.V., Uttam Kumar, B. Vaishnav and S.N. Prasad (2004), Geographic resources decision support system—an open source GIS, *Geospatial Today*, Vol. 3, issue 3, pp. 52-59.

Strobl, Josef (2005), GI Science and Technology—where next?, *Gis@development*, August (josef.strobl@sbg.ac.at).

APPENDICES

APPENDIX I

National Map Policy-2005

1. Preamble

All socio-economic developmental activities, conservation of natural resources, planning of disaster mitigation and infrastructure development require high quality spitial data. All advancements in digital technologies have now made it possible to use diverse spatial databases in an integrated manner. The responsibility for producing, maintaining and disseminating the topographic map database of the whole country, which is the foundation of all spatial data vests with the Survey of India (SOI). Recently, SOI has been mandated to take a leadership role in liberalizing access of spatial data to user groups without jeopardizing national security. To perform this role, the policy on dissemination of maps and spatial data needs to be clearly stated.

2. Objectives

- To provide, maintain and allow access and make available the National Topographic Database (NTDB) of the SOI conforming to national standards.
- To promote the use of geospatial knowledge and intelligence through partnerships and other mechanisms by all sections of the society and work towards a knowledge-based society.

3. Two Series of Maps

To ensure that in that furtherance of this policy, national security objectives are fully safeguarded, it has been decided that there will be two series of maps, namely:

(a) *Defence Series Maps (DSMs)*—These will be the topographical maps (on Everest/WGS-84 Datum and Polyconic/UTM Projection) on various scales (with heights, contours and full content without dilution of accuracy). These will mainly cater for defence and national security requirements.
The series of maps (in analogue or digital forms) for the entire country will be classified, an appropriate, and the guidelines regarding their use will be formulated by the Ministry of Defence.

(b) *Open Series Maps (OSMs)*—These will be brought out exclusively by SOI, primarily for supporting development activities in the country. OSMs shall bear different map sheet numbers and will be in UTM Projection on WGS-84 datum. Each of these OSMs (in both hard copy and digital form) will become "Unrestricted" after obtaining a one-time clearance of the Ministry of Defence. The content of the OSMs will be as given in Annexure 'B'. SOI will ensure that no civil and military Vulnerable Areas and Vulnerable Points (VA's/VP's) are shown on OSMs.

The SOI will issue from time to time detailed guidelines regarding all aspects of the OSMs like procedure for access by user agencies, further dissemination/sharing of OSMs amongst user agencies with or without value additions, ways and means of protecting business and commercial interests of SOI in the data and other incidental matters. Users will be allowed to publish maps on hard copy and web with or without GIS database. However, if the international boundary is depicted on the map, certification by SOI will be necessary.

In addition, the SOI is currently preparing City Maps. These City Maps will be on large scales in WGS-84 datum and in public domain. The contents of such maps will be decided by the SOI in consultation with Ministry of Defence.

4. National Topographical Database (NTDB)

SOI will continue to create, develop and maintain the National Topographical Database (NTDB) in analogue and digital forms consisting of following data sets:

(a) National Spatial Reference Frame,
(b) National Digitial Elevation Model,
(c) National Topographical Template,
(d) Administrative Boundaries, and
(e) Toponomy (place names).

Both the DSMs and OSMs will be derived from the NTDB.

5. Map Dissemination and Usage

- Open Series Maps of scales larger than 1:1 million either in analogue or digital formats can be disseminated by SOI by sale or through an agreement to any agency for specific end use. This transaction will be registered in the Registration database with details of the receiving agency, end use etc.
- Thrugh the agreement, SOI will allow a user to add value to the maps obtained (either in analogue or digital formats) and prepare his own value-added maps.
- The user should be able to share these maps with others—the information of all such sharing will also require to be logged in the Map Transaction Registry.

6. Applicability of Previous Instructions

The Ministry of Defence has from time to time issued detailed guidelines on various aspects of map access and use. These instructions shall continue to hold good but for the modifications cited herein.

GUIDELINES FOR IMPLEMENTING NATIONAL MAP POLICY

1. Background

The Central Government announced the National Map Policy (NMP) on 19th May 2005. The NMP document authorizes Survey of India (SOI) to issue detailed guidelines on the implementation of the NMP. These guidelines are issued in the light of the above cited authorization. The guidelines are clarificatory in nature and does not create any new rights nor abridge any existing rights which are enforceable in courts of law.

2. Information of Maps and Guidance of Indentors

(a) All up-to-date information on SOI maps shall be available at the various offices, map sales counters and Geospatial Data Centres (GDC) under the SOI. The information shall also be available on Survey of India website www.surveyofindia.gov.in created and maintained by the SOI. The information provided *inter alia* includes all relevant details of maps such as scale, information content, date of data capture, price, mode of data dissemination whether MOD clearance available for issue as Open Series Maps (OSMs) etc. The website also has a search engine by which the exact sheet number of a given locality can be traced by giving elementary details like name of district or any other prominent feature. SOI offices may be contacted for further information on OSMs.

(b) The lay out of OSM shees on UTM projection and WGS-84 datum is shown in Annexure 'A'. The annexure also shows the incidence of the existing topographical series on the OSM series.

(c) Data content: Approved data content of OSMs is given in Annexure 'B'.

(d) Export of all maps/digital data in 1:250K and larger scales through any means in prohibited *vide* Ministry of Finance (Department of Revenue) Notification No. 118-Cus./F.No. 21/5/62-Cus. I/VIII dt. 4th May 1963. Digital Topographical data will be licensed to only Indian individuals, organisations, firms or companies.

3. Sales/Data Dissemination

(a) *Analogue/Paper Maps*: of all scales shall be made available from Survey of India (SOI) offices and Map Sales Counters on payment at prices as may be fixed by SOI from time to time. The list of such outlets, their addresses with telephone numbers shall be made available in the website www.surveyofindia.gov.in. These maps can also be sold by any retailer. Digitisation of SOI analogue/paper maps is strictly forbidden.

(b) *Map Transaction Registry* (MTR): As stipulated in the National Map Policy, Survey of India will establish an on-line MTR for recording of all transactions relating to digital maps. Each user will be allotted a unique user ID and each transation with a unique transaction ID. For all future correspondence, this user ID and Transaction ID should be referred.

(c) *Digital Maps*: include both Raster and Vector forms. The ownership of all digital data vests solely with SOI and will be given only under licence against indent and on payment. Unauthorised copying and distribution of SOI digital data are strictly prohibited. All licences will be issued through the Map Transaction Registry (MTR). The format of the licences is available in SOI web site www.surveyofindia.gov.in. The indents may be made in the prescribed proforma.

4. Licensing of Digital Maps

Digital data will be available in single/multiple/commercial licensing for general use, value addition and marketing. All digital maps will be provided with encryptions/mechanisms which may corrupt the data while copying unauthorisedly or while attempting the same. Every such attempt shall attract criminal and civil liability from the user without prejudice to the corruption of data or software/hardware for which the SOI will not be liable.

SOI digital data will be licensed based on usage. Following are the categories:

- Digital Licence
- Publishing Licence
- Internet Licence
- Media Licence
- Value addition Licence

Terms and conditions governing each of the licence is available in SOI web site www.surveyofindia.gov.in.

5. Procedure for Value Added Products

Inasmuch as the National Map Policy is aimed at encouraging a flourishing geospatial industry and related knowledge services, there is much expectation from the industry to do value additions to the products of SOI. SOI encourages individuals/industry to do value addition. Intending value adders will submit a business model indicating the products to be generated and marketing strategy and also execute a Memorandum of Agreement (MOA) with SOI.

6. Pricing of Various Products

(i) The SOI shall notify from time to time the price list of its standard products.

(ii) The prices are subject to revision from time to time.

(iii) An indentor has to pay the amount at the rate prevailing on the data of supply unless the indenter has already paid at the earlier rates. In other words, the indenter will not be eligible for supply on payment on the plea that the rates prevailing on the date of indent were lower.

(iv) The SOI reserves its right for differential pricing, such as concessions to Universities, Research organizations, deserving NGOs on the basis of predetermined policies, different prices between single and multiple user licences etc.

7. e-Payment

Digital maps, on following the procedure, will be made available on variety of media like Compact Disks (CDs), DVD etc. Maps on 1:1M and smaller scale on compressed JPEG format will be available for download *against* online submission of licence agreement and on e-payment. The self-guide provided in www.surveyofindia.gov.in shall enable an indentor to make e-payment and download the requisite maps.

8. Settlement of Disputes

In case any dispute arises on the applicability or interpretation of these guidelines between the SOI and any other person, the matter shall be referred to the Secretary, Department of Science and Technology, Government of India, whose decision shall be binding on both the parties.

9. Applicability of Previous Instructions

The Ministry of Defence has from time to time issued detailed guidelines on various aspects of map access and use. These instructions shall continue to hold good but for the modifications cited in the NMP.

10. *General*

(i) Copyrights of all SOI maps (both digital and analogue) vest exclusively with the SOI. Any person resorting to unauthorized copying or use or attempts to do so shall make him liable to criminal and civil liability under existing laws.

(ii) While every effort will be made to ensure availability of maps of all areas containing all necessary details and with highest accuracy, SOI will not be responsible for any loss caused to any person on account of failure on any of the above factors or any other reason.

(iii) The SOI reserves the right to add, delete, modify amend any or all of these guidelines without notice and without assigning any reason.

Annexure 'A'

अनुबंध 'क'

Layout of Sheets

शीटों का विन्यास

Datum: World Geodetic System (WGS)-84

आधार : विश्व ज्योडीय पद्धति-84

Projection: Universal Transverse Mercator (UTM)

प्रक्षेप : यूनिवर्सल ट्रान्सवर्स मर्केटर

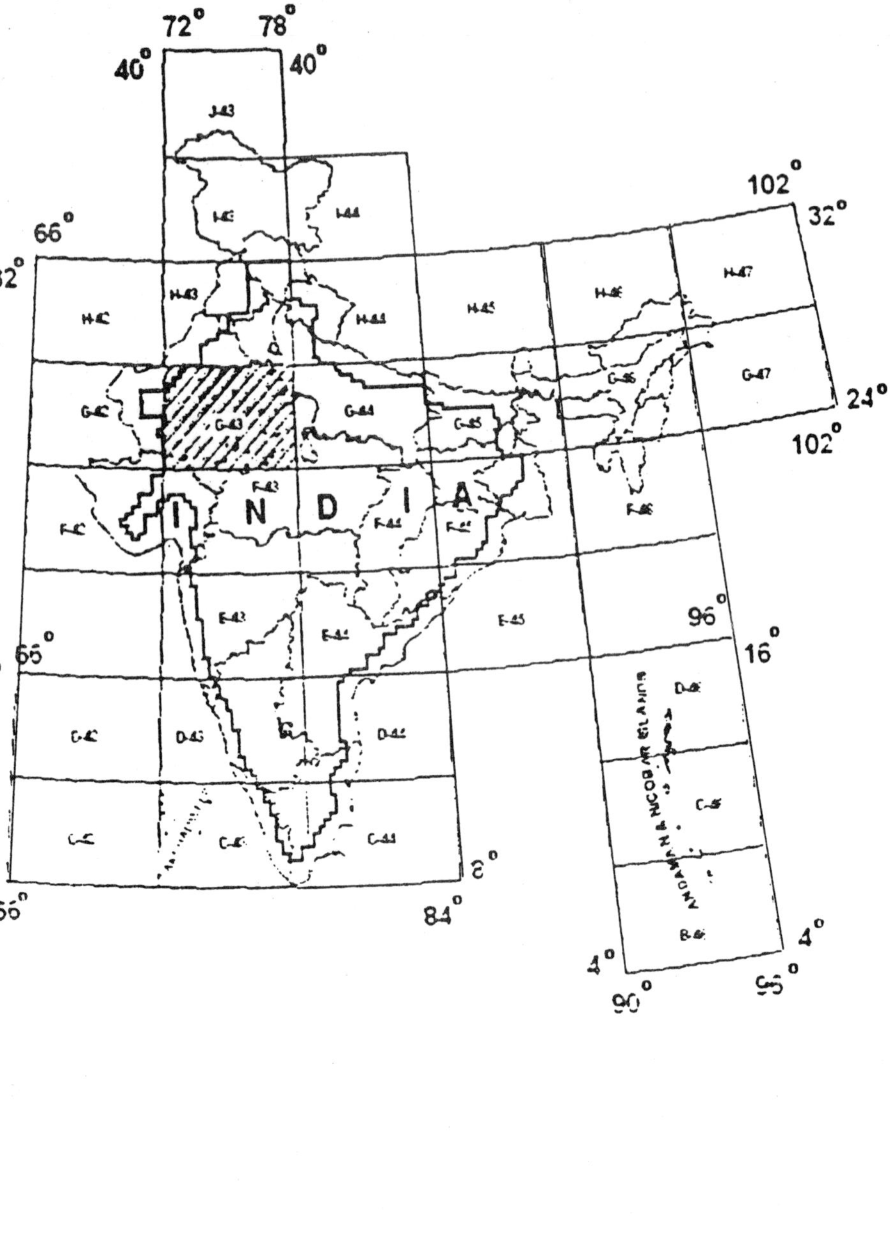

Annexure 'A' (contd.)

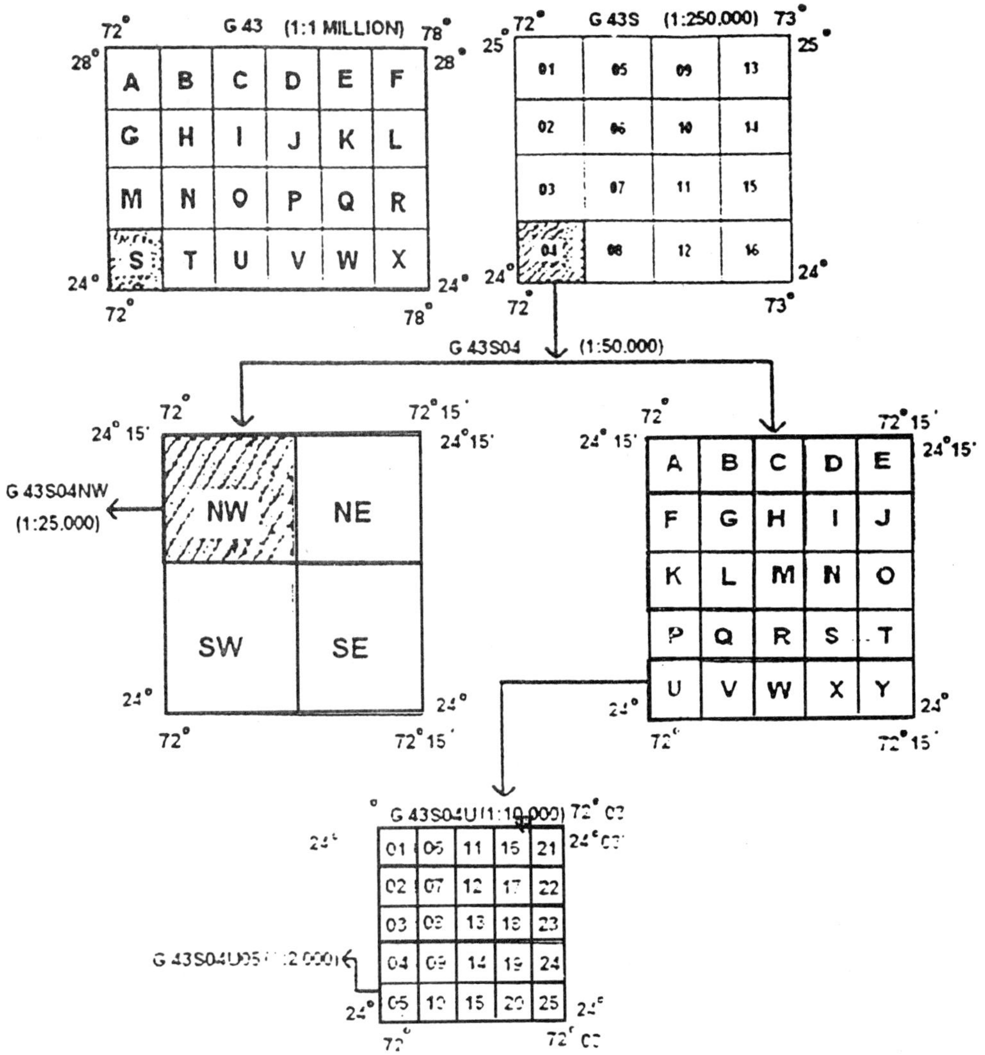

Annexure 'B'

Contents of OSM for Small Scale Maps

Sl.No.	*Category/Layer*		*Sub-details*
1.	General		Latitude/Longitude Name of State/District/Administrative index Topo sheet Number/Year of Survey/Edition No. Index to topo sheets Magnetic variation from true North direction Map reference Bar scale/Representative Factor
2.	Administrative Boundaries	Names Boundary Boundary Pillars	Administrative/Locality or tribal International to village, Forest, all boundary pillers, village trijunctions
3.	Communications/Roads	Roads Tracks Railways Embankments Other Lines	All Roads All Tracks, pass, footpath All gauges with stations, tunnels Light railways or tramway, all embankments, road/rail/tank
4.	Hydrology	Stream/Canals Dams Rivers and Banks Wells, Water Features	All streams/canals All earthwork dams All rivers with details, banks, islands All wells/tube wells/springs All Tanks (excluding overhead tanks), Lightship buoys, anchorages
5.	Settlement/Cultural Details	Towns Villages Offices Settlements	Village inhabited, deserted and forts Huts, Tower, Antiquities Religious places, tombs/grave All post/telegraphic/Police stations hut All Bungalows
6.	Transmission Line		
7.	Relief/Hypsography	Contours Sand Features Ice Forms Heights Benchmarks	Contours with sub-features All sand features Ice forms (all features) Spot height, Approximate height Benchmarks (Geodetic tertiary, canal)
8.	Vegetation	Plantations, Trees	All trees, Vine on Trellis, grass, scrub.
9.	Forest		Reserved/Protected

* Contours and heights will not be available in restricted zones as per MOD's instructions.

Annexure 'B' II

Contents for Large Scale Maps

1. Boundary (Polygon)

- Administrative Division
  - State
  - District
  - Tehsil
  - Village
  - Ward
  - Municipal/Corporation
- Locality
  - Hill Range
  - Tribal Area
  - Slum Area
- Boundary (Line)
  - State
  - Municipal/Corporation
  - District
  - Tehsil
  - Village
  - City
  - Colony/Locality with Pincode
  - Block/Sector/Sub Colony
  - Limits

2. Built Up

(A) Commercial
  - Bank Money Changers ATM
  - Cinema Theatres/Hall
  - District Centre
  - General Business Milk Booth

Hotel
  - Motels
  - Hostels
  - Dharamshalas
  - Guest Houses (Commercial)
  - Hotel with Star Categorisation
    - Restaurants

Non-Hierarchy Commercial Centre
  - Central/State and Other
  - Trade Centres—Shopping Area
  - Wholesale/Warehousing Depots

(B) Government Building

Courts
  - Supreme Court
  - High Court
  - Magistrate/Sessions Court
  - Munsif Court
  - Civil Court
  - Family Court
  - Customs and Excise Appellate Tribunal
  - Motor Accidents Claims Tribunal
  - University and School Tribunal
  - Consumer Courts

Jail

Office
  - Post and Telegraph Office
  - Rail Reservation Centre
  - Air Line Office
  - Check Post
  - Octroi Posts
  - Toll Points

- Manufacturing
  - Brick Kiln
  - Group Industry
  - Service Centre
- Public/Semi Public
  - Cremation and Burial
  - Electric Crematoriums
  - Cemeteries
  - Graveyards
  - Tower of Silence

Education and Research
  - School
  - College
  - University
  - Institution

Music/Dance/Drama/Martial Art School

Professional/Financial/Commercial/Philanthropic and Other Institutions

(contd.)

(contd.)

Facility Centre	Post HQ and Police Lines
Electricity Complaint Office	All Police Including State,
Sewer Complaint Office	Railway, Home Guards and Civil
Water Complaint Office	Defence, Traffice
Birth/Death Registration Office	Police Chowki
Marriage Registration Office	
Travel Agencies	Socio Cultural Institutions
Fire Station	Old Age Homes
Medical and Health	Art and Cultural Institutions
Hospital	Art Galleries
Dispensary	Marriage/Banquet Halls
Nursing Home	
Blood Bank	
Accident and Trauma Centre	
Ambulance	
Recreational	Major Parks—Gardens and Parks
Amusement or Water Parks	Playground
Places of Tourist Interest	Regional Parks
Film City	Religious Places
Auditorium	Temple
City Forest	Gopuram
Club—Libraries—Govt./Public	Mosques
District Parks	Idgah
Golf Course or Race Course	Gurudwara
Historical Monuments	Church
Forts	Synagogue
Place of Trourist Inerest	Fire Temple
View Points	National Park/Sanctuary
Statues	Sports Centre
Tombs	Sports Complex
	Stadium (Indoor/Open)
	Fountain
Residential	Residential
Foreign Mission	Build up Area
Embassies with their Office	House Number Information
Mixed Residential	Gate
Important Building	Wall—Fort Boundaries
Apartments/Flats/Quarters	
Govt. Colonies	
Bungalows of Chief Minister and All Ministers/Mayor	
3. Land Cover	
• Agriculture	• Road Circulation
Cultivation (Irrigated/Non-irregated)	Cart Track
Nursery	Divider
Farm Houses	Express Way—Arterial Road
Farms	Flyover

(contd.)

(contd.)

Vegetated Area Bound Tree Cover

- Vegetation
  - Forest
  - Reserved/Protected Forest
  - Forested Area/Wooded Area
  - Orchard
  - Plantation
- Govt. Land
  - Rented
  - Leased
  - Open
- Green Belt
  - Fallow Area
  - Green Belt
  - Poultry and Dairy Farming
- Parking and Other Open Space
  - Other Open Space
  - Parking
- Land
  - Rural Area
  - Barren Land
  - Marshy Land
  - Mud/Mangrove Swamp
- Water ways
  - Harbour
  - Jetty
- Airport
  - Domestic and International (Boundary wall and internal details will not be shown)
- Railway Line
  - Railway Line
    - Broad Gauge
  - Electrified/Single/Double/Multiple Line
    - Meter Gauge
    - Narrow Gauge
    - Metro Railway
- Railway Station
  - Railway terminal
  - Integrated Pass Terminal
  - Metro Pan Terminal
  - Goods Terminal
- Sewerage
  - Dust Bin Sanitary Landfill
- Solid Waste—Dumping ground
- Water
  - Overhead Tank
  - Water Swimming Pool

Bridges/Culverts
Level Crossing Rail/Road
Ford/Ferry Points (with periods)
Foot Path
Temporary Roads/
Bridges with Periods
Foot Bridges
Mountaineer Chock
    Route Tentative
National Highway
    With Number
State Highway—With Number
Other Roads
    Unmetalled Roads
    Lanes/Bylanes
    Subways
    One Way Road
Traffice Island/Light/Lamp Post
Traffic Restriction with Timing

- Bus Depot
  - Inter State Bus Terminus (ISBT)
  - Other Bus Depot—Bus
  - Shops/Bus Terminals
- Texi/Auto Stand
  - Other Auto Stand
  - Prepaid
- Truck Terminal—
  - Integrated Freight Complex—Parking
- Communication
  - Signal Point
- Drain
  - Drain—
    - Single and Double Line Drain
- Electricity
  - Lamp Post
  - Tower Pylons
- Fuel
  - CNG Station
  - LPG Distribution Centres
  - Petrol Pump—Petrol and Diesel/
    - CNG with Owner Information
  - Gas Pipeline Pole/
    - Service Station
- Water Bodies
- River—Flow Direction Arrows For River
  - River Island
    - Water Falls with Heights
    - Sand
- Tank/Well/Springs/Reservoirs/Salt Pan

(contd.)

(contd.)

• Other Tower Chimney	Water Limit Lakes Ponds Built Up Tanks Canals (Single and Double Line)
	Creek
	High Water Line Low Water Line Topographical Features Surveyed Tree

Note: Large Scale Maps of towns and cities falling in the restricted zone near the International boarder, Line of Control and Actual Line of Control will not be available in the public domain.
Authority: Ministry of Defence order no. 2(7)/2003-D(GS-III)/Vol. V dt. 14/21st March 2006.

Form 'A'/फार्म 'ए'

Survey of India/भारतीय सर्वेक्षण विभाग

Digital Products Indent Form/अंकिय उत्पाद मांग-पत्र फार्म

Organisation/संगठन

(Please select one from the following if you are an organisation)
(यदि आप एक संगठन (ऑर्गेनाइजेशन) हैं तो कृपया निम्नलिखित में से किसी एक का चयन करिए)

Government Organisation सरकारी संगठन	☐	Private Limited Company प्राइवेट लिमिटेड कम्पनी
Non-Government Organisation गैर सरकारी संगठन	☐	Educational Institution शैक्षणिक संस्थान
Others/अन्य	☐	
Student/छात्र	☐	
Others/अन्य	☐	

1. User Id (if allotted)/उपयोगकर्ता की पहचान (यदि आबंटित है) ☐☐☐☐☐☐☐☐☐
2. Organisation/Individual Name/संगठन/व्यक्ति का नाम
   (Strike out whichever is not applicable) (जो लागू न हो उसे काट दें) ☐☐☐☐☐☐☐☐☐☐☐☐☐☐☐☐☐☐☐☐☐☐☐☐☐
   ☐☐☐☐☐☐☐☐☐☐☐☐☐☐☐☐☐☐☐☐☐☐☐☐☐

3. Address/पता ☐☐☐☐☐☐☐☐☐☐☐☐☐☐☐☐☐☐☐☐☐☐☐☐☐
   ☐☐☐☐☐☐☐☐☐☐☐☐☐☐☐☐☐☐☐☐☐☐☐☐☐

3.1 City/शहर ☐☐☐☐☐☐☐☐☐☐☐☐☐☐☐☐☐☐☐☐☐☐☐☐☐

3.2 State/राज्य ☐☐☐☐☐☐☐☐☐☐☐☐☐☐☐☐☐☐☐☐☐☐☐☐☐

3.3. Pin Code/पिन कोड ☐☐☐☐☐☐☐☐☐

4. Phone/दूरभाष ☐☐☐☐☐☐☐☐☐☐☐☐☐☐
5. Fax/फैक्स ☐☐☐☐☐☐☐☐☐☐☐☐☐☐
6. E-mail/ई–मेल ☐☐☐☐☐☐☐☐☐☐☐☐☐☐☐☐☐☐☐☐☐☐☐☐☐
7. Website/वेबसाइट ☐☐☐☐☐☐☐☐☐☐☐☐☐☐☐☐☐☐☐☐☐☐☐☐☐
8. Field of work/कार्य क्षेत्र ☐

9. Purpose of Indenting
मांग–पत्र देने का प्रयोजन

10. Please atach a copy of proof/कृपया प्रमाण (प्रूफ) की प्रति संलग्न करें।

10.1 Proof of Identity/पहचान का प्रूफ

Passport/PAN/Driving Licence/Ration card/Voter identity card No. (Strike out whichever is not applicable)
पासपोर्ट/पैन/ड्राइविंग लाइसेंस/राशन कार्ड/मतदाता पहचान पत्र सं. (जो लागू न हो उसे काट दें)

□□□□□□□□□□□□□□□□□□□□□□□□□

10.2 Proof of Residence/निवास का प्रमाण

Tel. Bill/Electricity bill/House tax bill No./Notary affidavit (Strike out whichever is not applicable)
फोन का बिल/बिजली का बिल/गृह कर का बिल सं./नोटरी शपथ–पत्र (जो लागू न हो उसे काट दें)

□□□□□□□□□□□□□□□□□□□□□□□□□

11. Please fill the following details if you are an organization.
यदि आप संगठन (ऑरगेनाइजेशन) हैं तो कृपया निम्नलिखित ब्यौरे भरिए–

11.1 Organisation Registration No./संगठन का पंजीकरण (रजिस्ट्रेशन) सं.

□□□□□□□□□□□□□□□□□□□□□□□□□

11.2 Contact person name/संपर्क करने वाले व्यक्ति का नाम

□□□□□□□□□□□□□□□□□□□□□□□□□

11.3 Contact person Identification/संपर्क करने वाले व्यक्ति की पहचान

□□□□□□□□□□□□□□□□□□□□□□□□□

11.4 Contact person Address/संपर्क करने वाले व्यक्ति का पता

□□□□□□□□□□□□□□□□□□□□□□□□□
□□□□□□□□□□□□□□□□□□□□□□□□□

12. Please fill the following details if you are a student.
यदि आप छात्र हैं तो निम्नलिखित ब्यौरे भरिए–

12.1 Enrollment No./नामांकन संख्या

□□□□□□□□□□□□□□□□□□□□□□□□□

12.2 Course Name/पाठ्यक्रम (कोर्स) का नाम

□□□□□□□□□□□□□□□□□□□□□□□□□

12.3 Subject/विषय □□□□□□□□□□□□□□□□□□□□□□□□□

12.4 Research work/शोध कार्य □□□□□□□□□□□□□□□□□□□□□□□□□

12.5 College name/कॉलेज का नाम □□□□□□□□□□□□□□□□□□□□□□□□□

12.6 University name/विश्वविद्यालय का नाम

□□□□□□□□□□□□□□□□□□□□□□□□□

12.7 University/College address (Strike out whichever is not applicable)/विश्वविद्यालय/कॉलेज का पता (जो लागू न हो उसे काट दें)

□□□□□□□□□□□□□□□□□□□□□□□□□□

12.7.1 City/शहर □□□□□□□□□□□□□□□□□□□□□□□□□□

12.7.2. State/राज्य □□□□□□□□□□□□□□□□□□□□□□□□□□

12.8 Fax/फैक्स □□□□□□□□□□□□□□□□□□□□□□□□□□

12.9 E-mail/ई–मेल □□□□□□□□□□□□□□□□□□□□□□□□□□

12.10 Website/वेबसाइट □□□□□□□□□□□□□□□□□□□□□□□□□□

13. Details of Map Indent/मानचित्र मांग–पत्र का ब्यौरा

Please fill one or more information to indicate your requirements.

कृपया अपनी मांगों को दर्शाने के लिए एक या अधिक सूचना भरिए।

13.1 If you want to indent for Topographical Maps/यदि आप स्थलाकृतिक मानचित्रों के लिए मांग करना चाहते हैं।

(i) Sheet Nos./शीट सं. ______________________

____

____

(ii) Geographical Limits/भौगोलिक सीमाएं

Upper right (long/lat)/ऊपरी दाएं (देशान्तर/अक्षांश) □□□□□□□

Lower left (long/lat)/निचला बाएं (देशान्तर/अक्षांश) □□□□□□□

(iii) Place/District/State/स्थान/जिला/राज्य ______________________

____

13.2 For other maps (Please fill your requirement)/अन्य मानचित्रों के लिए (कृपया अपनी मांग भरिए)

14. Undertaking/वचनबंध

I solemnly affirm that the facts stated above are true to the best of my knowledge of belief/मैं सत्यनिष्ठा से प्रतिज्ञा करता हूँ कि मेरी जानकारी और विश्वास के अनुसार उपर्युक्त कथन सही है।

Date/दिनांक

Place/स्थान

Signature/हस्ताक्षर

National Spatial Data Infrastructure : Data Format

Table 1: Description and Format of First Two Records Common to All Files

Rec. No.	*Field*	*Contents*	*Type (Fortran)*	*Format*	*Starting Byte*	*Ending Byte*	*Remarks*
1	1	Map Sheet Code	Character	A40	1	50	Top Sheet Number as per Scale e.g. 56G/12
		Filler	—	—	51	52	2 Spaces
	2	NSDE Version	Character	A6	53	58	Version of the exchange format in which data is organized e.g. 1.0.1
		Filler	—	—	59	60	2 Spaces
	3	NSDI Layer	Integer*2	I4	61	64	NSDI Main Layer, e.g. Topography-001, Geology/Mining-002 etc.
	4	NSDI Sub-Layer	Integer*2	I4	65	68	Sub-layer Classification e.g. Coastal Geomorphology etc.
	5	Agency Code	Integer*2	I4	69	72	Unique Code of Data Generating Agency
2	1	File Name	Character	A7	1	7	7-Digit file name as per figure 1.
	2	Filler	—	—	8	9	2 Spaces
	3	Volume Code	Integer*4	16	10	15	6-Gidit Code, Unique for Data Set. Provided for future use.
	4	Data Type	Integer*2	12	16	17	Indicates type of Spatial Data 1. if data is base categorywise (Vector Format) 2. if data is Layerwise (Vector Format) 3. if raster multiband image (band-wise) 4. if Single band raster image values 5. if Single band coded raster image
		Filler	—	—	18	18	1 Space
	5	Serial Number of Media	Integer*2	I6	19	24	For data on Multiple Media Only. Otherwise this value is zero
	6	Total Number of Records in the file	Integer*4	I12	25	36	
	7	Base Category/ Layer serial No.	Integer*2	I6	37	42	Corresponds to field 1 of Record number 4 to 3+p of VOLDIR. Applicable only for files DATACAT and DATAFIL. FOR other files this value is zero.
		Filler	—	—	43	4	2 Spaces
	8	Data Classification	Integer*2	I2	45	46	0 Topological Non-structred 1 Topologically Structured
		Filler	—	—	47	72	26 Spaces

Table 2: Volume Directory (Voldir) Records

Rec. No.	*Field*	*Contents*	*Type (Fortran)*	*Format*	*Starting Byte*	*Ending Byte*	*Remarks*
3	1	Availability of Data as One Single file or Multiple Files Or No. of Layers Or No. of Bands (p)	Integer*2	I6	1	6	This value is zero for single file and non-zero (p) for multiple files, where p is the total number of base categories/Layers mentioned in the volumes. Record no. 4 onwards will appear only if this value is non-zero. Also the next field of this record is meaningful only if this value is non-zero. In case of Raster data, each file will contain data for a particular band. In case of Layer data, the file will contain data pertaining to a layer.
	2	Whether data is base category-wise/layerwise/ Image bandwise	Integer*2	I6	7	12	This value is 1. if data is base categorywise (Vector Format). 2. if data is Layerwise (Vector Format) 3. if raster multiband Image (band-wise) 4. if Single band raster image values 5. if Single band coded raster image
	3	Whether File TOPOINFO exists	Integer*2	I2	13	14	Relevant only for Topographic data 1 for yes 0 for no
	4	Whether File QUALINFO exists	Integer*2	I2	15	16	Flag for existence of file containing Quality processes and parameters 1 for yes 0 for no
		Filler	—	—	17	72	46 Spaces
4 to 3+p	1	Base Category/ Layer/Band Serial Number	Integer*2	I6	1	6	Order of listing of Base Categories, Layers, Bands
	—	Filler	—	—	7	7	1 Space
	2	Whether Attribute table Exist	Integer*2	I2	8	9	0 Indicates no attribute data attached in DATAFIL, ELSE number of attribute tables associated
	—	Filler	—	—	10	10	1 Space
	3	Description of Base Categories/ Layers/Bands	Character	A50	11	60	As per Section 3. Description of Base Categories, in case of vector topographic data Description of Layers, in case of user domain Layers in vector or raster format Description of Spectral Bands in case of multiple band image data (raster).

(contd.)

(contd.)

Rec. No.	Field	Contents	Type (Fortran)	Format	Starting Byte	Ending Byte	Remarks
	4	Whether Base Category/Layer/ Band included in the volume	Integer*2	I6	61	66	1, if includes Otherwise 0
	5	Serial Number of Media in which the Base Category/ Layer/Band is contained	Integer*2	I6	67	72	Applicable for Data on Media only

Table 3: General Information (GENINFO) Records

Rec. No.	Field	Contents	Type (Fortran)	Format	Starting Byte	Ending Byte	Remarks
3	1	Agency Acronym	Character	A10	1	10	Allotted by NSDI (ref NSDI metadata document 1.1.1.1)
	2	Layer Name	Character	A30	11	40	Allotted by NSDI (ref NSDI metadata document 1.1.1.2)
	3	Layer Type	Character	A20	41	60	Allotted by NSDI (ref NSDI metadata document 1.1.1.2)
		Filler	—	—	61	61	1 Space
	2	Number of States/ UT = (m)	Integer*2	12	62	63	As per Central Heading. Mostly applicable for SOI topographic data
	3	Source Type	Integer*2	19	64	65	0—Original 1—Compiled or Aggregated... *(Add new table)*
		Filler	—	—	66	72	7 Spaces
4	1	Scale	Integer*4	19	1	9	Scale of the map
		Filler	—	—	10	11	2 Space
	2	Edition Number	Integer*2	14	12	16	Edition of the source material
		Filler	—	—	17	18	2 Space
	3	Edition Year	Integer*2	18	19	26	Edition year e.g. 2001
		Filler	—	—	27	72	46 Spaces
5 to 5+m	1	StateName	Character	A68	1	64	One state per record (ref NSDI metadata document 1.1.1.4)
	2	Number of Districts (n)	Integer*2	I4	65	68	No. of Districts in the State
	3	Number of Location points(o)	Integer*2	I4	69	72	No. of unique reference locations
5+m to 5+m+n	1	DistName	Character	A72	1	72	Districts will be listed One District per record (ref NSDI metadata document 1.1.1.5)
5+m+n to 5+m +n+0	1	LocPtName	Character	A72	1	72	One Location name per record (ref NSDI metadata document 1.1.1.6)

(contd.)

(contd.)

Rec. No.	*Field*	*Contents*	*Type (Fortran)*	*Format*	*Starting Byte*	*Ending Byte*	*Remarks*
6+m+ n+o	1	Surveyed Year/ Data of acquisition	Character	A60	1	60	Year(s) of Survey of the source material e.g. 1968-69 Date of Acquisition in case of Satellite raster image
		Filler	—	—	61	68	8 Spaces
	2	Security classification of dataset	Integer*2	I4	69	72	0 Not Classified 1 Classified
7+m+ n+o	1	Data Type Code	Integer*2	I2	1	2	1. if data is base categorywise (Topographic Vector) 2. if data is Layerwise (Vector Format) 3. if raster multiband image (band-wise) 4. if Single band raster image values 5. if Single band coded raster image
	—	Filler	—	—	3	8	6. Spaces
	2	MapProjection	Character	A4	9	12	POLY—Polyconic, UTM LAMB—Lambert conical orthomorphic SOM—Space Oblique Mercater Any Other
		Filler	—	—	13	14	2 Spaces
	3	Units	Character	A6	15	20	Units of measure for ground planimetric co-ordinates. Its Value is METERS
		Filler	—	—	21	22	2 Spaces
	4	Grid Size (Resolution)	REAL*4	F8.2	23	30	The true ground distance in meters
	5	Number of Control (Tick) Points (q)	Integer*2	I6	31	36	Usually 4 corners of the sheet for Topographic data Could be more for User Domain data
		Filler	—	—	37	72	36 Spaces
8+m+ n+o	1	Central Longitude	Real*4	F12.5	1	12	Projection Parameters
	2	Central Latitude	Real*4	F12.5	13	24	Projection Parameters
	3	Standard Parallel 1	Real*4	F12.5	25	36	Projection Parameters
	4	Standard Parallel 2	Real*4	F12.5	37	48	Projection Parameters
	5	Scale Factor	Real*4	F6.4	49	54	Projection Parameters
		Filler	—	—	55	56	2 Spaces
	6	Spheroid	Character	A8	57	64	EVEREST, WGS84, WGS72 etc.
		Filler	—	—	65	72	8 Spaces
9+m+ n+0	1	Origin Latitude	Real*4	F12.5	1	12	Projection Parameters
	2	Origin Longitude	Real*4	F12.5	13	24	Projection Parameters
	3	Origin Easting	Real*4	F12.2	25	36	Projection Parameters
	4	Origin Northing	Real*4	F12.2	37	48	Projection Parameters
		Filler	—	—	49	72	24 Spaces

(contd.)

(contd.)

Rec. No.	*Field*	*Contents*	*Type (Fortran)*	*Format*	*Starting Byte*	*Ending Byte*	*Remarks*
10+m+ n+o to 10+m+ n+o+q	1	Control (Tick) point Label	Integer*2	I4	1	4	1 = SW/LL, 2 = SE/LR, 3 = NE/UR, 4 = NW/UL, 999 = Reserved for Identifying the origin of Raster Image (For Raster data) 5 to 998 can be used for identifying other control (Tick) points within the sheet/dataset
	2	Raster Image Orientation	Character	A2	5	6	Relevant for Control (Tick) Point Label = 999 (i.e. raster data) BL = Indicates raster image origin at Bottom left and rasterization is bottom up (row/scan lines) and Left-right (Columns/Pixels) TL = Indicates raster image origin at Top left and rasterization is Top-down (row/scan lines) and Left-right (Column/Pixels)
	3	Latitude	Real*4	F12.6	7	18	In Degrees of Tick Points
	4	Longitude	Real*4	F12.6	19	30	In Degrees of Tick Points
	—	Filler	—	—	31	36	6 Spaces
	5	X Co-ordinate	Real*4	F12.6	37	48	Projected in ground terms in meters
	6	Y Co-ordinate	Real*4	F12.6	49	60	Projected in ground terms in meters
	7	User Tick Label	Character	A10	61	70	User defined label for Tick point
	—	Filler	—	—	71	72	2 Spaces
11+m+ n+o+q	1	O.S. Name	Character	A18	1	18	Specifies name of source Operating System
	2	GIS Package	Character	A18	19	36	Specifies name of source GIS Software
	3	DBMS Name	Character	A18	37	54	Specifies name of source Database
	—	Filler	—	—	55	72	18 Spaces

Table 4: Quality Information (QUALINFO) Records

Rec. No.	*Field*	*Contents*	*Type (Fortran)*	*Format*	*Starting Byte*	*Ending Byte*	*Remarks*
3	1	Process Citation	Character	A69	1	69	General statement on data generation processes
	2	No. of Process stages (m)	Integer*2	I3	70	72	Number of distinct process stages
4 to 4+2*m	1	Process Description	Character	A62	1	62	Processes description
		Filler	—	—	63	64	2 Spaces
	2	Process Year	Integer*4	I8	65	72	Process/Source Data Year
	1	PAccHorX	Real	F12.2	1	12	Positional accuracy in X direction
	2	PAccHorY	Real	F12.2	13	24	Positional accuracy in Y direction
	3	PAccVer	Real	F12.2	25	36	Positional accuracy in Z direction
	4	TAcc	Real	F12.2	37	48	Thematic accuracy in %
	5	TAccConf	Real	F12.2	49	60	Confidence level for thematic accuracy
	6	TAccArea	Real	F12.2	61	72	Thematic Area accuracy in %
5+2*m	1	PACCLayer* (horizontal)	Real4	F12.2	1	12	Cumulative RMS Horizontal Positional accuracy (ref. NSDI metadata document 1.1.1.9)
	2	PACCLayer* (vertical)	Real	F12.2	13	24	Cumulative RMS vertical Positional accuracy (ref. NSDI metadata document 1.1.1.9)
	3	TACCLayer*	Real4	F12.2	25	36	Final % thematic accuracy (ref. NSDI metadata document 1.1.1.9)
	4	TACCLayer*(area)	Real4	F12.2	37	48	Final thematic area accuracy
		Filler	—	—	49	72	24 spaces

Table 5: General Topographic Information (TOPOINFO) Records

Rec. No.	*Field*	*Contents*	*Type (Fortran)*	*Format*	*Starting Byte*	*Ending Byte*	*Remarks*
3	1	Whether COMPIND Data Exists	Integer*2	I2	1	2	1 for yes, 0 for No
	2	Whether ADMIND Exists	Integer*2	I2	3	4	1 for yes, 0 for No
	3	Number of SPECIAL FOOTNOTES lines (m)	Integer*2	I2	5	6	
	4	Magnetic Variation E or W of True North	Character	A2	7	8	E = East of True North W = West of True North
	5	Annual Change in Magnetic Variation with sign	Integer*2	I4	9	12	+ for increasing, – for decreasing
		Filler	—	—	13	14	2 Spaces

(contd.)

(contd.)

Rec. No.	*Field*	*Contents*	*Type (Fortran)*	*Format*	*Starting Byte*	*Ending Byte*	*Remarks*
	6	Year of Magnetic Variation	Integer*2	I6	15	20	
	7	Copyright	Character	A24	21	44	Example: Govt. of India Copyright
	8	Year of Copyright	Integer*2	I6	45	50	Example: 1975
	9	Contour Interval	Integer*2	I6	51	56	Contour interval in meters
	8	Mean Grid North Alignment	Integer*2	2(13)	57	62	In degrees and minutes. For un-gridded sheets this field and the next are blank
		Filler	—	—	63	64	2 Spaces
	9	East and West of True North	Character	A1	65	65	E = East of True North W = West of True North
		Filler	—	—	66	67	2 Spaces
	10	Name of Grid	Character	A5	68	72	User Defined
4	1	Authority	Character	A64	1	64	Name of Authority under direction the data is published/generated
	2	Admn. Boundary Verification Remark	Integer*2	I2	64	66	1 = Adm. Boundaries are verfied 0 = Adm. Boundaries are not verified from appropriate Authority
	—	Filler	—	—	67	68	2 Spaces
	3	Triangulated heights and contours adjusted with spirit-leveled heights	Character	A1	69	69	Y = Adjusted N = Not Adjusted
	—	Fillers	—	—	70	71	2 Spaces
	4	Territorial water upto 12 Nautical miles from the coastline	Character	A1	72	72	Y in case of data having coastline; N in case of data not having coastline.
5to5+m	1	SPECIAL FOOTNOTES RECORES	Character	A72	1	72	

Table 6: Data Category (DATACAT) Records

Rec. No.	*Field*	*Contents*	*Type (Fortran)*	*Format*	*Starting Byte*	*Ending Byte*	*Remarks*
3	1	Base Category/ Layer/BAnd Serial Number	Integer*2	I6	1	6	Order of listing of Base Categories, Layers, Bands (Corresponds to field 1 of records 4 to 3+p ob Table 2)
	2	Number of major categories/Layer Categories (m)	Integer*2	I6	7	12	• Number of Base categories in case of Topographic Data • Number of Layer Categories in case of user domain thematic data in vector or raster form • Not relevant in case of Image data

(contd.)

(contd.)

Rec. No.	*Field*	*Contents*	*Type (Fortran)*	*Format*	*Starting Byte*	*Ending Byte*	*Remarks*
	3	Number of tables(nt)	Integer*2	I6	13	18	• Number of attribute tables attached to the layer
	—	Filler	—	—	19	72	54 Spaces
4 to 3 +m	1	Category Code	Integer*2	I2	1	2	Major Category/Layer Category Code
	—	Filler	—	—	3	8	6 Spaces
	2	Category Name	Character	A64	9	72	Full category name as per Appendix A Major Category/Layer category Name/Bandwidth
4+m	1	Number of Nodes (nn)	Integer*4	I8	1	8	Set to Zero for Raster Data
	2	No. of Lines (nl)	Integer*4	I8	9	16	Set to Zero for Raster Data
	3	Number of Areas (na)	Integer*4	I8	17	24	Set to Zero for Raster Data
	4	No. of Texts (nx)	Integer*4	I8	25	32	Set to Zero for Raster Data
	5	No. of Rows (nR)	Integer*4	I6	33	38	Set to Zero for Raster Data
	6	No. of Columns (nC)	Integer*4	I6	39	44	Set to Zero for Raster Data
	7	No. of Pixel Bytes (nBytes)	Integer*2	I2	45	46	Set to Zero for Vector Data, relevant to Image, Grid data 1. Single byte, binary for images 2. Two byte, binary for images/ noded grids 4. Four byte, binary for DEM
	8	Byte Order (depending upon architecture)	Character	A2	47	48	II/MM for more than one Byte meta II indicates Intel, Big Endian architecture MM indicates Motorola, Little Endian Architecture
	9	Record Type	Character	A6	49	54	RASTER or VECTOR
	10	Format of Value Records	Character	A10	55	64	Content would be LOGICAL*1 for Single byte, binary images INTEGER*2 for two byte, binary images/coded grids REAL*4 for four byte, binary DEM data (as per the details in field 7 above)
		Filler			65	72	8 spaces

Table 7: Raster (Grid) Data Records

Rec. No.	*Field*	*Contents*	*Type (Fortran)*	*Format*	*Starting Byte*	*Ending Byte*	*Remarks*
Value Records 3 to (nR*nC* nBytes)/ 72 + 3	1	Raster Values • Reflectance Values for image data • Elevation Values for DEM • Coded Values for Coded Raster Layer	Logical*1 or Integer*2 or Real*4 (As per record 4+m, field 7 in table 6)	Binary	1	NR*Nc *nByte	Order of Grid cell locations for Values Row 1: Col 1, Col 2,...... Col-nC Row 2:... Row-nr:....................................

Table 8: Node Records

Rec. No.	*Field*	*Contents*	*Type (Fortran)*	*Format*	*Starting Byte*	*Ending Byte*	*Remarks*
3	1	Record Type	Character	A4	1	4	NODE
	2	Element Internal Id No.	Integer*4	I8	5	12	This is Unique value for each Node
	3	Co-ordinates of Node Point	Real*4	2F12.2	13	36	X, Y co-ordinate in ground meters
	4	Angle	Real*4	F12.4	37	48	Orientation of point feature with X-axis in degrees (Relevant for Topographic Data)
	—	Filler	—	—	49	72	24 Space

Table 9: Line Records

Rec. No.	*Field*	*Contents*	*Type (Fortran)*	*Format*	*Starting Byte*	*Ending Byte*	*Remarks*
3	1	Record Type	Character	A4	1	4	LINE
	2	Element Internal ID	Integer*4	I8	5	12	This is Unique value for each Line
	3	Starting Node	Integer*4	I8	13	20	Internal ID Number This refers to data element 2 of the Node records
	4	Ending Node	Integer*4	I8	21	28	-do-
	5	Number of points in the line (p)	Integer*4	I8	29	36	Number of x-y Co-ordinate pairs (excluding the end points)
	6	Height	Real*4	F8.2	37	44	Height in Meters
	7	Number of Feaure Codes or Levels (q)	Integer*2	I4	45	48	
	8	Object ID	Integer*4	I8	49	56	
	—	Filler	—	—	57	72	16 Spaces
Co-ordinate records	1 to 2p	x, y co-ordinate string	Real*4	3	(2F 12.2)1	.....	Ending byte number is as per the number of points in the line
Code records	1 to 2q	Major and Minor List or	Integer*4	6(216)	1	....	• Major and Minor Codes for topographic data • Will extend to next line as per the number of feature codes
		Layer Category Code List	Integer*4	2(136)	1	.....	• Layer category Codes for User Domain Thematic data • Will extend to next line as per the number of feature codes

Note: All POINT features included in LINE type records. A POINT feature is distinguished from a LINE feature as follows:

- A POINT feature has the same start and end node ID (Data elements 3 and 4) and
- The number of points (Data element 5) for a POINT feature is zero (0)
- The key for linking with the text in the file can also be used for linking with non-graphic data outside the file.

Table 10: Area Records (Topologically Structured)

Rec. No.	*Field*	*Contents*	*Type (Fortran)*	*Format*	*Starting Byte*	*Ending Byte*	*Remarks*
3	1	Record Type	Character	A4	1	4	Area
	2	Element Internal ID No.	Integer*4	I8	5	12	This is Unique value within all areas (polygons)
	3	Co-ordinates of Area Point	Real*4	2F(12.2)	13	36	The area point (polygon label) is always within the pologon it represents
	5	Number of lines bounding the area (b)	Integer*2	I4	37	40	Number of x-y Co-ordinate pairs
	6	Number of Feature Codes (r)	Integer*2	I4	41	44	
	7	Object ID	Integer*4	I8	45	52	The key for linking with the text in the file, can also be used for linking with non-graphic data outside the file
		Filler	—	—	53	72	20 Spaces
Line ID records	1 to b	List of Line ID's	Integer*4	9(18)	1	.....	Internal ID number of lines bounding the area. Ending byte number is as per the number of lines in the area (polygon)
Code records	1 to 2r	Major and Minor Code List	Integer*4	216	1	12	• Major and Minor Codes in pairs for topographic data • Will extend to next line as per the number of feature codes
		Filler	—	—	13	18	6 Spaces
		Feature Name OR	Character	A54	19	72	
		Layer category Code List	Integer*4	2(136)	1	.....	• Layer category Codes for User Domain Thematic data • Will extend to next line as per the number of feature codes

Table 11: Area Records (Topologically Non-Structured)

Rec. No.	*Field*	*Contents*	*Type (Fortran)*	*Format*	*Starting Byte*	*Ending Byte*	*Remarks*
3	1	Record Type	Character	A4	1	4	Area
	2	Major Code	Integer*4	I6	5	10	Major Code of the Feature
	3	Minor Code	Integer*4	I6	11	16	Minor Code of the Feature
	6	Feature Type	Character	A52	17	68	
	5	Number of vertices (p) excluding end points	Integer*4	I4	69	72	Number of vertices
Co-ordinate records	1 to 2p	x, y coordinates of Area Points	Real*4	3(F12.2)	1	......	Ending bytes as per the number of vertices in the area feature

Table 12: Text Records

Rec. No.	*Field*	*Contents*	*Type (Fortran)*	*Format*	*Starting Byte*	*Ending Byte*	*Remarks*
3	1	Record Type	Character	A4	1	4	Text
	2	Element Internal ID	Integer*4	I8	5	12	This is Unique value for all text records
	3	Text Size	Integer*2	I4	13	16	In one hundredth of an inch
	4	Text Justification	Integer*2	I4	17	20	0 for left, 1 for right, 2 for centre justification
	5	Angle	Real*4	F12.4	21	32	Angle of Text in Degrees
	6	x, y Co-ordinates of reference Point	Real*4	2F(12.2)	33	56	
	7	Number (r) of attribute codes	Integer*2	I4	57	60	Number of x-y Co-ordinate pairs
	8	Number Text Character strings	Integer*2	I4	61	64	
	9	Object ID	Integer*4	I8	65	72	The key for linking with the object in the LINE or Area Records
	—	Filler	—	—	53	72	20 Spaces
Second	1	Text Characters	Character	A72	1	72	This record contains the text data describing Object ID
Third	1-2r	Major and Minor List	Integer*4	2I6	1	12	• Major and Minor Codes for topographic data • Will extend to next line as per the number of feature codes
		Filler	—	—	13	18	6 Spaces
		Feature Name OR	Character	A54	19	72	
		Layer category Code List	Integer*4	2(136)	1	.....	• Layer category Codes for User Domain Thematic data • Will extend to next line as per the number of feature codes

Table 13: Attribute Records

Rec. No.	*Field*	*Contents*	*Type (Fortran)*	*Format*	*Starting Byte*	*Ending Byte*	*Remarks*
3	1	Type	Character	A4	1	4	Indicates attribute table type (ATTR)
	2	Attribute Table ID.	Integer	I6	5	10	Attribute table ID.
	3	Name of Attribute Table	Character	AI2	1	24	Indicates name of the attribute table as per source specification.
	4	No. of fields (af)	Integer	I6	11	16	Indicates number of columns in attribute table
	5	No. of records (ar)	Integer	I6	17	22	Indicates number of records in attribute table
	—	Filler	—	—	23	72	50 Spaces
4 to 4+ af	1	Field name	Character	A12	1	12	Represents Field Item name
	2	Field type 1	Character	A12	13	24	Definition of Field giving Width, Output, Type, number of decimal places separated by comma.
	3	Field type 2	Character	A8	25	32	Represents ANSI standard format
	4	Key-Field	Character	I2	33	34	Indicates 1 if the field is Key-indexed, ELSE 0
	—	—	—	—	35	36	2 Spaces
	5	Remarks	Character	A36	37	72	Description/Remarks of the Field
Data/ Records 5+ af to 5+af+ar	1-af	List of field values	Character	2(A36)	1	....	• List of field data values • Will extend to next line as per the number of filled values in attribute records

Note: Records numbered 3 onwards... will be repeated as many times as the number of attribute tables indicated by number nt in datacat file.

List of GIS Vendors and Related Software Developers

- ACE Software *from PCI—USA*
- ARGUS Technologies (http://www.argusmap.com/)
- Autodesk *AutoCAD Map* (http://www3.autodesk.com/adsk/section/0,,693177-123112,00.html)
- Avenza (http://www.avenza.com/)
- Bentley Systems (http://dn1.bentley.com/products/geproducts) *Microstation Geographics*
- Blue Marble Geographics (http://www.bluemarblegeo.com/) *GeoView and other tools*
- Caliper (http://www.caliper.com/) *Maptitude, TransCAD, GIS+*
- DeLorme (http://www.delorme.com/)
- Digital Land Systems Research (http://www.dlsr.com.au/welcome.html) *SAGE*
- EarthResource Mapping (http://www.ermapper.com/) *ER Mapper*
- EPPL7 (http://www.mnplan.state.mn.us/EPPL7/)
- ESRI (http://www.esri.com/) *Arc/Info, ArcView, ArcCAD, SDE, MapObjects*
- ESRI (http://www.esri.ca/) *Canadian Distributor*
- Etak (http://www.etak.com/) *Digital Mapping*
- Genasys (http://www.genasys.com/)
- Geodyssey Limited of Calgary (http://www.cuug.ab.ca/~russellj) *Hipparchus*
- GEOsurv Inc.—10 Years (http://infoweb.magi.com/~geosurv/)
- Geosystems (http://www.geosys.com/) *GeoLocate*
- GRASS (http://www.cecer.army.mil/grass/GRASS.main.html)
- IBM's GIS (http://giswww.pok.ibm.com/gishome.html) *Several new software products*
- IDRISI (http://www.idrisi.clarku.edu/)
- Innogistic Software plc. (http://www.innogistic.co.uk/)
- INTERGRAPH (http://www.intergraph.com/) *MGE, FRAMME*
- INTERMAP (http://www.intermaptechnologies.com)
- MacGIS (http://shiva.uoregon.edu/macGIS/macGISHomePage.html)
- Mapcom (http://www.mapcom.com/) *M4*
- MapInfo (http://www.mapinfo.com/)
- MicroImages (http://tnt.microimages.com/) *TNT Familiy of products*
- Northwoood Geoscience LTD (http://www.northwoodgeo.com/)
- PCI (http://www.pci.on.ca/) *Remote Sensing Technologies*
- Planet One Software (http://www.planetonegis.com/)
- Ramas (http://www.ramas.com/) *Ecological Models*
- Scan/US (http://www.scanus.com/)
- Sedona Geo-Services (http://www.sedonageo.com/)

- SHL System House (http://www.gis.shl.com/) *Vision Solutions*
- Siemens-Nixdorf (http://www.mch.sni.de/public/sicad/sicad_us/sicad.htm) *SICAD, GDB-X, WinCAT*
- Smallworld (http://www.smallworld.co.uk/)
- SST Development Group (http://www.sstdevgroup.com/) *Consultants and Farm Managers*
- Star Informatic (http://www.star.be/) *Star Carto, Star Viewer*
- Tactician Corporation (http://www.tactician.com/)
- TEXO GIS (http://www.popnet.it/TEXO/texo_eng.html?)
- Thinkspace (http://www.thinkspace.com/) *Map Factory*
- TYDAC *SPANS* (http://www.tydac.com/)
- Universal Systems (http://www.universal.ca/) *CARIS*
- Vision International (http://www.gis.shl.com/) SQS
- Wolleysoft (http://www.woolleysoft.co.uk/) landscape visualization software

GIS Software Applications

At GeoPlan Center there was an attempt to make an effort to show people what their GIS software choices are. This Appendix will list any GIS software information which the Center has come across. They used quite a number of them at the Center, but they were interested in discovering what else is out there. If anybody know of any that aren't listed here, please write about them (e-mail us). According to the GeoPlan Center:

- We tried to list only the general-use GIS software here. There are lots of GIS software that are made for certain professions, such as hydrologic scientists, petroleum industry specialists, etc. that aren't listed here.
- Note that most of the software descriptions were written by the Software Makers/ Distributors.
- You may be able to get some of this software for less than indicated by the prices listed here. In all cases we listed the commercial price, not the educational/student, or similar discounted prices.

ArcCad
ESRI
www.esri.com
PLATFORM: Windows (95, NT)
"ESRI's ArcCAD software gives you full geographic information system (GIS) functionality within AutoCAD. ArcCAD provides powerful mapping, data management, and display tools that work directly with AutoCAD software's design and drafting tools. ArcCAD users can create, access, and exchange ARC/INFO data, the *de facto* standard for local, state, and national GIS installations."

ArcExplorer
ESRI
www.esri.com
PLATFORM: Windows
"ArcExplorer software is a lightweight GIS data explorer. ArcExplorer can work on its own with local data sets or as a client to Internet data and map servers. ArcExplorer is built using MapObjects, ESRI's powerful collection of GIS and mapping components for application developers. As a stand-alone application, ArcExplorer is a complete data explorer, allowing users to display and query a wide variety of standard data sources."

ArcFM
ESRI
www.esri.com
PLATFORM: Windows (NT)
"Arc Facilities Manager (ArcFM) is a powerful ARC/INFO-based application for the editing, maintenance, modelling, and data management of utility information. ArcFM is all-relational utility application software developed using ARC/INFO software's Open Development Environment and object-oriented programming."

Arc/Info
ESRI
www.esri.com
PLATFORM: Most Unix platforms, Windows (NT)
"Professional GIS offers advanced modelling and analysis capabilities"

ArcView Business Analyst
ESRI
www.esri.com
PLATFORM: Windows (3.51, 95, 98, NT)
"ArcView Business Analyst allows you to focus on getting the data analysis you need, rather than worrying about the details of the underlying technology. A wizard interface assists you with tasks ranging from basic, such as mapping customers, to complex, such as performing analyses or creating what-if scenarios."

ArcView GIS
ESRI
www.esri.com
PLATFORM: Most Unix platforms, Windows (95, 98, NT)
"The leading mapping and GIS software for the desktop."

Atlas GIS
ESRI
www.esri.com
PLATFORM: Windows (95, 98, NT)
"Business professionals worldwide rely on Atlas GIS to help them make decisions that improve the efficiency and profitability of their organizations. Atlas GIS is both powerful enough to perform sophisticated desktop mapping and intuitive enough that nontechnical professionals feel comfortable using it."

AutoCAD Map 2000
Autodesk
www.autodesk.com
PLATFORM: Windows (95, 98, NT)
"AutoCAD Map 2000 is the Autodesk software optimized for precision mapping and geographic

information system (GIS) analysis in the AutoCAD® environment. It provides the GIS analysis tools and productivity improvements you need to easily create, maintain, and produce maps and geographic information."

Autodesk MapGuide
Autodesk
www.autodesk.com
PLATFORM: Windows (95, NT)
"Autodesk MapGuide software delivers live maps and linked databases directly to your desktop, via your intranet or the Internet—anytime, anywhere."
Autodesk MapGuide Viewer
(A PC and a Java Edition)

Autodesk
www.autodesk.com
PLATFORM: Macintosh, Sun Solaris, Windows PC
"Works with your web browser and sites running Autodesk MapGuide"

Autodesk World
Autodesk
www.autodesk.com
PLATFORM: Windows
"Autodesk World 2.01 is a powerful Microsoft desktop application to integrate, analyse and display spatial data. With Autodesk World we are bringing GIS technology down to earth. It is a flexible, powerful spatial data information system that allows users to manage their data, their way. Autodesk World has an industry-standard interface that removes the complexity of GIS, while providing robust functionality."

BusinessMAP PRO
ESRI
www.esri.com
PLATFORM: Windows (3.1x, 95, NT)
"BusinessMAP PRO is an affordable mapping application that transforms information from databases, contact managers, spreadsheets, and electronic phone books to pushpin and color-filled maps. It will turn static data into meaningful images to help you understand your business better and help with sales automation and mission-critical decision making processes."

Cartalinx
Clark Labs, Clark University
www.clarklabs.org
PLATFORM: Windows (95, 98, NT)
Clark Labs' "first dedicated digitizing, database development and topological editing software product. CartaLinx is intended as a companion to a variety of popular Geographic Information System (GIS) and Desktop Mapping software products such as IDRISI, Arc/Info, ArcView, MapInfo, and so on.

CartaLinx is used to create coverages, consisting of the spatial definitions of features in vector format and associated attribute values files. These data are then typically exported to a GIS either as entire coverages or as a series of map layers. For example, CartaLinx might be used to develop a spatial frame of property parcel boundaries along with an attribute database recording the address, owner, area, property valuation, area and primary land use of each parcel. One might then choose to export land use and property valuation map layers to a GIS. Alternatively, one might wish to display and analyse the contents of a coverage using CartaLinx directly."

Data Automation Kit
ESRI
www.esri.com
PLATFORM: Windows (3.x, 95, NT)
"This software package perfectly complements ArcView GIS software and other desktop mapping software by allowing you to create fully intelligent (topological) data sets by digitizing or converting data from many formats, with functionality that previously existed only in a full ARC/INFO system."

Didger
Golden Software
www.golden.com
PLATFORM: Windows (3.51, 95, 98, NT)
"Didger is a highly accurate digitizing program that will be an invaluable addition to your software library. In seconds, Didger precisely transforms points, lines, or areas from your paper maps, graphs, aerial photos, or other plotted information to a versatile digital format you can use with your other software. You name it and Didger can handle it quickly, accurately, and usefully. And don't be fooled by the low $119 price. With Didger's multitude of features and ease-of-use, this is an unbelievable value considering the time and effort savings you will realize! You will soon wonder how you have done your job without this indispensible tool."

DLGV32
USGS
DLGv32 Page
PLATFORM: Windows 95/NT
"Windows 95/NT software for viewing a variety of digital cartographic data. It was originally developed in support of USGS Digital Line Graph (DLG) production activities. This software allows preview and evaluation of USGS data. It contains no editing capabilities, and is not a substitute for commercial Geographic Information System (GIS) software."

EPPL7
Minnesota Planning
EPPL7 Page
PLATFORM: Windows 95
"EPPL7 is a fast but powerful raster based GIS with a well stocked toolbox. EPPL performs a variety of complex map algebra functions as well as Boolean operations. There are powerful

filtering algorithms in EPPL7 as well as commands which perform interpolation, viewshed, slope, and aspect analysis. Users can view, query, and print data."

ERDAS IMAGINE Advantage
ERDAS
www.erdas.com
PLATFORM: Windows (95, NT), Most Unix platforms
A subset of ERDAS IMAGINE Professional
"Analyse data from your imagery via image mosaicking, surface interpolation, and advanced image interpretation and orthorectification tools."

ERDAS IMAGINE Essentials
ERDAS
www.erdas.com
PLATFORM: Windows (95, NT), Most Unix platforms
A subset of ERDAS IMAGINE Advantage and ERDAS IMAGINE Professional
"Our basic map production system enables you to produce quality maps through image visualization and analysis, information integration and geocorrection."

ERDAS IMAGINE Professional
ERDAS
www.erdas.com
PLATFORM: Windows (95, NT), Most Unix platforms
"Quick image testing and modelling allows you to automatically and quickly generate broad land cover maps. Perform graphical spatial data modelling and integrate radar imagery."

ERDAS MapSheets
ERDAS
www.erdas.com
PLATFORM: Windows
"All that work you put into gathering and analysing geographic data... without a quick and simple way to visually present the information to your boss, your client or your prospective customer.
Because its fully Microsoft Office Compatible, MapSheets gives you the same look and feel of other Microsoft applications, as well as the power to create a convincing PowerPoint presentation, an effective Word report or a compelling Web page."

ERDAS MapSheets Express
ERDAS
www.erdas.com
PLATFORM: Windows
"All that work you put into gathering and analysing geographic data... without a quick and simple way to visually present the information to your boss, your client or your prospective customer.

Because its fully Microsoft Office Compatible, MapSheets gives you the same look and feel of other Microsoft applications, as well as the power to create a convincing PowerPoint presentation, an effective Word report or a compelling Web page."

ER Mapper
Earth Resource Mapping
www.ermapper.com
PLATFORM: Windows, Unix
"Used in more than 142 countries worldwide and supported by 514 ER Mapper resellers around the globe, ER Mapper is the world's most popular desktop integrated image processing software."

ER Viewer
Earth Resource Mapping
www.ermapper.com
PLATFORM: Windows, Unix
"ER Viewer is an easy to use image viewer featuring interactive roaming and zooming with very large image files."

Geographic Calculator
Blue Marble Geographics
www.bluemarblegeo.com
PLATFORM: Windows
"The Geographic Calculator converts individual coordinates, point databases, and map files from virtually any coordinate system, datum, and map projection to any other."

Geographic Explorer
Blue Marble Geographics
www.bluemarblegeo.com
PLATFORM: Windows (95, NT)
"The Geographic Explorer extends the Windows 95 and NT Explorer to view, translate, and find map files. The Geographic Explorer provides a number of powerful utilities. You can explore map files directly on your desktop with the Geographic Explorer."

Geographic Tracker
Blue Marble Geographics
www.bluemarblegeo.com
PLATFORM: Windows
"The Geographic Tracker delivers all of the power and utility of GPS to the Windows environment by allowing you to link a GPS receiver to other Microsoft Windows applications through standard Windows DDE messaging.
This standard "plug-and-play" GPS interface component supports most of the leading GPS receivers on the market. Simply plug your GPS receiver into the serial port of a computer running Windows and the Geographic Tracker will GPS-enable your applications."

Geographic Transformer
Blue Marble Geographics
www.bluemarblegeo.com

GEMI Store
www.gemistore.com
PLATFORM: Windows
$799
upgrade: $299
"The Geographic Transformer allows you to easily establish an "image-to-world" relationship between image and map coordinates and reproject an image into a georeferenced image map."

Geographic Translator
Blue Marble Geographics
www.bluemarblegeo.com
PLATFORM: Windows
"The Geographic Translator supports comprehensive data and coordinate system translation of map files in the AutoCAD DWG & DXF, Microstation DGN, MapInfo MIF & TAB, and ESRI Shape formats. "Hands free" batch mode operation is just one compelling reason to add this new Blue Marble tool to your GIS toolbox."

Geomatica FreeView
PCI Geomatics
http://www.pcigeomatics.com
PLATFORM: Windows 95, 98, NT, 2000
"The Geomatica[tm] FreeView includes tools for fast image roaming and magnification, enhancements, numeric value display, and attributes table display."

Geomedia
Intergraph
http://www.intergraph.com
PLATFORM: Windows
"GeoMedia is your information integrator, serving as a visualization and analysis tool and a platform for custom GIS solutions. A universal GIS client, GeoMedia's live data connections allow you to maintain GIS data in native repositories, and ensure that your data is always up-to-date. You can integrate data automatically using on-the-fly coordinate transformation and feature definition. Perform simple and complex spatial analysis, such as incident and proximity analysis, to support decision-making activities."

Geomedia Viewer
Intergraph
http://www.intergraph.com
PLATFORM: Windows

"Create Thematic Maps... View and anlyze data in either Microsoft Access or Arc/View Shape format... Load a satellite raster image and easily manipulate your map with zoom-in, zoom-out capability... it's FREE!!!"

GRASS (Geographic Resources Analysis Support System)
U.S. Army Construction Engineering
Research Laboratories (USACERL)
www.cecer.army.mil

Baylor University (Official
Headquarters for GRASS support,
research and development-get GRASS here)
www.baylor.edu/~grass/
PLATFORM: Linux, Sun Solaris, Silicon Graphics Irix, HP-UX, DEC-Alpha and Windows 95/98/NT
GRASS is a raster-based GIS, vector GIS, image processing system, graphics production system, data management system, and spatial modelling system.
"The Geographic Resources Analysis Support System (GRASS) is a public domain image processing and Geographic Information System (GIS) that was initially developed by the US Army Construction Engineering Research Laboratories (USACERL) during the late 1980s and early 1990s. USACERL stopped developing GRASS, the last full USACERL release was in 1993. Many organizations have taken the public domain GRASS software code and enhanced its functionality beyond what was taken the public domain GRASS software code and enhanced its functionality beyond what was possible with federal dollars. Baylor University, Blackland Research Institute, GPZ Technologies, Logiciels et Applications Scientifiques, (LAS) Inc., and the National Oceanic and Atmospheric (NOAA) Geophysical Data Center have all released various public domain and commercial versions of GRASS on a variety of computer platforms."

IDRISI
Clark Labs, Clark University
www.clarklabs.org
PLATFORM: Windows
"IDRISI is the industry leader in raster analytical functionality covering the full spectrum of GIS and Remote Sensing needs from database query, to spatial modelling, to image enhancement and classification. Special facilities are included for environmental monitoring and natural resource management, including change and time series analysis, multi-criteria and multi-objective decision support, uncertainty analysis (including Bayesian and Fuzzy Set analysis) and simulation modelling (including force modelling and anisotropic friction analysis). Yet, despite the highly sophisticated nature of these capabilities, the system is as easy to use as it is inexpensive to buy."

Label-EZ
MapText, Inc.
www.maptext.com
PLATFORM: Windows 95, 98 and NT, UNIX
"A new productivity tool from MapText, Inc., Label-EZ places text automatically, without any human intervention, eliminating the need for manual text placement.The current version of the software works seamlessly with map data from: Arclnfo, ArcView, and Maplnfo."

LandForm C3
GEMI Store
www.gemistore.com
PLATFORM: Windows
"LandForm C3 allows users to fly through terrain data in real-time. It runs on Windows based desktop and laptop PCs, and reads many file types including DEM, DTED, TIFF, GeoTIFF, and others. LandForm has a unique capability to load DTED files automatically, based upon viewpoint and create a seamless land surface, at user-defined resolution. Aerial photographs, satellite images, and digital maps can be overlayed on the 3-D terrain. The viewpoint can be dynamically controlled by driving with the mouse, or the viewpoint can follow a flight track or ground track. At $995, this is the first product in its price range which allows users to combine topographic data with maps, satellite, aerial imagery, GPS and flight track data. Further, the program allows users to fuse these data into 3-D VRML models, which can be posted on Internet or intranet Web sites and viewed by Web browsers. The program was designed to be easy to use."

Manifold System
manifold.net
http://www.manifold.net
PLATFORM: Windows
"Manifold is a software package that combines database and mapping functionality. Manifold also includes a variety of statistics, logic, mathematics, and networking capabilities for use on maps, on databases, or all together at the same time."

Maplnfo Professional
Maplnfo
www.mapinfo.com/
PLATFORM: Windows, Macintosh, Sun, HP
"Maplnfo Professional is the world's leading desktop mapping software that helps you see patterns and trends in your data. Trends that may otherwise be impossible to identify."

Maplnfo ProViewer
Maplnfo
www.mapinfo.com/
PLATFORM: Windows
"Maplnfo's easy viewing tool designed for Maplnfo Professional users. With ProViewer, you

can share your MapInfo Professional maps and analyses with clients and co-workers, who may easily view and manipulate your creations even if they do not have MapInfo Professional."

Maplex
ESRI
www.esri.com
PLATFORM: Windows (NT)
"Maplex is a fully automated cartographic name placement software. Name placement can be one of the most time-consuming aspects of map production, often requiring considerable manual intervention to ensure that names do not overlap and are clearly associated with the features they annotate. Maplex provides the means to overcome this bottleneck in map production."

MapMarker and
MapMarker Plus
MapInfo
www.mapinfo.com/
PLATFORM: Windows
They both utilize "street maps to allow you to visually display all your geocoded data. MapMarker ships with MapMarker Streets—a five layer street display product based on TIGER data from U.S. Census Bureau. MapMarker Plus ships with MapMarker Plus Streets—a five-layer street display product based on data from GDT, Inc."

Maptitude
GEMI Store
www.gemistore.com
"MAPTITUDE Geographic Information System combines sophisticated geographic analysis capabilities, a powerful DBMS engine, and a tremendous amount of data all for $395. Maptitude lets you edit and update your maps, import maps from other products, or create your own maps from scratch. The MapWizard utility lets you create thematic maps with a single click of the mouse. Anyone can make beautiful maps with Maptitude. It includes two CD ROMs full of detailed maps and demographic data including every street in the U.S. It also has maps and data for the rest of the world with somewhat less detail. MAPTITUDE allows you to display and edit satellite images and digital orthophotos."

MAPublisher
Avenza
www.avenza.com
PLATFORM: MacOS PowerPC, Windows (95, 98, NT)
"MAPublisher 3.5 is a suite of Xtras for Macromedia FreeHand and Plug-ins for Adobe Illustrator 8.0.1 that bridges Geographic Information System (GIS) technology with high-end graphics software for high resolution printing and electronic publishing technology."

MAPublisher GPS
Avenza
www.avenza.com
PLATFORM: Windows (95, 98, NT)
"MAPublisher GPS allows you to publish your data quickly, easily and intuitively while speaking to a GPS receiver through a satellite/radio network from the Illustrator graphics environment. This data can efficiently be made available on the internet/intranet by using Avenza's pdfPLUS technology."

MAPublisher LT
Avenza
www.avenza.com
PLATFORM:MacOS PowerPC, Windows (95, 98, NT)
"MAPublisher Lt for Adobe Illustrator and Macromedia FreeHand imports raw GIS data files from Mac, PC or UNIX systems directly into the high end graphics environment with all attributes intact. MAPublisher Lt lets you import vector data (shapes, boundaries, routes, elevations, etc.) from MapInfo, ARC/INFO, ArcView, AutoCAD and others with all of the drawings data attributes intact. MAPublisher LT incorporates GIS object management tools such as feature selection and query logic tools for easy data selection. Use MAPublisher LT to GEOenable your websites or for simply importing your GIS/CAD files into high end graphics programs."

Map Viewer
Golden Software
www.golden.com
PLATFORM: Windows (95, 98, NT)
"MapViewer is a thematic mapping package that creates full-color maps by linking worksheet data to areas or points on a map. MapViewer includes numerous boundaries and data files for the United States and World to get you started, or you can create your own boundaries for use with MapViewer. By placing your data in the full-featured MapViewer worksheet, you can easily link your data to areas or points on the map, and generate instructive, publication quality maps in virtually no time. With the numerous map types, overlays, and drawing tools you will create the most informative maps possible."

Microdem
Bureau of Land Management
BLM ftp download site
"A program to manipulate gridded digital elevation models (DEMs), satellite imagery, and vector map data like DLG and TIGER."

MicroStation GeoGraphics
Bentley Systems, Incorporated
www.bentley.com
PLATFORM: Windows (3.1x, 95, NT), DOS. Releases are planned for Intergraph CLIX,

Hewlett Packard HP/UX, Sun Solaris, DEC Alpha NT, Silicon Graphics, IBM RS6000, and Macintosh Power
"MicroStation GeoGraphics is a comprehensive mapping toolbox that bridges the gap between CAD and GIS. MicroStation GeoGraphics combines MicroStation's industry-standard data capture and editing tools with a comprehensive database interface and a powerful spatial analysis engine. MicroStation GeoGraphics integrates seamlessly with MicroStation 95, and provides tools to input, validate, manage, analyse, and visualize geographical information. MicroStation GeoGraphics also has the ability to manage a diverse combination of vector data, raster data, and non-spatial attributes."

MrSID Stand Alone Viewer
LizardTech
www.lizardtech.com/
PLATFORM: Win95/NT, Macintosh & Unix(beta)
"View MrSID images on your personal computer."

NCAR Graphics
University Corporation for Atmospheric Research
ngwww.ucar.edu
PLATFORM: UNIX
"NCAR Graphics software continues to evolve its integrated environment for data processing and scientific visualization. Many tools come bundled with the distribution to perform such functions as viewing and editing metafiles, creating animations, converting between raster formats, resizing and compressing raster images, and zooming on images."

OziExplorer
ESRI
www.powerup.com.au/~lornew/oziexp.html
PLATFORM: WIN 95/98/NT4/NT5/WIN2000
"OziExplorer GPS Mapping Software will work with Magellan, Garmin, Lowrance and Eagle GPS receivers for the upload/download of data and most brand of GPS receivers for on screen tracking of position (Moving Map)."

PC Arc/Info
ESRI
www.esri.com
PLATFORM: DOS, Windows (3.x, 95, 98, NT)
"PC ARC/INFO is a series of six integrated software modules that combine fundamental geographic information system (GIS) tools and utilities for cartographic design and query, data entry and editing, data translation, polygon overlay and buffering, and network analysis and modelling."

pdfPLUS
Avenza
www.avenza.com
PLATFORM: MacOS PowerPC, Windows (95, 98, NT)
"pdfPLUS (formerly known as JAMBuddy) is an Adobe Acrobat 4.0 plug-in that allows you to read important database information embedded within a special PDF file. Save your MAPublisher creations and keep the data attributes alive and accessible for use on an intranet or the internet. Point and click on anything in the PDF and discover the hidden information. Query your data and watch pdfPLUS find the qualifiers."

Philcarto
Philippe WANIEZ
http://perso.club-internet.fr/philgeo/
PLATFORM: Macintosh or Power Macintosh
"Philcarto is a classical mapping software for spatialy agregated data (like census data). The input map is an Adobe Illustrator® file. The output map is an Adobe Illustrator® file too, with the same scale than the input map. Points and areas (polygons) are handled by the program. The data is a text-tabulated file containing an identifier, a name, and columns as variables. The different types of maps are:

- choropleth map
- proportional circles map
- proportional circles map coloured by a second variable
- proportional half-circles map
- choropleth associated with proportional circles map
- dot map
- links map."

Philexplo
Philippe WANIEZ
http://perso.club-internet.fr/philgeo/
PLATFORM: Macintosh or Power Macintosh
"Philexplo is a new software for cartographic exploratory data analysis. The purpose is to explore data and map using visualization tools. Data and map files are the same as with Philcarto. You can explore the statistical displays (histograms, box plots, bivariate plots and triangular diagrams) using tools like arrow, knife, lasso, etc.... The result (a selection of spatial units) can be stored in pict file or exported into a texte file compatible with Philcarto (to obtain a better cartographic output)."

Quad Companion
GEMI Store
www.gemistore.com
"Quad Companion was designed specifically to solve the problem of clipping collars and seaming DRGs together. Quad Companion clips off collars and seams DRGs together on-the-fly, just specify the region you are interested in and let the software do the rest."

QUADConvert
Applied Global Technologies, Inc.
www.appliedglobal.com/frames.htm
PLATFORM: DOS, Windows (3.1x, 95, NT), UNIX
Conversion Software for LTWG QUAD Format Satellite Scenes

R2V
Able Software Corp.
www.ablesw.com
PLATFORM: Windows (95/NT)
"The most advanced raster to vector conversion software for automated map digitizing and GIS data capture applications. R2V for Windows 95 & NT supports fully automatic vectorization and interactive line tracing from scanned maps and drawings. R2V inputs bi-level, grayscale and colour images in TIFF, GeoTIFF or BMP formats, exports vector data to ArcView Shapefile, Arc/Info Generate, MapInfo MIF/MID, AutoCAD DXF and MapGuide SDL formats. R2V supports complete on-screen image and vector editing, labelling, geo-referencing using user specified control points, image rubber sheeting, colour image classification and separation, text symbol detection and recognition (OCR), automatic polygon layer creation, 3D DEM and Grid creation, 3D display and image draping."

SAGE GIS for DOS Suite
DLSR (Digital Land Systems Research)
www.dlsr.com.au
PLATFORM: DOS
"The SAGE GIS for DOS system is a suite of three DOS programs which make up a functional and powerful GIS system. The system is comprised of: SAGE GIS (raster-based), SAGE Capture and SAGE Cartographer."

SAGE GIS for Windows
DLSR (Digital Land Systems Research)
www.dlsr.com.au
PLATFORM: Windows (95/NT)
"SAGE GIS support the analytics found in the SAGE GIS for DOS system, but with greatly improved speed and ease of use—perfect for training, teaching and consulting work."

Surfer
Golden Software
www.golden.com
PLATFORM: Windows (3.1, 95, 98, NT)
upgrade: $139
"Surfer is a contouring and 3D surface plotting program that runs under Microsoft Windows. Surfer quickly and easily converts your data into outstanding contour maps and surface plots. Virtually all aspects of your maps can be customized to produce exactly the presentation you want. Producing publication quality maps has never been quicker or easier. Both the 32-bit and 16-bit versions are included."

Terrain Professional
Maptech

GEMI Store
www.gemistore.com
PLATFORM: Windows (95, 98, NT)
"High-quality USGS topographic maps on CD-ROM. Upto 300 exportable maps per CD, at 160 dpi image resolution. Advanced features let you export and reproject full, georeferenced maps for use in GIS applications. Annotate and print customized maps. NGS (National Geodetic Survey) Data Sheets included."

Terrain Tools Freeware Edition
Softree
www.softree.com/betat e s t/t t ools.htm
PLATFORM: Windows 3.1, 95, 98 or NT
" Terrain ToolsTM—Freeware Edition is a quick and easy mapping package. Coordinates can be entered using the mouse, typed in, or imported from external files. Scanned topographic maps, air photos and other bitmap images can also be incorporated."

TNTatlas
MicroImages, Inc.
www.microimages.com
PLATFORM: Every Platform
"a free product that lets you publish and distribute your spatial materials on CD-ROM. TNTatlas CD's contain multiple versions of the software so that a single CD can be used on any popular computing platform. TNTatlas stacks are prepared with the HyperIndex feature in TNTmips."

TNTmips
MicroImages, Inc.
www.microimages.com
PLATFORM: Windows (3.1, 95, NT), Macintosh, UNIX
"TNTmips is the Map and Image Processing System, advanced software from MicroImages, Inc. for geospatial analysis. TNTmips is used in more than 70 nations around the world for geographic information analysis (GIS), advanced image processing, CAD, desktop cartography, electronic atlas preparation, and other spatial database management and visualization applications. TNTmips offers a level of data and process integration that is unmatched in any other professional system. And you can use TNTmips on PC, workstation, and Macintosh computers with exactly the same set of features and user interface on every platform. TNTmips is the most technically advanced and easy to use professional system on the market."

TNTview
MicroImages, Inc.
www.microimages.com
PLATFORM: Every Platform

"TNTview gives you the same powerful visualization and interpretation tools contained in MicroImages' industry-leading TNTmips product. TNTview is easy to learn and provides low cost visualization of your organization's project materials at sites where the users' scope of work does not include the preparation, processing, or final production of your spatial data."

WinCATS
Applied Global Technologies, Inc.
www.appliedglobal.com/frames.htm
PLATFORM: Windows
"WinCATS is the cutting edge intel tool for displaying database information over a vast host of geographic backgrounds. Advanced enough for the most complex cartographic demands, yet simple enough for the everyday software user, WinCATS is a serious mapping tool that remains user friendly."

Tcl/TK extensions
TK Geomap
www.tkgeomap.org
PLATFORM: Unix/Linux X11
Free set of extensions to Tcl/TK for map drawing.

WinCATS Viewer
Applied Global Technologies, Inc.
www.appliedglobal.com/frames.htm
PLATFORM: Windows (3.1x, 95)
The freeware version of WinCATS.

Source: www.geoplan.ufl.edu/software.html

Free GIS Software

1. **Free GIS** and CAD **Software**: geocomm.com—42k
2. **Free software**, data, and resources from ESRI ; www.esri.com/company/**free**.html—12k
3. FreeGIS.org: **freegis**.org—14k
4. **GIS Software**: www.**gis**development.net/downloads/**gi**
5. **Free GIS** mapping program **software** and resources: www.mapcruzin.com/**free_gis**.htm—22k
6. ArcExplorer — **GIS** Data Viewer: www.esri.com/**software**/arcexplorer—10k
7. EPIC 2001: **Free GIS software** for Minnesota Government: www.lmic.state.mn.us/EPPL7/EPIC—32k
8. **Free Software GIS**—Resources—PROCIG: www.procig.org/ing/resources-soft.htm—54k
9. **Free GIS Software**—GisWiki: www.**gis**.org.nz/index.php/**Free_GIS_Software**—21k
10. GRASS **GIS**—The World Leading **Free Software GIS**: grass.baylor.edu—9k
11. **Free** Maps and **GIS** Data: www.manifold.net/download/**free**maps.html—28k
12. **Free GIS Software**: www.**gis**.com/implementing_**gis**/**software**/**free_software**.html—11k
13. **Free GIS** Stuff: www.**gis**lounge.com/ll/**freegis**.shtml—23k
14. **Free GIS** Data: data.geocomm.com
15. **GIS** at CIP: www.cipotato.org/**gis**—6k
16. Christine **GIS** system: www.christine-**gis**.com—10k
17. ESRI Press: **GIS** for Everyone, Third Edition: **gis**.esri.com/esripress/display/index.cfm?...&websiteID=43&moduleID=1—15k
18. DIVA-**GIS** Homepage: www.diva-**gis**.org—8k
19. **GIS** Associations: www.reach.net/~maps/clearmaps/**free**.html—16k
20. **Free GIS** and Image Processing **Software**!: www.cof.orst.edu/cof/teach/for421/**Software**.html—6k
21. **Free** Tech Stuff: www.tenlinks.com/**free**.htm—21k
22. TNTlite for Archaeology: www.microimages.com/archaeol.htm—14k
23. SpatialNews, the leading **GIS**, LBS, and Geospatial technology news resource "spatialnews.geocomm.com/features/viewers2002 – 42k
24. **GIS** Data, Jobs, **Software**, News, and Fun—GeoCommunity: www.geocomm.com/channel/gps/**software**—49k
25. **free** downlaod **software**: www.spatialhydrology.com/**free**download.html—61k
26. GPS **Software**: www.**gis**development.net/downloads/gps—11k

27. [gislist] **Free GIS Software**—GISLIST: lists.directionsmag.com/discussion/read.php?f=20&i=34020&t=34020—78k
28. GISuser.com **GIS**, GPS, LBS, CAD, mapping news, jobs, **software**, data, community: RSS: View as XML: www.**gis**user.com
29. AGIS for Windows—simple mapping and **GIS** shareware: www.a**gis**map.com—12k
30. **GIS Software** and Resources: www.geoplan.ufl.edu/**software**.html—55k
31. **GIS Software**: home.earthlink.net/~**gis**webmaster/id14.html—20k
32. **Free Software**: www.hoslink.com/**freestuff/freesoftware**.htm—62k
33. **GIS** Dictionary: www.geo.ed.ac.uk/agidict/welcome.html—4k
34. GeoCommunity: The Premier Portal For Geospatial Technology Professionals: www.geocomm.com
35. USGS (U.S. Geological Survey) EROS, Sioux Falls, SD: edc.usgs.gov/geodata/public.html—8k
36. **Free GIS Software**: www.cmtinc.com/downloads/ftppc**gis**32.html—4k
37. **Free GIS Software**: www.cmtinc.com/downloads/pc**gis**.html—5k
38. **software**: www.yale.edu/**gis**/soft.htm—12k
39. Companies—Specializing—**Free Gis Software Software**: www.soft411.org/companies/**free-gis-software**.html—37k
40. Official GRASS **GIS** Homepage—The World Leading **Free Software GIS**: www.geog.uni-hannover.de/grass/index.php—9k
41. **Free GIS** Data Viewers: www.**gis**lounge.com/ll/**free**dataviewers.shtml—25k
42. Geographic information system—Wikipedia, the **free** encyclopedia: en.wikipedia.org/wiki/Geographic_information_system—66k
43. Open Directory—Science: Social Sciences: Geography: Geographic Information Systems : **Software**: dmoz.org/Science/Social_Sciences/Geography/**...**/**Software**—24k
44. **GIS** and Spatial Data Links: docs.lib.duke.edu/maps/guides/**gis**.html—28k
45. **GIS software**,arcgis,gwn-cogo,gwn-dtm,gwn-surf,mapviewer—Environmental **Software**: sci**software**.com/environmental_**software**/index.php?cPath=26_66&.
46. **Free** & Open Source **Software** Portal:**Software**/Productivity Tools/**GIS**—Geographical Information System: unesco.org/cgi-bin/webworld/portal_freesoftware/cgi/page.cgi?...&d=1
47. GInfoServer: **GIS-Software** + -Shareware: www.geo.uni-bonn.de/members/haack/**gis-software**.html—14k
48. Commercial **GIS Software**: www.**gis**.com/implementing_**gis/software**/commercial_**software**.html—11k
49. Download PC **Software** from NGS: www.ngs.noaa.gov/PC_PROD/pc_prod.shtml—29k
50. Main Page – GisWiki: www.**gis**.org.nz/index.php/Main_Page—9k
51. ArcLessons—Your Resource for **GIS** Educational Exercises: **gis**.esri.com/industries/education/arclessons/arclessons.cfm
52. Google Earth: www.keyhole.com/body.php?h=downloadKeyhole—13k
53. ESRI Canada-**Free**!: www.esricanada.com/english/bottom/**free**.asp—23k
54. **free** statistical **software**: gsociology.icaap.org/methods/soft.html—31k
55. Land Management Information Center Home Page: www.lmic.state.mn.us—33k

56. Jenness Enterprises—ArcView Extensions: www.jennessent.com/arcview/arcview_extensions.htm—50k
57. MicroImages: www.microimages.com—12k
58. **GIS** Data, Shaded Relief Images, DEM, Satellite Imagery, Geology of Madagascar, Global Mapping, Minerals, Fossils: www.gospatial.com—32k
59. Desktop **GIS** Web Publishing **software** products: www.collins**software**.com—8k
60. **Software** (RDMS/UD): www.rdms.udel.edu/rdms/main/**software**.html—31k
61. **Software** Ahoy!: www.ucalgary.ca/~clarko/mapsoft—24k
62. Ecological **Software** Solutions: www.ecostats.com/**software**/**software**.htm—11k
63. Yahoo! Groups: **free-gis**: groups.yahoo.com/group/**free-gis**
64. **GIS Software**: www.ies.wisc.edu/ltc/nap/**gis**vend.html—10k
65. **GIS** and Remote Sensing **Software**: www.indiana.edu/~dms/**gis**/**software**—9k
66. ESRI Conservation Program: supporting conservation geography, activism and multicultural social change: www.conservation**gis**.org/conservationb.html—26k
67. **Software** Resources—My Community, Our Earth: www.aag.org/sustainable/resources/**software**.html—10k
68. GISuser.com **GIS**, GPS, LBS, CAD, mapping news, jobs, **software**, data, community: www.**gis**user.com/content/view/3143/1—91k
69. **GIS** Internet Resources Body: www.tec.army.mil/**gis**/**gis**int_body1.html
70. **GIS software** downloads: www.**free**-download-soft.com/**Gis**+**software** – 30k
71. Map Maker—Desktop Mapping: www.mapmaker.com—14k
72. GISuser.com **GIS**, GPS, LBS, CAD, mapping news, jobs, **software**, data, community: www.**gis**user.com/content/view/4199/1—75k
73. SWEGIS, Consultant and **GIS Software** Provider—**Free** stuff: www.swe**gis**.com/english/download/**free**.asp—37k
74. WG VI.1/**Free software**: www.photogrammetry.ethz.ch/general/persons/jana/isprs/ed_**free**sw.html—21k
75. **Software**: **GIS**, remote sensing, hydrology and GPS **software**,**free** downlaod: spatial hydrology: www.spatialhydrology.com/**software**_**gis**rsgps.html—36k
76. **GIS** Internet Resources Body: www.tec.army.mil/**gis**/**gis**int_body1.html—47k
77. **Software** Resources—My Community, Our Earth: www.aag.org/sustainable/resources/**software**.cfm—10k
78. Central Region **GIS Software** and Utilities: cpg.cr.usgs.gov/pub/**software**.html—30k
79. ESRI **GIS Software**: www.esriindia.com/Products/ArcPad.htm—13k
80. Surveying **Software**: www.sonic.net/~trollhei/survsoft.html—4k
81. The Impoverished Social Scientist's Guide to **Free** Statistical **Software** and Resources: www.hmdc.harvard.edu/micah_altman/socsci.shtml—23k
82. **GIS** Links: www.acadweb.wwu.edu/**gis**/links/**gis**.htm—25k
83. PTTC Rockies—**Software** Fair: www.mines.edu/research/PTTC/**software**fair/**gis**.html—20k
84. Open Source **GIS**: www.opensource**gis**.org—118k
85. ALOV Map/TMJava. **Free** Java **GIS**: alov.org—10k
86. RockWare- earth science & **GIS software**: **free** catalog and trialware CD-ROM: www.rockware.com/catalog.html—18k

87. Forestry **GIS** (fGIS)—**Free GIS** Viewer and Shapefile Editor: www.digitalgrove.net/f**gis**.htm—43k
88. Manifold System—The ultimate **GIS** and Mapping Product!: www.manifold.net—102k
89. GRASS Geographic Information System Information: www.cecer.army.mil/grass/GRASS.main.html—2k
90. TatukGIS: www.tatuk**gis**.com—15k
91. Companies—Specializing—**Gis** Mapping **Software**: www.soft411.org/companies/**gis**-mapping.html—35k
92. **Free GIS** Data Viewers: **gis**.geog.ttu.edu/texastech_**gis**/viewers.htm—6k
93. **GIS** in Schools: Education Programme: Ministry of Education, Youth, and Culture – Jamaica: www.moec.gov.jm/projects/**gis**sep/techsupport.htm—11k
94. Quantitative Decisions: www.quantdec.com—16k
95. **Free Software**: www.geo.tu-freiberg.de/~apelm/miscsoft.htm—14k
96. Geographic information system—Wikipedia, the **free** encyclopedia: en.wikipedia.org/wiki/**Gis**
97. **Software**: www.explorationist.com/**Software**.htm—28k
98. PTTC Rockies: www.mines.edu/Research/PTTC/**Software** Fair/**gis**.shtml—22k
99. ESRI Data: maps.unomaha.edu/Workshops/Career/ESRI/index4.html—11k
100. The NITLE News Volume 1, Number 1, spring 2002: newsletter.nitle.org/v1_n1_spring2002/**gis**.php—4k
101. CIPE | **Software** and Images: www.evisual.org/www/**Software**/Soft.html—11k
102. http://www.unchs.org/guonet/documents/urban_observer/Urban_Observer_2-2003.txt: unchs.org/guonet/documents/urban_observer/Urban_Observer_2-2003.txt—7k
103. **Free** Desktop **Software** Viewers: chesapeake.towson.edu/mapping/**free**viewers.asp - 9k
104. Product Information on **GIS Software**: vnu.bitpipe.com/plist/**GIS-Software**.html—31k
105. ESRI INDIA Your Partner in **GIS**: www.esriindia.com/index/ESRIINDIA.htm—12k
106. **GIS** File Library—Directions Magazine: www.directionsmag.com/files—47k
107. [GRASSLIST:3906] Linux-based **GIS software**: grass.itc.it/pipermail/grassuser/2002-June/006754.html—7k
108. United States—**Free GIS** Data—**GIS** Data Depot: data.geocomm.com/catalog/US—28k
109. GISCafe—Downloads: www10.**gis**cafe.com/link/display_categories.php—24k
110. NCSU Sci-Link: www.ncsu.edu/scilink/workshops_2003.html—9k
111. Companies—Specializing—**Free** Gps **Software Software**: www.soft411.org/companies/**free**-gps-**software**.html—39k
112. **Free** Statistical **Software**: members.aol.com/johnp71/javasta2.html—118k
113. ArchNet: Map Libraries, **GIS**, and Mapping **Software**: archnet.asu.edu/archnet/topical/maps/maps.html—23k
114. **FGDL Software**: www.fgdl.org/**software**.html—26k
115. The Soft FAQ: **Software**, freeware and FAQ on **GIS** and spatial statistics: www.ai-geostats.org/**software**/**GIS**_Softfaq.htm—20k
116. **GIS** Data from the Washington Natural Heritage Program: www.dnr.wa.gov/nhp/refdesk/**gis**/wnhp**gis**.html—27k

117. **Software**: www.mngislis.org/oldsite/education/**software**.htm—7k
118. India Map—India Maps, India **Free** Map, India **GIS**, India **GIS** Data: www.maptell.com—19k
119. **GIS** Internet Links, UT Dallas: www.utdallas.edu/~briggs/other_**gis**.html—132k
120. http://ils.unc.edu/~ruvam/INLS110_111/fall03/slides/Lesson_2_rev.ppt (MICROSOFT POWERPOINT): ils.unc.edu/~ruvam/INLS110_111/fall03/slides/Lesson_2_rev.ppt—621k
121. **GIS**/GPS Link and **Free Software**: www.jttaylor.net/contents/3-**GIS**-GPS—975
122. The Soft Earth: www.df.lth.se/~micke/s_earth.html—71k
123. About this CD-ROM: www.csc.noaa.gov/products/sccoasts/html/about.htm—14k
124. **GIS** Resources: **Software**: www.geog.umn.edu/courses/**GIS**/**software**.htm
125. **Free Software** – Christian Singer, Solutions for Geographic Information: www.icsinger.de/**free**ware.htm—5k
126. ArcScripts Home—ESRI Support: arcscripts.esri.com—10k
127. **GIS**, CAD, and mapping daily news provided by Spatial News: spatialnews.geocomm.com/dailynews/2001/aug/20/news4.html—29k
128. @SoftTop—top text messaging **software**, download **free** text messaging **software**: www.**software**-top.com/softdesc.asp?LinkID=11450—24k
129. Clearinghouse Search: **GIS**: www.csiss.org/cgi- bin/texis/webinator/clearsearch/+BwwFqmHflxzmwwww—7k
130. Commercial **GIS** Providers (**Software** and Consultants): www.inweh.unu.edu/unuinweh/**gis**categories/commercial.htm—20k
131. Cultural Heritage **GIS** Resources: www-personal.umich.edu/~roberta/cult**GIS**.htm—13k
132. **GIS Software** Vendor Directory: www.bitpipe.com/olist/**GIS-Software**.html—30k
133. Infotiger search results: **gis** linux **free software**: www.infotiger.com search?qs=**gis**+linux+**free**+**software**&OK=++search+—25k
134. City of Monterey **GIS**—**Software** Requiremnts: **gis**web.ci.monterey.ca.us/pub/html/requirements.htm—7k
135. Geographical Information Systems (**GIS**) WWW Resource List: www.geo.ed.ac.uk/home/**gis**www.html—97k
136. STOMP **GIS Software** Reading: www.tec.army.mil/research/**software**/Stomp/swread.htm—16k
137. INSIDE Tutorial: **Software**: inside.uidaho.edu/tutorial/**software**/**software**.htm—12k
138. **GIS Software**: www.bakersfieldcity.us/**GIS**/notes/tech/**software**.htm—12k
139. CAD **GIS** Computer Aided Design—**Free** Guide: www.cad-directory.com—22k
140. ESRI—The **GIS Software** Leader: www.clas.ufl.edu/users/puhalla/ESRI
141. Geographic Information Service: volusia.org/**gis**—11k
142. Extract Systems: Home Page: www.icomap.com—10k
143. Components of a **GIS**: www.rst2.edu/ties/GENTOOLS/comp_**gis**.html—5k
144. **GIS Software** and Training for K-12 Schools—December 2004: www.nd.gov/**gis**/news/20041207.html—7k
145. User:Tschirl—Meta: meta.wikimedia.org/wiki/User:Tschirl—17k
146. Teachers Resources for **GIS**: www.columbia.edu/~ja681/**GIS**/rs3.html—11k

147. Welcome to **GIS**\Solutions — makers of **GIS**\Key, the premier **GIS software** for Windows: www.**gis**key.com/core.htm—21k
148. PASCO and myworld **GIS software**: www.pasco.com/myworld—41k
149. A series of four different labs across the three-day annual meeting presented by Charlie Fitzpatrick of ESRI (PDF): ncge.org/events/meetings/birmingham/.../PP-ComputerLabs.pdf—86k
150. d/freegis: lwn.net/1999/1104/a/**freegis**.html—8k

Other details

- Global Mapper—Affordable **GIS** At Last Global Mapper is an easy to use package that allows you to view, edit, merge, rectify, convert, and analyse a huge variety of map data formats. Download a **free** demo today.: www.globalmapper.com
- **GIS**/Mapping **Software** Business **software** information guide. See how it can help your business: www.techbusiness.ws
- **GIS** Information Guide A Directory On **GIS** Services Mapping **Software** and **GIS Software**:
www.advancedwebsearch.com
- **Free** 30-Day Trial of Autodesk Map 3D Increase geospatial design efficiency 25%. For **GIS**, engineering, mapping, survey professionals who want to create, manage, share spatial, design data. Integrate your CAD, **GIS** data: www.autodesk.com
- Mapping/**GIS Software** RockWare, Inc. supplies mapping **software** solutions including coordinate conversion, **GIS**, topo maps and more. We are licensed resellers of Arcview, Topo, Tralaine and more. **Free** demos: www.rockware.com
- **GIS** Information Guide: www.advancedwebsearch.com
- **Free** 30-Day Trial of Autodesk Map 3D: www.autodesk.com
- Mapping/**GIS Software**: www.rockware.com
- Find **Gis Software** Solutions: www.knowledgestorm.com
- **GIS** Information Guide A Directory On **GIS** Services Mapping **Software** and **GIS Software**: www.advancedwebsearch.com
- **Free** 30-Day Trial of Autodesk Map 3D Increase geospatial design efficiency 25%. For **GIS**, engineering, mapping, survey professionals who want to create, manage, share spatial, design data. Integrate your CAD, **GIS** data: www.autodesk.com
- Mapping/**GIS Software** RockWare, Inc. supplies mapping **software** solutions including coordinate conversion, **GIS**, topo maps and more. We are licensed resellers of Arcview®, Topo, Tralaine and more. **Free** demos: www.rockware.com

APPENDIX VI

2004 GIS Software Survey Notes

Program Name	Autodesk Raster Design 2005	Autodesk Survey 2005	Autodesk Land Desktop 2005	Autodesk Civil Design 2005	GEOPAK Survey	Power Survey	InRoads Survey	MicroStation GeoGraphics
Manufacturer/Distributor	Autodesk Inc.	Autodesk Inc.	Autodesk Inc.	Autodesk Inc.	Bentley Systems Inc.	Bentley Systems	Bentley Systems Inc.	Bentley Systems Inc.
Manufacturer's Phone Number	603/621-3270	603/621-3270	603/621-3270	603/621-3270	610/458-5000	610/458-5000	610/458-5000	610/458-5000
Cost Per Seat	Contact reseller	Contact reseller	Contact reseller	Contact reseller	Contact vendor	$1499	Contact vendor	Contact vendor
Date of First Release	1996	1998	1998	1998	1996	2003	2000	1995
SOFTWARE FEATURES								
Operating System/Network Support								
—Network Client-Server Support	No	Win NT/2000/XP	Win NT/2000/XP	Win NT/2000/XP	Yes	Yes	Yes	Yes
—Server Operating System	Win 2000/XP	No	No	No	Win NT 4.0/2000/XP/98/ME	Win NT 4.0/2000/XP/98/ME	Win NT 4.0/2000/XP/98/ME	WinXP/2000/NT 4.0/98
—Client Operating System	None	No	No	No	Win NT 4.0/2000/XP/98/ME	Win NT 4.0/2000/XP/98/ME	Win NT 4.0/2000/XP/98/ME	WinXP/2000/NT 4.0/98
—Internet Server Enabled	N/A	No	No	No	N/A	N/A	N/A	Via Bentley Publisher
Compatible Applications								
—Third Party Applications Designed to Run on Software	N/A	Yes, by vendor	[3]	Yes, by vendor	Yes	Yes	Yes	Yes
—Total Solution Packages Available	N/A	Yes	Yes	Yes	Yes	Yes	Yes	Yes
GIS Data Administration								
—Multi-user Edit Locking	Yes	Yes	Yes	Yes	N/A	N/A	Yes	[8]
—Versioning	No	No	No	No	N/A	N/A	Yes	[9]
—Metadata Maintenance	N/A	No	No	No	N/A	N/A	N/A	[10]
Database Management								
—Proprietary DBMS	No	No	No	No	No	No	No	No
—Relational Database Management System	No	No	No	No	Oracle, MS Access, SQL	Oracle, MS Access, SQL	Oracle, MS Access, SQL	[4]
—RDBMS Spatial Data Warehouse	No	No	No	No	Server, Sybase, Informix	Server, Sybase, Informix	Server, Sybase, Informix	Oracle 8i/9i Spatial and Oracle 9i Locator
Native Graphic Data Structure and Format								
—Vector-Spaghetti	No	No	No	No	N/A	N/A	N/A	DGN, DWG, DXF
—Vector-Topologic	No	No	No	No	DGN, DWG, DXF	DGN, DWG, DXF	DGN, DWG	Oracle Spatial/DGN post processing
—Parametric	No	Yes	Yes	Yes	N/A	N/A	N/A	Yes
—3-D	No	Yes	Yes	Yes	Yes	Yes	Yes	Yes
—TIN	No	Yes	Yes	Yes	Yes	Yes	Yes	Yes, via Bentley GEOPAK Site module
—Grid	No	Yes	Yes	Yes	Yes	Yes	Yes	N/A
—Raster Image	Yes							
GIS Data Import/Export Utilities		Yes	Yes	Yes	Yes	Yes	Yes	[2]
—Direct Import Formats	N/A	[1]	[1]	[1]	[2]	[2]	[2]	DXF, DWG, MID/MIF, SHP, E00
—Direct Export Formats	N/A	[1]	[1]	[1]	N/A	N/A	Yes	DXF, DWG, MID/MIF, SHP, E00
Readable Formats That Do Not Require Translation					N/A	N/A	Yes	Oracle Spatial, DGN, DWG
—Import	[6]	DWG, LandXML, ASCII, FBK	DWG, LandXML, ASCII	DWG, LandXML, ASCII	[6]	[6]	Yes	Oracle Spatial, DGN, DWG
—Export	Image Formats, CALS, TIFF, PNG, BMP, JPG	DWG, LandXML, ASCII, FBK	DWG, LandXML, ASCII	DWG, LandXML, ASCII	[6]	[6]	Yes	Yes
GIS Data Entry and Editing								
—Board Digitizing	Yes	Yes	Yes	Yes	Yes	Yes	Yes	Via Bentley GEOPAK Site
—Coordinate Geometry/Precision Entry	1-8 Decimal Places	0-8 decimal places	0-8 decimal places	0-8 decimal places	Yes	Yes	Yes	Via Bentley GEOPAK Site
—Electronic Survey Data Import	N/A	Yes	Yes, ASCII Coordinate files	Yes, with Survey Add-in	Yes	Yes	Yes	
—Heads-up Digitizing	Yes	Yes	Yes	Yes	Yes	Yes	Yes	Yes
—Vectorization	Yes	Yes	Yes	No	N/A	N/A	N/A	Via Bentley Descartes
—Map Rectification	Yes	Yes	No	No	N/A	N/A	Yes	Yes
—Graphic Error Check/Correction	Yes	Yes	No	No	Yes	Yes	Yes	Yes
—Field Data Entry	N/A	Yes	Yes	Yes	Yes	Yes	Yes	Yes
Map Design and Composition								
—Interactive Map Composition	Yes	Yes	Yes	Yes	Yes	Yes	Yes	Yes
—Annotation from Attributes	Yes	Yes	Yes	Yes	Yes	Yes	Yes	Yes
—Global Symbol Change	No	Yes	Yes	Yes	Yes	Yes	Yes	Yes
—Thematic Mapping	Yes	Yes	Yes	No	N/A	N/A	Yes	Yes
Geographic Query and Analysis Functions								
—Attribute Query and Selection	N/A	Yes	Yes	Yes	Yes	Yes	Yes	Yes
—Map Measurements	Yes	Yes	Yes	Yes	Yes	Yes	Yes	Yes
—Address Matching	N/A	No	No	No	N/A	N/A	N/A	Yes
—Buffer Generation	N/A	Yes	Yes	Yes	N/A	N/A	N/A	Yes
—Point/Line-in-Polygon Analysis	N/A	Yes	Yes	Yes	N/A	N/A	Yes	Yes
—Polygon Overlay	N/A	Yes	Yes	Yes	N/A	N/A	Yes	Yes
—Network Analysis	N/A	Yes	Yes	Yes	N/A	N/A	Yes	Yes
—Raster Document Query and Access	Yes	With add-on	With add-on	With add-on	N/A	N/A	N/A	Yes
—Direct Access to Other GIS Format	N/A	Yes	Yes	Yes	N/A	N/A	N/A	SHP, MID, MIF, E00
Terrain Data Processing and Analysis								
—DEM Generation	No	No	No	No	Yes	Yes	N/A	Via Bentley GEOPAK Site
—Contour Map Generation	Yes	Yes	Yes	Yes	Yes	Yes	Yes	Via Bentley GEOPAK Site
—3-D Display/Profile Generation	No	Yes	Yes	Yes	Yes	Yes	Yes	Yes
—Map Draping	No	No	No	No	N/A	N/A	Yes	Via Bentley Descartes
—Slope/Aspect Analysis	Yes	Yes	Yes	Yes	Yes			
Raster Image Capabilities						Yes	Yes	Via Bentley GEOPAK Site
—Geometric Rectification	Yes	With add-on	With add-on	With add-on	N/A	N/A	Yes	Via Bentley Descartes
—Orthoimage Generation	Yes	With add-on	With add-on	With add-on	N/A	N/A	N/A	N/A
—Image Enhancement	Yes	With add-on	With add-on	With add-on	N/A	N/A	N/A	Via Bentley Descartes
—Spectral Classification	Yes	With add-on	With add-on	With add-on	N/A	N/A	N/A	N/A
Application Development Language								
—Proprietary Application Development Language	No	No	No	No	MDL, JMDL, Basic	MDL, JMDL,Basic	MDL, JMDL, Basic, C++	MDL, JMDL, MicroStation Basic
—Industry Standard Programming Environment	Yes, ActiveX	No	ActiveX, VBA, C++, Lisp	ActiveX, VBA, C++	C, C++, Visual Basic, JAVA	C, C++, Visual Basic, JAVA	C, C++, Visual Basic, JAVA	C, C++, Visual Basic, JAVA
Extra Notes	[7]	[2]	[4]	[8]		[7]		[11]
Reader Service Number	1	2	3	4	5	6	7	8

N/A = Not Applicable (*Source*: www.pobonline.com)

2004 GIS Software Survey Notes

Program Name	Bentley Water	Bentley WasteWater	Bentley Publisher	Bentley Descartes	Bentley I/RAS B	Bentley PowerField	Bentley WorkPrint	Bentley PowerMap
Manufacturer/Distributor	Bentley Systems Inc.	Bentley Systems Inc.	Bentley Systems Inc.	Bentley Systems Inc.	Bentley Systems Inc.	Bentley Systems Inc.	Bentley Systems Inc.	Bentley Systems Inc.
Manufacturer's Phone Number	610/458-5000	610/458-5000	610/458-5000	610/458-5000	610/458-5000	610/458-5000	610/458-5000	610/458-5000
Cost Per Seat	Contact vendor	Contact vendor	Contact vendor	Contact vendor	Contact vendor	Contact vendor	Contact vendor	Contact vendor
Date of First Release	1999	1999	1996	1993	1992	2004	2004	2003
SOFTWARE FEATURES								
Operating System/Network Support								
—Network Client-Server Support	Yes	Yes	Yes	Yes	Yes	Yes	Yes	Yes
—Server Operating System	Win XP/2000/NT 4.0/98	Win XP/2000/NT 4.0/98	Win 2000/NT	Win XP/2000/NT 4.0/98	Win XP/2000/NT 4.0/98	Win XP/2000/NT 4.0/98/Me	Win XP/2000/NT 4.0/98/Me	Win XP/2000/NT 4.0/98/Me
—Client Operating System	Win XP/2000/NT 4.0/98	Win XP/2000/NT 4.0/98	Any with a WEB Browser	Win XP/2000/NT 4.0/98	Win XP/2000/NT 4.0/98	Win XP/2000/NT 4.0/98/Me	Win XP/2000/NT 4.0/98/Me	Win XP/2000/NT 4.0/98/Me
—Internet Server Enabled	Via Bentley Publisher	Via Bentley Publisher	Yes	Via Bentley Publisher	Via Bentley Publisher	Via Bentley Publisher	Via Bentley Publisher	Via Bentley Publisher
Compatible Applications								
—Third Party Applications Designed to Run on Software	N/A	N/A	Yes	Yes	Yes	Yes	Yes	Yes
—Total Solution Packages Available	Yes	Yes	Yes	Yes	Yes	Yes	Yes	Yes
GIS Data Administration								
—Multi-user Edit Locking	[8]	[8]	N/A	N/A	N/A	N/A	N/A	N/A
—Versioning	No	No	N/A	Yes, via MicroStation	N/A	N/A	N/A	N/A
—Metadata Maintenance	Yes	Yes	Yes	N/A	N/A	N/A	N/A	N/A
Database Management								
—Proprietary DBMS	No	No	No	No	No	No	No	No
—Relational Database Management System	[4]	[4]	[4]	No	N/A	[4]	[4]	[4]
—RDBMS Spatial Data Warehouse	Oracle 8i/9i Spatial and Oracle 9i Locator	Oracle 8i/9i Spatial and Oracle 9i Locator	Oracle 8i/9i Spatial and Oracle 9i Locator	No	N/A	[1]	Oracle 8i/9i, Spatial and Locator	Oracle 8i/9i, Spatial and Locator
Native Graphic Data Structure and Format								
—Vector-Spaghetti	DGN, DWG, DXF	DGN, DWG, DXF	DGN	DGN, DWG, DXF	DGN, DWG, DXF	DGN, DWG, DXF, MXD, SHP, LYR	DGN, DWG, DXF	DGN, DWG, DXF
—Vector-Topologic	Oracle Spatial/DGN post processing	Oracle Spatial/DGN post processing	Oracle 8i/9i Spatial and Oracle 9i Locator	N/A	N/A	Oracle Spatial (query-only)	[5]	[5]
—Parametric	Yes	Yes	Yes	Yes	Yes	Yes	Yes	Yes
—3-D	Yes	Yes	Yes	Yes	Yes	View-only	Yes	Yes
—TIN	N/A	N/A	N/A	Yes, via Bentley GEOPAK Site module	N/A	N/A	N/A	N/A
—Grid	N/A	N/A	N/A	N/A	N/A	N/A	N/A	N/A
—Raster Image	[2]	[2]	[14]	[14]	[17]	[2]	[2]	[2]
GIS Data Import/Export Utilities								
—Direct Import Formats	DXF, DWG, MID/MIF, SHP, E00	DXF, DWG, MID/MIF, SHP, E00	DXF/DWG	[14]	[17]	DGN, DWG, DXF, MXD, SHP, LYR	DGN, DWG, DXF	DGN, DWG, DXF, MXD, SHP, LYR
—Direct Export Formats	DXF, DWG, MID/MIF, SHP, E00	DXF, DWG, MID/MIF, SHP, E00	DXF/DWG	[14]	[17]	Read-only platform	DGN, DWG, DXF	DGN, DWG, DXF, SHP
Readable Formats That Do Not Require Translation								
—Import	Oracle Spatial, DGN, DWG	Oracle Spatial, DGN, DWG	Oracle Spatial, DGN, DWG	[14]	[17]	[3]	DGN, DWG, DXF	DGN, DWG, DXF, MXD, SHP, LYR
—Export	Oracle Spatial, DGN, DWG	Oracle Spatial, DGN, DWG	CGM, SVG, JPG	[14]	[17]	Read-only platform	DGN, DWG, DXF	DGN, DWG, DXF
GIS Data Entry and Editing								
—Board Digitizing	Yes	Yes	N/A	Yes	Yes	Yes	Yes	Yes
—Coordinate Geometry/Precision Entry	N/A	N/A	N/A	Yes	N/A	N/A	N/A	N/A
—Electronic Survey Data Import	N/A	N/A	N/A	N/A	N/A	N/A	N/A	N/A
—Heads-up Digitizing	Yes	Yes	N/A	Yes	Yes	N/A	Yes	Yes
—Vectorization	N/A	N/A	N/A	Yes	Yes	N/A	N/A	N/A
—Map Rectification	N/A	N/A	N/A	Yes	Yes	Yes	Yes	Yes
—Graphic Error Check/Correction	N/A	N/A	N/A	N/A	N/A	N/A	N/A	Yes
—Field Data Entry	N/A	N/A	N/A	N/A	N/A	Yes	Yes	Yes
Map Design and Composition								
—Interactive Map Composition	N/A	N/A	N/A	Yes	Yes	Yes	Yes	Yes
—Annotation from Attributes	Yes	Yes	Yes	No	Yes	Yes	Yes	Yes
—Global Symbol Change	Yes	Yes	Yes	Yes	N/A	Yes	Yes	Yes
—Thematic Mapping	Yes	Yes	Yes	Yes	Yes	Yes	Yes	Yes
Geographic Query and Analysis Functions								
—Attribute Query and Selection	Yes	Yes	Yes	Yes	Yes	Yes	Yes	Yes
—Map Measurements	Yes	Yes	Yes	Yes	Yes	Yes	Yes	Yes
—Address Matching	N/A	N/A	Available via additional services	N/A	N/A	N/A	Yes	Yes
—Buffer Generation	Yes	Yes	Yes	N/A	N/A	Yes	Yes	Yes
—Point/Line-in-Polygon Analysis	Yes	Yes	Yes	N/A	N/A	Yes	Yes	Yes
—Polygon Overlay	Yes	Yes	Yes	N/A	N/A	Yes	Yes	Yes
—Network Analysis	Yes	Yes	Yes	N/A	N/A	Yes	Yes	Yes
—Raster Document Query and Access	Yes	Yes	Yes	N/A	N/A	Yes	Yes	Yes
—Direct Access to Other GIS Format	SHP, MID, MIF, E00	SHP, MID, MIF, E00	N/A	N/A	N/A	Yes	Yes	Yes
Terrain Data Processing and Analysis								
—DEM Generation	N/A	N/A	N/A	N/A	N/A	N/A	N/A	N/A
—Contour Map Generation	N/A	N/A	N/A	Yes	N/A	N/A	N/A	Yes
—3-D Display/Profile Generation	N/A	N/A	N/A	Yes	N/A	Yes	Yes	Yes
—Map Draping	N/A	N/A	N/A	Yes	N/A	N/A	N/A	N/A
—Slope/Aspect Analysis	N/A	N/A	N/A	N/A	N/A	N/A	N/A	N/A
Raster Image Capabilities								
—Geometric Rectification	N/A	N/A	N/A	Yes	Yes	N/A	N/A	N/A
—Orthoimage Generation	N/A	N/A	N/A	N/A	N/A	N/A	N/A	N/A
—Image Enhancement	N/A	N/A	N/A	Yes	N/A	Yes	Yes	Yes
—Spectral Classification	N/A	N/A	N/A	N/A	N/A	N/A	N/A	N/A
Application Development Language								
—Proprietary Application Development Language	MDL, JMDL, Basic	MDL, JMDL, Basic	MDL, JMDL, Basic	MDL, JMDL, Basic	MDL	MDL, MicroStation Basic	MDL, MicroStation Basic	MDL, MicroStation Basic
—Industry Standard Programming Environment	C, C++, Visual Basic, JAVA, VBA	C, C++, Visual Basic, JAVA, VBA	ASP, JAVA Script, JAVA, Visual Basic, C, C++,	C, C++, Visual Basic, JAVA	Visual Basic	C, C++, Visual Basic, JAVA	C, C++, Visual Basic, JAVA	C, C++, Visual Basic, JAVA
Extra Notes	[12]	[13]	[15]	[16]	[18]			
Reader Service Number	9	10	11	12	13	14	15	16

N/A = Not Applicable

2004 GIS Software Survey Notes

Program Name	Geographic Transformer 5.0	Geographic Calculator 6.0	TransCAD 4.7	Maptitude 4.6	CARIS GIS	CARIS Carta	CEDRA AVseries [1]	ArcEditor	ArcView	ArcPad
Manufacturer/Distributor	Blue Marble Geographics	Blue Marble Geographics	Caliper Corporation	Caliper Corporation	CARIS	CARIS	The CEDRA Corporation	ESRI	ESRI	ESRI
Manufacturer's Phone Number	800/616-2725	800/616-2725	617/527-4700	617/527-4700	506/458-8533	506/458-8533	585/232-6998	800/447-9778	800/447-9778	800/447-9778
Cost Per Seat	$899	$499	$2995 (Base), $9995 (Standard)	$495	Contact vendor	Contact vendor	Contact vendor	Call ESRI	$1500	$495
Date of First Release	1996	1993	1988	1995	1983	2004	1985	2001	1992	2000
SOFTWARE FEATURES										
Operating System/Network Support										
—Network Client-Server Support	Yes	Yes	Win 98/NT and above	Win 98/NT and above	Yes	Yes	Yes	Yes	Yes	No
—Server Operating System	Win 98/NT and above	Win 98/NT and above	Win 98/NT and above	Win 98/NT and above	Win 95/98/NT/2000/XP/UNIX	Win NT/2000/XP	Win 95/98/NT/2000/XP/UNIX	Windows NT 4.0/2000/XP	Windows	No
—Client Operating System	Win 98/NT and above	Win 98/NT and above	Win 98/NT and above	Win 98/NT and above	Win 95/98/NT/2000/XP/UNIX	Win NT/2000/XP	Win 95/98/NT/2000/XP/UNIX	Windows NT 4.0/2000/XP	Windows	Windows CE
—Internet Server Enabled			With TransCAD for the Web	With Maptitude for the Web	No	No	Yes	Via ArcIMS	Via ArcIMS	Via ArcIMS
Compatible Applications										
—Third Party Applications Designed to Run on Software							ArcView or ArcGIS	www.esri.com/partners	www.esri.com/partners	www.esri.com/partners
—Total Solution Packages Available			2 [1]	3 [2]			12	Numerous	Numerous	Numerous
GIS Data Administration										
—Multi-user Edit Locking	Yes	Yes	Yes	Yes	No	No	Yes	Yes	No	No
—Versioning			No	No	No	No	Yes	Yes	No	No
—Metadata Maintenance			Yes	Yes	No	No	Yes	Yes	Yes	No
Database Management										
—Proprietary DBMS			No	No	No	No	No	No	No	No
—Relational Database Management System			Yes	Yes	Yes	Yes	Yes	Yes	Yes	Yes
—RDBMS Spatial Data Warehouse			Yes	Yes	No	Yes	Yes	[1]	[1]	No
Native Graphic Data Structure and Format										
—Vector-Spaghetti			Yes, can handle	Yes, can handle	Yes	Yes	Yes	Shapefile	Shapefile	Shapefile
—Vector-Topologic			Yes	Yes	Yes	Yes	Yes	Coverage, Geodatabase	Coverage, Geodatabase	No
—Parametric			No	No	Yes	No	Yes	Geodatabase	Geodatabase	No
—3-D			Yes (surfaces)	Yes (surfaces)	Yes	Yes	Yes	Shapefile, Geodatabase	Shapefile, Geodatabase	No
—TIN			Yes	Yes	Yes	Yes	Yes	ESRI TIN	ESRI TIN	No
—Grid	Yes	Yes	Yes	Yes	Yes	No	Yes	ESRI Grid	ESRI Grid	No
—Raster Image	Yes	Yes	Yes	Yes	Yes	Yes	Yes	[2]	[2]	Yes
GIS Data Import/Export Utilities										
—Direct Import Formats	[1]	[3]	16, including TIGER/Line, DWG, DGN	16, including TIGER/Line, DWG, DGN	Yes	Yes	Yes	Over 40 standard formats	Over 40 standard formats	N/A
—Direct Export Formats	[2]	[3]	8, including Shapefile, MID/MIF	8, including Shapefile, MID/MIF	Yes	Yes	Yes	Over 40 standard formats	Over 40 standard formats	N/A
Readable Formats That Do Not Require Translation										
—Import			Oracle Spatial, ESRI Shapefile, MapInfo TAB	Oracle Spatial, ESRI Shapefile, MapInfo TAB	Yes	Yes		Over 40 standard formats	Over 40 standard formats	Shapefiles, Mr SID, BMP, JPG
—Export			Oracle Spatial	Oracle Spatial	Yes	Yes		Over 40 standard formats	Over 40 standard formats	Shapefiles, Mr SID, BMP, JPG
GIS Data Entry and Editing										
—Board Digitizing			Yes	Yes	Yes	No	Yes	Yes	Yes	No
—Coordinate Geometry/Precision Entry	Yes	Yes	No	No	Yes	Yes	Yes	Yes	Yes	No
—Electronic Survey Data Import	Yes	Yes	GPS points	GPS points	Yes	No	Yes	Yes	Yes	No
—Heads-up Digitizing			Yes	Yes	Yes	No	Yes	Yes	Yes	Yes
—Vectorization			No	No	Yes	No	Yes	Yes	Yes	No
—Map Rectification	Yes	Yes	Yes	Yes	Yes	Yes	Yes	Yes	Yes	No
—Graphic Error Check/Correction	Yes	Yes	Yes	Yes	Yes	Yes	Yes	Yes	Yes	No
—Field Data Entry	Yes	Yes	Yes	Yes	Yes	Yes	Yes	Yes	Yes	Yes
Map Design and Composition										
—Interactive Map Composition			Yes	Yes	Yes	Yes	Yes	Yes	Yes	No
—Annotation from Attributes			Yes	Yes	Yes	Yes	Yes	Yes	Yes	No
—Global Symbol Change			Yes	Yes	Yes	Yes	Yes	Yes	Yes	Yes
—Thematic Mapping			Yes	Yes	Yes	Yes	Yes	Yes	Yes	No
Geographic Query and Analysis Functions										
—Attribute Query and Selection			Yes	Yes	Yes	Yes	Yes	Yes	Yes	Yes
—Map Measurements		Yes	Yes	Yes	Yes	Yes	Yes	Yes	Yes	Yes
—Address Matching			Yes	Yes	No	No	Yes	Yes	Yes	Yes
—Buffer Generation			Yes	Yes	Yes	Yes	Yes	Yes	Yes	No
—Point Line-in-Polygon Analysis			Yes	Yes	Yes	No	Yes	Yes	Yes	No
—Polygon Overlay			Yes	Yes	Yes	No	Yes	Yes	Yes	No
—Network Analysis			Yes	Yes (best route)	Yes	No	Yes	Yes	Yes	No
—Raster Document Query and Access	Yes		No	No	No	No	Yes	Yes	Yes	No
—Direct Access to Other GIS Format	Yes	Yes	Yes	Yes	No	Yes	Yes	Yes	Yes	No
Terrain Data Processing and Analysis										
—DEM Generation			Yes	Yes	Yes	Yes	Yes	Yes	Yes	No
—Contour Map Generation			Yes	Yes	Yes	Yes	Yes	Yes	Yes	No
—3-D Display/Profile Generation			Yes	Yes	Yes	No	Yes	Yes	Yes	No
—Map Draping			Yes	Yes	Yes	No	Yes	Yes	Yes	No
—Slope/Aspect Analysis			Yes	Yes	Yes	No	Yes	Yes	Yes	No
Raster Image Capabilities										
—Geometric Rectification	Yes		No	No	Yes	Yes	Yes	Yes	Yes	No
—Orthoimage Generation			No	No	Yes	No	Yes	Yes	Yes	No
—Image Enhancement	Yes		No	No	No	No	Yes	Yes	Yes	No
—Spectral Classification			No	No	Yes	No	Yes	Yes	Yes	No
Application Development Language										
—Proprietary Application Development Language	Yes	Yes	Yes (Caliper Script)	Yes (Caliper Script)	Yes	No		No	No	No
—Industry Standard Programming Environment			Yes (.NET)	Yes (.NET)	VB	No		[3]	[3]	VBScript, C/C++
Extra Notes			[37]	[40]				www.esri.com/arceditor	www.esri.com/arcview	[4]
Reader Service Number	17	18	19	20	21	22	23	24	25	26

N/A = Not Applicable

2004 GIS Software Survey Notes

Program Name	ArcGIS Survey Analyst	Surfer 8	Didger 3	MapViewer 5	GeoMedia/IntelliWhere [1]	ERDAS IMAGINE	MapInfo Professional	MapX Mobile
Manufacturer/Distributor	ESRI	Golden Software	Golden Software	Golden Software	Intergraph Mapping & Geospatial Solutions	Leica Geosystems GIS & Mapping	MapInfo Corporation	MapInfo Corporation
Manufacturer's Phone Number	800/447-9778	303/279-1021	303/279-1021	303/279-1021	256/730-7191 [2]	404/248-9000	800/327-8627	800/327-8627
Cost Per Seat	Call ESRI	$599	$329	$249	Call for pricing information	$2200+	Contact vendor	Contact vendor
Date of First Release	2003	2002	2001	2002	1997	1990	1985	2000
SOFTWARE FEATURES								
Operating System/Network Support								
—Network Client-Server Support	Yes			Yes	Yes	Yes	Yes	No
—Server Operating System	Win NT 4.0/2000/XP			Windows, Novell	[3]	Win XP/2000, Sun Solaris	MS Windows logo certified	No
—Client Operating System	Win NT 4.0/2000/XP	Win 98/Me/2000/XP	Win 98/Me/2000/XP	Win 98/Me/2000/XP	Win Me/XP/2000	Win XP/2000, Sun Solaris	MS Windows logo certified	[17]
—Internet Server Enabled	Via ArcIMS				Yes	No	Yes, via MapInfo Discovery	No
Compatible Applications								
—Third Party Applications Designed to Run on Software	www.esri.com/partners				Several	[7]	Numerous	Numerous
—T-tal Solution Packages Available	Numerous				Yes	Yes	Yes	
GIS Data Administration								
—Multi-user Edit Locking	Yes, with core ArcGIS				Yes	No	Yes, via database	No
—Versioning	Yes, with core ArcGIS				Yes	No	Yes, via database	No
—Metadata Maintenance	Yes, with core ArcGIS				Yes	Yes	Yes	No
Database Management								
—Proprietary DBMS	No			Yes	No	CellArray	[1]	No
—Relational Database Management System	Yes			Yes	[11]	[1]	Yes	Yes [6]
—RDBMS Spatial Data Warehouse	[1]			Yes	[11]	ArcSDE	[2]	No
Native Graphic Data Structure and Format								
—Vector-Spaghetti	Shapefile	Yes	Yes	Yes	[4]	Yes	Yes	Yes
—Vector-Topologic	Coverage, Geodatabase				Yes	Yes	No	No
—Parametric	Geodatabase				Yes	Yes	No	No
—3-D	Shapefile, Geodatabase	Yes			Yes	Yes	Yes [3]	No
—TIN	ESRI TIN				Yes	Yes	Yes, via Vertical Mapper	No
—Grid	ESRI Grid	Yes			Yes	Yes	Yes [16]	No
—Raster Image	[2]	Yes	Yes	Yes	Yes	Yes	Yes [16]	Yes
GIS Data Import/Export Utilities								
—Direct Import Formats	Over 40 standard formats	[1]	[3]	[5]	[5]	[2]	[16]	No
—Direct Export Formats	Over 40 standard formats	[2]	[4]	[6]	[6]	[3]	[16]	No
Readable Formats That Do Not Require Translation								
—Import	Over 40 standard formats	SRF, GSI, GSB, BLN	PJT, GSI, GSB, BLN	GSM, GSI, GSB, BLN	[7]	[4]	[16]	Shape, ASCII
—Export	Over 40 standard formats	SRF, GSI, GSB, BLN	PJT, GSI, GSB, BLN	GSM, GSI, GSB, BLN	[8]	[5]	[16]	No
GIS Data Entry and Editing								
—Board Digitizing	Yes		Yes		Yes	Yes	Yes	Yes
—Coordinate Geometry Precision Entry	Yes				Yes	No	Yes	Yes, with customization
—Electronic Survey Data Import	Yes		Yes		Yes	No	Yes	Yes
—Heads-up Digitizing	Yes, with ArcScan		Yes	Yes	Yes	Yes	Yes	Yes
—Vectorization	Yes, with ArcScan				Yes	Yes	Yes, via Vertical Mapper	No
—Map Rectification	Yes		Yes		Yes	Yes	No	No
—Graphic Error Check/Correction	Yes				Yes	Yes	Yes	No
—Field Data Entry	Yes				Yes	Yes	Yes	Yes
Map Design and Composition								
—Interactive Map Composition	Yes	Yes	Yes	Yes	Yes	Yes	Yes	Yes
—Annotation from Attributes	Yes		Yes	Yes	Yes	Yes	Yes	Yes
—Global Symbol Change	Yes	Yes	Yes	Yes	Yes	Yes	Yes	Yes
—Thematic Mapping	Yes			Yes	Yes	Yes	Yes	Yes
Geographic Query and Analysis Functions								
—Attribute Query and Selection	Yes, with core ArcGIS			Yes	Yes	Yes	Yes	Yes
—Map Measurements	Yes, with core ArcGIS		Yes	Yes	Yes	Yes	Yes	Yes
—Address Matching	Yes, with core ArcGIS				Yes	No [6]	Yes	No
—Buffer Generation	Yes, with core ArcGIS			Yes	Yes	Yes	Yes	No
—Point/Line-in-Polygon Analysis	Yes, with core ArcGIS			Yes	Yes	Yes	Yes	Yes
—Polygon Overlay	Yes, with core ArcGIS			Yes	Yes	Yes	Yes	No
—Network Analysis	Yes, with core ArcGIS				Yes	No	No	No
—Raster Document Query and Access	Yes, with core ArcGIS				Yes	Yes	Yes	No
—Direct Access to Other GIS Format	Yes, with core ArcGIS			Yes	Yes	Yes	Yes	Shape
Terrain Data Processing and Analysis								
—DEM Generation	Yes, with core ArcGIS				Yes	Yes	Yes	No
—Contour Map Generation	[5]	Yes			Yes	Yes	Yes, via Vertical Mapper	No
—3-D Display/Profile Generation	Yes, with ArcGIS 3D Analyst	Yes			Yes	Yes	Yes, via Vertical Mapper	No
—Map Draping	Yes, with core ArcGIS	Yes			Yes	Yes	Yes	No
—Slope/Aspect Analysis	Yes, with ArcGIS Spatial Analyst	Yes			Yes	Yes	Yes, via Vertical Mapper	No
Raster Image Capabilities								
—Geometric Rectification	Yes, with core ArcGIS		Yes		Yes	Yes	[4]	No
—Orthoimage Generation	Yes, with core ArcGIS				Yes	Yes	No	No
—Image Enhancement	Yes, with core ArcGIS		Yes	Yes	Yes	Yes	Yes	No
—Spectral Classification	Yes, with core ArcGIS				Yes	Yes	Yes, via Vertical Mapper	No
Application Development Language								
—Proprietary Application Development Language	No	Scripter		Scripter	[9]	IMAGINE Developers' Toolkit	Yes, MapBasic	No
—Industry Standard Programming Environment	[3]	C++, [illegible], Perl		C++, VB, Perl	Yes	C/C++ with some COM	[5]	[7]
Extra Notes	[6]				[10]			
Reader Service Number	27	28	29	30	31	32	33	34

N/A = Not Applicable

2004 GIS Software Survey Notes

Program Name	MapXtreme Java	MapXtreme 2004	Vulcan 4.5	Parcel Editor 8	ELT/5500	IMap	SOLO	TerraSync Professional Edition	GPS Pathfinder Office
Manufacturer/Distributor	MapInfo Corporation	MapInfo Corporation	Maptek Pty Ltd	NovaLIS Technologies	Paragon Imaging Inc.	SOKKIA	Tripod Data Systems	Trimble	Trimble
Manufacturer's Phone Number	800/327-8627	800/327-8627	303/763-4919	902/423-1800	800/937-4544	913/255-3913	541/753-9322	408/481-8000	408/481-8000
Cost Per Seat	Contact vendor	Contact vendor	Contact vendor	$2495 [1]	$3995	$1000 list	$1000	Contact vendor	Contact vendor
Date of First Release	1998	1998	2003	1996	2001	2000	1998	2000	1995
SOFTWARE FEATURES									
Operating System/Network Support									
—Network Client-Server Support	Yes	Yes	No	Terminal server	Yes, FlexLM	No	No	N/A	Yes
—Server Operating System	[8]	Win 2000/XP/2003	N/A	Unix, W2K, NT Server	Microsoft Windows	N/A		N/A	Win XP/2000/NT/Me/98/95
—Client Operating System	[13]	[13]	Win NT/2000/XP	Win XP/NT/2000	Microsoft Windows	Win CE/NT/98/2000/XP	Windows CE/Mobile	[1]	Win XP/2000/NT/Me/98/95
—Internet Server Enabled	Yes	Yes	No	N/A		No	No	Yes	Yes
Compatible Applications									
—Third Party Applications Designed to Run on Software	Numerous	Numerous	No	ArcGIS		No	RTI —RealTime Forest Inventory	N/A	N/A
—Total Solution Packages Available	Yes	Yes	Yes	Full Cadastral office suite		Yes	Yes	N/A	N/A
GIS Data Administration									
—Multi-user Edit Locking	Yes, via database	Yes, via database	Yes	Yes		No	N/A	No	No
—Versioning	Yes, via database	Yes, via database	No	Yes		No	No	No	No
—Metadata Maintenance	Yes	Yes	No	No		No	No	No	No
Database Management									
—Proprietary DBMS	[1]	[14]	Isis	Geodatabase		No	No	No	No
—Relational Database Management System	[9]	[9]	No	Geodatabase		No		No	No
—RDBMS Spatial Data Warehouse	[10]	[10]	Via ODBC	Geodatabase		No	No	No	No
Native Graphic Data Structure and Format									
—Vector-Spaghetti	Yes	Yes	Yes	Yes		Yes	Yes	Yes	Yes
—Vector-Topologic	No	No	No	Yes		Yes	Yes	No	No
—Parametric	Yes	Yes	No	No		No	No	No	No
—3-D	Yes, read-only	Yes, read-only from database	Yes	No		Yes	Yes	Yes	Yes
—TIN	No	No	Yes + solid triangulations	No		No	No	No	No
—Grid	Yes	Yes [16]	Yes	No		Yes	Yes	No	No
—Raster Image	Yes	Yes [16]	Yes	No	Native raster image support	Yes	Yes	Yes	Yes
GIS Data Import/Export Utilities									
—Direct Import Formats	[16]	[16]	[1]	[2]	[1]	[1]	[1]	ArcView Shapefile, Trimble SSF	[3]
—Direct Export Formats	[16]	[16]	[2]	[2]	NITF, GeoTiff and more	[2]	SHP, DXF, ASCII, CR5	ArcView Shapefile, Trimble SSF	[4]
Readable Formats That Do Not Require Translation									
—Import	[16]	[16]	[3]	[2]	[1]	[3]	UDF, SHP, MIF, DXF, DGN	Trimble SSF	Trimble SSF
—Export	[16]	[16]	[3]	[2]	[1]	[4]	UDF, CR5, SHP, DXF, ASCII	Trimble SSF	Trimble SSF
GIS Data Entry and Editing									
—Board Digitizing	No	Yes	Yes	Yes		No	No	No	No
—Coordinate Geometry/Precision Entry	No	Yes	Yes	Yes	[2]	Yes, no adjustments	Yes, no adjustments	N/A	Yes
—Electronic Survey Data Import	No	Yes	Yes	Yes		Yes	Yes	N/A	Yes
—Heads-up Digitizing	No	Yes	Yes	Yes		Yes	Yes	Yes	Yes
—Vectorization	No	No	No	Yes		No	No	No	No
—Map Rectification	No	No	Yes	Yes		No	No	No	No
—Graphic Error Check/Correction	No	No	No	Yes		No	No	No	No
—Field Data Entry	Yes	Yes	No	Yes		Primary function	Primary function	Yes	No
Map Design and Composition									
—Interactive Map Composition	Yes	Yes	Yes	Yes		Yes	Yes	Yes	Yes
—Annotation from Attributes	Yes	Yes	Yes	Yes		Yes	Yes	Yes	No
—Global Symbol Change	Yes	Yes	Yes	Yes		Yes	Yes	Yes	Yes
—Thematic Mapping	Yes	Yes	No	No		Yes	Yes	No	No
Geographic Query and Analysis Functions									
—Attribute Query and Selection	Yes	Yes	No	Yes		Yes	Yes	Yes	Yes
—Map Measurements	Yes	Yes	No	Yes	Image and map measurements	Yes	Yes	Yes	Yes
—Address Matching	Yes [11]	Yes	No	Yes		No	No	No	No
—Buffer Generation	Yes	Yes	No	No		No	Yes	No	No
—Point/Line-in-Polygon Analysis	Yes	Yes	No	Yes		No	No	No	No
—Polygon Overlay	No	Yes	No	Yes	Full polygon overlays	Yes	Yes	No	No
—Network Analysis	Yes [12]	No	Yes	Yes		No	No	No	No
—Raster Document Query and Access	Yes	No	No	No		No	No	No	No
—Direct Access to Other GIS Format	Yes	Yes (Oracle Spatial, Shape)	No	Yes		Yes	Yes	No	No
Terrain Data Processing and Analysis									
—DEM Generation	No	No	Yes	No		No	No	No	No
—Contour Map Generation	No	No	Yes	No		No	No	No	No
—3-D Display/Profile Generation	No	No	Yes + solid triangulations	No		No	No	No	No
—Map Draping	No	No	Yes	No		No	No	No	No
—Slope/Aspect Analysis	No	No	Yes	No		No	No	No	No
Raster Image Capabilities									
—Geometric Rectification	No	No	Yes	No	Yes	No	No	No	No
—Orthoimage Generation	No	No	No	No		No	No	No	No
—Image Enhancement	Yes	Yes	No	No	[3]	Yes	Yes	No	No
—Spectral Classification	No	No	No	No		No	No	No	No
Application Development Language									
—Proprietary Application Development Language	No	No	Lava	VB	Full API support with ELT/5500	Yes	Yes	No	No
—Industry Standard Programming Environment	Java and .Net languages	[15]	Perl	VB		Yes	Yes	No	No
Extra Notes							[2]	[2]	[5]
Reader Service Number	35	36	37	38	39	40	41	42	43

N/A = Not Applicable

2004 GIS Software Survey Notes

Summary of GIS Software Features

Software Feature Category	*Explanation*
Geographic Data Managemen	Database administration tools for managing data access by users, locking of data during edit and maintenance of metadata.
Tabular Attribute Data Management	Software environment and capabilities for storing and managing database attributes linked to map features in the GIS database. May involve use of a vendor-proprietary system for attribute storage or a commercial relational database management package.
GIS Data Import/ Export Utilities	Utility programs bundled with the GIS package for translation of GIS or CAD data to or from another format, including common industry-standard formats like DXF, SIF, DLG or SDTS.
GIS Data Entry and Editing	A range of interactive and batch processing functions for entry of map data through such means as board digitizing, coordinate geometry entry (COGO), scanning and heads-up digitizing, along with capabilities for editing GIS data, performing error checking and resolution, map rectification and transformation of coordinate systems and map projections.
Map Design and Composition	Interactive capabilities for the design of map plots and displays, automatic creation of thematic maps and legends, and modifying map symbology and annotation for custom map displays.
Basic Geographic Query and Analysis Functions	Basic tools for performing attribute or map-based queries and displays, basic distance and area measurements, query and access to scanned documents, buffer generation, polygon overlay operations and other query and analysis functions.
Network Analysis	Spatial analysis operations based on linear networks (e.g., road or pipeline systems), including such operations as shortest path tracing and region allocation. Network analysis capabilities in GIS packages often allow users to design network models based on attributes of network segments.
Terrain and 3-D Data Processing and Analysis	Capabilities for storing three-dimensional data normally in a grid or triangular integrated network (TIN) format with functions for 3-D analysis such as contour mapping, 3-D display, draping of map features over a 3-D display, slope and aspect analysis, etc.
Raster Image Processing Capabilities	Capabilities for the manipulation and processing of raster images (e.g., digital aerial photos or orthophotography, satellite images), including functions for the import and rectification of raw imagery, digital image enhancement and automated classification of multi-spectral imagery.
Application Development Languages	Programming environment for customizing applications accessing software functions provided by the package, including proprietary languages included with the GIS software package or industrystandard tools (e.g., C++, Visual Basic, Delphi) that may be used for application development.

Autodesk Inc.

1. ASCII, EOO, MID, MIF, ArcInfo coverage, ArcInfo Shape, DWG, LandXML, DGN, numerous data collectors.
2. Autodesk Survey is an add-in to Autodesk Land Desktop 2005. Survey adds bi-directional data collector support, traverse analysis, figures, basic coordinate geometry functions and annotation to Autodesk Land Desktop 2005.
3. Autodesk Raster Design. Autodesk Survey, Autodesk Civil Design.
4. Autodesk Land Desktop 2005 streamlines each phase of land development projects and provides a central location for project data that everbody on a team can access.
5. Autodesk Civil Design is an add-on for Autodesk Land Desktop 2005. The package includes tools to design roadways, grading plans, do hydrologic analysis, pond design and generate plan/profile sheets.
6. Image Formats, CALS, TIFF, GeoTIFF, SID, ECW, PNG, BMP, JPG, GeoSPOT, JPG2000, DOQ, DEM, satellite imagery.
7. Autodesk Raster Design 2005e is an add-on application to AutoCAD 2005, Autodesk Map 2005 and Autodesk Land Desktop 2004 that enables customers to easily edit, convert, and integrate scanned drawings and raster files into the digital design process.
8. Autodesk Civil Design is an add-on for Autodesk Land Desktop 2005. The package includes tools to design roadways, grading plans, do hydrologic analysis, pond design and generate plan/profile sheets.

Bentley Systems Inc.

1. Oracle (query only); any ODBD compliant database (read/write).
2. Yes, GeoTIFF, iTIFF, JPEG, TIFF, HMR, RGB, RLE, TGA, PNG, BMP, CAL, IMB, CIT, COT, TG4, PCX, RS, MrSID, ECW, JPG2000.
3. Oracle Spatial, DNG, DWG, SHP, MXD.
4. Oracle, any ODBC compliant database.
5. Oracle Spatial/DGN post processing.
6. ASCII XYZ, TDS, Sokkia, Leica, Topcon, SMI, Geodimeter, C&G.
7. Power Survey is a Standalone application with CADD included. Power Survey includes: DTM Editing, Volume Calculations,exportation of Machine Control data formats.
8. Yes, when Oracle 8i/9i Spatial/Locator is used.
9. Yes, with limitations (utilizing MicroStation Design History).
10. Yes, e.g. through links between geometry and a SQL database.
11. In the same way that MicroStation extends computer-aided design into an integrated engineering platform, MicroStation GeoGraphics transforms MicroStation into a comprehensive geospatial environment with premier mapping, data collection, editing, cleanup and analysis tools. MicroStation GeoGraphics allows users to integrate engineering and mapping applications on a single platform and extends the data capture and editing tools in MicroStation with a solid database interface and spatial analysis functionality. MicroStation GeoGraphics is a key engineering product that forms the foundation for geospatial applications. It significantly increases user productivity and offers an easy-to-use environment for geospatial users. With this discipline-specific platform, independent software developers can take advantage of highlevel functionality to increase application performance and reduce development time.
12. Bentley Water is a comprehensive, high-productivity engineering automation solution for water network design and management. All operations of a typical water supply network—from mapping and design to administrative functions such as maintenance and record-keeping—are addressed through an integrated environment. Municipal engineers can create a detailed system of water lines, valves, consumption points and other user-defined system components, while maintaining crucial network connectivity. Functions are provided for the placement, editing and attribution of facility graphics. Dynamic, database-driven facility annotation enables users to change existing annotation via a simple update function.
13. Bentley WasteWater is a comprehensive, highproductivity engineering solution for wastewater network design and management. All functions of a typical wastewater maintenance operation—from mapping and design to maintenance and record keeping—are addressed through an integrated environment. Municipal engineers can create a detailed model of storm water, sanitary and combined systems. The placement and edit functions create and maintain network connectivity, as well as distance, depth and connectivity of pipes and manholes.
14. GeoTIFF, iTIFF, JPEG, TIFF, HMR, RGB, RLE, TGA, PNG, BMP, CAL, IMG, CIT, COT, TG4, PCX, RS, MrSID, E.
15. Bentley Publisher is a powerful server-based application and a one-stop platform that allows users and implementers to publish their engineering information over the Internet. Bentley Publisher allows dynamic viewing and redlining of DGN, DWG and MicroStation GeoGraphics projects and images from Web browsers in raster, vector or hybrid raster/vector format. It provides instant and universal access to valuable data, without requiring any change to existing workflows. Bentley Publisher offers multiple publishing capabilities, from simple queries to enhanced navigation using serverside processing, to full features using the Bentley Publisher Viewer.
16. Bentley Descartes is a comprehensive imaging solution for image mapping, visualization and raster-to-vector conversion projects. Designed for both production and visualization, Bentley Descartes is used worldwide by engineers and planners in many disciplines, such as civil engineering, plant design, mapping and architectural design. Part of the Geospatial product line, Bentley Descartes is used by governments and utilities for mapping, imaging and spatial analysis. Bentley Descartes adds high-performance raster operations to MicroStation, allowing warp, reference, resample, mosaic creation and 3D terrain draping to produce outstanding image scenes. Additionally, Bentley Descartes provides tools such as an integrated Optical Character Recognition (OCR) and Line Following engines, plus several tools to enhance and repair legacy drawings. Because Bentley Descartes is tightly integrated into MicroStation, it can be customized to integrate in workflows where automation and interface modification is required. MicroStation Descartes adds high-performance raster operations to allow warp, reference, resample, create mosaics and drape and produce outstanding image scenes. Additionally, MicroStation Descartes provides tools such as an integrated OCR and Line Following engines, plus several tools to enhance and repair legacy drawings. Offering an API of over 1,700 synchronous keyins and a fully customizable interface, MicroStation Descartes can be customized to integrate in workflows where automation and interface customization is required.
17. RLE, CIT, TG4, GP4, RT4, TIF, CAL, CRL, T29, MPF, RLC, LRD, HMR, iTiff.
18. Bentley I/RAS B is a powerful tool for organizations that work with scanned raster data. A proven raster editor, it offers a complete set of tools to meet the needs of engineers, architects, facilities managers and other professionals. Bentley I/RAS B is an essential element in any MicroStation workflow, as it is the ideal product to digitally archive, clean up or convert scanned drawings. With Bentley I/RAS B tools, you can restore and revise legacy drawings in raster format. Bentley I/RAS B includes a full suite of engineering tools to help in the conversion from raster to vector. A powerful Optical Character Recognition engine allows seamless conversion of raster text-to-text elements and automated conversion for lines and symbols. Bentley I/RAS B is ProjectWise compliant and fully integrated into MicroStation.

Blue Marble Geographics

1. JPG, JP2, BMP, PNG, TIFF, GeoTIFF, MrSID, ECW, BIL, BSQ, BSB, ADRG, CADRG.
2. JPG, JP2, BMP, PNG, TIFF, GeoTIFF, ECW.
3. DXF, DWG thru R2000, SHP, TAB, MIF, ASCII, SEG P1, dBase, XLS, LOTUS.

Caliper Corporation

1. TransCAD for the Web, TransCAD for Great Britain.
2. Maptitude for the Web, Maptitude for Redistricting, Maptitude for Great Britain.

CEDRA Corporation

1. Cedra AVseries includes AVland, AVsand, AVparcel, AVwater and AVcogo.

ESRI

1. With ArcSDE: Oracle, MS SQL Server, IBM DB2, and Informix.
2. ESRI Grid, ESRI SDE raster, ESRI Raster Catalogs.
3. VBA (included) and COM compliant languages.
4. ArcPad is a field data collection technology for use on Windows CE handheld devices. For more information visit www.esri.com/arcpad.
5. Yes, with ArcGIS Geostatistical Analyst & Spatial Analyst.
6. Since our extensions have to run on a version of our core GIS product you receive a huge array of functionality. ArcInfo, ArcEditor, or ArcView can be used in combination with any extension to increase your productivity no matter what industry you are in.

Golden Software

1. EMF, E00, DXF, DLG, LGO, LGS, BNA, CLP, WMF, SHP, MIF, BMP, JPG, PNG, TIF, TGA, PCX, DCX, WPG, PCT, GIF, and USGS SDTS TVP (DDF), XLS, DAT, SLK, WKx, WRx, CSV, TXT, BNA, GRD, DEM, HDR, DDF, DT0, DT1, DT2.
2. EMF, MIF, DXF, SHP, WMF, CLP, CGM, TIF, BMP, JPG, TGA, PCX, DCX, WPG, PNG, BNA, GIF, PCT, XLS, SLK, CSV, TXT, DAT.
3. GeoTiFF, TFW, RSF, EMF, BNA, DLG, LGO, LGS, DXF, PLT, CLP, WMF, DDF, SHP, MIF, E00, TIF, BMP, TGA, PCX, GIF, WPG, DCX, EPS, JPG, PCT, PNG, XLS, SLK, DAT, CSV, TXT, BNA, WKx, WRx.
4. LAS, EMF, DAT, DXF, SHP, BNA, GIF, CGM, MIF, WMF, CLP, TIF, BMP, TGA, PCX, WPG, PNG, JPG, PCT and DCX.
5. GeoTIFF, SHP, MIF, DDF, E00, EMF, WPS, PNG, DLG, LGS, WPG, DXF, LGO, BNA, WMF, CLP, TIF, BMP, JPG, GIF, TGA, PCX, DCX, PCT, XLS, SLK, WK1, WK3, WK4, WKS, CSV, WRK, WR1, DBF, MDB, DB, TXT, DAT.
6. SHP, MIF, BLN, EMF, PNG, DXF, WMF, CLP, CGM, TIF, BMP, JPG, GIF, TGA, PCX, WPG, PCT, DCX, BNA, XLS, CSV, TXT, DAT, SLK.

Intergraph Mapping and Geospatial Solutions

1. GeoMedia Suite of Products and IntelliWhere Suite of Products.
2. US toll free phone number: 800/791-3357.
3. Win NTS 4.0/NTW 4.0/2000 Server/2000 Professional/XP Professional.
4. Yes, pseudo in core products. Complete with Dynnamo or 3rd party—radius topology.
5. OGC WMS, WFS, GML, ESRI shape, ArcInfo binary, MapInfo binary, Microstation DGN V7, Microstation DGN V8, AutoCAD DWG, DXF, MGE, MGDM, Oracle Object, Framme, SQL Server, SmartStore, Text/ASCII , Access.
6. Microstation DGN, MapInfo MIF, Oracle Object, SQL Server, AutoCAD DXF, AutoCAD DWG, GML, ESRI shape.
7. OGC WMS, WFS, GML, ESRI shape, ArcInfo binary, MapInfo binary, Microstation DGN V7, Microstation DGN V8, AutoCAD DWG, DXF, MGE, MGDM, Oracle Object, Framme, SQL Server, SmartStore, Text/ASCII , Access, all industry standard raster formats including Intergraph raster formats, JFIF, PCX, TIFF, GeoTIFF, MrSID, ECW, ESRI bip and bil, bitmap, CALS 24, Hitachi, IGS, USGS DOQ,DTED., USGS DTM.
8. Can export from any supported format.
9. No, can use any OLE/COM compatible language.
10. GeoMedia and IntelliWhere Products: GeoMedia, GeoMedia Grid, GeoMedia Image, GeoMedia Image Light Table, GeoMedia Objects, GeoMedia WebMap, GeoMedia WebMap Professional, GeoMedia WebMap Publisher, GeoMedia Parcel Manager, GeoMedia Professional, GeoMedia PublicWorks Manager, GeoMedia SMMS, GeoMedia Terrain, GeoMedia Transaction Manager, GeoMedia Transportation Analyst, GeoMedia Transportation Manager, GeoMedia Viewer (FREE), GeoMedia-Hansen Interface, SMMS, SMMS GeoConnect, IntelliWhere OnDemand, IntelliWhere LocationServer, IntelliWhere TrackForce.
11. Oracle, SQL Server, IBMDB2, Access.

Leica Geosystems GIS and Mapping

1. Only through customization with Developers' Toolkit.
2. 100+ standard raster and vector formats.
3. 30+ standard raster and vector formats.
4. 50+ standard raster and vector formats.
5. ERDAS IMAGINE .img, TIFF, GeoTIFF, ESRI GRID, Shapefile, Arc/Info coverage, Simple Personal Geodatabase.
6. Can be done using Developers' Toolkit.
7. ESRI ArcGIS, VLS Feature Analyst.

MapInfo Corporation

1. Spatial Data support for MS SQL server and IBM Informix.
2. MS Access, MS SQL Server, Oracle, IBM Informix.
3. Read-only via database.
4. Yes, via Vertical Mapper and Geographic Transformer.
5. Integrated mapping via Visual Basic.
6. Pocket PC MS SQL support.
7. Embedded Visual C and embedded Visual Basic.
8. Win 2000, 2003, Unix, Linux, or any OS with Java VM jdk1.4 or higher.
9. Oracle, Microsoft SQL Server, IBM Informix.
10. Oracle Spatial, SpatialWare for SQL Server and IBM Informix.
11. Via MapInfo MapMarker.
12. Via MapInfo Routing J Server.
13. Any system with a web browser.
14. SpatialWare for SQL Server and IBM Informix.
15. All Microsoft .NET languages are supported.
16. Numerous formats supported.
17. Microsoft Pocket PC PDA systems.

Maptek

1. ESRI shape, TIN and GRID; DGN; DXF; ASCII; Datamine strings, triangulations and databases; Micromine strings and databases; Surpac strings and triangulations; Easimine triangulations; Geocomp strings; Phodat; Minex strings and triangulations; Petrosys DIG; Whittle 4X.
2. ESRI shape, TIN and GRID; DGN; DXF; ASCII; Datamine strings, triangulations; Micromine strings; Easimine triangulations; Pitboss; Eclipse.
3. Vulcan databases (.DGD & Isis) and ASCII.

NovaLIS Technologies

1. Subsequent seats $1495.
2. DXF, DWG, ShapeFile, GDB, Coverage.

Paragon Imaging Inc.

1. NITF, GeoTiff, Shape, JPEG and more.
2. Accurate coordinate precision entry.
3. Advanced image enhancement functions.

SOKKIA

1. UDF, SHP, MIF, DXF, DGN, SIF, TIF, JPQ, DOQ.
2. DXF, ASCII, CR5, SHP, MIF.
3. UDF, SHP, MIF, DXF, DGN.
4. UDF, CR5, SHP, DXF, ASCII.

Tripod Data Systems

1. ESRI Shape, MapInfo MIF, AutoCAD DXF, Intergraph DGN.
2. RTI Application available through LandMark Systems.

Trimble

1. Windows XP/2000/NT/Me/98/95/CEv2.11/2.12/3.0/4.0.
2. The TerraSync software is a flexible GIS field data collection tool that can be used on a variety of field computers running Windows or Windows CE. The software is designed to integrate seamlessly with industry-standard GIS systems, and Trimble's GPS Pathfinder receiver range.
3. ArcView Shapefile, AutoCAD DXF, dBase, MapInfo MIF and Microsoft Access MDB.
4. ARC/INFO Generate, ArcView Shapefile, AutoCAD DXF, ASCII, MIF, MDB, DGN.
5. Trimble's GPS Pathfinder Office software supports all aspects of GIS data collection and data maintenance for Trimble's GPS Pathfinder and GeoExplorer systems.

Useful Web Addresses and Linkages

A

- **AGIS mapping/GIS shareware for Windows 3.x/95/NT** (http://www.agismap.com/) Free evaluation version of Windows mapping/GIS shareware. World map and demonstration examples provided.
- American Association of Geographers (AAG) **GIS Specialty Group** (http://www.cla.sc.edu/gis/aaggis)—online copies of their newsletter.
- **American Society for Photogrammetry and Remote Sensing** (http://www.asprs.org/asprs/) is to advance knowledge and improve understanding of mapping sciences to promote the responsible application of photogrammetry, remote sensing, geographic information systems and supporting technologies.
- **AM/FM Services for Electric Utilities** (http://srimap.com/) SRI utilizes the latest in GPS, Laser Rangefinders, and Pen Based Computers to complete a comprehensive inventory on power systems throughout North America.
- **AM/FM,GIS—Automated Digital Mapping** (http://www.4amfmgis.com/) Sentinel USA, Inc. specializes in AM/FM,GIS services, providing major utilities with Automated Digital Distribution Mapping, Field Inventory, and Digital Conversion Services.
- **AnGIS Software Home Page** (http://www.angis.com/) AnGIS Software is the developer of the AnGIS GIS system, as well as the Datamorpher engine. We support data conversions including ESRI SHPfiles as well as Landmark Argus GIS transfers.
- **ARGUS Technologies Inc.** (http://www.argusmap.com/) produces a Census Mapping GIS for Windows. Product information and downloadable data are available.
- **Array Systems** (http://www.array.ca/) Remote Sensing Image Processing and Archiving Systems.
- The **Association for Geographic Information** (http://www.geo.ed.ac.uk/agi/agi.html) The national coordinating body for GIS activities in the United Kingdom.
- **Atlas of Estonia** (http://atlas.ibs.ee/) Map server of Estonia with search by location or name.
- **Australian Centre of the Asian Spatial Information and Analysis Network (ACASIAN)** (http://www.asian.gu.edu.au/) is an academic and applied research institution specialising in Geographical Information System (GIS) databases for Asia and the former Soviet Union.
- **Australian Geodynamics on-line GIS** (http://www.ned.dem.csiro.au/AGCRC/4dgm/grasslinks/) Interactive GIS including information about Australian geology and geophysics.

- **Australian National University—Bioinformatics.** (http://life.anu.edu.au/) Biodiversity, Bio-information, Complex systems, and other material on complex systems research.
- **AutoDesk GIS Software** (http://www.autodesk.com/solution/gis/gis.htm) Descriptions of GIS solutions at AutoDesk, AutoCAD Map 2.0, AutoDesk World, and AutoDesk MapGuide.

B

- **Baltic Sea Drainage Basin GIS, Map and Statistical Database** (http://www.grida.no/baltic/) Europe's most comprehensive trans-boundary and multi-thematic GIS database. GIS data provided in PCARC/INFO and IDRISI formats. Thematic focus upon land cover and population. Links to related information.
- **Baltic Universities GIS (BUGIS) Project** (http://www.hydroconsult.se/UGIS/UGIS.html). Information resources and Data for the Baltic Countries.
- Bentley Systems **MicroStation Server** (http://www.bentley.com/) (see also Intergraph)
- The **Department of Geography** (http://www.geog.buffalo.edu/) at the State University of New York at Buffalo, and their **Geographic Information and Analysis Laboratory (GIAL)**. (http://www.geog.buffalo.edu/GIAL/welcome.html)
- **British Historical Geographical Information System Program** (http://www.qmw.ac.uk/~gbhgis) This major research programme is constructing a historical GIS covering the changing administrative units of Great Britain over the last 200 years combining computerised boundary information with a wide range of statistics.
- **BOSS International** (http://www.bossintl.com/) A recognized world leader in the field of civil and environmental engineering software.
- **Business Geographics Online** (http://www.geoweb.co.uk/) Comprehensive list of Census, geodemographic and lifestyle datasets. Locale on-line profiling system for generation of 1991 Census and Super Profiles local area reports. Description of Business Geographics Limited's activities, software and services. On-line click able maps and forms for finding UK property. An Internet GIS.

C

- **Caliper Corporation** (http://www.caliper.com/default.htm) Caliper Corporation performs GIS consulting and custom software development, and offers three Microsoft Windows-based GISs: Maptitude, TransCAD and GIS+.
- **CARIS GIS** (http://www.universal.ca/) Universal Systems Ltd.—home of CARIS GIS, CARIS Marine Information Systems and CARIS++ Spatial Objects, our new Application Development Toolkit.
- **Cartographic Technologies, Inc.** (http://www.sover.net/~cti) Cartographic Technologies, Inc. (CTI) provides a wide range of Geographic Information System (GIS), mapping, and information technology services to help clients gather and evaluate the information needed to make sound decisions.

- **Center for Advanced Spatial Technologies** (http://www.cast.uark.edu/) GIS Research, Professional Development Courses, Digital Data Catalog On-line Interactive Mapper, Other GIS links.
- **Southern California Traffic Report** (http://www.scubed.com/caltrans/transnet.html). This case study provides real-time mapping of the current traffic situation for commuters.
- Cartographic Communication research.
- **CIEG—Federal University of Parana, Brazil** (http://www.cieg.ufpr.br/) CIEG (GIS center) it's one of the most important center for research, education and projects in South Brazil. Links to Brazilian GIS sites, Projects Going on, News.
- **Community 2020 (tm) GIS for Community Organizations** (http://www.hud.gov/cpd/2020soft.html)
- **GIS data for Connecticut, USA** (http://magic.lib.uconn.edu/) MAGIC, the Map and Geographic Information Centre at the University of Connecticut, is the geodata library for the State of Connecticut.
- **Consortium for International Earth Science Information Network (CIESIN)** (http://www.ciesin.org/) are a gateway to data and other information.
- **Consulting company for GIS, remote sensing and informatics—GISL** (http://www.gisl.co.uk/) Company profile, description of services and expertise, summaries of previous projects, links to other GIS sites and much more.
- **Convergent Group** (http://www.convergentgroup.com/) GIS systems integration and consulting, including Smallworld, ESRI, Intergraph, MapInfo, Genasys. Related technologies including Outage Management, CIS, Field Data, Document Imaging, Data Conversion, hardware and database management vendors.
- **Institute of Geography, University of Copenhagen** (http://www.geogr.ku.dk/) Research, educational and staff information at the Institute of Geography, University of Copenhagen. Also the homepage of WinCHIPS (a remote sensing software package developed at the institute) and NOAA satellite images.

D

- **Data-on-the-Map 3.0—Desktop GIS/mapping software** (http://www.chartwrite.se/) An innovative, cost-effective and user-friendly GIS/mapping software based on object oriented programming.
- **DDS Digital Data Services GmbH** (http://www.dds.ptv.de/) Geographic and Demographic data for Germany and Europe. Offers a single data source for geomarkting applications primarily for Germany, but also for many other countries.
- **Department of Geology and Geography at Hunter College, NY** *(http://www.geo.hunter.cuny.edu/)*
- **Deutches Klimarechenzentrum** (http://www.dkrz.de/index-eng.html)(German Climatology Computer Centre). Climatic data and simulations.
- **Drake University Department of Geography and Geology** (http://www.geo.drake.edu/) Info about Drake's GIS Lab, courses, student projects and department of Geography and Geology (Des Moines, Iowa, USA).

E

- **EarthData International** (http://www.earthdata.com/) The Photo Science group specializes in the development of spatial data technologies to support the engineering, environmental and land management applications of its clients.
- **EarthWatch Incorporated** (http://www.digitalglobe.com/) is launching the world's first commercial, high-resolution satellite. Browse and order worldwide satellite imagery and geographic data from the Digital Globe database. The site also contains 1m and 3m satellite imagery.
- **Edinburgh University, Geography Department.** (http://www.geo.ed.ac.uk/home/gishome.html) The home of this page and other GIS Resources such as the AGI Dictionary, the Digital Elevation Data Catalogue and research projects including the first Arc/Info <-> WWW interface and the World-Wide Earthquake Locator.
- Edinburgh University, **Data Library,** (http://datalib.ed.ac.uk/) spatial metadata research, UKBORDERS digital data and ESRC Research Publications Database (incl. some GIS research).
- **Environmental Data Center** (http://www.edc.uri.edu/rreapage/edc/)—GPS data and Rhode Island Geographic Information System Data. The site contains information about the Environmental Data Center, Rhode Island GIS data, and GPS Base Station data.
- **Environmental Management Earth Observation Group** (http://www.unn.ac.uk/~evz9/eog/eog.htm) at University of Northumbria, UK.
- **Environmental Systems Research Institute (ESRI)** (http://www.esri.com/). The vendors of Arc/Info and related GIS software and data. Includes Arc View V1.0 for free.
- **ERDAS, Inc.** (http://www.erdas.com/) Visualization and analysis tools to help you make more informed business decisions involving the geography that surrounds you.
- **ER Mapper** (http://www.ctv.es/ermapper) site for Southern Europe and North Africa. ER Mapper is a powerful yet easy to use geographical data viewer and processor.
- **ESRI Canada Ltd.** (http://www.esri.ca/) Official Canadian Distributor of ESRI, world leader in Geographic Information System technology. Resources include GIS products, consulting, training, support, partner program, US.
- **ESRI Canada—Schools and Libraries Program** (http://www.esri.ca/k-12/) ESRI Canada is dedicated to promoting GIS awareness in schools!
- **ETAK—GIS Software and Digital Mapping Information** (http://www.etak.com/) A number of online demos for using maps on the internet, geocoding records over the internet, product descriptions.
- **Eurimage** (www.eurimage.com.) a major commercial course of remote sensing imagery
- **EUROGI** (http://www.eurogi.org/) the European Umbrella Organisation for GIS
- European Science Foundation, **GISDATA Programme** (http://www.shef.ac.uk/uni/academic/D-H/gis/gisdata.html) includes specialist meetings, GISDATA directory, news/outputs.

F

- **GIS in Production Agriculture** (http://www.farmsoft.com/) Farmer's Software Association (FSA) supports information tools for mapping and management of site-specific farming data in a GIS framework. Trainings are also held to educate agriculturalists in the application of GIS technology to farming.

- **Fenstermaker Online** (http://www.fenstermaker.com/) is your access gateway for interacting with our company. Learn about our GIS (Geographical Information Systems), Smart Maps, GPS (Global Positioning System) Surveying services, Engineering capabilities, comprehensive Environmental services, and some of our history.
- National Land Survey of Finland (the Finnish National Mapping Agency), **Geographic Information Centre** (http://www.nls.fi/index_e.html) includes details of their products and standards.
- **Institut Geographique National** (IGN) (http://www.ign.fr/)—the French national mapping agency.
- **Friends of the Earth** (UK) (http://www.foe.co.uk/cri/html/postcode.html) Describes their use of GIS with examples. Of particular interest is their Chemical Release Inventory.

G

- **Genasys II Inc.** (http://www.genasys.com/)—Information about their products.
- **Geodetic Survey of Canada** (http://www.geod.emr.ca/) Information about the organisation and the data products on offer.
- **GEODYNAMIC SOLUTIONS, INC.** (http://www.compassnet.com/gsi) Our company specializes in GIS applications and software solutions for the petroleum industry and for market, demographic and retail site selection analysis.
- **GEO/Graphics** (http://www.geograph.com/docs/html/homestart.html) GIS Consulting and Mapping
- **Geographical Institute (Slovenia)** (http://www.zrc-sazu.si/gi/) Geographic Institute has been established in 1948. It has been divided into 5 organisational units since 1993: Department for Geoecology, Department for Regional Geography, Department for Natural Disasters, Department for Geographical Information System and Department for Thematic Cartography. Activities have been organised into four research projects (basic, evolutional, applicational and aiming).
- **Geological and Geophysical Databases, Information and GIS Map Server, Cornell University** (http://atlas.geo.cornell.edu/) This page is designed to provide information, data, and maps for scientists, students and for those who are interested in the geological sciences.
- **Geomatics International Inc Web** (http://www.geomatics.com/) providing consulting services to private sector institutions, non-governmental organisations and all levels of government using leading edge GIS and web technology.
- **GeoNorth** (http://www.geonorth.com/) GeoNorth is a company owned by professionals dedicated to the development and support of GIS, database, and CAD systems. GeoNorth offers a full range of services including custom GIS and database application programming, GIS mapping and data conversion, and computer network administration. GeoNorth also offers a full line of the world's most popular software products including packages from ESRI, Sybase, and AutoDesk. We are also known for the development and marketing of out own, CD-ROM based products, and for developing software products for third parties.

- **GeoSoft Ltd.** (http://www.geosoft.co.uk/) are the U.K. based suppliers of Map Server and Map Server, which allow turnkey systems to be produced in Visual Basic.
- **GeoSolutions Consulting Inc.** (http://www.geosolutions.com/) GeoSolutions Consulting Inc. provides consulting and technical services in the area of Geographic Information System (GIS).
- **GeoSolutions, GIS Consultants and Application Developers** (http://www.geosolutions.ie/) GeoSolutions provide GIS services internationally. We focus on solutions and services specific to user requirements. GeoSolutions an ESRI Business Partner.
- **GeoSearch** (http://www.geosearch.com/) are a US-based GIS employment agency.
- **GEO/SQL** (http://www.geosearch.com/) Geo/SQL Corporation is an international company committed to providing a high level of consulting expertise in the design, implementation and maintenance of GIS systems.
- **GeoWeb—On-Line Resources for GIS/GPS/Remote Sensing** (http://www.ggrweb.com/) WWW Online Resources For Geographic Information System (GIS), Global Positioning System (GPS), and Remote Sensing (RS) Industries.
- GIS-L **Frequently Asked Questions (FAQ)** (http://www.census.gov/geo/www/faq-index.html) Page, which provides answers to many basic GIS questions.
- **GIS Internet Resources List** (http://www.lib.berkeley.edu/UCBGIS/gisnet.html) Extensive, annotated list of pointers, broken down into subject areas. Also has information on GIS education, research and facilities at UC Berkeley. This is part of the UCBGIS web site at: http://www.lib.berkeley.edu/UCBGIS/
- **GISLinx—complete GIS listing** (http://www.gislinx.com/) GISLinx is an Internet site devoted to listing any/all GIS related web sites. GISLinx contains links to GIS related software and hardware vendors, GeoMatics, GIS Events, and GIS Data.
- **GISL (UK) Ltd.—GIS, Remote Sensing and Informatics Consultants** (http://www.gisl.co.uk/gisl.htm) GISL is a UK firm specialising in the applications of GIS and remote sensing. Working worldwide on muti-disciplinary projects involving natural resources, rural development, environmental and urban planning. GISL provides support and training in most GIS and image processing software packages. We also have a register for freelance consultants with experience in GIS and/or remote sensing.
- **GISplan research group at Aalborg University, Denmark** (http://www.i4.auc.dk/gisplan/) A brief introduction to GISplan and the idea behind our research. There is also useful links to projects and partners around the world. We also have links to our own homepages. Expect to see updates every 3 month. We are committed to use the web for communicating our research and other information.
- **GIS World** (http://www.gisworld.com/), publishers of GIS World magazine and GIS Asia Pacific. Includes GIS conferences, books and other information.
- **Global Land Information System (GLIS)** (http://edcwww.cr.usgs.gov/glis/glis.html) Metadata and samples for many data products, such as AVHRR, TM, DEMs, Aerial Photographs, SLAR, World Ecosystems, and World Soil Series.
- **Global Positioning System Software (GPSS)** (http://www.gpss.co.uk/) Free GPS Software for Windows called GPSS—a mobile speech GIS. Information on what GPSS is, when it will next be on TV, and how to get a free copy, e.g. Personal Computer World CD-ROM Now available for whole World. Displays map and speak.

- **GRID-Geneva (Global Resource Information Database)** (http://www.grid.unep.ch/) GRID-Geneva is part of UNEP (United Nations Environment Programme). It provides GIS service and distributes free geo-referenced environmental data sets.
- **GTI (GIS Technology, Inc.)** (http://www.gistech.com/) All the GIS you can stand! Description of our products and services, articles, white papers, and other GIS links.
- **Geo InSight International, Inc.** (http://www.geoin.com/) GIS Training consulting and software development for private sector, state, federal govt, and environmental orgs.

H

- **Harlow Report Geographic Information Systems** (http://www.geoint.com/) Monthly newsletter on GIS since 1978. Free trail subscription available.
- **University of Hong Kong, GIS/LIS Research Centre** (http://www.hku.hk/cupem/hkugis)

I

- The **IDRISI Project** (http://www.idrisi.clarku.edu/) home page.
- An **IDRISI WWW Tutorial** (http://www.sbg.ac.at/geo/idrisi/wwwtutor/tuthome.htm) from the at the University of Salzburg.
- **ImageNet** (http://www.coresw.com/) from Core Software Technology—Search and Preview geospatial data (sat. imagery, air photography, vector data). There is a European node at **Eurimage** (http://www.eurimage.com/) in Italy.
- **Infomine Maps and GIS** (http://lib-www.ucr.edu/mapsinfo.html) University of California index to internet resource collections—access by subject, title, and keyword. Strong focus on California and the western United States.
- **Informed Solutions GIS and Spatial Data Consultants** (http://www.informed.co.uk/) A Group of Consultants specializing in the development of all leading GIS and RDBMS products. The leading independent consultancy in the country.
- **Institute of Advanced Technologies (Kyiv, Ukraine)** (http://www.iat-ant.com/) is the only organization in Ukraine, which is designed to develop geographical maps and atlases with help of exceptional GIS-technologies.
- **Intergraph Corporation** (http://www.intergraph.com), GIS product applications and Industry Solutions for Business and Governments.
- Iowa Department of Natural Resources, **Geological Survey Bureau** (http://samuel.igsb.uiowa.edu/). Information on geological and water resources in Iowa, including their Natural Resources GIS (with >400 documented databases) and publications.
- **International Institute for Aerospace Survey and Earth Sciences** (http://www.itc.nl/) The International Institute for Aerospace Survey and Earth Sciences (ITC) is the largest institute for international higher education in the Netherlands. The main objective of ITC is to assist developing countries in human resources development in aerospace surveys, remote sensing applications, the establishment of geoinformation systems and the management of geoinformation. To this end ITC concentrates on three main activities: education, research and advisory services.

- **IntelliGIS Inc.** (http://www.igis.com/) provides the Energy Industry with GIS Software, including a data navigator for geoscience databases and a system for the protection of underground utilities.
- **ISO/TC 211 Geographic Information/Geomatics** (http://www.isotc211.org/) This is the web-server of the ISO committee for geographic information/geomatics. Contains work programme, organization including working groups and experts, important information, calendar, list of documents, etc.

J

- **James W. Sewall Company** (http://www.jws.com/) Mapping, GIS, and resource management consultants to government, utilities, and private industry. Expertise in aerial photography, surveying/GPS, photogrammetric mapping, and orthophotography.
- The **Japan GIS/Mapping Sciences Resource Guide** (http://www.cast.uark.edu/jpgis/), which is a comprehensive reference for Japan's GIS, map and geospatial data products, activities and information sources.
- **GIS Jobs and Resources** (http://www.geojobsource.com/) An international listing of career opportunities in the spatial data professions and a growing directory of professional development resources.
- **GISLinx Job Department** (http://www.gislinx.com/) The GJD is a rapidly growing listing of GIS Job openings and GIS resumes from around the world. This listing is searchable, or may be viewed by categories.
- **JShape—Publish GIS to WWW** (http://shiuhlin.ml.org/) JShape is a Java applet to allow users to publish their GIS shapefile to WWW. JShape is a vector-based shapefile browser, which supports zoom/pan/query/select/label/... most GIS functions.

K

- **Land Info** (http://www.landinfo.com/) Provider of map data products. Land Info's Web site describes the company and its products, and includes a number of free downloadable sample map data product files.
- **Laser-Scan Ltd.** (http://www.lsl.co.uk/) are a U.K. GIS systems vendor. Details on products, case studies and technical papers.

M

- **MAPCOM Systems, Inc** (http://www.mapcom.com/) AM/FM-GIS Windows based software for utilities that Works! Descriptions of M4, Windows based AM/FM-GIS Software, applications and services.
- **Map*Factory—Raster Based GIS Modelling for the Macintosh** (http://www.keigansystems.com/) Keigan Systems specializes in the development of Macintosh spatial modelling, image processing, and GIS mapping products. Software tools include Map*Factory, Map*Factory Developer's Kit and MAP II.
- **MapInfo Corporation** (http://www.mapinfo.com/) MapInfo Corp website

- **Map Maker** (www.mapmaker.com) is a user friendly GIS for Windows.
- **MAPublisher** (http://www.avenza.com/) by Avenza Software Marketing Inc. MAPublisher is a software program that works with Adobe Illustrator and Macromedia Freehand to allow the import of GIS data into the graphics programs.
- **Massachusetts Electronic Atlas** (http://icg.harvard.edu/~maps/maatlas.htm) This is an interactive atlas showing data by minor civil division with over 200 thematic layers.
- **McGill University Department of Earth Sciences** (http://stoner.eps.mcgill.ca/) Some departmental information, together with an ftp server and a gopher server.
- **Meridian GIS—Consultants for Mineral Exploration** (http://www.meridian-gis.com.au/) Meridian GIS was formed in January 1997 to service the mapping and GIS community in Australia. We aim to supply MicroStation support in the mineral exploration and cartographic areas and more specifically the integration with GIS.
- **MicroImages, Inc.** (http://www.microimages.com/) Free TNTlite spatial data viewing and analysis software, plus the whole line of TNT professional GIS, Image Processing, and desktop cartography products.
- **Minnesota Legislative GIS Office** (http://www.commissions.leg.state.mn.us/). Information and mapping for Minnesota. Not much of great interest as yet.
- **Mississippi Automated Resource Information System** (http://www.maris.state.ms.us/) GIS clearing house for Mississippi. Online data, extensive metadata and links to other GIS sites around the country.
- **GIS/Census data for Missouri** (http://msdis.missouri.edu/) GIS datasets in ARC/INFO.e00 format for Missouri at the county level as well as the state level. Home pages for MGISAC and Missouri GIS user groups also available.
- **MIT Earth Resources Laboratory** (http://eaps.mit.edu/erl/) Environmental Engineering, Applied Geophysics, Centre for Advanced Geophysical Computing, and GPS information.
- **MultiGIS** (http://ourworld.compuserve.com/homepages/mgis) 3D photo realistic terrain visualization and DEM and planimetric digitisation. We offer FREE datum and coordinate software. Article preprints.
- **Municipal Software Corporation** (http://www.islandnet.com/municipal/) has been providing superior software solutions to local government for over 12 years.

N

- The U.S. **National Center for Geographic Information and Analysis (NCGIA)** (http://www.ncgia.ucsb.edu/)—Projects, References, Reports and Bibliographies relating to a range of GIS themes
- **State University of New York (SUNY) at Buffalo** (http://www.geog.buffalo.edu/ncgia/)
- **National Key Centre for Social Applications of GIS, Australia** (http://www.gisca.adelaide.edu.au/)
- **Natural History Museum** (http://www.nhm.ac.uk/) (U.K.)—Not much on GIS as yet, but we suspect they are working on it.
- The U.S. State of North Carolina's **Center for Geographic Information and Analysis** (http://cgia.cgia.state.nc.us/)—Spatial data and information about the state GIS in North Carolina.

- **Natural Resources Development Centre** (http://www2.tcd.ie/Natural_Resources/) The NRDC brings to environmental management and resource development an understanding of not only natural and human processes but also an expertise in information systems. By combining these abilities its develops new tools and provides new solutions for the optimal use of the environment.
- **NOAA National Geophysical Data Center** (http://www.ngdc.noaa.gov/) A major provider of environmental data (international data in addition to USA).

O

- Oak Ridge National Laboratory (U.S.) **Environmental Science Division,** (http://www.esd.ornl.gov/ern/embyr/embyr.html) including their **Fire Simulation Model (EMBYR)** (http://www.esd.ornl.gov/ern/embyr/embyr.html)
- **Object/FX Corp.—Embedded Tools and Interactive Internet Spatial Applications** (http://www.objectfx.com/) A software company offering spatial/mapping visualization tools and templates for viewing and interacting with enterprise data in either client/server or web environments.
- **Ohio State University—Center for Mapping** (http://www.cfm.ohio-state.edu/)
- Ohio State University, **Department of Geography** (http://thoth.sbs.ohio-state.edu/) provides a major GIS On-Line Bibliography.
- **On Target Mapping** (http://www.otmapping.com/) is a leading provider of desktop mapping solutions for service-intensive industries and is a leader in telecommunications mapping, routing, logistics and scheduling technologies.
- **Open GIS Consortium** (http://ogis.org/). Information on the OpenGIS Project, meetings, testbed and publications. OGIS also have pages at the **University of California at Berkeley** (http://www.regis.berkeley.edu/ogis/ogis.html) which include information on the public domain GIS.
- The **Ordnance Survey** (http://www.ordsvy.gov.uk/)—the U.K. National Mapping Agency. Including a query interface to the Spatial Information Enquiry Service (SINES).

P

- **PCI** (http://www.pci.on.ca/) produces software for Remote Sensing, GIS and Terrain Analysis, including EASI/PACE. Cartographic editing software in WYSIWYG environment reads 45+GIS file types. Produces high quality maps. Designed by Cartographers for cartographers.
- **PCI Pacific GeoSolutions, developers of PAMAP GIS** (http://www.pci-pacific.com/)
- **Pennsylvania Spatial Data Access** (http://www.pasda.psu.edu/) PASDA is PA's node on the NSDI. PASDA data includes Digital Elevation Models, Digital Raster Graphics, and Digital Ortho Photo Quadrangles as well as watersheds, streams, roads, etc.
- **Professional GEO Systems (PGS) Product Info (Java GIS Browser)** (http://www.pgs.nl/) PGS specializes in the development of Geographic Information Systems for Internet. We have developed a 100% Pure Java GIS browser.
- **PROGIS—Professional GIS software for MS Windows** (http://www.progis.com/) Professional GIS platform for Windows95/NT, add-on modules for GPS, Routing,

Multimedia, large Raster handling (TurboRaster) and Isoline generation. GIS downsizing with SDK, connection to any database.

R

- **GIS Conversion and Software Development** (http://www.ramtech-corp.com/) GIS Conversion and GIS Software Development Service Provider for Telecommunications, Electric, Gas, Water, Government, Mapping, Municipalities etc.
- **Retail Profit Management GIS Resources** (http://home.earthlink.net/~rpminfonet/gis.html) RPM, a provider of Atlas GIS solutions sponsors this comprehensive listing of GIS data sites, news and discussion groups, GIS and mapping related search engines, and more.
- **GIS data for Rhode Island** (http://www.edc.uri.edu/) Wetland, land use, images, roads, soils and utilities.
- The U.K. Regional Research Laboratory Initiative: **RRL Scotland** (http://www.geo.ed.ac.uk/home/research/rrl.html) Server includes information about RRL Scotland and links to the other RRLs.
- **RouteSmart Technologies (TM)** (http://www2.ari.net/distinct/) RouteSmart is a cutting edge vehicle routing and scheduling system for organizations that demand reductions in vehicle use, labour time and operating expenses.

S

- **SAFE Software Inc.** (http://www.safe.com/fme) Universal Spatial Data Translator—These pages contain a description, live demo, and downloadable version of the Feature Manipulation Engine (FME), which is a configurable, 'universal' Spatial and Attribute Data Translator. The FME supports most popular GIS formats, and more are being added all the time.
- **Sandia National Labs (USA) GIS Site** (http://www.sandia.gov/) Environmental, security, risk, transportation applications. ESRI ARC/INFO and ArcView technical information and code. MapObjects information.
- **Society of Cartographers** (http://www.shef.ac.uk/uni/projects/sc/index.html) Cartographic site organised by CARTOGRAPHERS for Cartographers and those interested in cartography and maps. Also has a listserver "CARTO-SOC" used by cartographers as an information and problem sharing resource.
- **South Carolina Dept. of Natural Resources GIS Data Clearinghouse** (http://www.dnr.state.sc.us/) Provides GIS data (soils, wetlands/land use, DLGs, DEMs, and DRGS) for public use in natural resource decision-making. All data is in Arc/Info EXPORT format and 1:24,000 scale.
- **State of Kansas GIS Data Access and Support Center** (http://gisdasc.kgs.ukans.edu/dasc.html)
- **The UK Street Map Web Site** (http://www.streetmap.co.uk/) Provides Street Maps of Greater London and Road Atlas coverage for the rest of the UK. Search by Post Code, Street Name, OS Grid & Landranger Grid (Lat/Long coming soon). Link to maps by Post Code or OS Grid.

- **Subway Navigator** (http://metro.jussieu.fr:10001/), An clever resource for navigating your way through the subway/underground systems of the world.
- Swiss Federal Institute of Technology—**GERMINAL Research Project** (http://dgrwww.epfl.ch/GERMINAL/Germinal.html), designing an Integrated Approach to Land Use Planning and Environmental Management.

T

- **TCI Software** (http://www.tcicorp.com/) An AutoDesk Registered Application Develope specializing in tools for creation, conversion and maintenance of AutoCAD basemaps since 1988.
- **Thomas Bros. Maps** (http://www.thomas.com/) of Irvine, California. Details on their PC GIS called GEOFINDER and their digital data.
- **TYDAC Research Inc.** (http://www.tydac.com/)

U

- **UNEP/GRID-Sioux Falls** (http://grid2.cr.usgs.gov/) Global Population Data Sets, Environmental Data-Sets Available On-Line, Digital Topographic Data, EDC's Landsat Archive and Landsat Pathfinder, Gap Analysis, Desertification, Land Cover Characterization.
- **The University of Arizona ARC/INFO User Group** (http://www.srnr.arizona.edu/UAAIUG/) The University of Arizona ARC/INFO User Group (UAAIUG) was established in 1994 to develop a forum for faculty, staff and students at the University of Arizona to share knowledge and ideas pertaining to Environmental Systems Research Institute (ESRI) software applications.
- United Nations Environment Programme (UNEP) **Global Resource Information Database (GRID) at Arendal, Norway** (http://www.grida.no/). Provide access to a number of environmental GIS databases.
- **The Center for Advanced Spatial Technologies (CAST)** (http://www.cast.uark.edu/) at the University of Arkansas.
- University of California at their **Research Program in Environmental Planning and Geographic Information Systems** (http://www.regis.berkeley.edu/) (REGIS). Information is available on the San Francisco Bay/Sacramento Area, also **GRASSLinks** (http://www.regis.berkeley.edu/grasslinks/index.html)—a WWW-GRASS interface.
- **University of California, Riverside Extension GIS Certificate** (http://www.unex.ucr.edu/GIS/GIS.html) Pages contain information regarding the curriculum of the innovative Certificate in Geographic Information Systems at UCR, the proposed Spring 1996 Institute and the current course offerings in GIS.
- **University College London, Dept. Photogrammetry and Surveying** (http://www.ps.ucl.ac.uk/) Information about research projects and master degrees in Photogrammetry, Surveying, GIS and Remote Sensing at UCL.
- **University of East Anglia, School of Environmental Sciences** (http://enva2.env.uea.ac.uk/)

- **University of Extremadura. Center of Teledetection and GIS.** (http://geot.unex.es/) Working group in GIS projects and digital image treatment, from satellite data, inside of the University of Extremadura in Spain.
- **University of Glasgow, Dept. of Geography and Topographic Science** (http://www.geog.gla.ac.uk/) Department Information, Undergraduate & Postgraduate courses, Staff Information, Specialist Department for cartography, digital mapping, GIS/LIS, remote sensing, photogrammetry, geodesy and surveying, Research Activities and Prospectus Information.
- **University of Leicester Department of Geography Research** (http://www.geog.le.ac.uk/argus/) Project Argus—Visualization for Spatial Scientists. These pages demonstrate the use of interactive computer visualization of spatially distributed data as a technique for helping teaching, learning and research in a variety of subjects, such as geography, geology, meteorology and the social sciences.
- **University of Maine, Spatial Information Science and Engineering** (http://www.spatial.maine.edu/) provides bibliographies to major GIS proceedings, full text of a GIS law proceedings, describes ongoing research and academic programs.
- **Department of Geography, University of Minnesota** (http://www.geog.umn.edu/) Complete description of degree programs in Geography at the University of Minnesota. Information for prospective and current students.
- **University of Minnesota Remote Sensing Laboratory** (http://www.gis.umn.edu/) for AML programs, general GIS/RS information, and GIS Jobs Clearinghouse.
- **University of Newcastle** (http://www.ncl.ac.uk/~ngraphic/) Information from the University Computing Service on Graphics and GIS at Newcastle University, including data, software, hardware and usage issues.
- **University of Salzburg, Institute for Geography** (http://www.sbg.ac.at/geo/home.htm) provide information which includes the UNIGIS German distance-learning diploma, and are an IDRISI Resource Center (http://www.sbg.ac.at/geo/idrisi/irchome.htm) (German language) which hosts an **IDRISI WWW Tutorial** (http://www.sbg.ac.at/geo/idrisi/wwwtutor/tuthome.htm) in English.
- University of South Carolina, teaches GIS at the **College of Liberal Arts** (http://www.cla.sc.edu/index.html#GIS)
- **University of Southampton, GeoData Institute** (http://www.geodata.soton.ac.uk/) is a research organisation and contract agency which provides tutorials and training aids.
- **University of Stuttgart, Institute for Photogrammetry** (http://www.ifp.uni-stuttgart.de/) Image processing, GIS projects and interests, and a GIS bibliography.
- **University of Virginia, Social Sciences Data Center** (**http://fisher.lib.virginia.edu/**) Virginia On-Line Atlas, Virginia Census Data.
- **State University of Utrecht, Faculty of Geographical Sciences** (http://www.frw.ruu.nl/) (The Netherlands) and their list of **"Nice" Geography/GIS/RS Servers** (http://www.frw.ruu.nl/nicegeo.html).
- **Urban Planning Job Links** (http://www.geocities.com/CollegePark/7442/jobs.html) This is the most complete collection of Urban Planning Job Links on the net. It includes a GIS section.
- **U.S. Army—Construction Engineering Research Laboratory (CERL)** (http://www.cecer.army.mil/) The home of the **GRASS** (http://www.cecer.army.mil/grass/

GRASS.main.html) public-domain GIS, the Open-GIS (OGIS) foundation, also spatial modelling information.

- **U.S. Bureau of the Census** (http://www.census.gov/)—General information and access TIGER census data and much other statistical information for the U.S.A.
- **U.S. Bureau Land Management** (http://www.blm.gov/)—includes a GIS page with data and information relevant to land management.
- **U.S. Environmental Protection Agency** (http://www.epa.gov/) includes information on projects using GIS such as their **National Estuary Program** (http://www.epa.gov/nep/nep.html). Also some information on software and data.
- The **U.S. Fish and Wildlife Service** (http://www.fws.gov/) supports the U.S. **National Wetlands Inventory** (http://www.nwi.fws.gov/) which provides access to GIS data relating to US Wetlands in a number of different formats.
- **U.S. Geological Survey (USGS)** (http://www.usgs.gov/), for general GIS information, application examples and lots more.
- **USGS Mid-Continent Mapping Center at Rolla, Missouri** (http://mcmcweb.er.usgs.gov/) USGS Map Products and Digital Data.
- **USGS EROS Data Center** (http://edcwww.cr.usgs.gov/)—Major Data Archive for world data, satellite imagery and etc.
- U.S. National Biological Service **Environmental Management Technical Center (EMTC)** (http://www.emtc.nbs.gov/http_data/image_maps/umrsys.html)—GIS Data Browsing and Retrieval System for the Upper Mississippi River System.
- **Universal Systems Ltd.** (http://www.universal.ca/) A software and systems integration company that develops, supports and markets, the CARIS GIS software.

V

- **Vermont Geographic Information System** (http://www.vcgi.org) The Vermont Center for Geographic Information offers GIS data for the state of Vermont.
- **Plug-in Visual WebMap for CAD/MAPS/GIS on WWW** (http://hem.passagen.se/project/npwebmap.htm) WWW Visual WebMap Plug-in for IT/CAD/GIS A Client/Server Concept for Intergraph or Bentley Systems vector and raster files or vector in database with API for user development.
- **Vysor Integration Inc.** (http://www.synapse.net/~vysor/)—V-Image Remote Sensing Software for Windows and downloadable Satellite Imagery.

W

- **Wageningen Agricultural University Department of Geographic Information Processing and Remote Sensing (GIRS)** (http://www.gis.wau.nl/girs/)
- Oliver Weatherbee's **GIS Software Listing** (http://triton.cms.udel.edu/~vinton/gis_gip/gis_gip_list.html) which contains contact information and short overviews of the majority of major GIS software packages currently available.
- **WWW Geospatial Metadata Server (GMS)** (http://www.geo.ed.ac.uk/~anp/gms/main.htm) On-line Demonstration software that includes a CSDGM metadata database,

metadata upload/parse/database entry, editing of database rows, data conversion utilities, a software robot for data mining and a coverage plotting utility.

- **WinCATS—GIS and Engineering Services** (http://www.appliedglobal.com/) WinCATS—Windows Cartographic Analysis Tool Set Powerful/Easy to use GIS for Windows.
- **Wireless Solutions through GIS** (http://www.geotranswireless.com/) Accelerated telecommunications PCS build-out solutions using GIS to site antenna facility identification, acquisition and construction management.
- **WorldSat International—satellite underlay for GIS** (http://www.worldsat.ca/) Complete cloud free images of the entire earth available as prints for display or in digital format for a GIS layer.

Additional Links

- **Geographic Engineering** (http://www.fleximage.fr/) FLEXIMAGE is a subsidiary of the French AEROSPATIALE MATRA Group. FLEXIMAGE is specialized in Image Processing software development and new technologies in GIS for Defence applications
- **MAPIX technologies Ltd** (http://www.mapix.com/) Hydrographic GIS applications using ESRI products, Oceanographic Databases for use in IS.
- **Update Description of Geodan** (http://www.geodan.nl/) Please skip the last part of the description: "Also on our site: Road planner, the planning tool for service engineers."
- **Versamap Digital Mapping—Cartography and Outline Maps** (http://www.versamap.com/index.htm) Home of the Versamap mapping program for drawing outline maps on many map projections, and of digital map databases designed for use with Versamap.
- **GIS Data for Australia and the World** (http://www.geomap.com.au/) Our extensive experience with GIS projects enables us to help clients design a GIS for their specific requirements and where necessary integrate GIS into existing operating procedures.
- **GIS Data Depot** (http://www.gisdatadepot.com/) Provides FREE GIS data downloads from over a terabyte of GIS data and creates custom spatial data CD-ROMs as an alternative.
- **Low-cost Desktop Digital Photogrammetry System (DDPS) for Windows** (http://www.3dmapper.com/) The DDPS is a fully featured, cost-effective, desktop soft photogrammetry solution that enables users to quickly and efficiently capture accurate 3D vector data and ortho-images...
- **Mapping Software Developer** (http://www.mapframe.com/) MapFrame Corporation develops smart simple solutions for field automation. The company's FieldSmart Solutions include map viewing, data collection, pole inspection, and mobile distribution design.
- **Innogistic Software plc** (http://www.innogistic.co.uk/) Developers and Suppliers of Geographical Information Systems, Land and Property Systems and AVL with two way data messaging.
- **The GeoCommunity** (http://www.geocomm.com/) Central location for GIS Data, News, Discussions, and much more.

- **Demographic and Spatial data products** (http://www.geolytics.com/) Data products on CD-Rom: CensusCD, CensusCD+Maps, CensusCD 1980, CensusCD Blocks, StreetCD (complete TIGER98)
- **Immense GIS software and GIS data site.** (http://www.manifold.net/)
- **Corporate GIS Consultants Web Site** (http://www.corp-gis.com.au/) An overview of Corporate GIS Consultants.
- **The only Interactive Webmap of Indonesia** (http://www.indomap.com/) If you want to find islands, mountains, straits, bays and street maps of Indonesian cities, find it here. And you can have your building/house (if it in Indonesia) to have a FREE page with the location map of course.
- **GIS Data Conversion** (http://www.kineticsolutions.net/)
- **Infrasoft Solutions Ltd—GIS and Engineering Solutions** (http://www.infrasoft-civil.com/) Company specialising in 3D data, OS products, 3D modelling and software for road, rail, air and GIS applications used in 62 countries.
- **ER Mapper Image Processing System** (http://www.ermapper.com/) Information on ER Mapper, ECW compression and Image Web Server Free GIS and other software plugins, Free software such as ER Viewer and Compressor and a free version of Image Web Server.
- **Image Web Server by Earth Resource Mapping** (http://www.earthetc.com/) Roam and zoom in realtime over terabytes of imagery using this revolutionary Image Web Server. Imagery ranges from scanned maps, aerial photo's and satellite data to traditional photography.
- **NYPIRG's Community Mapping Assistance Project** (http://www.cmap.nypirg.org/)
- **UNISIG Ecuador** (http://unisig.epn.edu.ec/) Unidad de Inteligencía Artificial y Sistemas de Información Geográfica (UNISIG) parte de la Facultad de Ingeniería de Sistemas de la Escuela Politécnica Nacional, Quito, Ecuador.
- **Discover the Benefits of a Proven GIS Solution.** (http://www.geosql.com/) Geo/SQL provides a cost effective way to create maps, integrate information, present powerful ideas, and develop effective solutions like never before.
- **GIS Software development recruitment consultants** (http://www.cityprojects.com/) Positions with the leading developers of GIS systems within the UK.
- **AM/FM/GIS for Telecommunications and Utility Companies** (http://www.enghouse.com/)

  Provide suites of specialist GIS software under the brand names TelcoWORKS, CATVWORKS, PowerWORKS & GasWORKS. For design and management of network infrastructures.
- **GIS applied to Marine Geology and Oceanography** (http://www.marinegis.com/) Resources for GIS systems applied to marine geology and oceanography (links to softwares, data, jobs, universities pages, etc.).
- **Remote Sensing Data and Geographic Information Systems** (http://lee1.en.a.u-tokyo.ac.jp/Rs&gis/Engl/Read_me.htm)
- **ENVI—The Environment for Visualizing Images** (http://www.rsinc.com/envi/index.cfm) Research Systems' remote sensing and image processing software specialising in hyperspectral analysis.

- **RiverTools—Surface Water and DEM modelling software** (http://www.rsinc.com/rivertools/index.cfm) Research Systems' hydrologic modelling software that can handle very large DEM's.
- **CgiMapScript** (http://imedia.king.ac.uk/) CGI script driven interactive web mapping using vector data to drive GIF & HTML generators or Java 'mapplets' derived from research into public transport information systems.
- **CAD/GIS/RemoteSensing/Digital Photogrammetry Solutions for 2000** (http://www.tricadinfo.com/) Total end to end Geomatics solutions.
- **The Engineers Resource** (http://engineersupply.com/) GIS Books, Engineering Reference, Products, Services, and Forums (discussion and chat) for Engineers of all types.
- **Devel-Tech Inc.** (http://www.develtech.ca/) Integrated Services in Geomatics and Information Management.
- **Smallworld** (http://www.gepower.com/dhtml/network_solutions/)
  Smallworld creates, markets and delivers spatial resource planning (SRP) software products to the Utility and Public Systems and Communications markets globally.
- **Planet Earth Home Page** (http://www.tidusa.com/PEHP2000/Planet_Earth/info.html) Connections to everywhere on every topic.
- **GIS Staff Consultants** (http://www.cityprojects.com/)
- **Mapping Services** (http://www.specksystems.com/)
  We offer services in mapping and GIS. We offer total solutions starting from data acquisition to the making of a final map
- **GIS and Mapping Services** (http://www.specksystems.com/)
  Facilities and capabilities of Speck Systems Limited, Hyderabad, India in the areas of GIS and mapping.
- **Florida Dept. of Environmental Protection GIS Section** (http://www.dep.state.fl.us/gis)
  Florida GIS data, satellite imagery, links to other GIS clearinghouses.
- **MapMart** (http://www.mapmart.com/) Online ordering of Digital Mapping and Aerial Photography. Specializing in USGS products as well as customized mapping. Provides aerial photography acquisition, photogrammetry services, digital mapping, ortho rectification, scanning and custom GIS solutions for public and private sector clients.
- **GIS Research Center Inasitute of Geograhy Russian AS** (http://geocnt.geonet.ru/)
  GIS—software, maps, data and projects, consultancy in Russia The most popular GIS software in Russia GeoDraw/GeoGraph/GeoConstructor ActiveX (2500 installations).
- **Developer of Mapping and Geographic Data Technologies** (http://www.chicagomap.com/) Developer of mapping and geographic data technologies. Key disciplines include vector map displays, geocoding, data compression, and other related software efforts.
- **Journalism and GIS** (http://groups.yahoo.com/group/JAGIS-L/)
  A resource site for journalists interested in using GIS for the analysis of a variety of phenomena and communicating of the results of that analysis.
- **GIS and GPS products and online support** (http://www.geoposition.com/)
  Products from ESRI, Trimble, Juniper, EDS. Support lists, data resources, forums, etc...

Recommended GIS and Related Books

Title	*Author*	*Description*
	1. Business GIS	
ArcView GIS Means Business: Geographic Information Systems Solutions for Business	Harder, Christian	Written for business professionals, this book presents success stories chronicling the application of desktop geographic information systems (GIS) to a variety of common business problems.
GIS for Business and Service Planning	Paul Longley (Editor),Graham Clarke (Editor)	Renowned contributors assess the links between technological change, analytical information and data customization which are now beginning to stimulate the wider adoption of GIS as a management and applied research tool.
GIS in Real Estate: Integrating, Analysing, and Presenting Locational Information	Gilbert H. Castle (Editor),Foreword by Joseph R. Stanfield	Advocates the use of Global Information. Systems (GIS) by real estate professionals, and demonstrates how this use can be achieved.
	2. GIS Developer Tutorials	
Focus on GIS Component Software: Featuring ESRI's Mapobjects	Hartman, Robert	Explaining what GIS component technology means for managers and developers, this text illustrates, through hands-on tutorials using Visual Basic and ESRI's MapObjects product, how to use GIS software components to create, display, and query geographic data. The CD-ROM includes an evaluation copy of Map Objects and tutorial code.
ArcView GIS/Avenue Developer's Guide, Third Edition	Razavi, Amir H.	ArcView GIS/Avenue Developer's Guide continues to offer readers one of the most complete introductions to Avenue, the programming language of ArcView GIS.

(contd.)

(contd.)

Title	Author	Description
		3. Cartography
How to Lie with Maps	Mark Monmonier	Originally published to wide acclaim, this lively, cleverly illustrated essay on the use and abuse of maps teaches us how to evaluate maps critically and promotes a healthy skepticism about these easy-to-manipulate models of reality.
AAA's Atlas of Population and Environment	Paul Harris, Fred Pearce	The AAAS Atlas of Population and Environment is an important analysis of the relationships between human population and the environment.
Longitude: The True Story of a Lone Genius Who Solved the Greatest Scientific Problem of his Time	Dava Sobel	Longitude is the dramatic human story of an epic scientific quest, and of John Harrison's forty-year obsession with building his perfect timekeeper, known today as the chronometer.
Cartographic Design Using ArcView GIS	Madej, Ed	Both an effective primer on digital map design, as well as a classic software tutorial, this book makes particular reference to cartographic methods available in one of today's most popular GIS applications software packages: ArcView GIS from ESRI, Inc. Structured around fundamental concepts of map design, each stand—a lone chapter introduces the reader to a general map design theory followed by opportunities to apply these theories.
Thematic Cartography and Visualization	Slocum, Terry A.	This comprehensive book bridges the evolution in cartography by presenting both traditional methods and recently developed computer-based cartographic techniques.
Mapping It Out: Expository Cartography for the Humanities and Social Sciences	Monmonier, Mark	Writers know only too well how long it can take—and how awkward it can be—to describe spatial relationships with words alone. And while a map might not always be worth a thousand words, a good one can help writers communicate an argument or explanation clearly, succinctly, and effectively.
Elements of Cartography	Robinson, Arthur H., Muehrcke, Phillip C. (Unk)	Recognized as the classic resource in cartography, this text continues to integrate the latest modern technology with traditional cartographic principles.

(contd.)

(contd.)

Title	Author	Description
Geographical Information Systems and Computer Cartography	Jones, Christopher B.	Presents the fundamental concepts in Geographical Information Systems (GIS), emphasizing the techniques in management, analysis, and graphic display of spatial information. (Paper)
	4. Data Conversion	
GIS Data Conversion: Strategies, Techniques, and Management	Hohl, Pat	This book provides an overview of the issues involved in GIS data conversion projects, including understanding and locating data; allocating personnel, budgets, and time; documenting the process; becoming familiar with the target system, and safeguarding data quality.
	5. Environment	
Air Apparent; How Meteorologists Learned to Map, Predict, and Dramatize Weather	Monmonier, Mark	Air Apparent is the singular history of the weather map, one that has developed around the twin poles of weather's many facets and the public's varied needs.
Cartographies of Danger: Mapping Hazards of America	Mark S. Monmonier	No place is perfectly safe, but some places are more dangerous than others. As Mark Monmonier shows in this entertaining and immensely informative book, maps can tell us a lot about where we can anticipate certain risks, but maps can also be dangerously misleading.
Environmentalism Unbound: Exploring New Pathways for Change	Robert Gottlieb	In Environmentalism Unbound, Robert Gottlieb proposes a new strategy for social and environmental change that involves reframing and linking the movements for environmental justice and pollution prevention.
Resource Wars: The New Landscape of Global Conflict	Michael T. Klare	This sobering look at the future of warfare predicts that conflicts will now be fought over diminishing supplies of our most precious natural resources.
	6. GIS/GPS Primers	
Understanding GPS: Principles and Applications (Artech	Kaplan, Elliott D.(Edt)	This practical design guide provides an in-depth description of GPS modern digital receiver, operation and explores future applications in air,

(contd.)

(contd.)

Title	*Author*	*Description*
House Telecommunications Library)		land, and marine navigation.
Fundamentals of Geographic Information Systems	DeMers, Michael N.	This comprehensive look at GIS doesn't overwhelm with details, equations, or computer codes.
The GIS Bcok 5th Edition	Korte, George	Discover what a geographic information system (GIS) is, how it applies to a variety of professions, and the benefits of using GIS for a variety of applications.
Principles of Geographical Information Systems	Burrough, P.A., Burrough, P.A.	Spatial Information Systems (Paper)
The GIS Book	Korte, George B.	Proven through three highly praised editions, The GIS Book, 4th Ed. is a completely revised and greatly expanded resource for anyone who needs to understand what a geographic system is, how it applies to their profession, and what it can do.
Extending ArcView GIS: With Network Analyst, Spatial Analyst, and 3D Analyst	Ormsby, Tim, Alvi, Jonell	Extending ArcView GIS is the companion workbook to ESRI's best-selling Getting to Know ArcView GIS. Taking up where that book leaves off, it explores the three most popular ESRI ArcView GIS extensions—ArcView Network Analyst, ArcView Spatial Analyst, and ArcView 3D Analyst.
Getting to Know ArcView GIS: The Geographic Information System (GIS) for Everyone	ESRI Press	Here's your opportunity to test drive ArcView GIS, the desktop geographic information system from ESRI, the leading global provider of GIS software. The CD packaged with this introductory book includes a working demonstration copy of ArcView software, a hands-on ArcView tutorial, and informative multimedia presentations.
GIS for Everyone—Exploring Your Neighbourhood and Your World with a Geographic Information System	Davis, David E.	Exploring your neighbourhood and your world with a Geographic Information System... This book with its illustrated exercises and CD-ROM support material, explain how GIS takes geographic information and makes it visually understandable to users.

(contd.)

(contd.)

Title	Author	Description
		7. GIS Tutorials
Map basic Developer's Guide	Whitener, Angela., Ryker, Breck	MapBasic Developer's Guide is a handbook for customizing MapInfo Professional. The book begins with a tutorial on MapBasic elements, the MapBasic development environment, and program building basics.
Inside MapInfo Professional: The Friendly User Guide to MapInfo Professional	Daniel, Larry/ Loree, Paula	Now fully updated to the software's latest features and functions, INSIDE MapInfo Professional, 2nd Edition continues to set the standard for desktop mapping tutorials.
Getting to Know ArcGIS Desktop: The Basics of ArcView, Arceditor, and Arcinfo	Tim Ormsby, Eileen Napoleon, Carolyn Groessl	ArcView is the world's most widely used Geographic Information Systems (GIS) software. Version 8 is the most significant upgrade to ArcView since its inception—it has been completely redesigned and engineered to be an easy-to-use, fast, modern, and powerful GIS, and requires a new guidebook for all users.
The ESRI Press Dictionary of GIS Terminology	ESRI Press	The ESRI Press Dictionary of GIS Terminology gives clear, concise definitions of over 1200 terms from GIS and related fields such as geography, GPS, cartography, remote sensing, photogrammetry, and geodesy.
Inside MapInfo Professional	Larry Daniel, Paula Loree, Angela Whitener	Inside MapInfo Professional is for desktop mapping neophytes who wish to quickly become proficient with MapInfo Professional software, MapInfo Corporation's desktop mapping market leader.
		8. GIS-Related Programming
Professional Java E-Commerce	Ronald Ashri, Robert Flenner, Tracie Karsjens, Ma Kerzner,Ronald Ashri,Chad Darby	This book has been designed for professional Java programmers who wish to leverage their in-depth knowledge of J2EE technologies to develop e-commerce solutions, either for consumer sites, or business-to-business scenarios.
Palm OS Programming for Dummies	Liz O'Hara, John Schettino	A one-stop reference for writing applications for the Palm platform used in the popular hand-held organizers.

(contd.)

(contd.)

Title	*Author*	*Description*
	9. Census	
Ecological Census Techniques: A Handbook	William J. Sutherland (Editor)	Virtually any exercise in ecology will require some knowledge of the techniques for carrying out a census of population numbers. This practical text outlines clearly, with worked examples, the main techniques used by field ecologists to enumerate plants and animals.
State Census Records	Ann Smith Lainhart	State censuses rank with federal censuses as a major genealogical resource, but, because they were taken randomly, remain a much under-utilized resource in American genealogy.
Counting on the Census?: Race, Group Identity and the Evasion of Politics	Peter Skerry	Skerry provides a skilful and very timely historicy of the various controversies involving the census and then lays out, and attemps to assess, the contemporary debate.
American Census: A Social History	Margo J. Anderson	This work examines the "history of the American census...how it was taken as well as an analysis of those... who made the decisions of how to gather, compile, and use the statistics.
	10. Adventure—With or Without GIS	
What the Ice Gets: Shackleton's Antarctic Expedition, 1914-1916	Melinda Mueller, Foreword by Gary Holthaus	Blending historical, scientific, and literary scholarship with an impressive range of poetic forms, Melinda Mueller masterfully brings us the legendary tale of Sir Ernest Shackleton's Trans-Antarctic Expedition.
Wilderness Navigation: Finding Your Way Using Map, Compass, Altimeter, and GPS	Bob Burns, Mike Burns,Paul Hughes	Wilderness Navigation answer your questions reading maps; buying a will all about compass; using a compass, a clinometer, an altimeter, and a GPS; orientation with map and compass; how to avoid getting lost, and what to do if that happens.
Power of Passion: Achieve Your Own Everests	Hobson, Alan, Clarke, Jamie	In this book, Jamie and Alan share the true story of their struggle to climb Everest. It is far more than a mountaineering tale, it's a real-life story—a story of hopes and dreams, of tragedy and tenacity to which everyone can relate. Mostly, it's a story about triumph—a triumph of the human spirit. It will help you climb your

(contd.)

(contd.)

Title	Author	Description
		own Everests—whatever mountains you are climbing in your own life.
		11. GPS
Fundamentals of Global Positioning System Receivers: A Software Approach	James Bao-Yen Tsui,James B.Y. Tsui	Provides engineers who use and design GPS systems with a reference on the operational principles guiding this technology, with an emphasis on software-based signal processing. Covers all aspects of receiver technology as well as relevant navigation schemes, concentrating on civilian C/A code used by commercial GPS receivers, rather than military code. Describes GPS basics and satellites, examines signal structure and tracking, and presents mathematical formulas for calculating a user's position.
Using GPS: Finding Your Way with Global Positioning System	Bruce Grubbs	
Integrating GIS and the Global Positioning System with CD-ROM	Karen Steede-Terry, Jack Dangermond	The coupling of the Global Positioning System (GPS) with Geographic Information System (GIS) is an information revolution that will give unprecedented accuracy in measurement and completeness of coverage to anyone who works with geographic data of any kind.
Understanding GPS: Principles and Applications (Artech House Telecommunications Library)	Kaplan, Elliott D. (Edt)	This practical design guide provides an in-depth description of GPS modern digital receiver operation and explores future applications in air, land, and marine navigation.
		12. Government
GIS in Public Policy	R. W. Greene	This case-study book shows how progressive policy makers have put GIS to use-the National Partnership for Reinventing Government being one example
Disaster Response: GIS for Public Safety	Gary Amdahl, Preface by Jack Dangermond	Local, state, and federal government units and the private sector need to learn how to best use the GIS technology now in the hands of those on the front lines fighting fires, managing

(contd.)

(contd.)

Title	*Author*	*Description*
		evacuations, and cleaning toxic spills. These case studies illustrate the best new strategies for mitigation, response, and recovery from both natural and human disasters. A wide variety of disasters and scenarios are represented, including a lethal mudslide and wildfires.
Opening Access: GIS in E-Government	R.W. Greene	As a result of the geographic aspect of most all government services, map-based information systems are critical to government performance. This guide shows how GIS is becoming an essential tool in the new world of e-government, demonstrating how GIS and the digital revolution are doing for government what they have done for business-making it more efficient in areas such as economic development and environmental protection.
Urban Planning and Development Applications of GIS	Said Easa, American Society of Civil Engineers Geographic Information Syste Yupo Chan (Editor)	Presents urban planning and development applications of Geographic Information Systems (GIS). The 13 chapters deal with GIS technology implementation, remote sensing, trends in spatial databases.
	13. Internet Mapping	
GIS Online: Information Retrieval, Mapping, and The Internet	Plewe, Brandon	Synopsis for people who want to build a Web site based around GIS and mapping technology or to include maps on their Web-sites, this text describes the concept of geographic information retrieval and data sharing, outlines motivations for sharing geographic data, and includes step-by-step instructions through the planning and development stages of the Web-site.
Serving Maps on the Internet: Geographic Information on the World Wide Web	Harder, Christian	Internet-based mapping applications are changing the way that people use and manage geographic information. Twelve fascinating case studies reveal how prominent organizations use Web-based geographic information systems (GIS) to improve customer service, increase employee efficiency, and inform the community at large.

(contd.)

(contd.)

Title	*Author*	*Description*
	14. Natural Resources	
Managing Natural Resources with GIS	Laura Lang, Environmental Systems Research Institute	As global population continues to grow, mankind must learn to balance consumption and conservation of the planet's scarece commodities. Geographic information system (GIS) technology brings a new perspective to the challenges faced by natural resource managers.
Geodestinies: The Inevitable Influence of Earth Resources on Nations and Individuals	Walter Lewellyn Youngquist	Geologists: Where is the search for energy and minerals today? How is the world Earth resource base changing? What new roles will alternative energy sources play?
Handbook of Hydrology	Maidment, David R.	Introducing Hydrology's New Benchmark Reference Here's the first book in nearly 30 years to provide comprehensive coverage of the current state of hydrologic knowledge and practice—saving you hours of time tracking down the latest techniques in professional journals.
Land Administration	Peter F. Dale, John D. McLaughlin	Land Administration provides a high level overview of recent advances in building formal property systems throughout the world, and reviews the role of property in advancing a society's economic and social agenda.
Flattening the Earth; Two Thousand Years of Map Projections	John P. Snyder	As long as there have been maps, cartographers have grappled with the impossibility of portraying the earth in two dimensions. To solve this problem mapmakers have created hundreds of map projections, mathematical methods for drawing the round earth on a flat surface.
	15. Non-Traditional GIS	
GIS for Landscape Architects	Hanna, Karen C.	Geographic information system (GIS) technology is an increasingly important software tool for organizing digital spatial data in the field of landscape architecture.
GIS for Health Organizations with CD-ROM	Laura Lang	Health care executives and managers will get a closer look at how to use geographic information systems (GIS) to improve patient care, better manage their facilities, and increase their market share with this book.

(contd.)

(contd.)

Title	*Author*	*Description*
GIS for Health Organizations with CD-ROM	Laura Lang	Health care executives and managers will get a closer look at how to use geographic information systems (GIS) to improve patient care, better manage their facilities, and increase their market share with this book. GIS is already used to monitor the spread of disease, assign patients to the nearest facility, and display injuries on a map of the human body.
Spatial Analysis, GIS, and Remote Sensing Applications in the Health Sciences	Donald Patrick Albert (Editor), Wilbert M. Gesler (Editor),Barbara Levergood (Editor)	Geographers and epidemiologists explain some of the ways geographical information systems can be applied in human and environmental health. Their topics include how spatial analysis can be used in medical geography, health services research, infectious disease, and a historical perspective on the development of remotely sensed data as applied to medical geography.
Remote Sensing and Geographical Information Systems in Epidemiology	Simon I. Hay (Editor),Sarah E. Randolph (Editor),Preface by J. R. Baker	With ever more sophisticated control methods being developed for infectious diseases, our ability to map spatial and temporal variation in risk is more important than ever. Only with global information may we plan control campaigns and deliver novel interventions and remedies where the need is greatest, and sustainable success is most likely.
Statistical Methods in Spatial Epidemiology	Andrew B. Lawson, A.B. Lawson	Epidemiology is the study of disease patterns. Spatial epidemiology considers the statistical aspects of the development of disease not only in relationship to time but in particular to geographical location.
Medical Geography, (Second Edition)	Melinda S. Meade,Robert J. Earickson	...this revised and expanded 2nd edition surveys the perspectives, methodologies and theories that geographers use to address the subject of human health and disease with clearly presented vignettes, tables, figures and maps
	16. Remote Sensing	
Remote Sensing of the Environment and Earth Resource Perspective	John R.. R. Jensen	This book introduces the principles of remote sensing from an Earth resource perspective.

(contd.)

(contd.)

Title	*Author*	*Description*
An Introductory Digital Image Processing: A Remote Sensing Perspective	John R. Jensen	Explains how to apply digital image processing techniques to remotely sensed imagery, and how to obtain useful results and information during this process.
Computer Processing of Remotely Sensed Images: An Introduction	Paul M. Mather, Paul Mather	Remotely sensed images provide a valuable source of information concerning the nature and variability of the land and sea surfaces of the world.
Remote Sensing and Geographical Information Systems in Epidemiology	Simon I. Hay (Editor),Sarah E. Randolph (Editor),Preface by J. R. Baker	With ever more sophisticated control methods being developed for infectious diseases, our ability to map spatial and temporal variation in risk is more important than ever.
Laser Remote Sensing of the Ocean: Methods and Applications	Konstantin I. Voliak, Alexey F. Bunkin	Recent advances in lasers and measuring devices, as well as new capabilities in automated data acquisition, have made possible the design and practical implementation of self-sufficient lidar systems.
Remote Sensing and Image Interpretation	Thomas M. Lillesand, Ralph W. Kiefer	Extensively illustrated, this updated edition provides a balance between classical visual image interpretation and digital image processing techniques.
GIS: A Visual Approach	Bruce Ellsworth Davis, Bruce Davis	This fully updated edition of Geographic Information Systems: A Visual Approach offers a comprehensive introduction to the application of GIS concepts.
	17. Spatial Data Analysis	
Spatial Analysis, GIS, and Remote Sensing Applications in the Health Sciences	Donald Patrick Albert (Editor), Wilbert M. Gesler (Editor), Barbara Levergood (Editor)	Geographers and epidemiologists explain some of the ways geographical information systems can be applied in human and environmental health.
The ESRI Guide to GIS Analysis, Volume 1: Geographic Patterns and Relationships	Mitchell, Andy	By the author of the best-selling GIS classic Zeroing In: GIS at Work in the Community comes an important new book about how to do real analysis with a geographic information system. The ESRI Guide to GIS Analysis, Volume 1: Geographic Patterns and

(contd.)

(contd.)

Title	*Author*	*Description*
		Relationships focuses on six of the most common geographic analysis tasks.
Spatial Analysis: Modelling in a GIS Environment	Longley, Paul/ Batty, Mike	This book addresses spatial analysis with an emphasis on the integration of different spatial analysis functions within GIS.
Casebook for Spatial Statistical Data Analysis: A Compilation of Analyses of Different Thematic Data Sets	Griffith, Daniel A./ Sone, Akio	This volume compiles geostatistical and spatial autoregressive data analyses involving geo-referenced socioeconomic, natural resources, agricultural, pollution, and epidemiological variables.
Interpolation of Spatial Data: Some Theory for Kriging	Stein, Michael Leonard	Kriging, which is the geostatistical term for optimal linear prediction of spatial processes, is widely used in geology, hydrology, environmental monitoring and other fields to interpolate spatial data.
Modelling Our World: The Esri Guide to Geodatabase Design	Zieler, Michael	This conceptual overview introduces and explains the concept of geodatabases-object-oriented data models—which are introduced in ArcInfo 8, the world's most advanced Geographic Information System (GIS) software package.
Beyond Mapping: Concepts, Algorithms, and Issues in GIS	Berry, Joseph K.	This book discusses maps as data with data structure implications; roving windows; spatial data modelling and management; assessing variability, shape, and pattern of map features; overlaying maps; cost-benefit analysis; algorithms involved in slope, distance, and connectivity; and cartographic and spatial modelling.
Cartography: Visualization of Spatial Data	Kraak, Menno-Jan/ Ormeling, Ferjan	Provides a comprehensive account of the acquisition of data, map design, topographic and statistical mapping and ends with examples of advanced mapping environments. (Paper)
Exploring Spatial Analysis in GIS	Chou, Yue-Hong	Written for Geographic Information Systems (GIS) professionals and students, this book provides an introduction to spatial analysis concepts and applications along with numerous examples and exercises.

(contd.)

(contd.)

Title	*Author*	*Description*
Geographic Information Systems and Science	John Wiley & Sons	There are many books integrating GIS technologies with disciplines such as remote sensing, cartography, geography, and computer science. This book, written by an authoritative and experienced author team, this book offers peerless coverage of issues related to geographical information science (GISc), GIS management, and real-world applications.
GIS in Telecommunications Management with CD-ROM	Lisa Winget Godin, Lisa Godin	Telecommunications companies use geographic infoımation systems (GIS) to plan, build, and maintain their networks. Techniques for merging spatial data from different sources into an integrated system are included, and the case studies profiled present innovative techniques for such daunting tasks as managing outside plant engineering processes
18. Transportation		
Transportation GIS	Laura Lang	This richly illustrated volume is an excellent introduction to GIS in the transportation industry. It's easy-to-read style and relevent case studies will appeal to industry professionals, students, and lay people alike.

Bibliography

Aangeenbrug, R.T. (1992) "A Review of 5 GIS Textbooks" *IJGIS*, Vol. 6, Iss 4, pp. 341-350.

Aangeenbrug, R.T. (1992) "Geographic Information Systems and Management Perspectives, by S. Aronoff" *IJGIS,* Vol. 6, Iss 4, pp. 344-345.

Aangeenbrug, R.T. (1992) "Geographic Information Systems—A Guide to the Technology, by J.C. Antenucci, K. Brown, P.L. Crosswell, M.J. Kevany, H. Archer" *IJGIS*, Vol. 6, Iss 4, pp. 348-350.

Aangeenbrug, R.T. (1992) "GIS in Education" *IJGIS*, Vol. 6, Iss 4, pp. 263-265.

Aangeenbrug, R.T. (1992) "Introduction to Urban Geographic Information Systems, by W.E. Huxhold" *IJGIS,* Vol. 6, Iss 4, pp. 346-348.

Aangeenbrug, R.T. (1992) "Principles of Geographic Information Systems for Land Resources Assessment", by P.A. Burrough *IJGIS*, Vol. 6, Iss 4, pp. 342-343.

Abel, D.J. (1985) "Some Elemental Operations on Linear Quadtrees for Geographic Information Systems " *Computer Journal*, Vol. 28, Iss 1, pp. 73-77.

Abel, D.J., Yap, S.K., Ackland, R. Cameron, M.A., Smith, D.F., Walker, G. (1992) "Environmental Decision Support System Project—An Exploration of Alternative Architectures for Geographical Information Systems" *IJGIS*, Vol. 6, Iss 3, pp. 193-204.

Abler, R.F. (1993) "Everything in Its Place—GPS, GIS, and Geography in the 1990s" *PG*, Vol. 45, Iss 2, pp. 131-139.

Adamus, C.L., Bergman, M.J. (1995) "Estimating Nonpoint-Source Pollution Loads with a GIS Screening Model" *Water Resources Bulletin*, Vol. 31, Iss 4, pp. 647-655.

Adinarayana, J., Flach, J.D., Collins, W.G. (1994) "Mapping Land-Use Patterns in a River Catchment Using Geographical Information Systems", *Journal of Environemntal Management*, Vol. 42, Iss 1, pp. 55-61.

Aitken, S., Stutz, F., Prosser, R., Chandler, R., (1993) "Neighbourhood Integrity and Residents Familiarity—Using a Geographic Information System to Investigate Place Identity", *Tijdschrift voor Economische en Scciale Geografie*, Vol. 84, Iss 1, pp. 2-12.

Alankary, K.M. (1991) "An Incremental Approach for Establishing a Geographical Information System in a Developing-Country—Saudi-Arabia", *IJGIS*, Vol. 5, Iss 1, pp. 85-98.

Albertson, P.E., Bourne, S.G., Hennington, G.W. (1992) "An Integrated Groundwater and Land-Use GIS for Impact Assessment", *PERS*, 1992, Vol. 58, Iss 8, pp. 1203-1207.

Albrecht, J. (1997) "GIS—A Computing Perspective, by M. Worboys" *IJGISC,* Vol. 11, Iss 4, pp. 405-406.

Ali, C.Q., Ross, L.G., Beveridge, M.C.M. (1991) "Microcomputer Spreadsheets for the Implementation of Geographic Information Systems in Aquaculture—A Case-Study on Carp in Pakistan", *Aquaculture*, Vol. 92, Iss 2-3, pp. 199-205.

Almond, S., (1991) "Geographic Information Systems—The Microcomputer and Modern Cartography", by D.R.F. Taylor, *Cartographic Journal*, Vol. 28, Iss 2, pp. 269-270.

Almond, S. (1994) "Mountain Environments and Geographic Information Systems", by M. Price, I. Heywood, *Cartographic Journal*, Vol. 31, Iss 2, pp. 145-145.

Alshuwaikhat, H.M., Nassef, K. (1996) "A GIS-Based Spatial Decision-Support System for Suitability Assessment and Land-Use Allocation", *Arabian Journal for Science and Engineering*, Vol. 21, Iss 4A, pp. 525-543.

Andrianarivo, J. (1993) "Using GIS to Evaluate the Crown-Line Intersect Sampling Method in Forest Survey" *Forest Ecology and Management*, Vol. 59, Iss 1-2, pp. 87-103.

Andrle, R. (1991) "Atlas GIS" *PG*, Vol. 43, Iss 3, pp. 362-363.

Anselin, L., Getis, A. (1992) "Spatial Statistical-Analysis and Geographic Information Systems", *Annals of Regional Science*, Vol. 26, Iss 1, pp. 19-33.

Ansoult, M.M., Soille, P.J., Loodts, J.A. (1990) "Mathematical Morphology—A Tool for Automated GIS Data Acquisition from Scanned Thematic Maps", *PERS*, Vol. 56, Iss 9, pp. 1263-1271.

Anthony, S.J., Corr, D.G. (1988) "Data-Structures in an Integrated Geographical Information System", *ESA Journal-European Space Agency*, Vol. 12, Iss 1, pp. 69-72.

Any, J., Treitz, P.M., and Howarth, P.J. (1992), Road network detection from SPOT imagery for updating geographical information systems in the rural-urban fringe, *International Journal on Geographic Information Systems*, Vol. 6, No. 2, pp. 141-57.

Apan, A.A. (1996) "Tropical Landscape Characterization and Analysis for Forest Rehabilitation Planning Using Satellite Data and GIS", *Landscape and Urban Planning*, Vol. 34, Iss 1, pp. 45-54.

Arbeit, D. (1992) "Geographic Information Systems—A Guide to the Technology", by J.C. Antenucci, K. Brown, P.L. Croswell, M.J. Kevany, H.A.V. Reinhold, *Journal of the Americam Planning Association*, Vol. 58, Iss 4, pp. 531-532.

Arbia, G. (1993) "The Use of GIS in Spatial Statistical Surveys", *International Statistical Review*, Vol. 61, Iss 2, pp. 339-359.

Archer, H., Croswell, P.L. (1989) "Public-Access to Geographic Information Systems—An Emerging Legal Issue", *PERS*, Vol. 55, Iss 11, pp. 1575-1581.

Archer, J.C. (1992) "Atlas-Asterisk-GIS", *CGIS*, Vol. 19, Iss 2, pp. 111-116.

Ardito, A., Iorio, R., Santagostino, G., Porrino, A., Fujimoto, N. (1992), "Accurate Modelling of Capacitively Graded Bushings for Calculation of Fast Transient Overvoltages in GIS", *IEEE Transactions on Power Delivery*, 1992, Vol. 7, Iss 3, pp. 1316-1327.

Arend, R.B. (1990) "GIS—Useful Tool or Expensive Toy", *Journal of Surveying Engineering-ASCE*, Vol. 116, Iss 2, pp. 131-138.

Arentze, T.A., Borgers, A.W.J., Timmermans, H.J.P. (1994) "Multistop-Based Measurements of Accessibility in a GIS Environment", *IJGIS*, Vol. 8, Iss 4, pp. 343-356.

Arlinghaus, S.L. (1996) "Geographical Record—A Path Through Recent GIS Literature", *Geographical Review*, Vol. 86, Iss 1, pp. 101-107.

Armstrong, M.P. (1994) "Requirements for the Development of GIS-Based Group Decision-Support Systems" *Journal of the Americal Society for Information Science*, Vol. 45, Iss 9, pp. 669-677.

Arthur, J.L., Nalle, D.J. (1997) "Clarification on the Use of Linear-Programming and GIS for Land-Use Modelling" *IJGISC*, Vol. 11, Iss 4, pp. 397-402.

Aschbacher, J., Ofren, R., Delsol, J.P., Suselo, T.B., Vibulsresth, S., Charrupat, T. (1995) "An Integrated Comparative Approach to Mangrove Vegetation Mapping Using Advanced Remote-Sensing and GIS Technologies—Preliminary-Results", *Hydrobiologia,* Vol. 295, Iss 1-3, pp. 285-294.

Aspinall, R., Matthews, K. (1994) "Climate-Change Impact on Distribution and Abundance of Wildlife Species—An Analytical Approach Using GIS", *Environmental Pollution*, 1994, Vol. 86, Iss 2, pp. 217-223.

Aspinall, R., Veitch, N. (1993) "Habitat Mapping from Satellite Imagery and Wildlife Survey Data Using a Bayesian Modelling Procedure in a GIS", *PERS*, Vol. 59, Iss 4, pp. 537-543.

Aspinall, R.J., Miller, D.R., Birnie, R.V. (1993) "Geographical Information Systemsfor Rural Land-Use Planning", *Applied Geogrpahy,* Vol. 13, Iss 1, pp. 54-66.

Astroth, J.H., Trujillo, J., Johnson, G.E. (1990) "A Retrospective Analysis of GIS Performance—The Umatilla Basin Revisited", *PERS*, Vol. 56, Iss 3, pp. 359-363.

Aultmanhall, L., Roorda, M., Baetz, B.W. (1997) "Using GIS for Evaluation of Neighbourhood Pedestrian Accessibility", *Journal of Urban Planning and Development-ASCE*, Vol. 123, Iss 1, pp. 10-17.

Bailey, R.G. (1988) "Problems with Using Overlay Mapping for Planning and Their Implications for Geographic Information Systems", *Environemntal Management,* Vol. 12, Iss 1, pp. 11-17.

Bailey, T.C. (1990) "GIS and Simple Systems for Visual, Interactive, Spatial-Analysis", *Cartographic Journal,* Vol. 27, Iss 2, pp. 79-84.

Baker, C.P., Bradley, M.D., Bobiak, S.M.K. (1993) "Wellhead Protection Area Delineation—Linking Flow Model with GIS", *Journal of Water Resources Planning and Management-ASCE*, Vol. 119, Iss 2, pp. 275-287.

Baker, W.L., Honaker, J.J., Weisberg, P.J. (1995) "Using Aerial-Photography and GIS to Map the Forest-Tundra Ecotone in Rocky-Mountain National-Park, Colorado, for Global Change Research", *PERS*, Vol. 61, Iss 3, pp. 313-320.

Bakker, S.A., Vandenberg, N.J., Speleers, B.P. (1994) "Vegetation Transitions of Floating Wetlands in a Complex of Turbaries Between 1937 and 1989 as Determined from Aerial Photographs with GIS", *Vegetatio,* Vol. 114, Iss 2, pp. 161-167.

Ball, G.L. (1994) "Ecosystem Modelling with GIS", *Environmental Management*, 1994, Vol. 18, Iss 3, pp. 345-349.

Banai, R. (1993) "Fuzziness in Geographical Information Systems—Contributions from the Analytic Hierarchy Process", *IJGIS*, Vol. 7, Iss 4, pp. 315-329.

Barrs, H.D., Prathapar, S.A. (1994) "An Inexpensive and Effective Basis for Monitoring Rice Areas Using GIS and Remote-Sensing", *Austrian Journal of Experimental Agriculture*, Vol. 34, Iss 7, pp. 1079-1083.

Battaglin, W.A., Hay, L.E., Parker, R.S., Leavesley, G.H. (1994) "Applications of a GIS for Modelling the Sensitivity of Water-Resources to Alterations in Climate in the Gunnison River Basin, Colorado", *Water Resources Bulletin*, 1993, Vol. 29, Iss 6, pp. 1021-1028.

Batty, M. (1992) "Geographic Information Systems—Development and Applications", by L. Worrall, *Urban Studies*, Vol. 29, Iss 1, pp. 132-134.

Batty, M. (1992)"Urban Modelling in Computer-Graphic and Geographic Information System Environments", *Environment and Planning B-Planning and Design*, Vol. 19, Iss 6, pp. 663-688.

Batty, M., Xie, Y.C. (1994) "Modelling Inside GIS.1. Model Structures, Exploratory Spatial Data-Analysis and Aggregation", *IJGIS*, Vol. 8, Iss 3, pp. 291-307.

Batty, M., Xie, Y.C. (1994) "Modelling Inside GIS.2. Selecting and Calibrating Urban Models Using Arc-Info", *IJGIS*, Vol. 8, Iss 5, pp. 451-470.

Baumgartner, M.F., Apfl, G. (1994) "Towards an Integrated Geographic Analysis System with Remote-Sensing, GIS and Consecutive Modelling for Snow Cover Monitoring", *IJRS*, 1994, Vol. 15, Iss 7, pp. 1507-1517.

Beard, R. (1992) "Geographic Information Systems and Their Socio-economic Applications", by D. Martin, *Cartographic Journal,* Vol. 29, Iss 1, pp. 80-81.

Bedient, P.B. (1992) "GIS and the Environment" *Water Environment Research,* 1992, Vol. 64, Iss 3, pp. 195-195.

Belyaeva, T., Higgins, J.M., Kirpichnikova, N., Lanzova, I., Hagerman, J.R. (1993) "GIS Application to Water-Quality Management in the Upper Volga River Basin—Joint TVA Russia Project", *Water Science and Technology*, Vol. 28, Iss 3-5, pp. 119-127.

Berry, J.K. (1986) "Geographic Information Systems.3. Learning Computer-Assisted Map Analysis", *JF*, Vol. 84, Iss 10, pp. 39-43.

Beyers, N., Gie, R.P., Zietsman, H.L., Kunneke, M., Hauman, J., Tatley, M., Donald, P.R. (1996) "The Use of a Geographical Information System (GIS) to Evaluate the Distribution of Tuberculosis in a High-Incidence Community", *South African Medical Journal*, Vol. 86, Iss 1, pp. 40-44.

Bishop, I. (1991) "MacGIS—A Geographic Information System for the Macintosh", *Journal of Architectural and Planning Reserach*, Vol. 8, Iss 1, pp. 78-80.

Bishop, I.D., Hulse, D.W. (1994) "Prediction of Scenic Beauty Using Mapped Data and Geographic Information Systems", *Landscape and Urban Planning*, Vol. 30, Iss 1-2, pp. 59-70.

Bissonette, J.A., Sherburne, S.S., Ramsey, R.D. (1994) "Analysing Telemetry Data with a GIS-Based Vector Structure", *IJGIS*, Vol. 8, Iss 6, pp. 533-543.

Blaise, P., Gessler, C. (1991) "Geobase—A Simple Geographical Information System on a Personal Computer" *Computer Applications in the Biosciences*, Vol. 7, Iss 2, pp. 155-160.

Blakemore, M. (1984), Generalization and error in spatial data bases, *Cartographia,* vol. 21, pp. 131-9.

Blakemore, M. (1985) "Cartography and Geographic Information Systems", *Progress in Human Geography*, Vol. 9, Iss 4, pp. 566-574.

Blakemore, M. (1986) "Cartography and Geographic Information Systems", *Progress in Human Geograph*, Vol. 10, Iss 4, pp. 553-563.

Blakemore, M. (1987) "Cartography and Geographic Information Systems", *Progress in Human Geography*, Vol. 11, Iss 4, pp. 590-606.

Blakemore, M. (1988) "Cartography and Geographic Information Systems", *Progress in Human Geography*, Vol. 12, Iss 4, pp. 525-537.

Blat, J., Delgado, A., Ruiz, M., Segui, J.M. (1995) "Designing Multimedia GIS for Territorial Planning—The Parcbit Case", *Environment and Planning B-Planning & Design*, Vol. 22, Iss 6, pp. 665-678.

Boeck, W., Kruger, K., (1992) "Arc Motion and Burn Through in GIS", *IEEE Transcations on Power Delivery*, 1992, Vol. 7, Iss 1, pp. 254-261.

Boggs, S.A. (1990) "Issues in Defining a Type Test for GIS Disconnectors", *IEEE Transcations on Power Delivery*, Vol. 5, Iss 4, pp. 1839-1845.

Boggs, S.A., Petterson, K., Diessner, A., Fujimoto, N., Leibold, A.A. (1992) "Partial Discharge Testing of Gas Insulated Substations—A Discussion Paper from the IEEE Substations Committee Working Group K4, GIS Diagnostic Methods", *IEEE Transcations on Power Delivery,* 1992, Vol. 7, Iss 2, pp. 499-506.

Boggs, S.A., Wang, Y. (1995) "Trapped Charge-Induced Field Distortion on GIS Spacers", *IEEE Transcations on Power Delivery*, Vol. 10, Iss 3, pp. 1270-1275.

Bojorqueztapia, L.A., Azuara, I., Ezcurra, E., Floresvillela, O. (1995) "Identifying Conservation Priorities in Mexico Through Geographic Information Systems and Modelling", *Ecological Applications*, Vol. 5, Iss 1, pp. 215-231.

Bolstad, Paul and Lillesand, T.M. (1991), Automated GIS integration in land cover classification, in Technical papers of the 1991, ACSM/ASPRS Annual Convention, vol. 3, pp. 23-32.

Bolstad, P.V. (1992) "Geometric Errors in Natural-Resource GIS Data—Tilt and Terrain Effects in Aerial Photographs", *Forest Science,* 1992, Vol. 38, Iss 2, pp. 367-380.

Bowler, I. (1994) "Landscape Ecology and Geographic Information Systems", by R. Hainesyoung *Geographical Journal*, Vol. 160, Iss Nov., pp. 345-345.

Bracken, I. (1992) "Geographic Information Systems—Developments and Applications", by L. Worrall *Progress in Human Geography,* Vol. 16, Iss 4, pp. 663-664.

Bracken, I. (1993) "An Introduction to Urban Geographic Information Systems", by W.E. Huxhold *Urban Studies,* Vol. 30, Iss 1, pp. 203-206.

Bracken, I. (1993) "Geographic Information Systems and Their Socioeconomic Applications, by D. Martin" *IEEE Transcations on Power Delivery,* Vol. 30, Iss 1, pp. 203-206.

Braddock, M., Lapidus, G., Cromley, E., Cromley, R., Burke, G., Banco, L. (1994) "Using a Geographic Information System to Understand Child Pedestrian Injury" *American Journal of Public Health,* Vol. 84, Iss 7, pp. 1158-1161.

Brainard, J., Lovett, A., Parfitt, J. (1996) "Assessing Hazardous-Waste Transport Risks Using a GIS" *IJGIS*, Vol. 10, Iss 7, pp. 831-849.

Braman, S. (1992) "Geographic Information Systems and Their Socioeconomic Applications", by D. Martin" *Journal of Communication*, Vol. 42, Iss 3, pp. 203-203.

Brandt, R., Groenewoudt, B.J., Kvamme, K.L. (1992) "An Experiment in Archaeological Site Location—Modelling in the Netherlands Using GIS Techniques" *World Archaeology,* Vol. 24, Iss 2, pp. 268-282.

Brannon, D.P., Irish, G.J. (1985) "Satellite Remote-Sensing and Geographic Information Systems Applied to Biogeochemical Flux Estimates" *Estuaries*, Vol. 8, Iss 2B, pp. A7-A7.

Braun, J.M., Ford, G.L., Fujimoto, N., Rizzetto, S., Stone, G.C., Boggs, S. (1993) "Reliability of GIS EHV Epoxy Insulators—The Need and Prospects for More Stringen—Acceptance Criteria" *IEEE Transactions on Power Delivery,* Vol. 8, Iss 1, pp. 121-131.

Braun, J.M., Rizzetto, S., Fujimoto, N., Ford, G.L. (1991) "Modulation of Partial Discharge Activity in GIS Insulators by X-Ray-Irradiation" IEEE *Transactions on Electrical Insulation* 1991, Vol. 26, Iss 3, pp. 460-468.

Briggs, D., Mounsey, H. (1989) "Integrating Land Resource Data into a European Geographical Information System—Practicalities and Problems" *Applied Geography,* Vol. 9, Iss 1, pp. 5-20.

Briggs, D.J., Collins, S., Elliott, P., Fischer, P., Kingham, S., Lebret, E., Pryl, K., Vanreeuwijk, H. (1997) "Smallbone-K Vanderveen-AMapping Urban Air-Pollution Using GIS—A Regression-Based Approach" *IJGISC,* Vol. 11, Iss 7, pp. 699-718.

Brimcombe, A.J., (1993) "Time in Geographic Information Systems, by G. Lanegran" *AREA,* Vol. 25, Iss 4, pp. 408-408.

Brinker, R.W., Jackson, B.D. (1991) "Using a Geographic Information System to Study a Regional Wood Procurement Problem" *Forest Science,* Vol. 37, Iss 6, pp. 1614-1631.

Bromley, R.D.F., Selman, J.M. (1993) "Geographic Information Usage in a Local-Authority—The Human Environment for GIS" *Area,* Vol. 25, Iss 3, pp. 228-236.

Brooks, K.R., London, J.B., Henry, M.S., Singletary, L.A. (1993) "Analysis and Simulation of Employment and Income Impacts of Infrastructure Investments in a State-Wide GIS Framework" *Computers Environment and Urban Systems,* Vol. 17, Iss 2, pp. 129-151.

Browne, T.J. (1992) "The Great Divide—Central Versus Departmental Support for Teaching GIS.1." *Journal of Geography in Higher Education,* Vol. 16, Iss 2, pp. 225-231.

Brown, S., Schreier, H., Thompson, W.A., Vertinsky, I. (1994) "Linking Multiple Accounts with GIS as Decision-Support System to Resolve Forestry Wildlife Conflicts" *Journal of Environmental Management,* Vol. 42, Iss 4, pp. 349-364.

Brunsdon, C. (1994) "Spatial-Analysis and GIS", by S. Fotheringham, P. Rogerson *Cartographic Journal*, Vol. 31, Iss 1, pp. 85-85.

Brunsdon, C. (1997) "International GIS Dictionary", by R. Mcdonnell, K. Kemp, *Environment and Planning B-Planning and Design,* Vol. 24, Iss 6, pp. 937-938.

Buck, K.E., Kleinetebbe, A., Rees, V. (1993) "Verification of the Seismic Resistance of Gas-Insulated Switchgear (GIS)" *IEEE Transactions on Power Delivery,* Vol. 8, Iss 1, pp. 114-120.

Budic, Z.D. (1994) "Effectiveness of Geographic Information Systems in Local-Planning" *Journal of the American Planning Association,* Vol. 60, Iss 2, pp. 244-263.

Budic, Z.D., Godschalk, D.R. (1994) "Implementation and Management Effectiveness in Adoption of GIS Technology in Local Governments" *Computers Environment and Urban Systems,* Vol. 18, Iss 5, pp. 285-304.

Bullen, N., Moon, G., Jones, K. (1996) "Defining Localities for Health-Planning—A GIS Approach" *Social Science and Medicine,* Vol. 42, Iss 6, pp. 801-816.

Burrough, P.A. (1986), *Principals of Geographical Information System for Land Resource Assessment,* Oxford University Press.

Burrough, P.A. (1992), Development of intelligent geographical information system, *International Journal of Geographical Information System,* Vol. 6:1, pp. 1-12.

Burrough, P.A. (1992) "Development of Intelligent Geographic Information Systems" *IJGIS,* Vol. 6, Iss 1, pp. 1-11.

Burrough, P.A. (1992) "Geographic Information Systems and Cartographic Modelling", by C.D. Tomlin, *IJGIS,* Vol. 6, Iss 5, pp. 449-450.

Burrough, P.A., Frank, A.U. (1995) "Concepts and Paradigms in Spatial Information—Are Current Geographical Information SystemsTruly Generic" *IJGIS,* Vol. 9, Iss 2, pp. 101-116.

Butler, W.I., Stehn, R.A., Balogh, G.R. (1995) "GIS for Mapping Waterfowl Density and Distribution from Aerial Surveys" *Wildlife Society Bulletin,* Vol. 23, Iss 2, pp. 140-147.

Buttenfield, B.P. (1992) "Geographic Information Systems and Cartographic Modelling", by C.D. Tomlin, *CGIS,* Vol. 19, Iss 4, pp. 262-263.

Cadoux-Hudson, J. and Heywood, I. (1991) (eds), *Geographic Information 1991,* Taylor and Francis, London.

Camm, J.D., Chorman, T.E., Dill, F.A., Evans, J.R., Sweeney, D.J., Wegryn, G.W. (1997) "Blending or/MS, Judgment, and GIS—Restructuring P-and-GS Supply Chain" *Interfaces,* Vol. 27, Iss 1, pp. 128-142.

Campbell, H. (1992) "The Impact of Geographic Information Systems on British Local-Government" *Computers Environment and Urban Systems,* Vol. 16, Iss 6, pp. 531-541.

Campbell, H. (1994) "How Effective Are GIS in Practice—A Case-Study of British Local-Government" *IJGIS,* Vol. 8, Iss 3, pp. 309-325.

Campbell, H., Masser, I. (1992) "GIS in Local-Government—Some Findings from Great-Britain", *IJGIS,* Vol. 6, Iss 6, pp. 529-546.

Campbell, H. (1992) "Organizational Issues and the Implementation of GIS in Massachusetts and Vermont—Some Lessons for the United-Kingdom" *Environment and Planning B-Planning and Design,* Vol. 19, Iss 1, pp. 85-95.

Campbell, J.C., Radke, J., Gless, J.T., Wirtshafter, R.M. (1992) "An Application of Linear-Programming and Geographic Information Systems—Cropland Allocation in Antigua", *Environment and Planning A,* Vol. 24, Iss 4, pp. 535-549.

Can, A. (1992) "Residential Quality Assessment—Alternative Approaches Using GIS", *Annals of Regional Science,* Vol. 26, Iss 1, pp. 97-110.

Can, A. (1993) "Tiger Line Files in Teaching GIS" *IJGIS,* Vol. 7, Iss 6, pp. 561-572.

Can, A. (1994) "Tiger Line Files in Teaching GIS (Vol. 7, Pg 561, 1993)" *IJGIS,* Vol. 8, Iss 4, pp. 403-403.

Cannistra, J.R., Leadbeater, R., Humphries, R. (1992) "Washington Suburban Sanitary Commission Implements GIS" *Journal American Water Works Association,* Vol. 84, Iss 7, pp. 62-67.

Carpentier, L.C., Bosch, D.J. (1994) "Adaptive Management and Geographic Information System to Overcome Informational Barriers to Pollution Reduction Trading" *American Journal of Agricultural Economics,* Vol. 76, Iss 5, pp. 1272-1272.

Carrara, A., Cardinali, M., Detti, R., Guzzetti, F., Pasqui, V., Reichenbach, P. (1991) "GIS Techniques and Statistical-Models in Evaluating Landslide Hazard" *Earth Surface Processes and Landforms 1991,* Vol. 16, Iss 5, pp. 427-445.

Carstensen, Jr., Laurence W. and Campbell, James B. (1991), Desktop scanning for cartographic digitization and spatial analysis, *Photogrammertic Engineering and Remote Sensing,* Vol. 57, No.11, pp. 1437-46.

Carstensen, L.W., Shaffer, C.A., Morrill, R.W., Fox, E.A. (1993) "Geosim—A GIS-Based Simulation Laboratory for Introductory Geography" *Journal of Geography,* Vol. 92, Iss 5, pp. 217-222.

Carter, J.R. (1985) *Geographic Names Information-System,* by F.R.L. Payne PG, Vol. 37, Iss 2, pp. 241-242.

Carver, S. (1997) "Innovations in GIS-3—Selected Papers from the 3rd National Conference on GIS-Research-UK (GISruk)", by D. Parker *IJGISC,* Vol. 11, Iss 5, pp. 521-521.

Carver, S.J. (1991) "Integrating Multicriteria Evaluation with Geographical Information Systems" *IJGIS,* Vol. 5, Iss 3, pp. 321-339.

Cassettari, S. (1991) "Geographic Information Systemsat Kingston" *Cartographic Journal,* Vol. 28, Iss 1, pp. 71-75.

Cassettari, S. (1992) "Time in Geographical Information Systems, by G. Langran" *Cartographic Journal,* Vol. 29, Iss 2, pp. 193-194.

Champion, A.G. (1988) "Population Information-Systems—Report of a Population Geography Study-Group Conference Held at Mansfield-College, Oxford, 7-8 September 1987" *AREA,* Vol. 20, Iss 1, pp. 82-83.

Chang, N.B., Lu-HY Wei,Y.L. (1997) "GIS Technology for Vehicle-Routing and Scheduling in Solid-Waste Collection Systems" *Journal of Environmental Engineering, ASCE,* Vol. 123, Iss 9, pp. 901-910.

Chen, J. Newkirk, R.T. Davidson, G. Gong, P. (1994) "The Development of a Knowledge-Based Geographical Information System for the Zoning of Rural-Areas", *Environment and Planning B-Planning and Design,* Vol. 21, Iss 2, pp. 179-190.

Chou, Y.H. (1992), Management of wildfires with a geographical information system, *International Journal on Geographic Information Systems,* Vol. 6, No. 2, pp. 123-40.

Chou, Y.H. (1992) "Slope-Line Detection in a Vector-Based GIS" *PERS* 1992, Vol. 58, Iss 2, pp. 227-233.

Chou, Y.H. (1992) "Management of Wildfires with a Geographical Information System" *IJGIS,* Vol. 6, Iss 2, pp. 123-140.

Chrisman, N.R. (1987) "Design of Geographic Information Systems Based on Social and Cultural Goals" *PERS,* Vol. 53, Iss 10, pp. 1367-1370.

Chuvieco, E. (1993) "Integration of Linear-Programming and GIS for Land-Use Modelling" *IJGIS,* Vol. 7, Iss 1, pp. 71-83.

Chuvieco, E. (1997) "Clarification on the Use of Linear-Programming and GIS for Land-Use Modelling—Reply" *IJGISC,* Vol. 11, Iss 4, pp. 403-403.

Chuvieco, E., Salas, J. (1996) "Mapping the Spatial-Distribution of Forest-Fire Danger Using GIS" *IJGIS,* Vol. 10, Iss 3, pp. 333-345.

Civco, D. (1994) "Erdas Imagine Remote-Sensing, Image-Processing, and GIS Software" *PERS* 1994, Vol. 60, Iss 1, pp. 35-39.

Civco, D.L. (1992) "VGA-Erdas Image-Processing and GIS Software" *PERS* 1992, Vol. 58, Iss 9, pp. 1350-1354.

Clarke, G. (1996) "Bringing GIS into Business", by D. Grimshaw *IJGIS,* Vol. 10, Iss 1, pp. 138-139.

Clarke, G.P., Langley, R. (1996) "A Review of the Potential of GIS and Spatial Modelling for Planning in the New Education Market" *Environment and Planning C-Government and Policy,* Vol. 14, Iss 3, pp. 301-323.

Clark, G.L. (1992) "GIS—What Crisis" *Environment and Planning A,* Vol. 24, Iss 3, pp. 321-322.

Clematis, A., Falcidieno, B., Spagnuolo, M., (1996) "Parallel-Processing on Heterogeneous Networks for GIS Applications" *IJGIS,* Vol. 10, Iss 6, pp. 747-767.

Collier, P., Fontana, D., Pearson, A. (1995) "GIS Mapping of Langstone-Harbor for an Integrated Ecological and Archaeological Study" *Cartographic Journal,* Vol. 32, Iss 2, pp. 137-142.

Coombes, M. Openshaw, S., Wong, C., Raybould, S. (1993)"Community Boundary Definition—A GIS Design Specification" *Regional Studies,* Vol. 27, Iss 3, pp. 280-286.

Coombes, M., Saunders, C., Raybould, S. (1993) "Critical Problems in Developing Databases for Upland Areas—The Potential Value of GIS Techniques" *Journal of Environmental Management,* Vol. 39, Iss 4, pp. 271-280.

Coppock, J.T. (1992) "GIS Education in Europe" *IJGIS,* Vol. 6, Iss 4, pp. 333-335.

Coppock, T. (1992) "GIS in Education—Foreword" *IJGIS,* Vol. 6, Iss 4, pp. R1-R1.

Coppock, T. (1993) "International Journal of Geographical Information Systems" *IJGIS,* Vol. 7, Iss 2, pp. 101-101.

Coppock, T., Guptill, S. (1991) "GIS in Developing-Nations" *IJGIS,* Vol. 5, Iss 1, pp. 1-1.

Cornelius, S.C., Sear, D.A., Carver, S.J., Heywood, D.I. (1994) "GPS, GIS and Geomorphological Field Work" *Earth Surface Processes and Landforms,* Vol. 19, Iss 9, pp. 777-787.

Coroza, O., Evans, D., Bishop, I. (1997) "Enhancing Runoff Modelling with GIS" *Landscape and Urban Planning,* Vol. 38, Iss 1-2, pp. 13-23.

Cosetino, Brian L. and Lillesand, Thomas M. (1991), Towards automated statewise land cover mapping in Wisconsin using satellite remote sensing and GIS Techniques, 1991 ACSM/ ASPRS Technical Papers, 3, 93-102.

Coulson, M.R.C. (1991) "Tactile-Map Output from Geographical Information Systems—The Challenge and Its Importance" *IJGIS,* Vol. 5, Iss 3, pp. 353-360.

Cova, T.J., Church, R.L. (1997) "Modelling Community Evacuation Vulnerability Using GIS" *IJGISC,* Vol. 11, Iss 8, pp. 763-784.

Cracknell, A.P. (1986) "Editorial—Geographical Information Systems" *IJRS,* Vol. 7, Iss 6, pp. 721-722.

Cracknell, A.P. (1993) "Icorg-92—Remote-Sensing Applications and Geographic Information Systems—Recent Trends—Preface" *IJRS,* Vol. 14, Iss 17, pp. 3077-3077.

Craglia, M. (1994) "Geographical Information Systemsin Italian Municipalities" *Computers Environment and Urban Systems,* Vol. 18, Iss 6, pp. 381-&.

Craglia, M., Raper, J. (1995) "GIS and Multimedia" *Environment and Planning B-Planning and Design,* Vol. 22, Iss 6, pp. 634-636.

Craul, P.J., Herrington, L.P. (1992) "GIS Applications—A Wide Spectrum Not Without Problems" *PERS* 1992, Vol. 58, Iss 8, pp. 1092-1094.

Cress, J.J., Deister, R.R.P. (1990) "Development and Implementation of a Knowledge-Based GIS Geological Engineering Map Production System" *PERS* Vol. 56, Iss 11, pp. 1529-1535.

Croswell, P.L., Clark, S.R. (1988) "Trends in Automated Mapping and Geographic Information System Hardware" *PERS,* Vol. 54, Iss 11, pp. 1571-1576.

Curran, P.J. (1984) "Geographic Information Systems" *AREA,* Vol. 16, Iss 2, pp. 153-158.

Curry, M.R. (1994) "Image, Practice and the Hidden Impacts of Geographic Information Systems" *Progress in Human Geography,* Vol. 18, Iss 4, pp. 441-459.

Curtis, S.E. (1989) "The Development of Geographical Information Systems for Locality Planning in Health-Care" *AREA,* Vol. 21, Iss 4, pp. 391-399.

Dahlberg, R.E. (1991) "Development of an Integrated Cartography-Remote Sensing-Geographic Information Systems Program at Northern-Illinois-University" *CGIS,* Vol. 18, Iss 3, pp. 171-175.

Dai, Q., Fletcher, J., Phipps, T., Yuill, C. (1996) "Development of a GIS Decision-Support System for Water-Quality Policy Analysis" *American Journal of Agricultural Economics,* Vol. 77, Iss 5, pp. 1372-1372.

Dale, P.F., (1991) "Information Technology in Geography and Planning—Including Principles of GIS, by I. Bracken, C. Webster" *IJGIS,* Vol. 5, Iss 2, pp. 256-258.

Dangermond, J. (1989) "Trends in Geographic Information Systems" *EKISTICS-The Problems and Science of Human settlements,* Vol. 56, Iss 338-, pp. 318-330.

Dangermond, J., Smith, L.K., (1988) "Geographic Information Systems and the Revolution in Cartography—The Nature of the Role Played by a Commercial Organization" *American Cartographer,* Vol. 15, Iss 3, pp. 301-310.

Dasgupta, A.R. (2005), Standards and interoperability for a seamless world, *Geospatial Today,* Vol.4, issue 2, pp. 22-4.

Davidson, D.A., Theocharopoulos, S.P., Bloksma, R.J. (1994) "A Land Evaluation Project in Greece Using GIS and Based on Boolean and Fuzzy Set Methodologies" *IJGIS,* Vol. 8, Iss 4, pp. 369-384.

Davidson, D.A., Theocharopoulos, S.P., Bloksma, R.J. (1996) "A Land Evaluation Project in Greece Using GIS and Based on Boolean and Fuzzy Set Methodologies (Vol. 8, Pg 369, 1994)" *IJGIS,* Vol. 10, Iss 2, pp. 245-245.

Davies, C., Medyckyjscott, D. (1994) "GIS Usability—Recommendations Based on the Users View" *IJGIS,* Vol. 8, Iss 2, pp. 175-189.

Davies, C., Medyckyjscott, D. (1996) "GIS Users Observed" *IJGIS,* Vol. 10, Iss 4, pp. 363-384.

Davis, B.A., Carr, H.V., Schmidt, N., Hickerson, L. (1993) "Opportunities in the Commercial Uses of Remote-Sensing and GIS Technologies—An Overview of NASA Visiting Investigator Program at Stennis-Space-Center" *PERS* Vol. 59, Iss 6, pp. 935-939.

Davis, F.W., Quattrochi, D.A., Ridd, M.K., Lam, N.S.N., Walsh, S.J., Michaelsen, J.C., Franklin, J., Stow, D.A., Johannsen, C.J., Johnston, C.A. (1991) "Environmental-Analysis Using Integrated GIS and Remotely Sensed Data—Some Research Needs and Priorities" *PERS* 1991, Vol. 57, Iss 6, pp. 689-697.

Davis, J.L. (1993) "GIS Approaches to Regional Analysis—A Case Study of the Island of Hvar, by V. Gaffney, Z. Stancic" *American Journal of Archaeology,* Vol. 97, Iss 4, pp. 799-800.

Davis, J.L. (1993) "Interpreting Space—GIS and Archaeology", by K.M.S. Allen, S.W. Green, E.B.W. Zubrow, *American Journal of Archaeology,* Vol. 97, Iss 2, pp. 357-359.

Deane, G.C. (1994) "The Role of GIS in the Management of Natural Resources" *ASLIB Proceedings,* Vol. 46, Iss 6, pp. 157-161.

Defusco, R.P. (1993) "Modelling Bird Hazards to Aircraft—A GIS Application Study" *PERS* Vol. 59, Iss 10, pp. 1481+.

Degg, M. (1994) "International Workshop on Geographical Information Systems in Assessing Natural Hazards, University-for-Foreigners, Perugia, Italy, 20-22 September, 1993" *Disasters*, Vol. 18, Iss 2, pp. 177-178.

Demers, M.N. (1993) "Geographical Information Systems, Vol. 1—Principles, Vol. 2—Applications", by D.J. Maguire, M.F. Goodchild, D.W. Rhind, *Geographical Review,* Vol. 83, Iss 3, pp. 330-332.

Demers, M.N., Fisher, P.F. (1991) "Comparative Evolution of Statewide Geographic Information Systems in Ohio" *IJGIS,* Vol. 5, Iss 4, pp. 469-485.

Denning, J. (1993) "Small-Government GIS" *Civil Engineering,* Vol. 63, Iss 6, pp. 52-54.

Derenyi, E., Pollock, R. (1990) "Extending a GIS to Support Image-Based Map Revision", *PERS* Vol. 56, Iss 11, pp. 1493-1496.

Deroo, A.P.J., Burrough, P.A., Hazelhoff, L. (1989) "Soil-Erosion Modelling Using Answers and Geographical Information Systems" *Earth Surface Processes and Landforms,* Vol. 14, Iss 6-7, pp. 517-532.

Desmet, P.J.J., Govers, G. (1995) "GIS-Based Simulation of Erosion and Deposition Patterns in an Agricultural Landscape—A Comparison of Model Results with Soil Map Information" *CATENA* Vol. 25, Iss 1-4, pp. 389-401.

Desmet, P.J.J., Govers, G. (1997) "Modelling Topographic Potential for Erosion and Deposition Using GIS—Comment" *IJGISC,* Vol. 11, Iss 6, pp. 603-610.

Devantier, B.A. (1993) "Review of GIS Applications in Hydrologic Modelling (Vol. 119, Pg 248, 1993)" *Journal of Water Resources Planning and Management-ASCE* Vol. 119, Iss 4, pp. 504-504.

Devantier, B.A., Feldman, A.D. (1993) "Review of GIS Applications in Hydrologic Modelling" *Journal of Water Resources Planning and Management-ASCE* Vol. 119, Iss 2, pp. 246-261.

Dillworth, Mary E. (1991), Geographic windows in remote sensing: does window size matter? *ACSM/ASPRS Technical Papers,* Vol. 3, pp.122-28.

Dillworth, M.E., Stieg, R.T. (1992) "GIS in the Memphis, Tennessee Region—Working Together Saves Time" *PERS* 1992, Vol. 58, Iss 8, pp. 1213-1216.

Djokic, D., Maidment, D.R. (1993) "Application of GIS Network Routines for Water-Flow and Transport" *Journal of Water Resources Planning and Management-ASCE* Vol. 119, Iss 2, pp. 229-245.

Dobson, J.E. (1992) "Geographical Information Systems—Principles and Applications", by D.J. Maguire, M.F. Goodchild, D.W. Rhind, PG, Vol. 44, Iss 3, pp. 358-359.

Dobson, J.E. (1993) "Commentary—A Conceptual Framework for Integrating Remote Sensing, GIS, and Geography" *PERS* Vol. 59, Iss 10, pp. 1491-1496.

Dockrill, S.J. (1992) "GIS Approaches to Regional-Analysis—A Case Study of the Island of Hvar", by V. Gaffney, Z. Stancic, *Antiquity,* Vol. 66, Iss 250, pp. 266-268.

Dockrill, S.J. (1992) "Interpreting Space—GIS and Archaeology", by K.M.S. Allen, S.W. Green, E.B.W. Zubrow, *Antiquity,* Vol. 66, Iss 250, pp. 266-268.

Dockrill, S.J. (1992) "Introductory Readings in Geographic Information Systems", by D.J. Pequet, D.F. Marble *Antiquity,* Vol. 66, Iss 250, pp. 266-268.

Dorling, D.F.L. (1992) "Geographical Information Systems and Their Socio-economic Applications", by D. Martin, *IJGIS,* Vol. 6, Iss 5, pp. 447-448.

Duecker, K.J., Delacy, P.B. (1990) "GIS in the Land-Development Planning Process—Balancing the Needs of Land-Use Planners and Real-Estate Developers" *Journal of the American Planning Association,* Vol. 56, Iss 4, pp. 483-491.

Dueker, K.J. (1984) "Proceedings, United-States Australia Workshop on Design and Implementation of Computer-Based Geographic Information Systems", by D. Peuquet, J. Ocallaghan *PG,* Vol. 36, Iss 4, pp. 518-518.

Dueker, K.J. (1987) "Geographic Information Systems and Computer-Aided Mapping" *Journal of the American Planning Association,* Vol. 53, Iss 3, pp. 383-390.

Duguay, C.R., Beaudoin, M., Leroux, D. (1992) "Fundamentals of Geographic Information Systems—A Compendium", by W.J. Ripple, *Canadian Geographer-Geograhe Canadien,* Vol. 36, Iss 1, pp. 103-108.

Duguay, C.R., Beaudoin, M., Leroux, D. (1992) "Geographic Information Systems—An Introduction", by J. Star, J. Estes, *Canadian Geographer-Geograhe Canadien,* Vol. 36, Iss 1, pp. 103-108.

Duguay, C.R., Beaudoin, M., Leroux, D. (1992) "Geographic Information Systems for Urban and Regional Planning", by H.J. Scholten, J.C.H. Stillwell, *Canadian Geographer-Geograhe Canadien,* Vol. 36, Iss 1, pp. 103-108.

Duguay, C.R., Beaudoin, M., Leroux, D. (1992) "Geographic Information Systems—Developments and Applications", by L. Worrall, *Canadian Geographer-Geograhe Canadien,* Vol. 36, Iss 1, pp. 103-108.

Duguay, C.R., Beaudoin, M., Leroux, D. (1992) "Geographic Information Systems and Cartographic Modelling", by C.D. Tomlin, *Canadian Geographer-Geograhe Canadien,* Vol. 36, Iss 1, pp. 103-108.

Duncan, C.J., Epps, W.R. (1993) "GIS and the Role of the State Down Under—Comment" *Political Geography,* Vol. 12, Iss 1, pp. 3-7.

Dunn, C.E., Atkins, P.J. Townsend, J.G. (1997) "GIS for Development—A Contradiction in Terms" *AREA,* Vol. 29, Iss 2, pp. 151-159.

Dunn, C.E., Newton, D. (1992) "Optimal Routes in GIS and Emergency Planning Applications" *AREA,* Vol. 24, Iss 3, pp. 259-267.

Dunn, R. (1991) "Geographic Information Systems—Developments and Applications", by L. Worrall, *Local Government Studies,* Vol. 17, Iss 3, pp. 101-104.

Dunn, R. (1992) "Geographic Information Systems and Their Socioeconomic Applications", by D. Martin, *Environment and Planning A,* Vol. 24, Iss 7, pp. 1068-1069.

Eckhardt, D.W., Verdin, J.P., Lyford, G.R. (1990) "Automated Update of an Irrigated Lands GIS Using Spot HRV Imagery" *PERS* Vol. 56, Iss 11, pp. 1515-1522.

Edralin, J.S. (1991) "International-Conference on Geographic Information Systems Applications for Urban and Regional-Planning—Ciloto, Puncak, Indonesia, 3-6 October 1989" *IJGIS,* Vol. 5, Iss 1, pp. 147-154.

Egenhofer, M.J. (1992), Why not SQL!, *International Journal of Geographic Information Systems,* Vol. 6, No. 2, pp. 71-85.

Egenhofer, M.J., Frank, A.U. (1990) "Lobster—Combining AI and Database Techniques for GIS", *PERS* Vol. 56, Iss 6, pp. 919-926.

Ehlers, M., Greenlee, D., Smith, T., Star, J. (1991) "Integration of Remote-Sensing and GIS—Data and Data Access" *PERS* 1991, Vol. 57, Iss 6, pp. 669-675.

Ellis, J.R., Casavant, K.L., Jessup, E.L. (1996) "A GIS and Transportation Optimization Model Approach to Determining Highway and Rural Road Commodity Flows" *American Journal of Agricultural Economics,* Vol. 78, Iss 5, pp. 1394-1394.

Elston, D.A., Buckland, S.T. (1993) "Statistical Modelling of Regional GIS Data—An Overview" *Ecological Modelling,* Vol. 67, Iss 1, pp. 81-102.

Elston, D.A., Jayasinghe, G., Buckland, S.T., Macmillan, D.C., Aspinall, R.J. (1997) "Adapting Regression Equations to Minimize the Mean Squared Error of Predictions Made Using Covariate Data from a GIS" *IJGISC,* Vol. 11, Iss 3, pp. 265-280.

Elvis, M., Matsuoka, M., Siemiginowska, A., Fiore, F., Mihara, T., Brinkmann, W. (1994) "An Asca GIS Spectrum of S5-0014+813 at Z=3.384", *Altrophysical Journal,* 1994, Vol. 436, Iss 1, pp. L55-L58.

Engel, B.A., Srinivasan, R., Arnold, J., Rewerts, C., Brown, S.J. (1993) "Nonpoint-Source (Nps) Pollution Modelling Using Models Integrated with Geographic Information Systems (GIS)", *Water Science and Technology,* Vol. 28, Iss 3-5, pp. 685-690.

Esnard, A.M., Macdougall, E.B. (1997) "Common Ground for Integrating Planning-Theory and GIS Topics" *Journal of Planning Education and Research,* Vol. 17, Iss 1, pp. 55-62.

ESRI (2001), *Map Book,* Environmental System research Institute, Redlands.

Etzelmuller, B., Vatne, G., Odegard, R.S., Sollid, J.L. (1994) "Mass-Balance and Changes of Surface Slope, Crevasse and Flow Pattern of Erikbreen, Northern Spitsbergen—An Application of a Geographical Information System (GIS)" *Polar Research,* 1993, Vol. 12, Iss 2, pp. 131-146.

Evans, B.M., Myers, W.L. (1990) "A GIS-Based Approach to Evaluating Regional Groundwater Pollution Potential with Drastic" *Journal of Soil and Water Conservation,* Vol. 45, Iss 2, pp. 242-245.

Fairbairn, D. (1992) "Geographical Information Systems—Principles and Applications, Vol. 1, Vol. 2", by D. Maguire, M. Goodchild, D. Rhind, *Cartographic Journal,* Vol. 29, Iss 1, pp. 79-80.

Fairbairn, D., Openshaw, S. (1991)"Cartographic and GIS Activity at the University-of-Newcastle-upon-Tyne 1987-1991", *Cartographic Journal,* Vol. 28, Iss 1, pp. 79-82.

Farrell, T. (1995) "Closing with the Enemy—How GIS Fought the War in Europe, 1944-45", by M.D. Doubler, *Journal of Strategic Studies,* Vol. 18, Iss 4, pp. 139-140.

Farthing, K. (1996) "The GIS Book, 3rd Edition", by G.B. Korte, *IJGIS,* Vol. 10, Iss 1, pp. 139-141.

Farthing, S. (1991) "Information Technology in Geography and Planning—Including Principles of GIS", by I. Bracken, C. Webster, *Regional Studies,* Vol. 25, Iss 2, pp. 192-192.

Faust, N.L. (1987) "Automated Data Capture for Geographic Information- Systems—A Commentary" *PERS,* Vol. 53, Iss 10, pp. 1389-1390.

Faust, N.L., Anderson, W.H., Star, J.L. (1991) "Geographic Information Systems and Remote-Sensing Future Computing Environment" *PERS,* Vol. 57, Iss 6, pp. 655-668.

Felleman, J. (1990) "Theres a GIS in Your Future", *Government Information Quarterly,* Vol. 7, Iss 3, pp. 261-267.

Ferenth, M. (1991) "Gimms-(GIS)-LTD—Products and Platforms" *Cartographic Journal,* Vol. 28, Iss 1, pp. 47-48.

Fernandez, R.N., Lozanogarcia, D.F., Johannsen, C.J. (1989) "Development of Microcomputer Based Geo-Referenced Information-Systems (GIS)—Case Studies in Argentina and Mexico", *Interciencia,* Vol. 14, Iss 5, pp. 247-253.

Fernandez, R.N., Rusinkiewicz, M. (1993) "A Conceptual Design of a Soil Database for a Geographical Information System" *IJGIS,* Vol. 7, Iss 6, pp. 525-539.

Ferndndez, R. Norberto, D. Fabidn Lozano-Garcia, Greg Deeds and Cgria IJ. Johannsen (1991), Accuracy assessment of map coordinate retrieval, *Photogrammetric Engineering & Remote Sensing,* Vol. 37, No.11, pp. 1447-52.

Ferrier, G., Wadge, G. (1997) "An Integrated GIS and Knowledge-Based System as an Aid for the Geological Analysis of Sedimentary Basins" *IJGISC,* Vol. 11, Iss 3, pp. 281-297.

Feser, K., Sun, R., Eriksson, A., Frohlich, K. (1991) "On-Site Dielectric Testing of GIS—Theoretical and Practical Considerations" *IEEE Transactions on Power Delivery,* Vol. 6, Iss 2, pp. 615-625.

Fischer, M.M., Nijkamp, P. (1992) "Geographic Information Systems and Spatial-Analysis" *Annals of Regional Science,* Vol. 26, Iss 1, pp. 3-17.

Fisher, P. (1996) "Geographic Information Systems for Geoscientists—Modelling with GIS", by G.F. Bonhamcarter, *Environment and Planning B-Planning and Design,* Vol. 23, Iss 3, pp. 384-385.

Fisher, P. (1996) "Spatial-Analysis and GIS", by S. Fotheringham, P. Rogerson, *Environment and Planning B-Planning and Design,* Vol. 23, Iss 2, pp. 245-246 Fisher-P Strachan-A Stuart-N (1996) "UK Developments in Environmental GIS—In Honor of Coppock,T" *IJGIS,* Vol. 10, Iss 1, pp. 3+.

Fisher, P.F. (1989) "Geographical Information System Software for* Teaching", *Journal of Geography in Higher Education,* Vol. 13, Iss 1, pp. 69-80.

Fitzsimons, D. (1991), "Cartography Geographic Information Systems at Southwest-Texas-State-University" *CGIS,* Vol. 18, Iss 3, pp. 176-178.

Flowerdew, R. (1988) "Rrl Workshop on Geographical Information Systems—Report of a 2-Day Workshop Sponsored by Esrc for the Regional Research Laboratory Initiative at the University-of-Edinburgh, 18-19 September 1987", *AREA,* Vol. 20, Iss 1, pp. 89-90.

Flowerdew, R. (1990) "Population and Geographical Information Systems" *AREA,* Vol. 22, Iss 2, pp. 207-208.

Flowerdew, R., Green, M. (1992) "Developments in Areal Interpolation Methods and GIS" *Annals of Regional Science*, Vol. 26, Iss 1, pp. 67-78.

Fogarty, B.W. (1994) "Grand Challenges for GIS and Whats Really Relevant to Social-Science" *Social Science Computer Review,* Vol. 12, Iss 2, pp. 193-201.

Foody, G. (1991) "3-Dimensional Applications in Geographic Information Systems", by J. Raper, *AREA,* Vol. 23, Iss 3, pp. 278-279.

Forer, P. (1993) "An Introduction to Urban Geographic Information Systems", by W.E. Huxhold, *Australian Geographer,* Vol. 24, Iss 2, pp. 89-90.

Forrest, D. (1990) "3-Dimensional Applications in Geographic Information Systems", by J. Raper, *Cartographic Journal,* Vol. 27, Iss 1, pp. 70-72.

Fotheringham, A.S. (1991) "GIS and Spatial-Analysis—An Ncgia Research Initiative" *Environment and Planning A,* Vol. 23, Iss 10, pp. 1390-1391.

Fotheringham, A.S. (1992) "Exploratory Spatial Data-Analysis and GIS" *Environment and Planning A,* Vol. 24, Iss 12, pp. 1675-1678.

Fotheringham, A.S. (1993) "On the Future of Spatial-Analysis—The Role of GIS" *Environment and Planning A,* Iss NSI, pp. 30-34.

Fotheringham, A.S. (1994) "What Does Doing a PhD in GIS Mean", *Environment and Planning A,* Vol. 26, Iss 1, pp. 6-8.

Fotheringham, A.S. (1996) "1994 International GIS Sourcebook, by GIS-Wcrld", *Progress in Human Geography,* Vol. 20, Iss 3, pp. 426-426.

Fotheringham, A.S., Rogerson, P.A. (1993) "GIS and Spatial Analytical Problems", *IJGIS,* Vol. 7, Iss 1, pp. 3-19.

Fox, L., Carlson, S.A., Campbell, H. (1992) "Integrating Geographic Information Systems Education into the Curriculum at Humboldt-State-University" *PERS,* Vol. 58, Iss 8, pp. 1158-1158.

Frank, Andrew U., Egenhofer, Max J. and Kuhn, Werner (1992), A perspective on GIS Technology in the nineties, *Photogrammetric Engineering and Remote Sensing,* Vol. 57, No.11, pp. 1431-6.

Frank, A.U., Egenhofer, M.J., Kuhn, W. (1991) "A Perspective on GIS Technology in the Nineties" *PERS* Vol. 57, Iss 11, pp. 1431-1436.

Franklin, C. (1992) "An Introduction to Geographic Information Systems—Linking Maps to Databases", *Database-The Magazine of Database Reference and Review,* Vol. 15, Iss 2, pp. 12.

Franklin, C. (1993) "Using Geographic Information Systems (GIS) Software to Improve the Quality of Access to Us Census-Data on Cd-ROM" *Online aand CDRom* Review, Vol. 17, Iss 3, pp. 190-190.

French, S.P., Wiggins, L.L. (1990) "California Planning Agency Experiences with Automated Mapping and Geographic Information Systems" *Environment and Planning B-Planning and Design,* Vol. 17, Iss 4, pp. 441-450.

Furdada, G., Marti, G., Oller, P., Garcia, C., Mases, M., Vilaplana, J.M. (1995) "Avalanche Mapping and Related GIS Applications in the Catalan-Pyrenees", *Surveys in Geophysics,* Vol. 16, Iss 5-6, pp. 681-693.

Gaile, G.L. (1996) "Possibilities and Constants or GIS Applications in Developing-Countries", by P. Vanteeffelen, L. Vangrunsven, O. Verkoren, *Annals of the Association of American Geographers,* Vol. 86, Iss 4, pp. 802-803.

Gainey, Kevin and Tim Johnson (2000), Faster, flexible redistricting applications built with MapObjects, *ArcUser,* April-June, pp. 10-3.

Ganter, J.H. (1990) "3-Dimensional Applications in Geographical Information Systems, by J. Raper" *PG,* Vol. 42, Iss 4, pp. 508-509.

Garg, P.K., Harrison, A.R. (1992) "Land Degradation and Erosion Risk Analysis in S.E. Spain—A Geographic Information System Approach" *CATENA,* Vol. 19, Iss 5, pp. 411-425.

Gastelluetchegorry, J.P., Estreguil, C., Mougin, E., Laumonier, Y. (1993) "A GIS Based Methodology for Small-Scale Monitoring of Tropical Forests—A Case-Study in Sumatra" *IJRS* Vol. 14, Iss 12, pp. 2349-2368.

Gastelluetchegorry, J.P., Sinulingga, A.B. (1988) "Designing a GIS for the Study of Forest Evolution in Central Java" *TIJDSCHRIFT VOOR Economische en Sociale Geografie,* Vol. 79, Iss 2, pp. 93-103.

Gatrell, A. (1991) "Information Technology in Geography and Planning—Including Principles of GIS", by I. Bracken, C. Webster, *Urban Studies,* Vol. 28, Iss 6, pp. 1050-1052.

Gatrell, A. (1994) "Geographic Information Systems, Spatial Modelling and Policy Evaluation", by M.M. Fischer, P. Nijkamp, *Environment and Planning B-Planning and Design,* Vol. 21, Iss 5, pp. 647-648.

Gatrell, A.C. (1988) "A UK Geographic Information System for Environmental Monitoring, Resource Planning and Management Capable of Integrating and Using Satellite Remotely Sensed Data, by J.A.T. Young" *Regional Studies,* Vol. 22, Iss 2, pp. 172-173.

Gatrell, A.C., Dunn, C.E., Boyle, P.J. (1991) "The Relative Utility of the Central Postcode Directory and Pinpoint Address Code in Applications of Geographical Information Systems" *Environment and Planning A,* Vol. 23, Iss 10, pp. 1447-1458.

Gatrell, T. (1991) "Geographical Information Systems for Urban and Regional Planning", by H.J. Scholten, J.C.H. Stillwell, *Regional Studies,* Vol. 25, Iss 4, pp. 371-371.

Gatrell, T. (1997) "Spatial-Analysis—Modelling in a GIS Environment, by P. Longley, M. Batty" *Geography,* Vol. 82, Iss 356, pp. 288-288.

Gauthier, M.J. (1994) "Geographic Information System in Images—French", by C. Collet, *Canadian Geographer-Geograhe Canadien,* Vol. 38, Iss 2, pp. 191-192.

Geertman, S.C.M., Vaneck, J.R.R. (1995) "GIS and Models of Accessibility Potential—An Application in Planning" *IJGIS,* Vol. 9, Iss 1, pp. 67-80.

Gentile, J. (1993) "GIS Applications in Natural-Resources", by M. Heit, A. Shortreid, *PG,* Vol. 45, Iss 1, pp. 113-113.

Ghosh, R., Goel, R.K., Lole, B.S., Singh, T.P., Sastry, K.L.N., Patel, J.G., Vanikar, Y.V., Thakker, P.S., Navalgund, R.R. (1993) "District Level Planning—A Case-Study for the Panchmahals District Using Remote-Sensing and Geographic Information System Techniques," *International Journal of Remote Sensing,* Vol. 14, Iss 17, pp. 3163-3168.

Gifford, J.A. (1993) "Videography and Geographical Information Systems for Recording the Excavation of a Prehistoric Underwater Site," *International Journal of Nautical Archaeology,* Vol. 22, Iss 2, pp. 167-172.

Gilliland, M.W., Baxterpotter, W. (1987) "A Geographic Information System to Predict Non-Point Source Pollution Potential" *Water Resources Bulletin,* Vol. 23, Iss 2, pp. 281-291.

Gilmartin, P., Cowen, D. (1991) "Educational Essentials for Todays and Tomorrows Jobs in Cartography and Geographic Information Systems" *CGIS,* Vol. 18, Iss 4, pp. 262-267.

Gleave, D. (1992) "Spatial-Analysis and Spatial Policy Using Geographic Information Systems", by L. Worrall, *Regional Studies,* Vol. 26, Iss 6, pp. 600-601.

Gluck, M. (1992) "Prospecting for User-Based GIS Processes," *Proceedings, of the ASIS Annual Meeting,* Vol. 29, pp. 242-248.

Godschalk, D.R., Mcmahon, G., Kaplan, A., Qin, W. (1992) "Using GIS for Computer-Assisted Dispute Resolution" *PERS* 1992, Vol. 58, Iss 8, pp. 1209-1212.

Golledge, R.G., Loomis, J.M., Klatzky, R.L., Flury, A., Yang, X.L. (1991)"Designing a Personal Guidance-System to Aid Navigation Without Sight—Progress on the GIS Component" *IJGIS,* Vol. 5, Iss 4, pp. 373-395.

Gonzalez, C., Ortigosa, L., Marti, C., Garciaruiz, J.M. (1995) "The Study of the Spatial-Organization of Geomorphic Processes in Mountain Areas Using GIS," *Mountain Research and Development,* Vol. 15, Iss 3, pp. 241-249.

Goodchild, M.F. (1991), Geographical information science, *International Journal on Geographical Information System,* Vol. 6, No. 1, pp. 31-45.

Goodchild, M., Haining, R., Wise, S., Arbia, G., Anselin, L., Bossard, E., Brunsdon, C., Diggle, P., Flowerdew, R., Green, M., Griffith, D., Hepple, L., Krug, T., Martin, R., Openshaw, S. (1992) "Integrating GIS and Spatial Data-Analysis—Problems and Possibilities" *International Journal of Geographic Information Systems,* Vol. 6, Iss 5, pp. 407-423.

Goodchild, M.F. (1988) "Geographic Information Systems," *Progress in Human Geography,* Vol. 12, Iss 4, pp. 560-566.

Goodchild, M.F. (1990) "GIS and Basic Research—The National Center for Geographic Information and Analysis," *Government Information Quarterly,* Vol. 7, Iss 3, pp. 343-355.

Goodchild, M.F., (1991) "Geographic Information Systems", *Journal of Retailing,* Vol. 67, Iss 1, pp. 3-15.

Goodchild, M.F. (1991) "Geographic Information Systems,"*Progress in Human Geography,* Vol. 15, Iss 2, pp. 194-200.

Goodchild, M.F., Mark, D.M. (1993) "GIS, Geography, and Ncgia—Response," *PG,* Vol. 45, Iss 2, pp. 216-220.

Goodchild, M.F., Yang, S.R. (1992) "A Hierarchical Spatial Data Structure for Global Geographic Information Systems," *CVGIP-Graphical Model and Image Processing,* Vol. 54, Iss 1, pp. 31-44.

Gordon, W.R. (1994) "A Role for Comprehensive Planning, Geographical Information System (GIS) Technologies and Program-Evaluation in Aquatic Habitat Development," *Bulletin of Marine Science,* Vol. 55, Iss 2-3, pp. 995-1013.

Gordon, W.R., Soubra, N.M. (1992) "Geographical Information Systems and Planning in the USA—Selected Municipal Adoption Trends and Educational Concerns" *IJGIS,* Vol. 6, Iss 4, pp. 267-278.

Gould, M.D. (1994) "GIS Design—A Hermeneutic View" *PERS* 1994, Vol. 60, Iss 9, pp. 1105-1116.

Gould, M.I. (1992) "The Use of GIS and Cac by Health Authorities—Results from a Postal Questionnaire" *AREA,* Vol. 24, Iss 4, pp. 391-401.

Goulter, I.C., Forrest, D. (1987) "Use of Geographic Information Systems (GIS) in River Basin Management," *Water Science and Technology,* Vol. 19, Iss 9, pp. 81-86.

Goulter, I.C., Forrest, D. (1987) "Use of Geographic Information Systems (GIS) in River Basin Management," *Water Science and Technology,* Vol. 19, Iss 9, pp. 81-86.

Graham, D.F., Bonhamcarter, G.F. (1993) "Airborne Radiometric Data—A Tool for Reconnaissance Geological Mapping Using a GIS" *PERS* Vol. 59, Iss 8, pp. 1243-1249.

Gratton, D.J., Howarth, P.J., Marceau, D.J. (1990) "Combining dem Parameters with LANDSAT MSS and TM Imagery in a GIS for Mountain Glacier Characterization" *IEEE Transactions on Geoscience and Remote Sensing,* Vol. 28, Iss 4, pp. 766-769.

Green, N.P., Finch, S. and Wiggins, J. (1986), The state of art in geographic information system, *Area,* Vol. 17.No. 4, pp. 295-301.

Green, K., Finney, M., Campbell, J., Weinstein, D., Landrum, V. (1995) "Fire—Using GIS to Predict Fire Behavior" *JF* Vol. 93, Iss 5, pp. 21-25.

Green, N.P., Finch, S., Wiggins, J. (1985) "The State of the Art in Geographical Information Systems" *AREA,* Vol. 17, Iss 4, pp. 295-301.

Green, R. (1990) "Geographic Information Systems in Europe" *Cartographic Journal,* Vol. 27, Iss 1, pp. 40-42.

Greer, J.D. (1994) "GIS and Remote-Sensing for Wildland Fire Suppression and Burned Area Restoration" *PERS* 1994, Vol. 60, Iss 9, pp. 1059-1064.

Grelot, J.P. (1994)"Cartography in the GIS Age" *Cartographic Journal,* Vol. 31, Iss 1, pp. 56-60.

Greve, C.W., Kelmelis, J.A., Fegeas, R., Guptill, S.C., Mouat, N, (1993) "Investigating United-States-Geological-Survey Needs for the Management of Temporal GIS Data" *PERS* Vol. 59, Iss 10, pp. 1503-1507.

Griffiths, A. (1989) "Sagis—A Proposal for a Sardinian Geographical Information System and an Assessment of Alternative Implementation Strategies," *Journal of Information Science,* Vol. 15, Iss 4-5, pp. 261-267.

Grimshaw, D.J. (1989) "Geographical Information Systems—A Tool for Business and Industry", *International Journal of Information Management,* Vol. 9, Iss 2, pp. 119-126.

Grimshaw, D.J. (1991) "Geographical Information Systems as Part of the Corporate Information Strategy," *International Journal of Information Management,* Vol. 11, Iss 4, pp. 292-297.

Griner, A.J. (1994) "Development of a Water-Supply Protection Model in a GIS," *Water Resources Bulletin* 1993, Vol. 29, Iss 6, pp. 965-971.

Grossmann, W.D., Eberhardt, S. (1992) "Geographical Information Systems and Dynamic Modelling—Potentials of a New Approach," *Annals of Regional Science,* Vol. 26, Iss 1, pp. 53-66.

Grove, M., Hohmann, M. (1992) "Social Forestry and GIS" *JF* 1992, Vo!. 90, Iss 12, pp. 10-15.

Grunblatt, J., Ottichilo, W.K., Sinange, R.K., (1992) "A GIS Approach to Desertification Assessment and Mapping," *Journal of Arid Environments,* Vol. 23, Iss 1, pp. 81-102.

Gumbricht, T. (1996) "Application of GIS in-Training for Environmental-Management," *Journal of Environmental Management,* Vol. 46, Iss 1, pp. 17-30.

Guthe, W.G., Tucker, R.K., Murphy, E.A., England, R., Stevenson, E., Luckhardt, J.C. (1993) "Reassessment of Lead-Exposure in New-Jersey Using GIS Technology" *Environmental Research* 1992, Vol. 59, Iss 2, pp. 318-325.

Haining, R., Ma, J.S., Wise, S. (1996) "Design of a Software System for Interactive Spatial Statistical-Analysis Linked to a GIS," *Computation*al Statistics, Vol. 11, Iss 4, pp. 449-466.

Hall, G.B., Wang, F., Subaryono (1992) "Comparison of Boolean and Fuzzy Classification Methods in Land Suitability Analysis by Using Geographical Information Systems," *Environment and Planning A,* Vol. 24, Iss 4, pp. 497-516.

Healey, R.G. and Desa, G.B. (1989), Transputer based parallel processing for GIS analysis: problems and potentialities, *Auto-Carto, 9 Proceedings,* April, pp. 90-99.

Hendrix, W.G., Price, J.E., Fabos, J.G. (1988) "An Ecological Approach to Landscape Planning Using Geographic Information System Technology" *Landscape and Urban Planning,* Vol. 15, Iss 3-4, pp. 211-225.

Hernon, P. (1997) "GIS as a Service Option," *Journal of Academic Librarianship,* Vol. 23, Iss 3, pp. 235-237.

Herold, P. (1996) "Moving Geospatial Data to the Web—GIS at Mann-Library," *LIbrary Hi Tech,* Vol. 14, Iss 4, pp. 86-87.

Herr, A.M., Queen, L.P. (1993) "Crane Habitat Evaluation Using GIS and Remote-Sensing" *PERS* Vol. 59, Iss 10, pp. 1531-1538.

Herring, G.C. (1986) "And a Hard Rain Fell—A GIS True Story of the War in Vietnam", by J. Ketwig, *Journal of American History,* Vol. 73, Iss 1, pp. 152-164.

Hession, W.C., Shanholtz, V.O. (1988) "A Geographic Information System for Targeting Nonpoint Source Agricultural Pollution," *Journal of Soil and Water Conservation,* Vol. 43, Iss 3, pp. 264-266.

Heywood, I. (1990) "Geographic Information Systems in the Social-Sciences—Introduction" *Environment and Planning A,* Vol. 22, Iss 7, pp. 849-852.

Hickey, R. Jankowski, P. (1997) "GIS and Environmental Decision-Making to Aid Smelter Reclamation Planning," *Environment and Planning A,* Vol. 29, Iss 1, pp. 5-19.

Hicks, D.R. (1992) "The Role of Auburn-University Department of Geography in Remote-Sensing GIS for International Programs" *PERS* 1992, Vol. 58, Iss 8, pp. 1155-1157.

Higgs. G. (1994) "Landscape Ecology and GIS", by R.H. Hainesyoung, D.R. Green, S.H. Cousins, *Environment and Planning B-Planning and Design,* Vol. 21, Iss 3, pp. 383-385.

Higgs, G. (1996) "Environmental Modelling with GIS", by M.F. Goodchild, B.O. Parks, L.T. Steyaert, *Environment and Planning B-Planning and Design,* Vol. 23, Iss 4, pp. 501-503.

Hinton, J.C. (1996) "GIS and Remote-Sensing Integration for Environmental Applications" *IJGIS,* Vol. 10, Iss 7, pp. 877-890.

Hirschfield, A., Brown, P., Todd, P. (1995) "GIS and the Analysis of Spatially-Referenced Crime Data -Experiences in Merseyside, UK" *IJGIS,* Vol. 9, Iss 2, pp. 191-210.

Hirschfield, A., Brown, P.J.B., Bundred, P. (1995) "The Spatial-Analysis of Community-Health Services on Wirral Using Geographic Information Systems," *Journal of the Operational Research Society,* Vol. 46, Iss 2, pp. 147-159.

Hiscock, K.M., Lovett, A.A., Brainerd, J.S., Parfitt, J.P. (1995) "Groundwater Vulnerability Assessment—2 Case-Studies Using GIS Methodology," *Quarterly Journal of Engineering Geology* Vol. 28, Iss MAY, pp. 179-194.

Hodgson, M.E. (1992) "Geographic Information Systems—An Introduction", by J. Star, J. Estes *CGIS,* Vol. 19, Iss 3, pp. 181-182.

Hodgson, M.E., Jensen, J.R., Coulter, M.C., Mackey, H.E. (1988)"Monitoring Wood Stork Foraging Habitat Using Remote-Sensing and Geographic Information Systems" *PERS,* Vol. 54, Iss 11, pp. 1601-1607.

Hodgson, M.E., Price, K.P. (1993) "Special GIS Issue—Foreword" *PERS* Vol. 59, Iss 10, pp. 1489-1490.

Holden, R., Turner, T. (1997) "Western-Europe, Current City Expansion and the Use of GIS" *Landscape and Urban Planning,* Vol. 36, Iss 4, pp. 315-326.

Horowitz, A.J. (1997) "Integrating GIS Concepts into Transportation Network Data-Structures," *Transportation Planning and Technology,* Vol. 21, Iss 1-2, pp. 139-153.

Howard, D.C. (1996) "Landscape Ecology and GIS", by R.H. Young, D.R. Green, S.H. Cousins, *IJGIS,* Vol. 10, Iss 1, pp. 137-138.

Hrkal, Z., Trouillard, J.M. (1994) "Use of GIS for Optimalization of Human Activity in a Catchment-Area—An Example of the Beauce Region (France)," *Environmental Geology* 1994, Vol. 24, Iss 1, pp. 22-27.

Huber, T.P. (1993) "Integrated Remote-Sensing and GIS Techniques for Elk Habitat Management," *Journal of Environmental Systems,* Vol. 22, Iss 4, pp. 325-339.

Hultquist, N.B. (1993) "Geographic Information Systems—Developments and Applications", by L. Worrall, *PG,* Vol. 45, Iss 4, pp. 497-498.

Hultquist, N.B. (1994) "Understanding GIS Tutorial GIS Software" *PERS* 1994, Vol. 60, Iss 10, pp. 1203-1205.

Hunt, E.D. (1992) "Upgrading Site-Catchment Analyses with the Use of GIS—Investigating the Settlement-Patterns of Horticulturalists," *World Archaeology,* Vol. 24, Iss 2, pp. 283-309.

Hunter, A.A. (1996) "Modelling Prehistoric Agricultural Productivity in Southwestern Colorado—A GIS Approach", by C.R. Vanwest, *American Antiquity,* Vol. 61, Iss 2, pp. 441-442.

Hutchinson, C.F., Toledano, J. (1993) "Guidelines for Demonstrating Geographical Information Systems Based on Participatory Development," *IJGIS,* Vol. 7, Iss 5, pp. 453-461.

Hutchinson, M.F. (1988) "Principles of Geographical Information Systems for Land Resources Assessment", by P.A. Burrough, *Landscape and Urban Planning,* Vol. 15, Iss 3-4, pp. 357-358.

Huxhold, W.E. (1992) "Spatial-Analysis and Spatial Policy Using Geographic Information Systems", by L. Worrall, *Urban Studies,* Vol. 29, Iss 8, pp. 1352-1356.

Innes, J.E., Simpson, D.M. (1993) "Implementing GIS for Planning—Lessons from the History of Technological Innovation," *Journal of the American Planning Association,* Vol. 59, Iss 2, pp. 230-236.

Jackson, A. (1991) "Resource Assessment in Western Australia Using a Geographic Information System," *AAPG Bulletin-American Association of Petroleum Geologists,* Vol. 75, Iss 3, pp. 600-601.

Jackson, M.J. (1986) "Report of a Workshop on Geographical Information Systems Held in London on 13 September 1985," *IJRS,* Vol. 7, Iss 6, pp. 745-755.

Jackson, M.J. (1992) "Integrated Geographical Information Systems," *IJRS,* Vol. 13, Iss 6-7, pp. 1343-1351.

Jackson, M.J., James, W.J., Stevens, A. (1988) "The Design of Environmental Geographic Information Systems," *Philosophical Transactions of the Royal Society of London Series A-Mathematical and Physical Sciences,* Vol. 324, Iss 1579, pp. 373.

Jaga, R.M.N., Sundaram, A., Natarajan, T. (1993) "Wasteland Development Using Geographic Information System Techniques," *IJRS,* Vol. 14, Iss 17, pp. 3249-3257.

Jankowski, P., Richard, L. (1994) "Integration of GIS-Based Suitability Analysis and Multicriteria Evaluation in a Spatial Decision-Support System for Route Selection," *Environment and Planning B-Planning and Design,* Vol. 21, Iss 3, pp. 323-340.

Janssen, L.L.F., Molenaar, M. (1995) "Terrain Objects, Their Dynamics and Their Monitoring by the Integration of GIS and Remote-Sensing," *IEEE Transactions in Geoscience and Remote Sensing,* Vol. 33, Iss 3, pp. 749-758.

Jensen, J.R., Christensen, E.J. (1986) "Solid and Hazardous-Waste Disposal Site Selection Using Digital Geographic Information System Techniques," *Science of the Total Environment,* Vol. 56, Iss NOV, pp. 265-276.

Jensen, J.R., Cowen, D.J., Halls, J., Narumalani, S., Schmidt, N.J., Davis, B.A., Burgess, B. (1994) "Improved Urban Infrastructure Mapping and Forecasting for Bellsouth Using Remote-Sensing and GIS Technology," *PERS* 1994, Vol. 60, Iss 3, pp. 339-346.

Jensen, J.R., Narumalani, S., Weatherbee, O., Morris, K.S., Mackey, H.E. (1992) "Predictive Modelling of Cattail and Waterlily Distribution in a South-Carolina Reservoir Using GIS," *PERS* 1992, Vol. 58, Iss 11, pp. 1561-1568.

Jenson, S.K., Domingue, J.O. (1988) "Extracting Topographic Structure from Digital Elevation Data for Geographic Information System Analysis" *PERS*, Vol. 54, Iss 11, pp. 1593-1600.

Jing, T. (1995) "Surface-Charge Accumulation—An Inevitable Phenomenon in DC-GIS," *IEEE Transactions on Dielectrics and Electrical Insulation* Vol. 2, Iss 5, pp. 771-778.

John, S.A. (1993) "Data Integration in a GIS—The Question of Data Quality," *ASLIB Proceedings,* Vol. 45, Iss 4, pp. 109-119.

Johnson, D., Keys, L. (1991) "October 1990 GIS Observer," *PERS* Vol. 57, Iss 2, pp. 158-158.

Johnson, E., Klein, M., Mickus, K. (1993) "Assessment of the Feasibility of Utilizing Landsat for Detection and Monitoring of Landfills in a Statewide GIS," *Environmental Geology* Vol. 22, Iss 2, pp. 129-140.

Johnson, M.L. (1996) "GIS in Business—Issues to Consider in Curriculum Decision-Making," *Journal of Geography,* Vol. 95, Iss 3, pp. 98-105.

Johnson, R.M. (1989) "Scholars Guide to Washington, DC for Cartography and Remote-Sensing Imagery (Maps, Charts, Aerial Photographs, Satellite Images, Cartographic Literature and Geographic Information Systems)", by R.E. Ehrenberg, *Journal of American History,* Vol. 75, Iss 4, pp. 1393-1394.

Johnsson, K. (1995) "Fragmentation Index as a Region-Based GIS Operator," *IJGIS,* Vol. 9, Iss 2, pp. 211-220.

Johnston, C.A., Niemi, G.J., Detenbeck, N.E., Bonde, J.P. (1988)"Geographic Information Systems for Cumulative Impact Assessment," *PERS*, Vol. 54, Iss 11, pp. 1609-1615.

Johnston, K.M. (1987) "Natural-Resource Modelling in the Geographic Information System Environment," *PERS*, Vol. 53, Iss 10, pp. 1411-1415.

Jones, G.E. (1992) "Geographic Information Systems and Their Socioeconomic Applications", by D. Martin, *Geography,* Vol. 77, Iss 336, pp. 278-279.

Jones, G.E., Davidson, D.A. (1987) "A Computer-Based Land Resource Information-System—A Geographical Contribution to Land-Use Planning" *Town Planning Review,* Vol. 58, Iss 1, pp. 81-94.

Jones, K. (1996) "GIS for Business and Service Planning", by P. Longley, G. Clarke, *Cartographic Journal,* Vol. 33, Iss 2, pp. 157-157.

Jones, R.M., Copas, C.V., Edmonds, E.A. (1997) "GIS Support for Distributed Group-Work in Regional-Planning," *IJGISC,* Vol. 11, Iss 1, pp. 53-71.

Jordan, C., Mihalyfalvy, E., Garrett, M.K., Smith, R.V. (1994) "Modelling of Nitrate Leaching on a Regional-Scale Using a GIS," *Journal of Environmental Management,* Vol. 42, Iss 3, pp. 279-298.

Jordan, L.E., Rado, B.Q., Sperry, S.L. (1992) "Meeting the Needs of the GIS and Image-Processing Industry in the 1990s" *PERS* 1992, Vol. 58, Iss 8, pp. 1249-1251.

Jordan, L.E., Star, J. (1992) "A Call to Action—Standards for the GIS Community," *PERS* 1992, Vol. 58, Iss 6, pp. 863-864.

Junkin, B.G. (1982) "Development of 3-Dimensional Spatial Displays Using a Geographically Based Information-System," *PERS*, Vol. 48, Iss 4, pp. 577-586.

Kainz, W. (1989) "GIS Technology—Geographical Information Systems, Land Information-Systems and Their Fundamental Principles—German", by N. Bartelme, *Mitteilungen Der Osterreichischen Geographischen Gesellschaft,* Vol. 131, pp. 306-306.

Kainz, W. (1990) "Applications of Spatial Data-Structures—Computer Graphic, Image-Processing, and GIS", by H. Samet, *Mitteilungen Der Osterreichischen Geographischen Gesellschaft,* Vol. 132, pp. 325-325.

Kainz, W. (1991) "4th Vienna Symposium—Geographic Information Systems and Cartography—Theoretical Principles and Future Aspects—Vienna, July 8-9, 1991," *Mitteilungen Der Osterreichischen Geographischen Gesellschaft,* Vol. 133, pp. 293-294.

Kainz, W. (1991)"Applications of Spatial Data-Structures—Computer-Graphics, Image-Processing, and GIS", by H. Samet, *IJGIS,* Vol. 5, Iss 2, pp. 253-254.

Kapetsky, J.M., Worthy, L.D., Hill, J.M. (1988) "A Geographical Information System for Catfish Farming Development," *Aquaculture,* Vol. 68, Iss 4, pp. 311-320.

Karaki, S., Yamazaki, T., Nojima, K., Yokota, T., Murase, H., Takahashi, H., Kojima, S. (1995) "Transient Impedance of GIS Grounding Grid" *IEEE Transaction on Power Delivery,* Vol. 10, Iss 2, pp. 723-731.

Kay, Simon (1988), Geographic information systems: An integrated database for Badiyah, *The Journal of Oman Studies,* Special Report No. 3 (The Scientific results of the Royal Geographical Society's Oman Wahiba Sands Project 1985-1987), pp. 47-54.

Keating, T., Ingram, K., Phillips, W. (1987) "An Integrated Topologic Database Design for Geographic Information Systems" *PERS,* Vol. 53, Iss 10, pp. 1399-1402.

Kelk, B. (1990) "3-Dimensional Applications in Geographical Information Systems", by J. Raper" *Geographical Journal,* Vol. 156, Iss JUL, pp. 226-227.

Keller, C.P., Davis, T.J., Canessa, R.R. (1996) "A Hypertext Tutor for Teaching Principles and Techniques of GIS," *Journal of Geography in Higher Education* 1996, Vol. 20, Iss 2, pp. 193-207.

Kemp, K.K., Frank, A.U. (1996) "Toward Consensus on a European GIS Curriculum—The International Postgraduate Course on GIS," *IJGIS,* Vol. 10, Iss 4, pp. 477-497.

Kemp, K.K., Goodchild, F.M. (1992) "Evaluating a Major Innovation in Higher-Education—The Ncgia Core Curriculum in GIS," *Journal of Geography in Higher Education,* Vol. 16, Iss 1, pp. 21-35.

Kemp, K.K., Goodchild, M.F. (1991) "Developing a Curriculum in Geographic Information Systems—The National Center for Geographic Information and Analysis Core Curriculum Project," *Journal of Geography in Higher Education*, Vol. 15, Iss 2, pp. 123-134.

Kemp, K.K., Goodchild, M.F., Dodson, R.F. (1992) "Teaching GIS in Geography" *PG,* Vol. 44, Iss 2, pp. 181-191.

Kennedy, G.A., Dai, M., Henning, S.A., Vandeveer, L.R. (1996) "A GIS-Based Approach for Including Topographic and Locational Attributes in the Hedonic Analysis of Rural Land Values," *American Journal of Agricultural Economics,* Vol. 78, Iss 5, pp. 1419-1419.

Kennedy, M. (1994) "Geographical Information Systems—Principles and Applications", by D.J. Maguire, M.F. Goodchild, D.W. Rhind, *Annals of the Association of American Geographers,* Vol. 84, Iss 1, pp. 172-173.

Kent, M., Jones, A., Weaver, R., (1993) "Geographical Information Systems and Remote-Sensing in Land-Use Planning—An Introduction," *Applied Geography,* Vol. 13, Iss 1, pp. 5-8.

Kessler, B, (1995) "PC-Based GIS Grows on Foresters," *JF* Vol. 93, Iss 5, pp. 28-29.

Kessler, B. (1995) "PC-Based GIS Grows on Foresters," *JF* Vol. 93, Iss 8, pp. 37-37.

Khorram, S., Devine, H., Brockhaus, J., Cheshire, H. (1992) "Remote-Sensing and GIS Research and Training at North-Carolina-State-University," *PERS* 1992, Vol. 58, Iss 8, pp. 1171-1172.

Kienegger, E.H. (1992) "Assessment of a Waste-Water Service Charge by Integrating Aerial-Photography and GIS," *PERS* 1992, Vol. 58, Iss 11, pp. 1601-1606.

Kim, K.Y., Ventura, S.J., Harris, P.M., Thum, P.G., Prey, J. (1993)"Urban Non-Point-Source Pollution Assessment Using a Geographical Information System" *Journal of Environmental Management,* Vol. 39, Iss 3, pp. 157-170.

King, G.Q. (1991) "Geography and GIS Technology," *Journal of Geography,* Vol. 90, Iss 2, pp. 66-72.

Kirkby, S.D. (1996) "Integrating a GIS with Arm Expert-System to Identify and Manage Dryland Salinization," *Applied Geography,* Vol. 16, Iss 4, pp. 289-303.

Kitron, U., Bouseman, J.K., Jones, C.J. (1992) "Use of the Arc Info GIS to Study the Distribution of Lyme-Disease Ticks in an Illinois County," *Preventive Veterinary Medicine* 1991, Vol. 11, Iss 3-4, pp. 243-248.

Kloos, H.W. (1992) "Graphics Data-Processing and Geographical Information Systems in Municipal Administration," *Wirtschaftsinformatik,* Vol. 34, Iss 6, pp. 577-583.

Klosterman, R.E. (1991) "An Introduction to Urban Geographic Information Systems", by W.E. Huxhold, *Journal of the American Planning Association,* Vol. 57, Iss 2, pp. 236-239.

Klosterman, R.E. (1991) "Geographic Information Systems and Cartographic Modelling", by C.D. Tomlin, *Journal of the American Planning Association,* Vol. 57, Iss 2, pp. 236-239.

Klosterman, R.E. (1991) "Geographic Information Systems—A Management Perspective", by S. Aronoff, *Journal of the American Planning Association,* Vol. 57, Iss 2, pp. 236-239.

Klosterman, R.E. (1991) "Geographic Information Systems—An Introduction", by J. Star, J. Estes, *Journal of the American Planning Association,* Vol. 57, Iss 2, pp. 236-239.

Klosterman, R.E. (1991) "Ncgia Core Curriculum in GIS", by M.F. Goodchild, K.K. Kemp, *Journal of the American Planning Association,* Vol. 57, Iss 2, pp. 236-239.

Klosterman, R.E. (1991) "Principles of Geographic Information Systems for Land Resource Assessment", by P.A. Burrough, *Journal of the American Planning Association,* Vol. 57, Iss 2, pp. 236-239.

Klosterman, R.E. (1991) "Understanding GIS—The Arc/Info Method, by Environm-Syst-Res, Inst," *Journal of the American Planning Association,* Vol. 57, Iss 2, pp. 236-239.

Knapp, W.G.M. van der (1992), The vector to raster conversion: (mis)use in geographical information systems, *International Journal on Geographic Information Systems,* Vol. 6, No. 2, pp.159-70.

Knapp, P.A., Bishop, K.A. Lancaster, J.M., Taylor, R.L. (1993) "Use of GIS in Optimizing Timber Thinning Strategies in the Eastern Sierra-Nevada," *PG,* Vol. 45, Iss 3, pp. 323-331.

Kollias, V.J., Voliotis, A. (1991) "Fuzzy-Reasoning in the Development of Geographical Information Systems—FRSIS—A Prototype Soil Information-System with Fuzzy Retrieval Capabilities," *IJGIS,* Vol. 5, Iss 2, pp. 209-223.

Kontoes, C., Wilkinson, G.G., Burrill, A., Goffredo, S., Megier, J. (1993)"An Experimental System for the Integration of GIS Data in Knowledge-Based Image-Analysis for Remote-Sensing of Agriculture," *IJGIS,* Vol. 7, I ss 3, pp. 247-262.

Kontoes, C., Wilkinson, G.G., Burrill, A., Goffredo, S., Megier, J. (1994) "An Experimental System for the Integration of GIS Data in Knowledge-Based Image-Analysis for Remote-Sensing of Agriculture (Vol. 7, Pg 247, 1993)," *International Journal of Geographic Information Systems,* Vol. 8, Iss 5, pp. 489-489.

Korp, P., Mckenzie, B. (1990) "Federal-Land and Geographic Information System Activities" *PERS,* Vol. 56, Iss 5, pp. 631-631.

Kosters, G., Pagel, B.U., Six, H.W. (1997) "GIS-Application Development with Geoooa," *IJGISC,* Vol. 11, Iss 4, pp. 307-335.

Koutnik, M.A., Padilla, D.K. (1994) "Predicting the Spatial-Distribution of Dreissena-Polymorpha (Zebra Mussel) Among Inland Lakes of Wisconsin—Modelling with a GIS" *Canadian Journal of Fisheries of Aquatic Sciences* 1994, Vol. 51, Iss 5, pp. 1189-1196.

Krautheim, K.R., Aldrich, T.E. (1997) "Geographic Information System (GIS) Studies of Cancer Around NPL-Sites (1997) Reprinted from Hazardous-Waste, Impacts on Human and Ecological Health, pp. 459-464) *Toxicology and Industrial Health,* 1997, Vol. 13, Iss 2-3, pp. 357-362.

Krist, F.J., Brown, D.G. (1994) "GIS Modelling of Paleo-Indian Period Caribou Migrations and Viewsheds in Northeastern Lower Michigan" *PERS* 1994, Vol. 60, Iss 9, pp. 1129-1137.

Kubo, S. (1984) "ALIS—A Geographical Information System for Urban Research" *IEEE Computer Graphics and Applications,* Vol. 4, Iss 5, pp. 68-76.

Kuijperslinde, M. (1992) "Geographic Information Systems—Developments and Applications", by L. Worrall, *Environment and Planning B-Planning and Design,* Vol. 19, Iss 3, pp. 359-360.

Kumar, L., Skidmore, A.K., Knowles, E. (1997) "Modelling Topographic Variation in Solar-Radiation in a GIS Environment," *IJGISC,* Vol. 11, Iss 5, pp. 475-497.

Kumar, N.K., Evans, I.S. (1997) "Map Output Format from GIS—Optimizing Visual Quality of Paper and Electronic Atlases," *Cartographic Journal,* Vol. 34, Iss 1, pp. 37-41.

Kupiszewski, M. (1993) "Transfer of Geographical Information Systems to Poland as an Example of the Transfer of Information Technology," *International Journal of Information Management,* Vol. 13, Iss 1, pp. 41-50.

Liengsakul, M., Mekpaiboonwatana, S., Pramojanee, P., Bronsveld, K., Huizing, H. (1994) "Use of GIS and Remote-Sensing for Soil Mapping and for Locating New Sites for Permanent Cropland—A Case-Study in the Highlands of Northern Thailand," *Geoderma* 1993, Vol. 60, Iss 1-4, pp. 293-307.

Lachowski, H., Maus, P., Platt, B. (1992) "Integrating Remote Sensing with GIS" *JF,* Vol. 90, Iss 12, pp. 16-21.

Lachowski, H.M., Wirth, T., Maus, P., Avers, P. (1994) "Remote-Sensing and GIS—Their Role in Ecosystem Management," *JF,* Vol. 92, Iss 8, pp. 39-40.

Lackaff, B.B., Hunt, B.J., Vonessen, I.E. (1994) "The Development and Implementation of a GIS-Based Contaminant Source Inventory over the Spokane Aquifer, Spokane County, Washington," *Water Resources Bulletin* 1993, Vol. 29, Iss 6, pp. 949-955.

Lai, P.C. (1990) "Feasibility of Geographic Information Systems Approach for Natural-Resource Management," *Environmental Management,* Vol. 14, Iss 1, pp. 73-80.

Lai, P.C. (1991) "Issues Concerning the Technology-Transfer of Geographic Information Systems," *Environmental Management,* Vol. 15, Iss 5, pp. 595-601.

Lai, P.C., Gillies, C.F. (1991) "The Impact of Geographical Information Systems on the Role of Spatial Data Libraries," *IJGIS,* Vol. 5, Iss 2, pp. 241-251.

Lake, R.W. (1993) "Planning and Applied Geography—Positivism, Ethics, and Geographic Information Systems," *Progress in Human Geography,* Vol. 17, Iss 3, pp. 404-413.

Lal, H., Hoogenboom, G., Calixte, J.P., Jones, J.W., Beinroth, F.H. (1993) "Using Crop Simulation-Models and GIS for Regional Productivity Analysis," *Transactions of the ASAE* Vol. 36, Iss 1, pp. 175-184.

Land Management Decisions, *Ecological Enomics,* Vol. 23, Iss 1, pp. 45-57.

Landis, J.D. (1990) "Microcomputer-Based GIS Systems," *Journal of the American Planning Association,* Vol. 56, Iss 3, pp. 373-376.

Lanfear, K.J. (1989) "Geographic Information Systems and Water-Resources Applications," *Water Resources Bulletin,* Vol. 25, Iss 3, pp. R5-R6.

Langford, M., Unwin, D.J. (1994) "Generating and Mapping Population Density Surfaces Within a Geographical Information System" *Cartographic Journal,* Vol. 31, Iss 1, pp. 21-26.

Lang, R., Muller, A., Lenz, R., Selige, T., Forster, E.M. (1997) "Landscape Modelling and GIS-Applications in the Munich-Research-Association-for-Agricultural-Ecosystems" *Landscape and Urban Planning,* Vol. 37, Iss 1-2, pp. 11-18.

Lanter, D.P. (1991) "Design of a Lineage-Based Meta-Data Base for GIS," *CGIS,* Vol. 18, Iss 4, pp. 255-261.

Lanter, D.P., Veregin, H. (1992) "A Research Paradigm for Propagating Error in Layer-Based GIS," *PERS* Vol. 58, Iss 6, pp. 825-833.

Lao, Y. (1996) "Innovations in GIS 1, by M.F. Worboys," *PG,* Vol. 48, Iss 2, pp. 235-236.

Lapalme, G., Rousseau, J.M., Chapleau, S., Cormier, M., Cossette, P., Roy, S. (1992) "Georoute—A Geographic Information System for Transportation Applications," *Communications of the ACM,* Vol. 35, Iss 1, pp. 80-88.

Lauer, D.T., Estes, J.E., Jensen, J.R., Greenlee, D.D. (1991) "Institutional Issues Affecting the Integration and Use of Remotely Sensed Data and Geographic Information Systems," *PERS,* Vol. 57, Iss 6, pp. 647-654.

Laxton, J.L. (1996) "Geographic Information Systems for Geoscientists—Modelling with GIS", by G.F. Bonhamcarter, *IJGIS,* Vol. 10, Iss 3, pp. 355-356.

Lee, C.T., Marsh, S.E. (1995) "The Use of Archival Landsat MSS and Ancillary Data in a GIS Environment to Map Historical Change in an Urban Riparian Habitat," *PERS* Vol. 61, Iss 8, pp. 999-1008.

Lee, J., Kretzschmar, W.A. (1993) "Spatial-Analysis of Linguistic Data with GIS Functions," *IJGIS,* Vol. 7, Iss 6, pp. 541-560.

Lees, B.G., Ritman, K., (1991) "Decision-Tree and Rule-Induction Approach to Integration of Remotely Sensed and GIS Data in Mapping Vegetation in Disturbed or Hilly Environments," *Environmental Management* Vol. 15, Iss 6, pp. 823-831.

Lee, Y.C. (1990) "Geographic Information Systems for Urban Applications Problems and Solutions," *Environment and Planning B-Planning and Design,* Vol. 17, Iss 4, pp. 463-473.

Lee, Y.C., Chin, F.L. (1995) "An Iconic Query Language for Topological Relationships in GIS," *IJGIS,* Vol. 9, Iss 1, pp. 25-46.

Leggett, D.J., Jones, A. (1996) "The Application of GIS for Flood Defense in the Anglian Region—Developing for the Future," *IJGIS,* Vol. 10, Iss 1, pp. 103-116.

Lehmann, A., Jaquet, J.M., Lachavanne, J.B. (1994) "Contribution of GIS to Submerged Macrophyte Biomass Estimation and Community Structure Modelling, Lake Geneva, Switzerland" *Aquati Botany* Vol. 47, Iss 2, pp. 99-117.

Leipnik, M.R., Kemp, K.K., Loaiciga, H.A. (1993) "Implementation of GIS for Water-Resources Planning and Management" *Journal of Water Resources Planning and Management-ASCE* Vol. 119, Iss 2, pp. 184-205.

Leung, Y., Leung, K.S. (1993) "An Intelligent Expert-System Shell for Knowledge-Based Geographical Information Systems.1. The Tools" *IJGIS,* Vol. 7, Iss 3, pp. 189-199.

Leung, Y., Leung, K.S. (1993) "An Intelligent Expert-System Shell for Knowledge-Based Geographical Information Systems.2. Some Applications," *IJGIS,* Vol. 7, Iss 3, pp. 201-213.

Levine, J., Landis, J.D. (1989) "Geographic Information Systems for Local-Planning," *Journal of the American Planning Association,* Vol. 55, Iss 2, pp. 209-220.

Levine, N. (1996) "Spatial Statistics and GIS—Software Tools to Quantify Spatial Patterns" *Journal of the American Planning Association,* Vol. 62, Iss 3, pp. 381-391.

Lew, A.A. (1992) "Geographic Information Systems and Their Socioeconomic Applications", by D. Martin, *PG,* Vol. 44, Iss 4, pp. 488-489.

Lewis, DM (1995) "Importance of GIS to Community-Based Management of Wildlife—Lessons from Zambia," *Ecological Applications* Vol. 5, Iss 4, pp. 861-871.

Lillesand, J.M. (1990), Satellite Remote Sensing: Its Evolution and Synergism with GIS technology, *Government Information Quarterly,* Vol. 7:3, pp. 307-27.

Lillesand, T. (1990) Satellite Remote-Sensing—Its Evolution and Synergism with GIS Technology," *Government Information Quarterly,* Vol. 7, Iss 3, pp. 307-327.

Limp, W.F. (1987) "Integrating Database and Geographic Information Systems for Archaeologists," *American Journal of Archaeology,* Vol. 91, Iss 2, pp. 323-323.

Linden, G. (1987) "An Integrated Study of Desertification—Applications of Remote-Sensing, GIS and Spatial Models in Semi-Arid Sudan", by L. Olsson, *Third World Planning Review,* Vol. 9, Iss 2, pp. 195-196.

Lindholm, M., Sarjakoski, T. (1992) "User Models and Information-Theory in the Design of a Query Interface for GIS," *Lecture Notes in Comptuer Science,* Vol. 639, pp. 328-347.

Lindhult, M.S., Price, N., Brown, P., Fabos, J. (1988) "Using Geographic Information Systems to Assess Conflicts Between Agriculture and Development," *Landscape and Urban Planning,* Vol. 16, Iss 4, pp. 333-343.

Lin, H., Wan, Q., Li, X., Chen, J., Kong,Y. (1997) "GIS-Based Multicriteria Evaluation for Investment Environment," *Environment and Planning B-Planning and Design,* Vol. 24, Iss 3, pp. 403-414.

Linsey, T.K., Raper, J.F. (1993) "Hyperarc—A Task-Oriented Hypertext GIS Interface," *IJGIS,* Vol. 7, Iss 5, pp. 435-452.

Lipschultz, M.S. Glaser, D.E. (1992) "Rainy Season Flooding Being Tracked by GIS" *Water-Engineering and Management,* Vol. 139, Iss 6, pp. 22-25.

Liu, J.G., Dunning, J.B., Pulliam, H.R. (1995) "Potential Effects of a Forest Management Plan on Bachman Sparrows (Aimophila-Aestivalis)—Linking a Spatially Explicit Model with GIS," *Conservation Biology* Vol. 9, Iss 1, pp. 62-75.

Llobera, M. (1996) "Exploring the Topography of Mind—GIS, Social Space and Archaeology" *Antiquity,* Vol. 70, Iss 269, pp. 612-622.

Lo, C.P. (1995) "Automated Population and Dwelling Unit Estimation from High-Resolution Satellite Images—A GIS Approach," *IJRS* Vol. 16, Iss 1, pp. 17-34.

Lo, C.P., Shipman, R.L. (1990) "A GIS Approach to Land-Use Change Dynamics Detection" *PERS* Vol. 56, Iss 11, pp. 1483-1491.

Loh, D.K., Rykiel, E.J. (1992) "Integrated Resource-Management Systems—Coupling Expert Systems with Database-Management and Geographic Information Systems," *Environmental Management,* Vol. 16, Iss 2, pp. 167-177.

Lombard, A.T. (1993) "Multispecies Conservation, Advanced Computer Architecture and GIS—Where Are We Today," *South African Journal of Science* Vol. 89, Iss 9, pp. 415-418.

Lombard, A.T. August, P.V., Siegfried, W.R. (1992) "A Proposed Geographic Information System for Assessing the Optimal Dispersion of Protected Areas in South-Africa," *South African Journal of Science*, Vol. 88, Iss 3, pp. 136-140.

Longley, P. (1990) "Geographical Information Systems and the Regional Research Laboratories," *AREA,* Vol. 22, Iss 2, pp. 199-200.

Longley, P. (1991) "Geographic Information Systems and Cartographic Modelling, by C.D. Tomlin," *Environment and Planning B-Planning and Design,* Vol. 18, Iss 2, pp. 256-257.

Longley, P. (1992) "Introductory Readings in Geographic Information Systems", by D.J. Peuquet, D.F. Marble, *Environment and Planning B-Planning and Design,* Vol. 19, Iss 1, pp. 116-117.

Longley, P. (1993) "Quantitative-Analysis, Technological Change, and the Dissemination of GIS Research—Some Reflections and Prospects," *Environment and Planning A,* Iss NSI, pp. 35-37.

Longley, P. (1995) "The 1995 European GIS Yearbook", by P.J. Shand, P.J. Ireland *Environment and Planning B-Planning and Design,* Vol. 22, Iss 6, pp. 772-773.

Longley, P., Higgs, G., Martin, D., (1994) "The Predictive Use of GIS to Model Property Valuations" *IJGIS,* Vol. 8, Iss 2, pp. 217-235.

Longley, P.A. (1989) "Applied Geographical Information Systems," *AREA,* Vol. 21, Iss 3, pp. 337-338.

Lopezblanco, J. Villersruiz, L. (1995) "Delineating Boundaries of Environmental Units for Land Management Using a Geomorphological Approach and GIS—A Study in Baja-California, Mexico," *Remote Sensing of Environment* Vol. 53, Iss 2, pp. 109-117.

Lopez, X.R. (1993) "Database Copyright Issues in the European GIS Community," *Government Information Quarterly,* Vol. 10, Iss 3, pp. 305-318.

Loukes, D.K., Mclaughlin, J. (1991) "GIS and Transportation—Canadian Perspective" *Journal of Surveying Engineering-ASCE,* Vol. 117, Iss 3, pp. 123-133.

Love, D., Lindquist, P. (1995) "The Geographical Accessibility of Hospitals to the Aged-A Geographic Information Systems Analysis Within Illinois," *Health Services Research,* Vol. 29, Iss 6, pp. 629-651.

Loveland, Thomas R., James W. Merchant, Donald O. Ohlen and Jesslyn F. Brown (1991), Development of a land-cover characteristics database for the conterminous U.S., *Photogrammertic Engineering and Remote Sensing,* Vol. 57, No. 11, pp. 1453-63.

Lovett, A.A., Parfitt, J.P., Brainard, J.S. (1997) "Using GIS in Risk Analysis—A Case-Study of Hazardous-Waste Transport" *Risk Analysis,* Vol. 17, Iss 5, pp. 625-633.

Lovett, J. (1993) "GIS Promise Applied," *SEARCH* Vol. 24, Iss 1, pp. 15-15.

Low, Cost Housing Development in Kuala-Lumpur, (Malaysia1990)" *Environment and Planning B-Planning and Design,* Vol. 17, Iss 4, pp. 385-393.

Lowell, K. (1991) "Utilizing Discriminant Function-Analysis with a Geographical Information System to Model Ecological Succession Spatially" *IJGIS,* Vol. 5, Iss 2, pp. 175-191.

Lowell, K.E. (1990) "Differences Between Ecological Land Type Maps Produced Using GIS or Manual Cartographic Methods," *PERS* Vol. 56, Iss 2, pp. 169-173.

Lowell, K.E. (1994), "Probabilistic Temporal GIS Modelling Involving More Than 2 Map Classes," *IJGIS,* Vol. 8, Iss 1, pp. 73-93.

Lowry, J.H., Miller, H.J., Hepner, G.F. (1995) "A GIS-Based Sensitivity Analysis of Community Vulnerability to Hazardous Contaminants on the Mexico/Us Border" *PERS* Vol. 61, Iss 11, pp. 1347-1359.

Ludeke, A.K., Maggio, R.C., Reid, L.M. (1990) "An Analysis of Anthropogenic Deforestation Using Logistic-Regression and GIS," *Journal of Environmental Management,* Vol. 31, Iss 3, pp. 247-259.

Lui, C.Y., Hiley, J. (1994) "Computational Study of Very Fast Transients in GIS with Special Reference to Effects of Trapped Charge and Risetime on Overvoltage Amplitude" *IEE Proceedings-Generation and Distribution,* Vol. 141, Iss 5, pp. 485-490.

Lull, K.J., Tindall, J.A., Potts, D.F. (1995) "Assessing Nonpoint-Source Pollution Risk—A GIS Application" JF Vol. 93, Iss 1, pp. 35-40.

Lundgaard, L.E., Tangen, G., Skyberg, B., Faugstad, K. (1992) "Acoustic Diagnoses of GIS—Field Experience and Development of Expert System" *IEEE Transactions of Power Delivery,* Vol. 7, Iss 1, pp. 287-294.

Lunetta, R.S., Congalton, R.G., Fenstermaker, L.K., Jensen, J.R., Mcgwire, K.C., Tinney, L.R. (1991) "Remote-Sensing and Geographic Information System Data Integration—Error Sources and Research Issues," *PERS,* Vol. 57, Iss 6, pp. 677 -687.

Lupien, A.E., Dangermond, J. Moreland, W.H. (1987) "Network Analysis in Geographic Information Systems," *PERS,* Vol. 53, Iss 10, pp. 1417-1421.

Lupton, M. Mather, C. (1997) "The Antipolitics Machine—GIS and the Reconstruction of the Johannesburg Local State," *Political Geography,* Vol. 16, Iss 7, pp. 565-580.

Macdougall, E.B. (1981) "Harvard Papers on Geographic Information Systems—1st International Advanced Study Symposium on Topological Data-Structures for Geographic Information Systems", by G. Dutton, *Environment and Planning A,* Vol. 13, Iss 5, pp. 650-651.

Macdougall, E.B. (1992), "Exploratory Analysis, Dynamic Statistical Visualization, and Geographic Information Systems," *CGIS,* Vol. 19, Iss 4, pp. 237-246.

Maceachren, A.M. (1988) "Principles of Geographical Information Systems for Land Resources Assessment", by P.A. Burrough, *PG,* Vol. 40, Iss 2, pp. 249-250.

Mace, T. (1991) "Special Issue—Integration of Remote-Sensing and GIS and Featuring the 1991 Asprs Directory of the Mapping Sciences—Foreword," *PERS* Vol. 57, Iss 6, pp. 641-641.

Macgill, S.M. (1990) "GIS in the 1990s" *Environment and Planning A,* Vol. 22, Iss 12, pp. 1559-1560.

Maclean, A.L. Davello, T.P. Shetron, S.G. (1993) "The Use of Variability Diagrams to Improve the Interpretation of Digital Soil Maps in a GIS," *PERS* Vol. 59, Iss 2, pp. 223-228.

Maclean, A.L., Reed, D.D., Mroz, G.D., Lyon, G.W. Edison, T. (1992) "Using GIS to Estimate Forest Resource Changes," *JF,* Vol. 90, Iss 12, pp. 22-25.

Madon, S., Sahay, S. (1997) "Managing Natural-Resources Using GIS—Experiences in India" *Information and Management,* Vol. 32, Iss 1, pp. 45-53.

Maguire, D.J., Goodchild, M.F. and Rhind. D, (1991), *Geographic Information System,* Longman, London.

Maguire, D. (1997) "GIS and Organizations—How Effective Are GIS in Practice", by H. Campbell, I. Masser, *AREA,* Vol. 29, Iss 2, pp. 189-191.

Maguire, D.J. (1991) "Information Technology in Geography and Planning—Including the Principles of GIS", by I. Bracken, C. Webster, *Transactions-Institute of British Geographers,* Vol. 16, Iss 2, pp. 245-246.

Maguire, D.J. (1992) "Geographic Information Systems—An Introduction", by L. Star, J.E. Estes, *Progress in Human Geography,* Vol. 16, Iss 3, pp. 486-487.

Malczewski, J. (1996) "A GIS-Based Approach to Multiple Criteria Group Decision-Making" *IJGIS,* Vol. 10, Iss 8, pp. 955-971.

Manning, E.W. (1990) "Geographic Information Systems and Sustainable Development" *Government Information Quarterly,* Vol. 7, Iss 3, pp. 329-342.

Marble, D.F. (1991) "Geographical Information Systems for Urban and Regional-Planning", by H.J. Scholten, J.C.H. Stillwell, *Geographical Analysis,* Vol. 23, Iss 4, pp. 371-373.

Marble, D.F. Amundson, S.E. (1988) "Microcomputer-Based Geographic Information Systems and Their Role in Urban and Regional-Planning," *Environment and Planning B-Planning and Design,* Vol. 15, Iss 3, pp. 305-324.

Mark, David M. and Gould, Michael D. (1991), Interacting with Geographic information: A commentary, *Photogrammertic Engineering and Remote Sensing,* Vol. 57, No.11, pp. 1427-30.

Mark, D.M. Calkins, H.W. (1991) "Cartography and Geographic Information Systems at the State-University-of-New-York at Buffalo," *CGIS,* Vol. 18, Iss 3, pp. 198-200.

Marsh, S.E., Walsh, J.L., Hutchinson, C.F. (1990) "Development of an Agricultural Land-Use GIS for Senegal Derived from Multispectral Video and Photographic Data" *PERS* Vol. 56, Iss 3, pp. 351-357.

Martin, D. (1988) "Principles of Geographical Information Systems for Land Resources Assessment", by P.A. Burrough, *Environment and Planning B-Planning and Design,* Vol. 15, Iss 2, pp. 231-232.

Martin, D (1991) "Geographic Information Systems—Developments and Applications", by L. Worrall" *Regional Studies,* Vol. 25, Iss 6, pp. 573-574.

Martin, D. (1994) "Recent Development in GIS" *AREA,* Vol. 26, Iss 2, pp. 199-200.

Martin, D. Longley, P. Higgs, G. (1994) "The Use of GIS in the Analysis of Diverse Urban Databases," *Computers Environment and Urban Systems,* Vol. 18, Iss 1, pp. 55-66.

Martin, F.C. (1985) "Using a Geographic Information System for Forest Land Mapping and Management" *PERS,* Vol. 51, Iss 11, pp. 1753-1759.

Masaki, K., Sakakibara, T., Murase, H., Akazaki, M., Uehara, K. Menju, S. (1994) "On-Site Measurement for the Development of Online Partial Discharge Monitoring-System in GIS" *IEEE Transactions of Power Delivery* Vol. 9, Iss 2, pp. 805-810.

Maslia, M.L., Aral, M.M., Williams, R.C. Susten, A.S., Heitgerd, J.L. (1994) "Exposure Assessment of Populations Using Environmental Modelling, Demographic-Analysis, and GIS," *Water Resources Bulletin,* Vol. 30, Iss 6, pp. 1025-1041.

Mason, D.C., Oconaill, M.A., Bell, S.B.M. (1994), "Handling 4-Dimensional Geo-Referenced Data in Environmental GIS," *IJGIS,* Vol. 8, Iss 2, pp. 191-215.

Masser, Ian (1998), *Governments and Geographic Information,* Taylor & Francis, London.

Masser, I. (1988) "The Development of Geographic Information Systems in Britain—The Chorley Report in Perspective," *Environment and Planning B-Planning and Design,* Vol. 15, Iss 4, pp. 489-494.

Masser, I. (1990) "Information Technology in Geography and Planning Including Principles of GIS", by I. Bracken, C. Webster, *Environment and Planning B-Planning and Design,* Vol. 17, Iss 3, pp. 367-368.

Masser, I. (1992) "Geographical Information Systems for Urban and Regional Planning", by H.J. Scholten, J.C.H. Stillwell, *IJGIS,* Vol. 6, Iss 1, pp. 59-60.

Mather, P.M. (1992) "Introductory Readings in Geographic Information Systems", by D.J. Peuquet, D.F. Marble, *Geographical Journal*, Vol. 158, Iss MAR, pp. 95-96.

Matsuyama, T., Hao, L.V., Nagao, M. (1984) "A File-Organization for Geographic Information Systems Based on Spatial Proximity," *Computer Vision Graphics and Image Processing,* Vol. 26, Iss 3, pp. 303-318.

Matthews, K.B., Macdonald, A., Aspinall, R.J., Hudson, G., Law, A.N.R., Paterson, E. (1994) "Climatic Soil-Moisture Deficit—Climate and Soil Data Integration in a GIS," *Climatic Change,* Vol. 28, Iss 3, pp. 273-287.

Matthews, S.A. (1992) "Geographic Information Systems and Their Socioeconomic Applications", by D. Martin, *Environment and Planning B-Planning and Design,* Vol. 19, Iss 2, pp. 238-239.

Mattson, M.D., Godfrey, P.J. (1994) "Identification of Road Salt Contamination Using Multiple-Regression and GIS," *Environmental Management,* Vol. 18, Iss 5, pp. 767-773.

Maurel, P. (1993) "Satellite Imagery and Geographical Information Systems," *RECHERCHE,* Vol. 24, Iss 258, pp. 1140-1142.

Mcclean, C. (1996) "Remote-Sensing and GIS in Ecosystem Management", by V.A. Sample, *Land Use Policy,* Vol. 13, Iss 3, pp. 237-238.

Mcdonald, G. (1991) "Map-II Map Processor—A Geographic Information System for the Macintosh," *Journal of Architectural and Planning Research*, Vol. 8, Iss 1, pp. 76-78.

Mcdonald, G.T. (1994) "Environmental Modelling with GIS", by M.F. Goodchild, B.O. Parks, L.T. Steyaert, *PG,* Vol. 46, Iss 3, pp. 393-394.

Mcglamery, P., Lamont, M. (1994) "New Opportunities and Challenges—Geographic Information Systems in Libraries," *Database-the Mgazine of Electronic Database Reviews,* Vol. 17, Iss 6, pp. 35-&.

Mclaren, R. (1997) "GIS and Organizations", by H. Campbell, I. Masser" *IJGISC,* Vol. 11, Iss 5, pp. 521-522.

Mcnaughton, R.B. (1992) "Geographic Information Systems—The Microcomputer and Modern Cartography", by D.R.F. Taylor, *Canadian Geographer-Geograhe Canadien,* Vol. 36, Iss 4, pp. 392-393.

Meehan, K.T. (1991) "Integrating Geologic and Economic Data Using a Geographic Information System (GIS)—Offshore Gulf of Mexico," *AAPG Bulletin-American Association of Petroleum Geologists,* Vol. 75, Iss 3, pp. 633-633.

Meijer, E. (1989) "From High-End to Low-End Geographic Information Systems—A Continuum," *EKISTICS-the Problems and Science of Human Settlements,* Vol. 56, Iss 338-, pp. 300-303.

Mellerowicz, K.T., Rees, H.W., Chow, T.L., Ghanem, I. (1994) "Soil Conservation Planning at the Watershed Level Using the Universal Soil Loss Equation with GIS and Microcomputer Technologies—A Case-Study," *Journal of Soil and Water Conservation* Vol. 49, Iss 2, pp. 194-200.

Merchant, J.W. (1988) "Special Issue—Geographic Information Systems (GIS)—Foreword" *PERS*, Vol. 54, Iss 11, pp. 1545-1545.

Merchant, J.W. (1994) "GIS-Based Groundwater Pollution Hazard Assessment—A Critical-Review of the Drastic Model," *PERS* Vol. 60, Iss 9, pp. 1117-1128.

Merchant, J.W. (1994) "Landscape Ecology and GIS", by R. Hainesyoung, D.R. Green, S. Cousins, *PG,* Vol. 46, Iss 4, pp. 516-517.

Michalak, W.Z. (1993) "GIS in Land-Use Change Analysis—Integration of Remotely Sensed Data into GIS," *Applied Geography,* Vol. 13, Iss 1, pp. 28-44.

Mikhova, D., Pickles, J. (1994) "GIS in Bulgaria—Development and Perspectives," *IJGIS,* Vol. 8, Iss 5, pp. 471-477.

Miller, H.J. (1991) "Modelling Accessibility Using Space-Time Prism Concepts Within Geographical Information Systems," *IJGIS,* Vol. 5, Iss 3, pp. 287-301.

Miller, H.J. (1996) "GIS and Geometric Representation in Facility Location Problems," *IJGIS,* Vol. 10, Iss 7, pp. 791-816.

Millette, T.L. (1994) "Geographic Information Systems, Spatial Modelling, and Policy Evaluation", by M.M. Fischer, P. Nijkamp, *Journal of Regional Science,* Vol. 34, Iss 3, pp. 465-468.

Milne, J.A., Sear, D.A. (1997) "Modelling River Channel Topography Using GIS," *IJGISC,* Vol. 11, Iss 5, pp. 499-519.

Mires, P.B. (1993) "Relationships of Louisiana Colonial Land Claims with Potential Natural Vegetation and Historic Standing Structures—A GIS Approach," *PG,* Vol. 45, Iss 3, pp. 342-350.

Missotten, R., Han, Q., Aureli, A. (1994) "Remote-Sensing and GIS Activities in UNESCO" *IJRS* Vol. 15, Iss 15, pp. 3111-3118.

Mitasova, H., Hofierka, J., Zlocha, M., Iverson, L. (1997) "Modelling Topographic Potential for Erosion and Deposition Using GIS—Reply," *IJGISC,* Vol. 11, Iss 6, pp. 611-618.

Mitasova, H., Hofierka, J., Zlocha, M., Iverson, L.R. (1996) "Modelling Topographic Potential for Erosion and Deposition Using GIS," *IJGIS,* Vol. 10, Iss 5, pp. 629-641.

Mitchell, D. (1991) "3-Dimensional Applications in Geographical Information Systems", by J. Raper" *IJGIS,* Vol. 5, Iss 2, pp. 255-256.

Monmonier, M. (1993) "What a Friend We Have in GIS" *PG,* Vol. 45, Iss 4, pp. 448-450.

Morad, M., Chalmers, A.I., Oregan, P.R. (1996) "The Role of Root-Mean-Square Error in the Geo-Transformation of Images in GIS," *IJGIS,* Vol. 10, Iss 3, pp. 347-353.

Morgan, J.M. (1987) "Academic Geographic Information Systems Education—A Commentary" *PERS*, Vol. 53, Iss 10, pp. 1443-1445.

Morgan, K.M., Busbey, A.B., Donovan, R.N., Newland, L.W. (1992) "Image-Processing and GIS at Tcus Center for Remote-Sensing and Energy Research" *PERS* Vol. 58, Iss 8, pp. 1173-1174.

Morgan, K.M., Newland, L.W. (1990) "Geographic Information Systems (GIS) for Environmental Data-Base Management and Analysis," *Toxicological and Environmental Chemistry,* Vol. 26, Iss 1-4, pp. 129-136.

Morse, A., Zarriello, T.J., Kramber, W.J. (1990) "Using Remote-Sensing and GIS Technology to Help Adjudicate Idaho Water Rights," *PERS,* Vol. 56, Iss 3, pp. 365-370.

Moskos, C.C. (1985) "Female GIS in the Field," *SOCIETY,* Vol. 22, Iss 6, pp. 28-33.

Mounsey, H.M. and Tomlinson,R.F., (1988), *Building Data Bases for Global Science,* Taylor and Francis, London.

Moyer, D.D. (1991) "Introductory Readings in Geographic Information Systems", by D.J. Peuquet, D.F. Marble", *Journal of Agricultural Economics Research,* Vol. 43, Iss 4, pp. 28-29.

Mozeto, A.A., Stone, T.A., Brown, I.F., Skole, D.L. (1990) "The Use of a Remote Sensing Geographic Information System in the Evaluation of Environmental-Impact in the Ecological Station of Uhe Samuel, Rondonia, Brazil," *Interciencia,* Vol. 15, Iss 5, pp. 265-271.

Muehrcke, P.C. (1990) "Cartography and Geographic Information Systems," *CGIS,* Vol. 17, Iss 1, pp. 7-15.

Mueller, M. (1988) "Geographic Information System—Small Area Applications," *Population Index,* Vol. 54, Iss 3, pp. 426-426.

Mukaiyama, Y., Takagi, I., Izumi, K., Sekiguchi, T., Kobayashi, A., Sumikawa, T. (1990) "Investigation on Abnormal Phenomena of Contacts Using Disconnecting Switch and Detachable Bus in 300Kv GIS" *IEEE Transactions of Power Delivery* Vol. 5, Iss 1, pp. 189-195.

Muller, J.C. (1993) "Latest Developments in GIS LIS," *IJGIS,* Vol. 7, Iss 4, pp. 293-303.

Muller, J.C. (1996) "Advances in GIS—Proceedings of the 6th International-Symposium on Spatial Data Handling", by R.G. Healey, T.C. Waugh, *IJGIS,* Vol. 10, Iss 2, pp. 238-241.

Muller, P.O. (1991) "An Introduction to Urban Geographic Information Systems", by W.E. Huxhold, *Urban Geography,* Vol. 12, Iss 4, p. 387.

Murty, C.V.S.S.B.R., Venkatachalam, P., (1992) "Delineation of Erosion Cells in a Watershed Using GIS," *Journal of Environmental Management,* Vol. 36, Iss 3, pp. 159-166.

Nag, P. and M. Kudrat (1998), *Digital Remote Sensing,* Concept Publishing Company, New Delhi.

Nag, P. (1986), A proposed base for geographical information system for India, *International Journal for Geographical Information System,* Vol. 1, No. 2, pp. 181-7.

Nag, P. (1987), A proposed geographical information system for India, *International Journal on Geographical Information System,* Vol. 1:2.

Namken, J.C. Stuth, J.W. (1997) "A Prototype Graphic Landscape Analysis System.1. Predicting Spatial Patterns of Grazing Pressure Using GIS" *IJGISC,* Vol. 11, Iss 8, pp. 785-798.

Neldner, V.J., Crossley, D.C., Cofinas, M. (1995) "Using Geographic Information Systems (GIS) to Determine the Adequacy of Sampling in Vegetation Surveys" *Biological Conservation,* Vol. 73, Iss 1, pp. 1-17.

Nellis, M.D. (1991) "Geographic Information Systems—The Microcomputer and Modern Cartography", by D.R.F. Taylor, *Journal of Rural Studies,* Vol. 7, Iss 3, pp. 352-353.

Nichol, J.E. (1994) "A GIS-Based Approach to Microclimate Monitoring in Singapore Highrise Housing Estates" *PERS* Vol. 60, Iss 10, pp. 1225-1232.

Nichol, J.E. (1994) "An Examination of Tropical Rain-Forest Microclimate Using GIS Modelling," *Global Ecology and Biogeography Letters,* Vol. 4, Iss 3, pp. 69-78.

Nielsen, G.A., Caprio, J.M., Mcdaniel, P.A., Snyder, R.D., Montagne, C. (1990) "Maps—A GIS for Land Resource-Management in Montana" *Journal of Soil and Water Conservation,* Vol. 45, Iss 4, pp. 450-453.

Nishiwaki, S., Nojima, K., Tatara, S., Kosakada, M., Tanabe, N., Yanabu, S. (1995) "Electromagnetic-Interference with Electronic Apparatus by Switching Surges in GIS—Cable System" *IEEE Transactions of Power Delivery* Vol. 10, Iss 2, pp. 739-746.

Nitta, T., Nakanishi, K. (1991) "Charge Accumulation on Insulating Spacers for HVDC GIS" *IEEE Transactions on Electrical Insulation,* Vol. 26, Iss 3, pp. 418-427.

Noronha, V.T. (1990) "Geographic Information Systems—A Management Approach", by S. Aronoff, *CGIS,* Vol. 17, Iss 2, pp. 178-179.

Novak, K. (1995) "Mobile Mapping Technology for GIS Data-Collection" *PERS* Vol. 61, Iss 5, pp. 493-501.

NSDI (197), *A Strategy for the National Spatial Data Infrastructure,* Federal Geographic Data Committee, US Geological Survey, Reston.

Nugent, J.L. (1995) "Quality-Control Techniques for a GIS Database Development Project" *PERS* Vol. 61, Iss 5, pp. 523-527.

Nyerges, T.L. (1991) "Geographical Information Systems for Urban and Regional-Planning", by H.J. Scholten, J.C.H. Stillwell, *Journal of Regional Science,* Vol. 31, Iss 2, pp. 231-233.

Nyerges, T.L., Chrisman, N.R. (1989) "A Framework for Model Curricula Development in Cartography and Geographic Information Systems," *PG,* Vol. 41, Iss 3, pp. 283-293.

Obermeyer, N.J. (1993) "Certification and Accreditation in GIS—Prospects and Pitfalls" *Computers Environment and Urban Systems,* Vol. 17, Iss 1, pp. 91-102.

Obermeyer, N.J. (1994) "Spatial-Analysis and Spatial Policy Using Geographic Information Systems", by L. Worral, *PG,* Vol. 46, Iss 2, pp. 276-276.

Obermeyer, N.J. (1994) "GIS—A New Profession" *PG,* Vol. 46, Iss 4, pp. 498-503.

Okabe, S., Kosakada, M., Toda, H., Suzuki, K., Ishikawa, M. (1993) "Investigations of Multiple Reignition Phenomena and Protection Scheme of Shunt Reactor Current Interruption in GIS Substations" *IEEE Transactions of Power Delivery* Vol. 8, Iss 1, pp. 197-202.

Oldak, A. (1992) "Geographical Information System Techniques in Land-Use Assessment Within Landscape Types in Eastern Poland," *Land Degradation and Rehabilitation,* Vol. 3, Iss 3, pp. 181-194.

Olivier, J.J., Greenwood, P.H., Cooper, A.K., Mcpherson, D.R., Engelbrecht, R. (1990) "Selecting a GIS for a National Water Management Authority" *PERS,* Vol. 56, Iss 11, pp. 1471-1475.

Ong, Y. (1994) "Mapping and Monitoring the Intertidal Zone of the East-Coast of England Using Remote-Sensing Techniques and a Coastal Monitoring GIS," *Marine Technology Society Journal* 1994, Vol. 28, Iss 2, pp. 19-29.

Openshaw, S. (1988) "An Evaluation of the BBC Domesday Interactive Videodisc As a GIS for Planners," *Environment and Planning B-Planning and Design,* Vol. 15, Iss 3, pp. 325-338.

Openshaw, S. (1991) "A View on the GIS Crisis in Geography, or, Using GIS to Put Humpty-Dumpty Back Together Again," *Environment and Planning A,* Vol. 23, Iss 5, pp. 621-628.

Openshaw, S. (1992) "Further Thoughts on Geography and GIS—A Reply" *Environment and Planning A,* Vol. 24, Iss 4, pp. 463-466.

Openshaw, S. (1992) "Some Suggestions Concerning the Development of Artificial-Intelligence Tools for Spatial Modelling and Analysis in GIS" *Annals of Regional Science,* Vol. 26, Iss 1, pp. 35-51.

Openshaw, S. (1993) "GIS Crime and GIS Criminality," *Environment and Planning A,* Vol. 25, Iss 4, pp. 451-458.

Openshaw, S., Charlton, M., Wymer, C. (1986) "A Geographical Information and Mapping System for the BBC Domesday Optical Disks," *Transactions-Institute of British Geographers,* Vol. 11, Iss 3, pp. 296-304.

Orland, B. (1994) "Visualization Techniques for Incorporation in Forest Planning Geographic Information Systems" *Landscape and Urban Planning,* Vol. 30, Iss 1-2, pp. 83-97.

Orr, R.A. (1991) "Bartholomew and the Giant GIS" *Cartographic Journal,* Vol. 28, Iss 1, pp. 30-33.

Osmokrovic, P., Krstic, S., Ljevak, M., Novakovic, D. (1992) "Influence of GIS Parameters on the Topler Constant" *IEEE Transactions on Electrical Insulation* Vol. 27, Iss 2, pp. 214-220.

Otoole, K. (1994) "GIS, Administrative Reform and Local-Government—The Issue of Privacy" *Australian Journal of Social Issues,* Vol. 29, Iss 4, pp. 345-354.

Ottens, H.F.L. (1991) "Geographical Information Systems in the Netherlands" *TIJDSCHRIFT Voor Economische en Sociale Geographie,* Vol. 82, Iss 4, pp. 306-309.

Oyama, M., Hanai, E., Aoyagi, H., Murase, H., Ohshima, I., Menju, S. (1994) "Development of Detection and Diagnostic-Techniques for Partial Discharges in GIS" *IEEE Transactions of Power Delivery* Vol. 9, Iss 2, pp. 811-818.

Ozernoy, V.M., Smith, D.R., Sicherman, A. (1981) "Evaluating Computerized Geographic Information Systems Using Decision Analysis" Interfaces, Vol. 11, Iss 5, pp. 92-100.

Padmavathy, A.S., Raj, K.G., Yogarajan, N., Thangavel, P., Chandrasekhar, M.G. (1993) "Checkdam Site Selection Using GIS Approach" *Advances in Scace Research,* Vol. 13, Iss 11, pp. 123-127.

Pal, Saroj K. and Dilip Kumar (2003), GIS and remote sensing in natural resources mapping in Ferozpur Jhirka Block, District Gurgaon, Harayana, Paper presented in the International Conference on National & Regional Atlases, Kolkata.

Parker, H. Dennison (1991), The role of private sector in GIS, *Photogrammetric Engineering and Remote Sensing,* vol. 37, No. 11, pp. 1465-6.

Parker, H. Dennison (1991), The role of the private sector in GIS, *Photogrammertic Engineering and Remote Sensing,* Vol. 57, No.11, pp. 1465-6.

Parker, H.D. (1988) "The Unique Qualities of a Geographic Information System-A Commentary," *PERS,* Vol. 54, Iss 11, pp. 1547-1549.

Parker, H.D. (1991) "The Role of the Private-Sector in GIS" *PERS* Vol. 57, Iss 11, pp. 1465-1466.

Parker, S., Cocklin, C. (1993) "The Use of Geographical Information Systems for Cumulative Environmental Effects Assessment," *Computers Environment and Urban Systems,* Vol. 17, Iss 5, pp. 393-407.

Pathan, S.K., Sastry, S.V.C., Dhinwa, P.S., Rao, M., Majumdar, K.L., Kumar, D.S., Patkar, V.N., Phatak, V.N. (1993) "Urban-Growth Trend Analysis Using GIS Techniques—A Case-Study of the Bombay Metropolitan Region," *IJRS,* Vol. 14, Iss 17, pp. 3169-3179.

Pearson, M., Wheaton, S. (1993) "GIS and Storm-Water Management,"*Civil Engineering,* Vol. 63, Iss 9, pp. 72-73.

Peled, A., Adler, R. (1993) "A Common Database for Digital Mapping and GIS," *IJGIS,* Vol. 7, Iss 5, pp. 425-434.

Pereira, J.M.C., Duckstein, L. (1993) "A Multiple Criteria Decision-Making Approach to GIS-Based Land Suitability Evaluation," *IJGIS,* Vol. 7, Iss 5, pp. 407-424.

Pereira, J.M.C., Itami, R.M. (1991) "GIS-Based Habitat Modelling Using Logistic Multiple-Regression—A Study of the Mt Graham Red Squirrel," *PERS* Vol. 57, Iss 11, pp. 1475-1486.

Perera, L.K., Tateishi, R. (1995) "Do Remote-Sensing and GIS Have a Practical Applicability in Developing-Countries—(Including Some Sri-Lankan Experiences)," *IJRS,* Vol. 16, Iss 1, pp. 35-51.

Perkins, C. (1994) "Quality in Map Librarianship and Documentation in the GIS Age," *Cartographic Journal,* Vol. 31, Iss 2, pp. 93-99.

Pesachowitz, A.M. (1992) "Using Geographic Information Systems (GIS) for Environmental Decision-Making," *Abstracts of Papers of the American Chemical Society* Vol. 204, Iss AUG, pp. 83-ENVR.

Petersen, J.F. (1987) "An Inexpensive Videotape About Geographic Information Systems" *Journal of Geography,* Vol. 86, Iss 1, pp. 34-34.

Petit, J. (1994) "Looking at Urban Forests with GIS Technology," *JF* Vol. 92, Iss 10, pp. 45-45.

Peuquet, D.J. (1994) "Its About Time—A Conceptual-Framework for the Representation of Temporal Dynamics in Geographic Information Systems," *Annals of the Association of American Geographers,* Vol. 84, Iss 3, pp. 441-461.

Peuquet, D.J., Bacastow, T. (1991) "Organizational Issues in the Development of Geographical Information Systems—A Case-Study of United States Army Topographic Information Automation," *IJGIS,* Vol. 5, Iss 3, pp. 303-319.

Pickles, J. (1992)"Geographic Information Systems and Their Socioeconomic Applications", by D. Martin, *Environment and Planning D-Society and Space,* Vol. 10, Iss 5, pp. 597-600.

Pickles, J. (1997) "Tool or Science—GIS, Technoscience, and the Theoretical Turn," *Annals of the Association of American Geographers,* Vol. 87, Iss 2, pp. 363-372.

Poiani, K.A., Bedford, B.L. (1995) "GIS Based Nonpoint-Source Pollution Modelling—Considerations for Wetlands," *Journal of Soil and Water Conservation,* Vol. 50, Iss 6, pp. 613-619.

Pollak, L.M., Corbett, J.D. (1994) "Using GIS Datasets to Classify Maize-Growing Regions in Mexico and Central-America," *Agronomy Journal,* 1993, Vol. 85, Iss 6, pp. 1133-1139.

Pollock, R.J., Mclaughlin, J.D. (1991) "Database Management System Technology and Geographic Information Systems," *Journal of Surveying Engineering-ASCE,* Vol. 117, Iss 1, pp. 9-26.

Polydorides, N.D. (1994) "GIS-Related Developments in Europe," *Social Science Computer Review,* Vol. 12, Iss 4, pp. 515-531.

Populus, J., Moreau, F., Coquelet, D., Xavier, J.P. (1995) "An Assessment of Environmental Sensitivity to Marine Pollutions—Solutions with Remote-Sensing and Geographical Information Systems (GIS)," *IJRS* Vol. 16, Iss 1, pp. 3-15.

Poulsen, M.F. (1994) "Human-Geography and Its Relationship to GIS—A Perspective from the Vantage Point of GIS," *Australian Geographer,* Vol. 25, Iss 2, pp. 170-177.

Pratt, Monica (2000), Works for labor information analysts, *ArcUser,* April-June, pp. 14-6.

Prevedel, D.A. (1995) "Project Sparkey—A Strategic Wildfire Monitoring Package Using AVHRR Satellites and GIS," *PERS* Vol. 61, Iss 3, pp. 271-278.

Priestnall, G. (1997) "GIS—A Computing Perspective", by M.F. Worboys, *Transactions of the Institute of British Geographers,* Vol. 22, Iss 2, pp. 271-272.

Probert, M. (1995) "Innovations in GIS-2—Selected Papers from the 2nd National Conference on GIS Research UK", by P. Fisher, *Cartographic Journal,* Vol. 32, Iss 2, pp. 187-188.

Prosperi, D.C. (1997) "GIS and Organizations", by H. Campbell, I. Masser, *Journal of the American Planning Association,* Vol. 63, Iss 3, pp. 414-414.

Puffer, R.L. (1994) "Closing with the Enemy—How GIS Fought the War in Europe, 1944-1945", by M.D. Doubler, *Library Journal,* Vol. 119, Iss 19, pp. 76-76.

Quattrochi, D.A. (1993) "The Need for a Lexicon of Scale Terms in Integrating Remote-Sensing Data with Geographic Information Systems," *Journal of Geography,* Vol. 92, Iss 5, pp. 206-212.

Qureshi, M.A., Khan, S. (1994) "Environmental Pollution Information System Using Remote-Sensing and GIS—Report on Training Program," *Journal of Scientific and Industrial Research,* Vol. 53, Iss 4, pp. 275-282 -R-.

Raafat, H.M., Xiao, Q.H., Gauthier, D.A. (1991) "An Extended Relational Database for Remotely Sensed Image Data Management Within GIS," *IEEE Transactions on Geoscience and Remote Sensing,* Vol. 29, Iss 4, pp. 651-655.

Ramasubramanian, L. (1995) "Building Communities—GIS and Participatory Decision-Making," *Journal of Urban Technology,* Vol. 3, Iss 1, pp. 67-79.

Rao, M.N. (1991) "Introductory Readings in Geographic Information Systems", by D.J. Peuquet, D.F. Marble, *PG,* Vol. 43, Iss 4, pp. 547-548.

Raper, J. (1993) "Geographical Information Systems," *Progress in Physical Geography,* Vol. 17, Iss 4, pp. 493-502.

Raper, J. (1993) "Time in Geographic Information Systems", by G. Langran, *Environment and Planning B-Planning and Design,* Vol. 20, Iss 6, pp. 734-735.

Raper, J., Green, N. (1989) "The Development of a Hypertext-Based Tutor for Geographical Information Systems," *British Journal of Educational Technology,* Vol. 20, Iss 3, pp. 164-172.

Raper, J., Green, N. (1992) "Teaching the Principles of GIS—Lessons from the GIStutor Project," *IJGIS,* Vol. 6, Iss 4, pp. 279-290.

Raper, J.F. (1992) "Map-II Map Processor—A Geographic Information System for the Macintosh, Vol. 1-2", by M. Pazner, K.C. Kirby, N. Thies, *Geographical Journal,* Vol. 158, Iss JUL, pp. 240-240.

Raubal, M., Gaupmann, B., Kuhn, W. (1997) "Teaching Raster GIS Operations with Spreadsheets,".*Journal of Geography,* Vol. 96, Iss 5, pp. 258-263.

Rhind, D. (1992), Data Access, Charging and Copyright and Their Implications for Geographic Information Systems, *International Journal on Geographical Information System,* Vol. 6:1, pp. 13-30.

Rhind, D. (1991) "Geographical Information Systems and Environmental Problems," *International Social Science Journal,* Vol. 43, Iss 4, pp. 649-668.

Rhind, D. (1992) "Data Access, Charging and Copyright and Their Implications for Geographic Information Systems," *IJGIS,* Vol. 6, Iss 1, pp. 13-30.

Rhind, D., Openshaw, S., Armstrong, P. (1988) "The Domesday Machine—A Nationwide Geographical Information System," *Geographical Journal,* Vol. 154, Iss MAR, pp. 56-68.

Rhind, D.W. (1986) "Remote-Sensing, Digital Mapping, and Geographical Information Systems—The Creation of National Policy in the United Kingdom," *Environment and Planning C-Government and Policy,* Vol. 4, Iss 1, pp. 91-102.

Rhoads, A.F., Thompson, L. (1992) "Integrating Herbarium Data into a Geographic Information System—Requirements for Spatial-Analysis," *TAXON,* Vol. 41, Iss 1, pp. 43-49.

Rice, K.W. (1992) "Introductory Readings in Geographic Information Systems", by D.J. Pevquet, D.F. Marble, *CGIS,* Vol. 19, Iss 2, pp. 105-106.

Richards, C., Host, G. (1994) "Examining Land-Use Influences on Stream Habitats and Macroinvertebrates—A GIS Approach," *Water Resources Bulletin,* Vol. 30, Iss 4, pp. 729-738.

Rifai, H.S., Hendricks, L.A., Kilborn, K., Bedient, P.B. (1993) "A Geographic Information System (GIS) User Interface for Delineating Wellhead Protection Areas," *Ground Water,* Vol. 31, Iss 3, pp. 480-488.

Rifai, H.S., Newell, C.J., Bedient, P.B. (1993) "Getting to the Nonpoint-Source with GIS," *Civil Engineering,* Vol. 63, Iss 6, pp. 44-46.

Ringrose, S., Vanderpost, C., Matheson, W. (1996) "The Use of Integrated Remotely-Sensed and GIS Data to Determine Causes of Vegetation Cover Change in Southern Botswana," *Applied Geography,* Vol. 16, Iss 3, pp. 225-242.

Robinson, G.M. (1992) "Geographical Information Systems (GIS) and Planning in Scotland—The Rluis Project and Advances in GIS," *Scottish Geographical Magazine,* Vol. 108, Iss 1, pp. 22-28.

Robinson, G.M. (1992) "Geographical Information Systems (GIS) and Planning in Scotland—The Rluis Project and Advances in GIS," *Scottish Geographical Magazine,* Vol. 108, Iss 1, pp. 22-28.

Robinson, G.M., Furley, P.A., Gray, D.A., Healey, R.G. (1989) "Developing a Geographical Information System (GIS) for Agricultural-Development in Belize, Central-America," *Applied Geography,* Vol. 9, Iss 2, pp. 81-94.

Robinson, G.M., Furley, P.A., Gray, D.A., Healey, R.G. (1989) "Developing a Geographical Information System (GIS) for Agricultural-Development in Belize, Central-America," *Applied Geography,* Vol. 9, Iss 2, pp. 81-94.

Robinson, V.B., Frank, A.U. (1987) "Expert Systems for Geographic Information Systems," *PERS,* Vol. 53, Iss 10, pp. 1435-1441.

Robinson, V.B., Frank, A.U., Blaze, M.A. (1986) "Expert Systems and Geographic Information Systems—Review and Prospects," *Journal of Surveying Engineering-ASCE,* Vol. 112, Iss 2, pp. 119-130.

Rocheleau, B. (1992) "An Introduction to Urban Geographic Information Systems, by W.E. Huxhold," *Public Administration Review,* Vol. 52, Iss 4, pp. 398-400.

Roche, S.C., Gittings, B.M. (1996) "Parallel Polygon Line Shading—The Quest for More Computational Power from an Existing GIS Algorithm," *IJGIS,* Vol. 10, Iss 6, pp. 731-746.

Rogerson, R.J. (1992) "Teaching Generic GIS Using Commercial Software," *IJGIS,* Vol. 6, Iss 4, pp. 321-331.

Rogowski, A.S. (1995) "Quantifying Soil Variability in GIS Applications.1. Estimates of Position," *IJGIS,* Vol. 9, Iss 1, pp. 81-94.

Rogowski, A.S. (1996) "Quantifying Soil Variability in GIS Applications.2. Spatial-Distribution of Soil Properties," *IJGIS,* Vol. 10, Iss 4, pp. 455-475.

Roseberry, J.L., Richards, B.J., Hollenhorst, T.P. (1994) "Assessing the Potential Impact of Conservation Reserve Program Lands on Bobwhite Habitat Using Remote-Sensing, GIS, and Habitat Modelling," *PERS* Vol. 60, Iss 9, pp. 1139-1143.

Rosenthal, W.D., Srinivasan, R., Arnold, J.G. (1995) "Alternative River Management Using a Linked GIS-Hydrology Model," *Transactions of the ASAE* Vol. 38, Iss 3, pp. 783-790.

Roy, P.S. (2005),. Standards and Interoperability for Creating a Common Code, *Geospatial Today,* Vol.4, issue 2, pp. 40-3.

Rutten, B.J.C.M., Vonbeek, A., Vanderhaspel, B. (1988) "GISwa—Geographical Information System Wind Resource," *Journal of Wind Engineering and Industrial Aerodynamics,* Vol. 27, Iss 1-3, pp. 103-112.

Sader, S.A., Ahl, D., Liou, W.S. (1995) "Accuracy of Landsat-TM and GIS Rule-Based Methods for Forest Wetland Classification in Maine," *Remote Sensing of Environment* Vol. 53, Iss 3, pp. 133-144.

Sadler, J. (1992) "Interpreting Space—GIS and Archaeology", by K.M.S. Allen, S.W. Green, E.B.W. Zubrow, *IJGIS,* Vol. 6, Iss 1, pp. 60-61.

Sahay, S., Walsham, G. (1996) "Implementation of GIS in India—Organizational Issues and Implications," *IJGIS,* Vol. 10, Iss 4, pp. 385-404.

Sanders, P.F. Thornton, L.L., Louis, J.B., Ivahnenko, T., Evans, B.M. (1991)"Using a Geographical Information System to Assess the Susceptibility of Surface Drinking-Water Supplies to Pesticide Contamination," *Abstracts of Paper of the American Chemical Society,* Vol. 2 02, Iss AUG, pp. 53.

Savabi, M.R., Flanagan, D.C., Hebel, B., Engel, B.A. (1995) "Application of Wepp and GIS-Grass to a Small Watershed in Indiana," *Journal of Soil and Water Conservation* Vol. 50, Iss 5, pp. 477-483.

Scholten, Henk J. and Stillwell, John C.H. (1990)(eds), *Geographical Information System for Urban and Regional Planning,* Kluwer Academic Publishers, Dordrechht.

Scholten, H.J., Padding, P. (1990) "Working with Geographic Information Systems in a Policy Environment" *Environment and Planning B-Planning and Design,* Vol. 17, Iss 4, pp. 405-416.

Scholten, H.J., Vandervlugt, M. (1989) "Application of Geographic Information Systems in Europe," *EKISTICS, The Problems and Science of Human Settlements,* Vol. 56, Iss 338-, pp. 304-311.

Schreier, H., Brown, S., Kennedy, G., Shah, P.B. (1991) "Food, Feed, and Fuelwood Resources of Nepal—A GIS Evaluation," *Environmental Management,* Vol. 15, Iss 6, pp. 815-822.

Schreier, H., Brown, S., Schmidt, M., Shah, P., Shrestha, B., Nakarmi, G., Subba, K., Wymann, S. (1994) "Gaining Forests But Losing Ground—A GIS Evaluation in a Himalayan Watershed," *Environmental Management,* Vol. 18, Iss 1, pp. 139-150.

Schreier, H., Shah, P.B., Kennedy, G. (1990) "Evaluating Mountain Watersheds in Nepal Using Micro-GIS," *Mountain Research and Development,* Vol. 10, Iss 2, pp. 151-159.

Scott, L.M. (1994) "Identification of GIS Attribute Error Using Exploratory Data-Analysis," *PG,* Vol. 46, Iss 3, pp. 378-386.

See, R.B., Naftz, D.L., Qualls, C.L. (1992) "GIS-Assisted Regression-Analysis to Identify Sources of Selenium in Streams," *Water Resources Bulletin* Vol. 28, Iss 2, pp. 315-330.

Sellars, A.G., Farish, O., Hampton, B.F. (1994) "Assessing the Risk of Failure Due to Particle Contamination of GIS Using the UHF Technique," *IEEE Transactions on Dielectruics and Electrical Insulation* Vol. 1, Iss 2, pp. 323-331.

Sellars, A.G., Farish, O., Hampton, B.F. (1994) "Characterizing the Discharge Development Due to Surface Contamination in GIS Using the UHF Technique," *IEE Proceedings-Science Measurement and Technology,* Vol. 141, Iss 2, pp. 118-122.

Shaffer, C.A. (1992) "Data Representations for Geographic Information Systems," *Annual Review of Information Science and Technology,* Vol. 27, pp. 135-172.

Sharma, P. (1993) "Geographic Information Systems—Developments and Applications", by L. Worrall, *Australian Geographer,* Vol. 24, Iss 1, pp. 86-88.

Sharma, S.K., Anjaneyulu, D. (1994) "Application of Remote Sensing and GIS in Water-Resource Management," *IJRS* 1993, Vol. 14, Iss 17, pp. 3209-3220.

Shaw, S.L. (1993) "Geography—A Place for GIS," *Applied Geography,* Vol. 13, Iss 2, pp. 107-110.

Shaw, S.L. (1993) "GIS for Urban Travel Demand Analysis—Requirements and Alternatives," *Computers Environment and Urban Systems,* Vol. 17, Iss 1, pp. 15-29.

Shea, C., Grayman, W., Darden, D., Males, R.M., Sushinsky, P. (1993) "Integrated GIS and Hydrologic Modelling for Countywide Drainage Study," *Journal of Water Resources Planning and Management-ASCE* Vol. 119, Iss 2, pp. 112-128.

Shelton, R.L., Estes, J.E. (1981) "Remote-Sensing and Geographic Information Systems—An Unrealized Potential," *Geo-Processing,* Vol. 1, Iss 4, pp. 395-420.

Shennan, S.J. (1993) "Interpreting Space—GIS and Archaeology", by K.M.S. Allen, S.W. Green, E.B.W. Zubrow, *Journal of Archaeological Science,* Vol. 20, Iss 2, pp. 237-238.

Shih, S.F., Jordan, J.D. (1992) "Landsat Midinfrared Data and GIS in Regional Surface Soil-Moisture Assessment," *Water Resources Bulletin,* Vol. 28, Iss 4, pp. 713-719.

Shih, S.F., Jordan, J.D. (1993) "Landsat Midinfrared Data and GIS in Regional Surface Soil-Moisture Assessment—Discussion," *Water Resources Bulletin,* Vol. 29, Iss 2, pp. 313-314.

Shih, S.F., Jordan, J.D. (1993) "Use of Landsat Thermal-Ir Data and GIS in Soil-Moisture Assessment," *Journal of Irrigation and Drainage Engineering-ASCE* Vol. 119, Iss 5, pp. 868-879.

Short, N.M., Dickens, L. (1995) "Automatic-Generation of Products from Terabyte-Size Geographical Information Systems Using Planning and Scheduling" *IJGIS,* Vol. 9, Iss 1, pp. 47-65.

Siva Kumar, R. (2005), Standards and Interoperability for an Indian standard, *Geospatial Today,* Vol.4, issue 2, pp. 54-56.

Skidmore, A.K., Gauld, A., Walker, P. (1996) "Classification of Kangaroo Habitat Distribution Using 3 GIS Models," *IJGIS,* Vol. 10, Iss 4, pp. 441-454.

Skidmore, A.K., Ryan, P.J., Dawes, W., Short, D., Oloughlin, E. (1991) "Use of an Expert System to Map Forest Soils from a Geographical Information System," *IJGIS,* Vol. 5, Iss 4, pp. 431-445.

Sleight, P. (1996) "GIS for Business and Service Planning", by P. Longley, G. Clarke, *IJGIS,* Vol. 10, Iss 5, pp. 664-665.

Slocum, T.A. (1991) "Geographic Information Systems and Cartographic Modelling", by C.D. Tomlin, *PG,* Vol. 43, Iss 1, pp. 126-126.

Smith, C.L., Steel, B.S., List, P.C., Cordray, S. (1995) "Making Forest Policy—Integrating GIS with Social Processes," *JF* Vol. 93, Iss 5, pp. 31-36.

Smith, D.A., Tomlinson, R.F. (1992) "Assessing Costs and Benefits of Geographical Information Systems—Methodological and Implementation Issues," *IJGIS,* Vol. 6, Iss 3, pp. 247-256.

Smith, J.L., Logan, J.A., Gregoire, T.G. (1992) "Using Aerial-Photography and Geographic Information Systems to Develop Databases for Pesticide Evaluations," *PERS*, Vol. 58, Iss 10, pp. 1447-1452.

Smith, M.B., Brilly, M. (1992) "Automated Grid Element Ordering for GIS-Based Overland-Flow Modelling," *PERS* Vol. 58, Iss 5, pp. 579-585.

Smith, M.B., Vidmar, A. (1994) "Data Set Derivation for GIS-Based Urban Hydrological Modelling," *PERS* Vol. 60, Iss 1, pp. 67-76.

Sombroek, W.G., Antoine, J. (1994) "The Use of Geographic Information Systems (GIS) in Land Resources Appraisal," *Outlook on Agriculture,* Vol. 23, Iss 4, pp. 249-255.

Somers, R. (1987) "Geographic Information Systems in Local-Government—A Commentary," *PERS*, Vol. 53, Iss 10, pp. 1379-1382.

Sprunt, B. (1988) "A UK Geographic Information System for Environmental Monitoring, Resource Planning and Management Capable of Integrating and Using Satellite Remotely Sensed Data", by J.A.T. Young, *Cartographic Journal,* Vol. 25, Iss 1, pp. 69-70.

Srinivasan, A., Richards, J.A. (1993) "Analysis of GIS Spatial Data Using Knowledge-Based Methods," *IJGIS,* Vol. 7, Iss 6, pp. 479-500.

Srinivasan, R., Arnold, J.G. (1994) "Integration of a Basin-Scale Water-Quality Model with GIS," *Water Resources Bulletin,* Vol. 30, Iss 3, pp. 453-462.

Staebler, R.N. (1992) "Natural-Resource Management and GIS," *JF,* Vol. 90, Iss 12, pp. 2-2.

Star, J.L., Estes, J.E., Davis, F. (1991)"Improved Integration of Remote-Sensing and Geographic Information Systems—A Background to Ncgia Initiative 12," *PERS*, Vol. 57, Iss 6, pp. 643-645.

Steinke, T.R. (1990) "3-Dimensional Applications in Geographical Information Systems", by J. Raper, *CGIS,* Vol. 17, Iss 2, pp. 177-178.

Stillwell, J. (1992) "Geographic Information Systems—Developments and Applications", by L. Worrall, *IJGIS,* Vol. 6, Iss 1, pp. 56-59.

Stillwell, J. (1992) "Geographical Information Systems and Urban and Rural Planning", by T.W. Rideout, *IJGIS,* Vol. 6, Iss 5, pp. 445-447.

Stockwell, J.R., Sorensen, J.W., Eckert, J.W., Carreras, E.M. (1993) "The United-States EPA Geographic Information System for Mapping Environmental Releases of Toxic-Chemical Release Inventory (TRI) Chemicals," *Risk Analysis,* Vol. 13, Iss 2, pp. 155-164.

Stoms, D.M., Davis, F.W., Cogan, C.B. (1992) "Sensitivity of Wildlife Habitat Models to Uncertainties in GIS Data," *PERS* Vol. 58, Iss 6, pp. 843-850.

Stoorvogel, J.J. (1995) "Linking GIS and Models—Structure and Operationalization for a Costa-Rican Case Study," *Netherlands Journal of Agricultural Science* Vol. 43, Iss 1, pp. 19-29.

Strachan, A.J., Stuart, N. (1996) "UK Developments in Environmental GIS," *IJGIS,* Vol. 10, Iss 1, pp. 17-20.

Strand, E.J., Parker, H.D. (1992) "GIS Has a Bright Future in Business," *PERS* Vol. 58, Iss 8, pp. 1259-1262.

Strasser, T. (1997) "International GIS Dictionary", by R. Mcdonnell, K. Kemp, *Journal of Government Information,* Vol. 24, Iss 2, pp. 130-132.

Stuart, N. (1992) "Geographic Information Systems—Developments and Applications", by L. Worrall, *Scottish Geographical Magazine,* Vol. 108, Iss 2, pp. 134-135.

Stuart, N. (1992) "Time in Geographic Information Systems", by G. Langran, *IJGIS,* Vol. 6, Iss 6, pp. 547-547.

Stuart, N. (1995) "Visualization in Geographical Information Systems", by H. Hearnshaw, D.J. Unwin, *Applied Geography,* Vol. 15, Iss 1, pp. 91-92.

Stuebe, M.M., Johnston, D.M. (1990) "Runoff Volume Estimation Using GIS Techniques," *Water Resources Bulletin* Vol. 26, Iss 4, pp. 611-620.

Sui, D.Z. (1991) "Geographic Information System—A Management Perspective", by S. Aronoff, *PG,* Vol. 43, Iss 3, pp. 370-370.

Sui, D.Z. (1994) "GIS and Urban Studies—Positivism, Post-Positivism, and Beyond," *Urban Geography,* Vol. 15, Iss 3, pp. 258-278.

Sui, D.Z. (1997) "Reconstructing Urban Reality—From GIS to Electropolis," *Urban Geography,* Vol. 18, Iss 1, pp. 74-89.

Sussman, R. (1993) "Municipal GIS and the Enterprise Data Model," *IJGIS,* Vol. 7, Iss 4, pp. 367-377.

Takeyama, M., Couclelis, H. (1997) "Map Dynamics—Integrating Cellular-Automata and GIS Through Geo-Algebra," *IJGISC,* Vol. 11, Iss 1, pp. 73-91.

Tan, Y.R., Shih, S.F. (1990) "GIS in Monitoring Agricultural Land-Use Changes and Well Assessment," *Transactions of the ASAE* Vol. 33, Iss 4, pp. 1147-1152.

Tappan, G.G., Moore, D.G., Knausenberger, W.I. (1991) "Monitoring Grasshopper and Locust Habitats in Sahelian Africa Using GIS and Remote-Sensing Technology," *IJGIS,* Vol. 5, Iss 1, pp. 123-135.

Tarboton, K.C. (1992) "Interfacing GIS and Hydrological Modelling—Mgeni Case-Study," *Water SA* Vol. 18, Iss 4, pp. 273-278.

Taylor, D.R.F. (1991) "GIS in Developing-Nations," *IJGIS,* Vol. 5, Iss 1, pp. 3-3.

Taylor, P.J., Overton, M. (1991) "Further Thoughts on Geography and GIS—A Preemptive Strike," *Environment and Planning A,* Vol. 23, Iss 8, pp. 1087-1090.

Theocharopoulos, S.P., Davidson, D.A., Mcarthur, J.N., Tsouloucha, F. (1995) "GIS as an Aid to Soil Surveys and Land Evaluation in Greece," *Journal of Soil and Water Conservation* Vol. 50, Iss 2, pp. 118-124.

Thomas, A. (1992) "Agricultural Water-Balance of Yunnan Province, Pr China -Agroclimatic Zoning with a Geographical Information System," *Agricultural Water Management,* Vol. 21, Iss 4, pp. 249-263.

Thorpe, J.A., Gilbert, D.M. (1987) "AM/FM Requirements of Geographic Information Systems—A Commentary," *PERS,* Vol. 53, Iss 10, pp. 1403-1404.

Tim, U.S. (1995) "The Application of GIS in Environmental-Health Sciences—Opportunities and Limitations," *Environmental Research,* Vol. 71, Iss 2, pp. 75-88.

Tobin, R. (1989) "Introducing Geographic Information Systems to a Small Municipality," *EKISTICS-The Problems and Science of Human Settlements,* Vol. 56, Iss 338-, pp. 296-299.

Tobler, W. (1995) "The Resel-Based GIS," *IJGIS,* Vol. 9, Iss 1, pp. 95-100.

Tomlinson, R.F. (1989) "Geographic Information Systems and Geographers in the 1990s," *Canadian Geographer-Geograhe Canadien,* Vol. 33, Iss 4, pp. 290-298.

Towers, G. (1997) "GIS Versus the Community—Siting Power in Southern West-Virginia," *Applied Geography,* Vol. 17, Iss 2, pp. 111-125.

Treitz, P., Elliot, L., Howarth, P. (1993) "Differential Global Positioning System—Potential for Geographical Information System Database Management" *Environment and Planning A,* Vol. 25, Iss 6, pp. 883-898.

Treitz, P.M., Howarth, P.J., Gong, P. (1992) "Application of Satellite and GIS Technologies for Land-Cover and Land-Use Mapping at the Rural-Urban Fringe—A Case-Study," *PERS,* Vol. 58, Iss 4, pp. 439-448.

Trotter, C.M. (1991) "Remotely-Sensed Data as an Information Source for Geographical Information Systems in Natural-Resource Management—A Review," *IJGIS,* Vol. 5, Iss 2, pp. 225-239.

Tudor, G.S., Sugarbaker, L.J. (1993) "GIS Orthographic Digitizing of Aerial Photographs by Terrain Modelling," *PERS* Vol. 59, Iss 4, pp. 499-503.

Twigg, L. (1990) "Health Based Geographical Information Systems—Their Potential Examined in the Light of Existing Data Sources," *Social Science and Medicine,* Vol. 30, Iss 1, pp. 143-155.

Unwin, D. (1993) "Geographic Information Systems and Their Socioeconomic Applications", by D. Martin, *Progress in Human Geography,* Vol. 17, Iss 1, pp. 147-148.

Unwin, D. (1994) "Cartography, Visc and GIS," *Progress in Human Geography,* Vol. 18, Iss 4, pp. 516-522.

Unwin, D., Dale, P. (1990) "An Educationalists View of GIS—Some Educational and Sociological Concerns," *Journal of Geography in Higher Education*, Vol. 14, Iss 2, pp. 166-170.

Unwin, D.J. (1992) "Introductory Readings in Geographic Information Systems", by D.J. Peuquet, D.F. Marble, *AREA,* Vol. 24, Iss 2, pp. 189-190.

Unwin, D.J. (1996) "GIS, Spatial-Analysis and Spatial Statistics," *Progress in Human Geography,* Vol. 20, Iss 4, pp. 540-551.

Utsumi, T., Endo, F., Ishikawa, T., Iwaasa, S., Yamagiwa, T., Diessner A., Gorablenkow, J., Hashoff, L. (1993) "Preventive Maintenance System with a Different Gas Injecting Facility for GIS," *IEEE Transactions of Power Delivery* Vol. 8, Iss 3, pp. 1107-1113.

Valenzuela, C.R. (1989) "The Digital Geographic Information System As a Management Tool," *Impact of Science on Society,* Vol. 39, Iss 4, pp. 315-324.

Vancleynenbreugel, J., Fierens, F., Suetens, P., Oosterlinck, A. (1990) "Delineating Road Structures on Satellite Imagery by a GIS-Guided Technique," *PERS* Vol. 56, Iss 6, pp. 893-898.

Vanderknaap, W.G.M. (1992) "The Vector to Raster Conversion—(MIS) Use in Geographical Information Systems," *IJGIS,* Vol. 6, Iss 2, pp. 159-170.

Vangroenigen, J.W., Stein, A., Zuurbier, R. (1997) "Optimization of Environmental Sampling Using Interactive GIS," *Soil Technology,* Vol. 10, Iss 2, pp. 83-97.

Vanlanen, H.A.J., Vandiepen, C.A., Reinds, G.J., Dekoning, G.H.J., Bulens, J.D., Bregt, A.K. (1992) "Physical Land Evaluation Methods and GIS to Explore the Crop Growth-Potential and Its Effects Within the European Communities," *Agricultural Systems,* Vol. 39, Iss 3, pp. 307-328.

Vaughan, P.J., Corwin, D.L. (1994) "A Method of Modelling Vertical Fluid-Flow and Solute Transport in a GIS Context," *GEODERMA,* Vol. 64, Iss 1-2, pp. 139-154.

Verbyla, D.L. (1993) "Landsat Midinfrared Data and GIS in Regional Surface Soil-Moisture Assessment—Discussion," *Water Resources Bulletin* Vol. 29, Iss 2, pp. 309-311.

Veregin, H. (1994) "Integration of Simulation Modelling and Error Propagation for the Buffer Operation in GIS," *PERS* Vol. 60, Iss 4, pp. 427-435.

Vertinsky, I., Brown, S., Schreier, H., Thompson, W.A., Vankooten, G.C. (1994) "A Hierarchical-GIS-Based Decision-Model for Forest Management—The Systems-Approach," *Interfaces,* Vol. 24, Iss 4, pp. 38-53.

Viet, C.P., Phuong, M.N. (1993) "Natural Resources Evaluation by the Use of Remote Sensing and GIS Technology for Agricultural-Development," *Advances in Sapce Research,* Vol. 13, Iss 11, pp. 117-121.

Villa, F., Ceroni, M., Mazza, A. (1996) "A GIS-Based Method for Multiobjective Evaluation of Park Vegetation," *Landscape and Urban Planning,* Vol. 35, Iss 4, pp. 203-212.

Vincent, P. (1993) "Geographical Information Systems, Principles and Applications", by D.J. Maguire, M.F. Goodchild, D.W. Rhind, *Progress in Human Geography,* Vol. 17, Iss 3, pp. 439-440.

Visvalingam, M. (1989) "Cartography, GIS and Maps in Perspective," *Cartographic Journal,* Vol. 26, Iss 1, pp. 26-32.

Visvalingam, M. (1992) "GIS in Malaysia," *Cartographic Journal,* Vol. 29, Iss 1, pp. 55-56.

Wadge, G. (1992) "Geological Applications of GIS," *Journal of the Geological Society,* Vol. 149, Iss Jul., pp. 672-672.

Wagner, D.E. (1995) "Reactive Data-Structures for Geographic Information Systems", by P.J.M. Vanoosterom, *PG,* Vol. 47, Iss 1, pp. 113-114.

Walford, N., Lane, M., Shearman, J. (1989) "The Rural-Areas Database—A Geographical Information and Mapping System for the Management and Integration of Data on Rural Britain," *Transactions-Institute of British Geographers,* Vol. 14, Iss 2, pp. 221-230.

Walker, D.J., Painter, K.M. (1994) "Measuring Rate and Onsite Damage Using GIS—A Policy Application," *American Journal of Agricultural Economics,* Vol. 76, Iss 5, pp. 1258-1258.

Walker, D.R.F., Newman, I.A., Medyckyjscott, D.J., Ruggles, C.L.N. (1992) "A System for Identifying Datasets for GIS Users," *IJGIS,* Vol. 6, Iss 6, pp. 511-527.

Walker, G. (1992) "GIS and Urban and Rural Planning," *AREA,* Vol. 24, Iss 2, pp. 230-231.

Walpole, S.C., Sinden, J.A. (1997) "BCA and GIS—Integration of Economic and Environmental Indicators to Aid."

Walsh, S.J. (1985) "Geographic Information Systems for Natural Resource Management," *Journal of Soil and Water Conservation,* Vol. 40, Iss 2, pp. 202-205.

Walsh, S.J. (1988) "Geographic Information Systems—An Instructional Tool for Earth Science Educators," *Journal of Geography,* Vol. 87, Iss 1, pp. 17-25.

Walsh, S.J. (1992) "Spatial Education and Integrated Hands-on Training—Essential Foundations of GIS Instruction," *Journal of Geography,* Vol. 91, Iss 2, pp. 54-61.

Walsh, S.J., Cooper, J.W., Vonessen, I.E., Gallager, K.R. (1990) "Image-Enhancement of LANDSAT Thematic Mapper Data and GIS Data Integration for Evaluation of Resource Characteristics," *PERS* Vol. 56, Iss 8, pp. 1135-1141.

Walsh, S.J., Lightfoot, D.R., Butler, D.R. (1987) "Recognition and Assessment of Error in Geographic Information Systems," *PERS,* Vol. 53, Iss 10, pp. 1423-1430.

Wang, F. (1994) "The Use of Artificial Neural Networks in a Geographical Information System for Agricultural Land Suitability Assessment," *Environment and Planning A,* Vol. 26, Iss 2, pp. 265-284.

Wang, F.J., Hall, G.B. (1996) "Fuzzy Representation of Geographical Boundaries in GIS," *IJGIS,* Vol. 10, Iss 5, pp. 573-590.

Wang, J.F., Treitz, P.M., Howarth, P.J. (1992) "Road Network Detection from Spot Imagery for Updating Geographical Information Systems in the Rural Urban Fringe," *IJGIS,* Vol. 6, Iss 2, pp. 141-157.

Wanless, B. (1992) "GPS Positioned Digital Video for Airborne GIS Data Acquisition," *Journal of Surveying Engineering—ASCE* Vol. 118, Iss 3, pp. 80-89.

Warwick, J.J., Haness, S.J. (1994) "Efficacy of Arc/Info GIS Application to Hydrologic Modelling," *Journal of Water Resources Planning and Management-ASCE* Vol. 120, Iss 3, pp. 366-381.

Webster, C.J. (1993) "GIS and the Scientific Inputs to Urban-Planning.1. Description," *Environment and Planning B-Planning and Design,* Vol. 20, Iss 6, pp. 709-728.

Webster, C.J. (1994) "GIS and the Scientific Inputs to Planning.2. Prediction and Prescription," *Environment and Planning B-Planning and Design,* Vol. 21, Iss 2, pp. 145-157.

Webster, C.J. (1994) "Structured Methods for GIS Design.1. A Relational System for Physical Plan Monitoring," *Computers Environment and Urban Systems,* Vol. 18, Iss 1, pp. 1-18.

Webster, C.J. (1997) "GIS and Organizations—How Effective Are GIS in Practice", by H. Campbell, I. Masser, *Environment and Planning B-Planning and Design,* Vol. 24, Iss 4, pp. 636-638.

Wehmeyer, M.G. (1994) "Integrating SCADA Systems with Am/FM-GIS Technology," *Water-Engineering and Management* Vol. 141, Iss 7, pp. 18+.

Weibel, R., Buttenfield, B.P. (1992) "Improvement of GIS Graphics for Analysis and Decision-Making," *IJGIS,* Vol. 6, Iss 3, pp. 223-245.

Welborn, H., Wagner, J., Weber, D. (1990) "Trends in Choosing a Geographic Information System," *Water-Engineering and Management,* Vol. 137, Iss 11, pp. 20-23.

Welch, R., Remillard, M., Alberts, J. (1992) "Integration of GPS, Remote-Sensing, and GIS Techniques for Coastal Resource-Management," *PERS* Vol. 58, Iss 11, pp. 1571-1578.

Welch, R., Remillard, M., Doren, R.F. (1995) "GIS Database Development for South Florida National-Parks and Preserves," *PERS* Vol. 61, Iss 11, pp. 1371-1381.

Wentz, E.A. (1996) "Innovations in GIS—Selected Papers from the First National Conference on GIS Research UK", by M.F. Worboys, *Annals of the Association of American Geographers,* Vol. 86, Iss 2, pp. 376-377.

Wheeler, J.O. (1994) "GIS and Critical Social-Theory—Whither Urban-Geography," *Urban Geography,* Vol. 15, Iss 2, pp. 103-105.

Wheeler, J.O. (1997) "Quantitative-Analysis and GIS in Urban-Geography—Then and Now," *Urban Geography,* Vol. 18, Iss 2, pp. 95-99.

White, K.L., Simms, M. (1993) "Geographic Information Systems as an Educational-Tool," *Journal of Geography,* Vol. 92, Iss 2, pp. 80-85.

Wiggins, L.L., Ferreira, J. (1992) "MITS Computer Resource Lab—A Research and Educational Facility for GIS in Urban-Planning," *IJGIS,* Vol. 6, Iss 4, pp. 299-307.

Wiggins, L.L., French, S.P. (1990) "Desk-Top Mapping and GIS," *Journal of the American Planning Association,* Vol. 56, Iss 3, pp. 370-373.

Wilkinson, G.G. (1996) "A Review of Current Issues in the Integration of GIS and Remote-Sensing Data," *IJGIS,* Vol. 10, Iss 1, pp. 85-101.

Williams, D.C., Lyon, J.G. (1991) "Use of a Geographic Information System Data Base to Measure and Evaluate Wetland Changes in the St-Marys River, Michigan," *Hydrobiologia,* Vol. 219, Iss JUL, pp. 83-95.

Williams, G.J. (1995) "Templates for Spatial Reasoning in Responsive Geographical Information Systems," *IJGIS,* Vol. 9, Iss 2, pp. 117-131.

Williamson, A.N. (1992) "Map Scanning for GIS Applications," *PERS* Vol. 58, Iss 8, pp. 1199-1202.

Wise, S. (1992) "Geographic Information Systems—Developments and Applications", by L. Worrall, *Applied Geography,* Vol. 12, Iss 1, pp. 85-85.

Wise, S. (1992) "Introductory Readings in Geographical Information Systems", by D.J. Peuquet, D.F. Marble, *Progress in Human Geography,* Vol. 16, Iss 3, pp. 474-476.

Wise, S. (1993) "Geographic Information Systems and Their Socioeconomic Applications", by D. Martin, *Geographical Journal,* Vol. 159, Iss JUL, pp. 229-231.

Wise, S. (1993) "Geographical Information Systems, Vol. 1, Principles, Vol. 2, Applications", by D.J. Maguire, M.F. Goodchild, D.W. Rhind, *Geography,* Vol. 78, Iss 338, pp. 89-90.

Wise, S. (1993) "Geographical Information Systems—Principles and Applications", by D. Maguire, M. Goodchild, D. Rhind, *Geographical Journal,* Vol. 159, Iss JUL, pp. 229-231.

Wise, S. (1996) "Innovations in GIS-2—Selected Papers from the 2nd National Conference on GIS Research UK", by P. Fisher, *IJGIS,* Vol. 10, Iss 2, pp. 241-242.

Wise, S.M. (1993) "Geographic Information Systems—A Guide to the Technology", by A.N. Antenucci, K. Brown, P.L. Croswell, M.J. Kevany, *Applied Geography,* Vol. 13, Iss 1, pp. 100-101.

Wong, D. (1997) "Enhancing Segregation Studies Using GIS," *Computers Environment and Urban Systems,* Vol. 20, Iss 2, pp. 99-109.

Wood, B., Washino, R., Beck, L., Hibbard, K., Pitcairn, M., Roberts, D., Rejmankova, E., Paris, J., Hacker, C., Salute, J., Sebesta, P., Legters, L. (1992) "Distinguishing High and Low Anophéline-Producing Rice Fields Using Remote-Sensing and GIS Technologies," *Preventive Veterinary Medicine* 1991, Vol. 11, Iss 3-4, pp. 277-288.

Woodcock, C.E., Sham, C.H., Shaw, B. (1990) "Comments on Selecting a Geographic Information System for Environmental Management," *Environmental Management,* Vol. 14, Iss 3, pp. 307-315.

Woodsford, P.A. (1991) "Cartographic and GIS Developments at Laser-Scan, 1987-91," *Cartographic Journal,* Vol. 28, Iss 1, pp. 43.

World, L. (1994) "GIS Feels the Earth Move in Los-Angeles," *IEEE Computer Graphics and Applications,* Vol. 14, Iss 3, pp. 16-16.

Worrall, Les (1990)(ed), *Geographic Information System: Development and Applications,* Taylor and Francis, London.

Worrall, L. (1994) "Justifying Investment in GIS—A Local-Government Perspective," *IJGIS,* Vol. 8, Iss 6, pp. 545-565.

Worrall, L. (1994) "The Role of GIS-Based Spatial-Analysis in Strategic Management in Local-Government," *Computers Environment and Urban Systems,* Vol. 18, Iss 5, pp. 323-332.

Wright, D.J., Goodchild, M.F. (1997) "Data from the Deep—Implications for the GIS Community," *IJGISC,* Vol. 11, Iss 6, pp. 523-528.

Wright, D.J., Goodchild, M.F., Proctor, J.D. (1997) "Demystifying the Persistent Ambiguity of GIS as Tool Versus Science," *Annals of the Association of American Geographers,* Vol. 87, Iss 2, pp. 346-362.

Wright, J. (1988) "The Plain Fellows Guide to Geographic Information Systems," *Geographical Journal,* Vol. 154, Iss JUL, pp. 161-168.

Wrigley, N. (1991) "Market-Based Systems of Health-Care Provision, the NHS Bill, and Geographical Information Systems," *Environment and Planning A,* Vol. 23, Iss 1, pp. 5-8.

Wyatt, P.J. (1997) "The Development of a GIS-Based Property Information-System for Real-Estate Valuation," *IJGISC,* Vol. 11, Iss 5, pp. 435-450.

Xiang, W.N. (1993) "Application of a GIS-Based Stream Buffer Generation Model to Environmental-Policy Evaluation," *Environmental Management,* Vol. 17, Iss 6, pp. 817-827.

Xiang, W.N. (1993) "A GIS Method for Riparian Water-Quality Buffer Generation," *IJGIS,* Vol. 7, Iss 1, pp. 57-70.

Xiang, W.N. (1993) "A GIS MMP-Based Coordination Model and Its Application to Distributed Environmental Planning," *Environment and Planning B-Planning and Design,* Vol. 20, Iss 2, pp. 195-220.

Xiang, W.N. (1996) "A GIS Based Method for Trail Alignment Planning," *Landscape and Urban Planning,* Vol. 35, Iss 1, pp. 11-23.

Xiang, W.N. (1996) "GIS-Based Riparian Buffer Analysis—Injecting Geographic Information into Landscape Planning," *Landscape and Urban Planning,* Vol. 34, Iss 1, pp. 1-10.

Xiang, W.N., Stratton, W.L. (1996) "The B-Function and Variable Stream Buffer Mapping—A Note on a GIS Method for Riparian Water-Quality Buffer Generation," *IJGIS,* Vol. 10, Iss 4, pp. 499-510.

Xu, F., Prato, T., Fulcher, C. (1993) "Broiler Litter Application to Land in an Agricultural Watershed—A GIS Approach," *Water Science and Technology,* Vol. 28, Iss 3-5, pp. 111-118.

Xu, H., Bailey, J.O., Barrett, E.C., Kelly, R.E.J. (1994) "Monitoring Snow Area and Depth with Integration of Remote-Sensing and GIS," *IJRS* 1993, Vol. 14, Iss 17, pp. 3259-3268.

Xu, J.P., Lathrop, R.G. (1995) "Improving Simulation Accuracy of Spread Phenomena in a Raster-Based Geographic Information System," *IJGIS,* Vol. 9, Iss 2, pp. 153-168.

Yaakup, A.B., Healey, R.G. (1994) "A GIS Approach to Spatial Modelling for Squatter Settlement Planning in Kuala-Lumpur, Malaysia," *Environment and Planning B-Planning and Design,* Vol. 21, Iss 1, pp. 21-34.

Yapa, L.S. (1991) "Is GIS Appropriate Technology," *IJGIS,* Vol. 5, Iss 1, pp. 41-58.

Yeh, A.G. (1991) "The Development and Applications of Geographic Information Systems for Urban and Regional-Planning in the Developing Countries," *IJGIS,* Vol. 5, Iss 1, pp. 5-27.

Yeh, A.G.O., Batty, M. (1990) "Applications of Geographic Information Systems in Urban and Regional-Planning," *Environment and Planning B-Planning and Design,* Vol. 17, Iss 4, pp. 369-374.

Yi, G.C., Risley, D., Koneff, M., Davis, C. (1994) "Development of Ohio GIS-Based Wetlands Inventory," *Journal of Soil and Water Conservation* Vol. 49, Iss 1, pp. 23-28.

Young, J.A.T. (1986) "Remote-Sensing and an Experimental Geographic Information System for Environmental Monitoring, Resource Planning and Management," *IJRS,* Vol. 7, Iss 6, pp. 741-744.

Zack, J.A., Minnich, R.A. (1991) "Integration of Geographic Information Systems with a Diagnostic Wind-Field Model for Fire Management," *Forest Science,* Vol. 37, Iss 2, pp. 560-573.

Zaima, E., Okabe, S., Nishiwaki, S., Ishikawa, M., Nakamoto, T., Murayama, Y., Kawaguchi, Y. (1990) "Reignition Surges At Reactor Current Interruption in Cable-System GIS," *IEEE Transactions of Power Delivery* Vol. 5, Iss 2, pp. 947-954.

Zech, Y., Sillen, X., Debources, C., Vanhauwaert, A. (1994) "Rainfall-Runoff Modelling of Partly Urbanized Watersheds—Comparison Between and Distributed Model Using GIS and Other Models Sensitivity Analysis," *Water Science and Technology,* Vol. 29, Iss 1-2, pp. 163-170.

Zhou, H.Z., Macdonald, K.B., Moore, A. (1991) "Some Cautions on the Use of Geographic Information System (GIS) Technology to Integrate Soil Site and Area Data," *Canadian Journal of Soil Science,* Vol. 71, Iss 3, pp. 389-394.

Zhou, J., Civco, D.L. (1996) "Using Genetic Learning Neural Networks for Spatial Decision-Making in GIS," *PERS* 1996, Vol. 62, Iss 11, pp. 1287-1295.

Zhou, Q.M. (1993) "Arcview for Windows GIS Software," *PERS* Vol. 59, Iss 7, pp. 1103-1106.

Zhu, Z.L., Evans, D.L. (1992) "Mapping Midsouth Forest Distributions—AVHRR Satellite Data and GIS Help Meet RPA Mandate," *JF* Vol. 90, Iss 12, pp. 27-30.

Zumbrunnen, C. (1991) "GIS tutor (GIST)—The Geographical Information Systems Tutor," *PG,* Vol. 43, Iss 2, pp. 235-236.

Zwarenstein, M., Krige, D., Wolff, B. (1991) "The Use of a Geographical Information System for Hospital Catchment-Area Research in Natal Kwazulu," *South African Medical Journal,* Vol. 80, Iss 10, pp. 497-500.

Zybach, B., Barrington, M., Downey, T. (1995) "Converting Historical Information to GIS—Political Boundaries of the Douglas-Fir Region, 1788 to 1995," *JF* Vol. 93, Iss 5, pp. 15-20

Abbreviations Used

IJGIS	:	International Journal of Geographical Information Systems
IJGISC	:	International Journal of Geographic Information Science
IJRS	:	International Journal of Remote Sensing
PERS	:	Phogrammmetric Engineering and Remote Sensing
Pg	:	PG
JF	:	JF

Index